Wildland Fire Behaviour

Dynamics, Principles and Processes

Mark A Finney, Sara S McAllister, Torben P Grumstrup,
Jason M Forthofer

© CSIRO 2021

All rights reserved. Except under the conditions described in the *Australian Copyright Act 1968* and subsequent amendments, no part of this publication may be reproduced, stored in a retrieval system or transmitted in any form or by any means, electronic, mechanical, photocopying, recording, duplicating or otherwise, without the prior permission of the copyright owner. Contact CSIRO Publishing for all permission requests.

The authors assert their moral rights, including the right to be identified as an author.

A catalogue record for this book is available from the National Library of Australia.

ISBN: 9781486309085 (pbk)
ISBN: 9781486309092 (epdf)
ISBN: 9781486309108 (epub)

How to cite:
Finney MA, McAllister SS, Grumstrup TP, Forthofer JM (2021) *Wildland Fire Behaviour: Dynamics, Principles and Processes*. CSIRO Publishing, Melbourne.

Published by:

CSIRO Publishing
36 Gardiner Road, Clayton VIC 3168
Private Bag 10, Clayton South VIC 3169
Australia

Telephone: +61 3 9545 8400
Email: publishing.sales@csiro.au
Website: www.publish.csiro.au

Front cover: 'Buoyant dynamics of wildfire flames' by Brian Elling

Set in 10.5/14 Palatino & Optima
Edited by Joy Window (Living Language)
Cover design by Cath Pirret
Typeset by Envisage Information Technology
Index by Max McMaster
Printed by Ingram Lightning Source

CSIRO Publishing publishes and distributes scientific, technical and health science books, magazines and journals from Australia to a worldwide audience and conducts these activities autonomously from the research activities of the Commonwealth Scientific and Industrial Research Organisation (CSIRO). The views expressed in this publication are those of the author(s) and do not necessarily represent those of, and should not be attributed to, the publisher or CSIRO. The copyright owner shall not be liable for technical or other errors or omissions contained herein. The reader/user accepts all risks and responsibility for losses, damages, costs and other consequences resulting directly or indirectly from using this information.

Acknowledgement
CSIRO acknowledges the Traditional Owners of the lands that we live and work on across Australia and pays its respect to Elders past and present. CSIRO recognises that Aboriginal and Torres Strait Islander peoples have made and will continue to make extraordinary contributions to all aspects of Australian life including culture, economy and science. CSIRO is committed to reconciliation and demonstrating respect for Indigenous knowledge and science. The use of Western science in this publication should not be interpreted as diminishing the knowledge of plants, animals and environment from Indigenous ecological knowledge systems.

FOREWORD

Fire shares a long and complicated history with humankind. The taming of fire for heating, cooking and, over the last two centuries, combustion for mechanical work and electrical power has allowed for substantial advancement of civilised society that we recognise today. Fire, of course, has always existed naturally across many landscapes. Recognition of the physical mechanisms in fires notably began with Michael Faraday's *The Chemical History of a Candle* in 1848, which the authors use to help explain fire processes. The study of fire and combustion, however, often diverged from the natural processes we associate with wildfires. Humankind was not always ignorant of fire across the landscape, and we now recognise the importance of Indigenous practices on burning. While fire may not have always been studied with the mathematical tools we've become familiar with in the scientific literature, Indigenous people have long had an integral cultural and experiential understanding of fire and have used it to manage lands for their own needs in sustainable ways.

While many are working to maintain and restore Indigenous history and understanding of fire, we have still made significant advancements in understanding the physical processes of fire. Yet, today's wildfires do not appear like those of the past – around the world whole communities are destroyed, large areas are enveloped in smoke for days or weeks at a time, ecosystems are damaged, and both human life and property are destroyed at increasing rates. While landscape fires are a critical and necessary natural process, fire disasters which severely impact people and property have dramatically increased in frequency and severity. This increase can be attributed to a combination of land management, climate change and population movement, and thus, the path forward requires us to look back and find ways to better understand the processes that are affecting us. Unlike the use of fire confined as combustion within engines, wildland fires are a natural and unconfined process that we must learn to live with. By understanding the process, we may find better ways to abide, adapt, control and even use fire across the natural landscape once again.

Wildland Fire Behaviour advances a legacy of work from the USDA Forest Service's Missoula Fire Sciences Laboratory, which since its inception in 1960 has led the direction of wildfire research in the United States. As when it first opened, the focus of the laboratory has been to develop a greater understanding of wildland fire and the associated technology to translate that knowledge into practice. A crucial development in fire science came from the fire spread model developed by Richard (Dick) Rothermel and published in 1972. This model became the basis for wildland fire predictions and, while originally developed for one-dimensional applications (and placed onto slide rules of the day), led to a dramatic transition where the path of a fire could finally be quantitatively predicted. While other models have been developed since, it remains the most widely used in practice and even underlies two- and three-dimensional simulations which run on today's advanced supercomputing clusters. Fire behaviour, however, often deviates from the conditions under which this model was first developed. We understand that many fires are not steady, burn through heterogenous live and dead fuels, interact with the atmosphere, and burn under conditions that were not imagined or tested during the model's initial development. Despite these challenges, the model remains an essential component of fire practice, motivating the development of a next-generation replacement for the future.

The work of Missoula researchers has recently been devoted to developing this next-generation fire spread model and, through this process, advancing our fundamental understanding of how fires spread. Even within the physical sciences it is well understood that fire spread is a complex process, comprising fluid dynamics, heat transfer and chemical kinetics. Paired with wildland fuels that, as you'll learn reading this book, behave in unique ways when heated and burned, wildfire spread incorporates a vast array of disciplines which are hard to understand from any one perspective alone. The advancements described in this book are perhaps most impactful in the way in which they have combined the knowledge of disparate fields of science into a more comprehensive, holistic view of wildland fire behaviour.

One of the most exciting aspects of this book is that its primary audience is not scientists, although they may also be excited to learn from it, but practitioners who seek to further understand the fire processes that they use and manage. New discoveries, such as the way in which fires heat fuels and contribute to fire spread, have real applications to fire management and local understanding of fire behaviour through the vast array of natural lands covering our globe. This book, therefore, will be unique and influential in the ways that it shares this physical perspective to a wide audience who may use it in their practical applications, while still exciting scientists from all fields to expand their scope and knowledge on wildland fire behaviour.

It is fitting that the pair of us were asked to write the foreword. Stephens has known the first author, Finney, since graduate school and once taught McAllister's course on wildland fire at UC Berkeley, while Gollner has known McAllister since graduate school in California and holds her and Finney mostly responsible for his transition from fundamental combustion into wildland fire. UC Berkeley's shared history with the team who wrote this book, however, goes further. Both Finney and McAllister are Berkeley alumni but from different schools: Natural Resources and Engineering. This ties directly with a unique contribution of the book that makes its advancement so pivotal, a linking of knowledge between engineering and natural sciences. Advancements rarely occur alone, but are often the result of several developments that, when combined and applied to the right problem, can result in paradigm-changing discoveries. Here, it's the application of engineering and natural science knowledge to wildland fire that has led to major advances. This is not unique, as Richard Rothermel himself was an engineer who moved into the field of wildland fire. The University of California, Berkeley continues to be a renowned leader in fire science due to unique collaborations between faculty in wildland fire such as Harold Biswell and Bob Martin, and engineers like Patrick Pagni and Brady Williamson.

We hope that this text serves to not only educate its readers, but also to inspire the next generation of collaborative, transformative research in wildfire. The world is plagued with a new set of wildfire challenges which need sound science to inform solutions. Reducing wildfire risk to communities demands knowledge of both the built and natural environments, shifting land management agencies from reactive fire suppression to proactive fire management requires developing advanced expertise in strategic fire planning, and mitigating effects of climate change on forests and wildfire behaviour requires ambitious intervention at broad scales in the wildfire cycle. While humans are often aware of some fire risks, they rarely take action to mitigate them before disaster strikes. Each of these and many other problems could use a multidisciplinary approach to work towards tomorrow's solutions. We hope you enjoy this book as much as we did and take this knowledge with you into the field, forest or laboratory.

Michael J. Gollner
Department of Mechanical Engineering
University of California, Berkeley

Scott Stephens
Department of Environmental Science, Policy, and Management
University of California, Berkeley

CONTENTS

Foreword   iii
Preface and acknowledgements   x
About the authors   xii
Nomenclature   xiii

1   Introduction to wildfire science   1
    Fire science and the need for experiments   3
    Wildland fire science since 1900   4
    Modelling and field-scale research   7
    The challenge of validation   8
    Outline of the book   10
    References   12

2   Fire and wildland fire behaviour   15
    The burning candle as a fire process   15
       Igniting and burning a candle   17
       Flame shape   18
       Flame size   19
       Candles and wildfires as coupled systems   20
    Wildfire behaviour triangle: fuels, weather, topography   21
       Wildfire classification   23
    Initial fire growth   24
    Line fire concept   25
    Wildfire behaviour   28
       Fire spread rate   29
       Fire shapes   32
       Fire area and perimeter   33
       Heat release and fireline intensity   35
       Flame length   37
       Fire characteristics chart   39
       Fire acceleration   41
    Summary   43
    Supplementary calculations   43
    Endnote   44
    References   44

3 Thermodynamics, fluid mechanics and heat transfer — 46
- Basic concepts, material properties and terminology — 47
- Thermodynamics — 50
- Fluid mechanics — 53
  - Boundary layers — 54
  - Vortex flows — 58
  - Ember lofting — 61
- Heat transfer — 61
  - Conduction heat transfer — 61
  - Radiation heat transfer — 64
  - Convection heat transfer — 73
  - Combined heat transfer — 79
- Summary — 80
- Endnote — 80
- References — 80

4 Combustion — 82
- Fuels — 82
- Thermodynamics of combustion — 84
  - Combustion reactions — 84
  - Heat of combustion — 85
  - Flame temperatures — 88
- Brief discussion of chemical kinetics — 89
- Types of flames — 91
  - Premixed flames — 91
  - Non-premixed or diffusion flames — 93
- Smouldering and glowing — 106
- Summary — 112
- Endnotes — 112
- References — 112

5 Ignition — 114
- The ignition process — 114
- Flaming ignition criteria — 117
- Types of ignition — 119
- Critical heat flux for ignition — 119
- Predicting ignition times — 120
- Factors that affect ignition time — 122
- Live fuels — 127
- Summary — 130
- Endnotes — 130
- References — 131

6	**The environment in wildfire dynamics**	**132**
	Wildland fuel	132
	Fuel particles and fuel beds	136
	Fuel moisture	142
	Live fuels	152
	Implications for fuel characterisation and classification	155
	Weather	157
	Winds	157
	Solar radiation	172
	Topography	173
	Fire configurations	177
	Flame front width and shape	177
	Backing fires and flanking	179
	Multiple flame zones and air-flow interactions	182
	Summary	185
	References	185
7	**Wildfire spread**	**191**
	System behaviour	191
	Model framework	192
	Fuel particles	193
	Burning rate	195
	Flame radiation heat transfer	196
	Solid glowing radiation	197
	Ambient environment radiation heat transfer	197
	Convection heat transfer	198
	Model function	201
	Modelled fire spread and behaviour	202
	Simple fire spread dynamics	204
	Fuel particle heating and ignition	206
	Fuel loading	207
	Flame front width	208
	Effects of wind	209
	Non-steady wind	211
	Effects of slope	211
	Effects of dead fuel moisture	212
	Effects of fuel continuity	213
	Positive and negative feedbacks	216
	Model improvements	218
	Combustion	218
	Ignition	218
	Wind	219

Flame zone orientation	219
Heat transfer	219
Crown fire	219
Summary	219
References	219

8 Behaviours of large fires — 222

Crown fire	222
Spotting and spot fires	229
Fire shapes and growth patterns	234
Burn streets	238
Plumes and pyroconvective atmospheric storms	242
Vorticity	246
Pulsating or puffing	247
Fire whirls	250
Counter-rotating vortex pairs and wake vortices	253
Vorticity-driven lateral spread	255
Mass fires	256
Summary	259
References	260

9 Measurements in fire behaviour — 265

Sampling and experimental design	266
Fire measurements	266
Combustion and heat release	266
Fuel consumption	267
Heat release	281
Flame zone properties	282
Heat transfer	292
Radiation	292
Convection	294
Ignition	299
Rate of spread	301
One-dimensional spread rate	301
Two-dimensional spread rate and fire growth	303
Environmental measurements	305
Fuel moisture	305
Weather and wind	308
Temperature	310
Relative humidity	310
Summary	311
References	311

10 Ignition techniques for experimental burning 318
   Point ignition 319
   Single line ignition 320
      Heading fires 320
      Backing fires 323
      Flanking fires 323
      Other line ignitions 324
   Multiple line fires 324
      Strip head fire 325
      Flank fire 327
   Multiple spot ignitions 328
   Ring fire, centre fire, mass ignition 329
   Summary 331
   References 331

11 Conclusions 333
   Key principles and insights 333
   Principal value to researchers 335
   Principal value to managers 336

Appendix A: Physical quantities and units 338
Appendix B: Thermal and physical properties of air 343
Index 349

PREFACE AND ACKNOWLEDGEMENTS

Practical concerns over wildland fire and its management have inspired a century of worldwide scientific research. Studies in the laboratory, in the field and by modelling have probed physical, ecological and meteorological dimensions of wildfire phenomena across a range of spatial scales, but have also exposed contradictions and persistent mysteries, even for what appear to be the simplest and most observable behaviours. Consequently, the body of scientific observations, correlations, predictions and speculations cannot yet be claimed to constitute a coherent physical theory of how wildfires spread and behave.

This book is our attempt to assemble and organise information on what is known and unknown about wildfire spread and its behaviours. It is clear to any observer that wildfires are non-steady and ever-changing, and we have therefore approached the challenge of explaining wildfire behaviour as a dynamical system of interdependent physical processes. We rely primarily on the power of experiments to offer clear demonstrations of principles and concepts, making the subject more approachable to a broader audience. Most of the people directly involved with wildland fire are foresters, firefighters, land managers, social scientists or ecologists who work with wildfires in one way or another but who do not specialise in physics or engineering. Most physicists, chemists, atmospheric scientists and engineers have little experience with wildfire, prescribed burning, or natural resource sciences or management. The benefit of this book to both groups is that it offers a way to look beneath the empirical correlations and complexities of existing fire models for explanation of wildfire behaviour dynamics. By explaining how wildland fires spread and behave we hope to advance the curricula for firefighter training, increase the fidelity of predictive modelling and the data required to properly characterise the fire environment, and finally to improve the reliability of planning for fuel mitigation for proactive fire management. Hopefully, readers experienced with wildfires will gain insight into the processes that produce fire behaviours, and those who are experts in the physical sciences or engineering will gain appreciation for how the problems in wildland fire are distinct from those in well-controlled and measured engineered environments.

There remains much work to be done in order to understand wildfires. We offer this book as a means of establishing a foundation for research and teaching on wildfire behaviour, but progress will require explanations from many different disciplines to build the needed body of theory required for improved tools, training and reliable proactive wildland fire management. We do not expect readers to be experts in engineering, physics or combustion, and our discussion of relevant topics in these fields occasionally sacrifices some nuance to ensure broad comprehension. There are many textbooks and research papers in those fields that provide detailed treatment and we offer suggestions throughout for further study. Likewise, there are many texts that cover practical material on fire ecology, wildfire management and approaches to fire behaviour prediction. Our specific aim is to communicate the physical concepts underlying wildfire spread and behaviour. This has not been done before, and it certainly has not been attempted from a dynamical standpoint with experiments to illustrate the physical processes and relations.

To read and understand this book you will not need your calculator or computer. We have tried to minimise the mathematical content. Equations are presented principally to illustrate relations among the main variables that produce the physical effects and to familiarise readers with the primary physical processes in fire behaviour. The principles of wildfire spread are presented mostly using graphs and pictures, and demonstrations of simple experiments, some of which can be duplicated in a small laboratory or on your back porch. Like studies of most natural phenomena, experience with fire is necessary, but alone it is insufficient, if true understanding is desired.

This book is a cumulative product of the contributions from many colleagues and team members. All authors are employees of the US Forest Service in the Department of Agriculture (USDA Forest Service). The USDA Forest Service has responsibility for managing fire and ecosystems covering approximately 78 million hectares (193 million acres) in the United States. Complementary land management agencies in the Department of Interior and individual US states are entrusted with similar land and fire management responsibilities. We have a vested interest in seeking both fundamental and applied knowledge of wildland fire for application in accomplishing the mission of land management. Our offices are located at the Missoula Fire Sciences Laboratory of the Rocky Mountain Research Station, which has been engaged in wildland fire research since it opened in 1960. Much of our personal research and our perspectives presented here are the products of the experiences and achievements of our present colleagues and previous generations of scientists and staff at our facility. In particular, we are especially grateful to Jack Cohen and Don Latham for their notable career-long efforts to elucidate the physical processes of fire behaviour and their philosophical approaches to fire science. Their curiosity and collaborations in our own research have been invaluable.

Current and former members of our research group have been essential to the experimental and modelling research, including the design and fabrication of laboratory and field apparatus, developing and operating specialised instrumentation, obtaining and assembling sample materials, conducting laboratory and field experiments, and performing modelling and analyses. Thank you to Jon Bergroos, Bill Chatham, Josh Deering, Jay Fronden, Andrew Gorris, Isaac Grenfell, Chelsea Phillips, Randy Pryhorocki, Sophie Vernholm, Mark Vosburgh and Cyle Wold for all your contributions. Tanner and Trevor Finney assisted with graphics. Ian Grob has generously contributed his expertise in photography and videography in documenting many experiments, some specifically for this book. Brian Elling's talented artistry has provided vivid illustrations of complex fire and physical phenomena. Jane Kapler-Smith conducted detailed and demanding editing of the manuscript that has been indispensable to improving the clarity and readability of each chapter.

We also recognise the longstanding collaborations with Kozo Saito (University of Kentucky), Michael Gollner (University of California, Berkeley), Arnaud Trouve (University of Maryland) and their graduate students, who have contributed greatly to experimental research and interpretations of fire behaviours.

Our colleagues at SCION in New Zealand have made possible the opportunity to develop and employ field measurements to document and test theories of physical processes in fire behaviour in field experiments. Thank you to Grant Pearce, Tara Strand and Marwan Katurji in particular, and to Jessica Kerr, Veronica Clifford, Hugh Wallace and Katie Melnik for all your work.

Finally, the continuity of wildfire research would not be possible without the long-standing support by the USDA Forest Service. The Rocky Mountain Research Station has steadfastly sustained fire science efforts by recurrent hiring of personnel and maintaining and upgrading the physical infrastructure of the laboratory facility itself. Much of the current fire behaviour research has been generously supported for decades by Fire and Aviation Management of the USDA Forest Service through the National Fire Decision Support Center, and the individual commitment of John Phipps, the Deputy Chief of State and Private Forestry for all phases of research, development and application.

ABOUT THE AUTHORS

Dr Mark A Finney is a Senior Scientist and Research Forester. He began his career as a seasonal wildland firefighter with the Bureau of Land Management and worked as an ecologist for Sequoia National Park before joining the U.S. Forest Service at the Missoula Fire Sciences Laboratory. His research has involved fire history and ecology, prescribed burning, modelling of fire growth, landscape fuel treatment design, wildfire risk analysis, and laboratory and field experiments on the physics of wildland fire behaviour.

Dr Sara S McAllister is a Research Mechanical Engineer for the U.S. Forest Service at the Missoula Fire Sciences Laboratory. She earned her PhD in Mechanical Engineering from the University of California, Berkeley, where she studied material flammability in spacecraft and co-authored a textbook on combustion fundamentals. Her current research includes understanding the critical conditions for solid fuel ignition, flammability of live forest fuels, ignition due to convective heating, and fuel bed property effects on burning rate.

Dr Torben P Grumstrup is a Research Mechanical Engineer at the U.S. Forest Service, Missoula Fire Sciences Laboratory. His introduction to wildland fire was working as a Forest Service seasonal wildland firefighter on engine, helitack and helirappel crews. Torben has diverse engineering experience, ranging from particle accelerator engineering to laser spectroscopy of combustion emissions. His present research concerns laboratory and field experiments on heat transfer and fluid flow in wildland fire.

Jason M Forthofer began his career with the U.S. Forest Service as a seasonal firefighter and now works as a Mechanical Engineer at the Missoula Fire Sciences Laboratory. His research includes numerical, field and laboratory studies of heat transfer and fluid flow relating to wildland fires. Results of his work include improvements to fire shelters and safety zone size guidelines and development of fire tornado training material, operational wind models and a fire spread model. Jason is a qualified Division Supervisor for wildland fire suppression in the USA.

NOMENCLATURE

Latin symbols

Symbol	Units[i]	Description
A	See footnote [ii]	Pre-exponential factor
A_e	m²	Area of an elliptic fire
A_f	m²	Focal sector area
A_s	m²	Exposed surface area of fuel; surface area
A_v	m²	Area of vertical passages in a wood crib
A_1	m²	Area of surface 1
A_2	m²	Area of surface 2
a	–	Exponent in chemical reaction rate equation
B^*	–	Flame zone aspect ratio
b	–	Exponent in chemical reaction rate equation
C_n	–	Convection number
c_p	J kg⁻¹ K⁻¹	Specific heat of a gas in a constant-pressure process
c_s	J kg⁻¹ K⁻¹	Specific heat of a solid
D	m	Flame zone depth
D^*	–	Normalised flame zone depth
d	m	Diameter or thickness
E_a	J mol⁻¹	Activation energy per mole for a chemical reaction
$\dot{E}''_b$	W m⁻²	Radiant emissive flux of fuel bed surface
$\dot{E}''_f$	W m⁻²	Radiant emissive flux of flame surface
EMC	%	Equilibrium moisture content of fuel particle
e	–	Base of the natural logarithm: 2.71828…
F_{12}	–	View factor; radiation from 1 incident on 2
F_{21}	–	View factor; radiation from 2 incident on 1
f	s⁻¹, Hz	Frequency of flame pulsation
Gr	–	Grashof number: $Gr = g\tau\lvert T_s - T_\infty\rvert \ell^3 v^{-2}$

Symbol	Units[i]	Description
Gr_{crit}	–	Critical Grashof number
Gr_d	–	Grashof number with respect to diameter d
g	m s⁻²	Acceleration due to gravity: 9.807…
HRR	W	Heat release rate
H_c	J kg⁻¹	Heat of combustion
$H_{c,eff}$	J kg⁻¹	Effective heat of combustion
H_v	J kg⁻¹	Heat of vaporisation
h	W m⁻² K⁻¹	Convection heat transfer coefficient
h_1, h_2	m	Reference heights for logarithmic wind speed profile
I_B	W m⁻¹	Fireline intensity[iii]
j	–	Number of molecules in a chain
k	W m⁻¹ K⁻¹	Thermal conductivity
L	m	Length of a crib stick
l_x	–	Major axis of an elliptical fire shape
l_y	–	Minor axis of an elliptical fire shape
ℓ	m	Length or distance; characteristic length scale
ℓ^*	–	Normalised distance scale
ℓ^*_r	–	Characteristic roughness scale
ℓ_f	m	Flame length
ℓ_p	m	Characteristic plume length
ℓ_{th}	m	Thermal penetration depth
M	kg mol⁻¹	Molecular mass of a substance[iv]
M_{gas}	kg mol⁻¹	Molecular mass of a specific gas[iv]
MC	%	Moisture content (%)
MF	-	Moisture Fraction (MC)(100)⁻¹
m	kg	Mass of a substance
m_c	kg	Mass of fuel consumed by combustion

Symbol	Units[i]	Description
m_{dry}	kg	Mass of fuel with no moisture content
m_{water}	kg	Mass of water moisture in fuel
$\dot{m}_b$	kg s^{-1}	Fuel burning rate
$\dot{m}''_b$	kg m^{-2} s^{-1}	Fuel burning rate per surface area
m''_c	kg m^{-2}	Fuel consumed in the active flaming front
m''_d	kg m^{-2}	Dry mass of fuel per unit area; dry fuel loading
m''_f	kg m^{-2}	Mass of fuel per area; fuel loading
N	mol	Moles of a substance
$\dot{N}'''$	mol m^{-3} s^{-1}	Volumetric molar chemical reaction rate
Nu	–	Nusselt number: $Nu = h\ell k^{-1}$
n	–	Number of sticks per layer in a crib
P	m	Perimeter of an elliptical fire
P_f	kg s^{-3}	'Power of the fire' as defined by George M. Byram
P_w	kg s^{-3}	'Power of the wind' as defined by George M. Byram
Pr	–	Prandtl number: $Pr = \nu\alpha^{-1}$
p	Pa	Static pressure
p_d	Pa	Dynamic pressure
p_h	Pa	Hydrostatic pressure
q	J	Energy; thermal energy; heat; thermal radiation
q_b	J	Heat release from burning
q_{tot}	J	Total heat absorbed
q_v	J	Thermal energy required for vaporisation
$q_{1\to 2}$	J	Heat absorbed going from state 1 to state 2
$q_{2\to 3}$	J	Heat absorbed going from state 2 to state 3
$q_{3\to 4}$	J	Heat absorbed going from state 3 to state 4
$\dot{q}$	W	Heat transfer rate
$\dot{q}_{abs}$	W	Rate of radiation absorption
$\dot{q}_b$	W	Heat evolution rate due to burning
$\dot{q}_c$	W	Rate of convection heat transfer
$\dot{q}_{emit}$	W	Rate of radiation emission
$\dot{q}_{in}$	W	Thermal energy input rate
$\dot{q}_k$	W	Rate of conduction heat transfer
$\dot{q}_{out}$	W	Thermal energy output rate
$\dot{q}_{12}$	W	Net rate of heat transfer from surface 1 to surface 2
$\dot{q}_r$	W	Rate of radiation heat transfer
$\dot{q}_{ref}$	W	Rate of radiation reflection
$\dot{q}'$	W m^{-1}	Fireline intensity [iii]
$\dot{q}''$	W m^{-2}	Heat transfer flux
$\dot{q}''_c$	W m^{-2}	Convection heat transfer flux
$\dot{q}''_{r,b}$	W m^{-2}	Incident radiant flux from burning fuel bed to fuel particle
$\dot{q}''_{r,e}$	W m^{-2}	Net radiation heat transfer flux with the ambient environment
$\dot{q}''_{r,f}$	W m^{-2}	Incident radiant flux from flames to fuel particle
$\dot{q}''_s$	W m^{-2}	Net heat transfer flux at a surface
R	J mol^{-1} K^{-1}	Universal gas constant
R_{air}	J kg^{-1} K^{-1}	Specific gas constant for air
R_{gas}	J kg^{-1} K^{-1}	Specific gas constant for an arbitrary gas
Ra	–	Rayleigh number: $Ra = Pr\, Gr$
Ra_d	–	Rayleigh number with respect to diameter d
Re	–	Reynolds number: $Re = U\ell\nu^{-1}$
Re_{crit}	–	Critical Reynolds number
Re_d	–	Reynolds number with respect to diameter d
RH	%	Relative humidity of the air
Ri	–	Richardson number: $Ri = Gr\, Re^{-2}$
r	m s^{-1}	Fire spread rate
r_a	m s^{-1}	Forward expansion rate of an elliptical shaped fire
r_b	m s^{-1}	Lateral expansion rate of an elliptical shaped fire
r_c	m s^{-1}	Elliptical expansion rate relative to ellipse centroid
r_d	m s^{-1}	Backing spread rate; into the wind or downhill

NOMENCLATURE xv

Symbol	Units[i]	Description
S	m² m⁻³	Total fuel particle surface area per unit fuel bed volume
s	m	Stick spacing in cribs
s^*	–	Sine slope function
T	K, °C	Temperature
T_a	K, °C	Air temperature
T_b	K, °C	Radiating temperature of a burning fuel bed
T_e	K, °C	Ambient environment temperature
T_f	K, °C	Flame temperature; flame radiating temperature
T_{film}	K, °C	Film temperature; i.e. the average of T_s and T_∞
T_g	K, °C	Gas temperature
T_{ig}	K, °C	Ignition temperature
T_L	K, °C	Lower temperature
T_o	K, °C	Initial temperature
T_p	K, °C	Temperature of products
T_r	K, °C	Temperature of reactants
T_s	K, °C	Surface temperature
T_U	K, °C	Upper temperature
T_x	K, °C	Final temperature
T_∞	K, °C	Freestream temperature
T_1	K, °C	Temperature of surface 1
T_2	K, °C	Temperature of surface 2
t	s	Time
t_{ig}	s	Ignition time
t_p	s	Particle flaming time
t_r	s	Flame residence time
U	m s⁻¹	Wind speed; freestream flow speed
U_{mf}	m s⁻¹	Wind speed at mid-flame height
U_1	m s⁻¹	Wind speed at reference height
U_2	m s⁻¹	Wind speed at arbitrary height
U_{10}	m s⁻¹	10-metre open wind speed
U^*	–	Wind-flame momentum balance
u	m s⁻¹	Flow velocity in the x-direction
V	m³	Volume of gas
v	m s⁻¹	Flow velocity in the y-direction

Symbol	Units[i]	Description
W	m	Flame zone width; fuel bed width
WAF	–	Wind adjustment factor
w	m s⁻¹	Flow velocity in the z-direction
w_f	m s⁻¹	Gas velocity in the z-direction at the flame tip
x	m	Horizontal coordinate; streamwise direction
x_a	m	Forward ignition point on upper fuel bed surface
x_b	m	Flameout point of fuel bed upper surface
x_c	m	Flameout point of fuel bed lower surface
x_d	m	Flame detachment location
y	m	Transverse coordinate; spanwise direction
Z	m	Intermediate variable used in Chapter 7
z	m	Vertical coordinate; vertical direction
z_0	m	Roughness length for logarithmic wind speed profile

Greek symbols

Symbol	Units[i]	Description
α	m² s⁻¹	Thermal diffusivity
β	–	Packing ratio
γ	m	Boundary layer thickness
Δq_{th}	J	Change in thermal energy
ΔT	K, °C	Change in temperature
δ	m	Fuel bed depth
ε	–	Thermal radiation emissivity
ε_b	–	Thermal radiation emissivity of a burning bed
ε_e	–	Thermal radiation emissivity of the ambient environment
ε_f	–	Thermal radiation emissivity of a flame sheet
ε_p	–	Thermal radiation emissivity of a fuel particle
ε_1	–	Thermal radiation emissivity of surface 1

Symbol	Units[i]	Description
ε_2	–	Thermal radiation emissivity of surface 2
ζ	–	Thermal radiation absorptivity
ζ_b	–	Thermal radiation absorptivity of a fuel bed
ζ_2	–	Thermal radiation absorptivity of surface 2
η	–	Thermal radiation reflectivity
η_2	–	Thermal radiation reflectivity of surface 2
θ	°	Flame tilt angle measured from line normal to surface of fuel bed
θ_c	°	Angular position on a cylinder
θ_f	°	Focal angle
θ_h	°	Angle from heading direction
θ_s	°	Slope angle
θ_w	°	Wind-induced flame angle
λ	m	Porosity
λ_f	m	Wavelength of flame peaks
μ	Pa s	Dynamic viscosity
ν	m² s⁻¹	Kinematic viscosity
ξ	m	Radial distance
π	–	Ratio of a circle's circumference to diameter: 3.14159…

Symbol	Units[i]	Description
ρ	kg m⁻³	Density of a substance
ρ_a	kg m⁻³	Air density
ρ_b	kg m⁻³	Bulk density
ρ_f	kg m⁻³	Flame gas density
ρ_p	kg m⁻³	Density of a fuel particle/element
ρ_s	kg m⁻³	Density of a solid
σ	m⁻¹	Ratio of surface area to volume of a fuel particle
σ_B	W m⁻² K⁻⁴	Stefan-Boltzmann constant: 5.67 × 10⁻⁸
τ	K⁻¹	Volumetric thermal expansion rate
ϕ	m	Porosity of cribs
ψ	–	Eccentricity of ellipse

[i] All units of symbols are shown as SI base units and accepted derived units. Those symbols with a '–' in the units column are dimensionless.

[ii] Units depend on how the equation with which A is associated is applied.

[iii] Fireline intensity appears twice here with different notations. First, in a form consistent with the nomenclature of this book, $\dot{q}'$. Second, by widely accepted tradition, calculations pertaining to the theories of fire science pioneer, George M. Byram, use his notation, wherein fireline intensity is represented by I_B.

[iv] SI base units are shown, but molecular mass is traditionally in terms of grams per mol (g mol⁻¹).

1

Introduction to wildfire science

Humans are the fire species (Pyne 2010). For at least 400 000 years, humans have claimed mastery over domestic fires for heating, lighting, and cooking (Bowman *et al*. 2009). Our species employed widespread deliberate burning of vegetation to improve foraging and agriculture on every inhabited continent through the 19th century CE. Throughout aeons of changing climate and culture, fires from human as well as natural sources have developed and sustained the landscapes and ecosystems on which we depend. Wildland fires, whether intentional and carefully managed prescribed burns or free-spreading wildfires, will forever remain part of human existence.

For most of history, human proficiency with wildland fire was based on experience rather than scientific explanations. In recent millennia, western civilisation has sought explanations of fire, first through mysticism and alchemy (Williams 1992), then through experimental science – which is now accepted as the basis for our understanding of physical phenomena. But neither our experience nor our science has yet comprehensively answered these simple questions: how do wildland fires spread? How do fuel particles ignite in spreading fires? How long do wildfires burn? How does living vegetation ignite and burn? Are big wildfires just larger versions of small wildfires? In this book, we use scientific principles to address these questions and many others about wildland fire.

Wildlands supply people with essential natural resources, including clean air and water, timber and building materials, livestock forage, recreational amenities, and healthy habitat for wildlife. Many of the world's wildland ecosystems are *fire dependent*, meaning that fire is needed to maintain their function and productivity. It is impossible to completely prevent or suppress fire in these ecosystems. Wildfire science has been developed to increase understanding of fire and thus provide a foundation for managing it for the benefit of people and wildland resources.

Many fire-dependent ecosystems are near or adjacent to human communities, so we are faced with a conundrum: we need fire to sustain ecological processes, yet we need to prevent wildfire from endangering people and their communities. The escalating frequency and magnitude of wildfire disasters worldwide makes it clear that current practices are not meeting either objective. Perhaps we have inferior science, inadequate experience or insufficient technology; perhaps we are using these tools in the wrong fashion; or we are focusing on unrealistic objectives. Our intention in this book is to describe our current understanding of wildfire in terms of basic scientific principles, so scientists, technology experts, fire specialists and natural resource managers can develop objectives, tools and technology that harvest the benefits of wildland fire while protecting people and human communities from destruction. Our specific objectives are:

- Articulate a theory of fire spread and behaviour as a physical system specifically in a wildland context.
- Improve understanding of wildfire behaviour so we can ultimately make better predictions and identify opportunities for sustainable wildland management.
- Describe what is known and what remains unknown about the physical principles and processes of wildland fire.

This book is about improving wildfire science, not about implementation. Numerous cultural factors will control movement towards management solutions (Finney 2020). But only through increased understanding can we align our expectations, tools and technology with the reality that we must live with wildfire. At present, we live with and suffer only the most extreme fires, which are immune to control. We are consequently afraid of wildfires and their destruction. We react by expending vast resources in attempts (and failures) to eliminate wildfire. By our success in suppressing fires under mild and moderate conditions, we are deceived into thinking that it is possible to eliminate fire from our wildlands. Paradoxically, attempts to do this lead to more extreme fires (Brown and Arno 1991; Calkin *et al*. 2015). Reality does not give us the choice to *not* have fires, but it does give us the choice of *when* to have them and *what kind* of fires to have. In this book, we provide a scientific basis for understanding fire in wildlands so people can ultimately learn again to live with fire, as they have done for most of human history, not by removing it (which nature proves every year is impossible) but by using it deliberately to meet the current needs of our modern societies and the ecosystems that support us. Greater science-based understanding will enable us to:

- interpret our observations of wildfire behaviour and expand opportunities for accomplishing prescribed goals
- recognise prospects for precisely mitigating unwanted impacts of wildfire on our natural resources and developed infrastructure
- replace the illusions of knowledge offered by black-box fire models with actual understanding of wildfire processes
- substitute fear with acceptance.

Basic questions about wildfire behaviour are difficult to answer for two reasons. First, our human perspectives are poorly attuned to physically important dimensions of the problem; our intuition about fires is frequently wrong – not just a little wrong, but *exactly* wrong. For example, fires feel hot to us because our skin is very sensitive, and we tend to interpret phenomena in terms of our sensory experiences. Anyone standing too close to a campfire will soon move farther away to reduce the incoming heat from thermal radiation, so our intuition suggests that radiation must be critical to wildfire spread. As will be seen in **Chapter 5**, this is demonstrably incorrect for many wildland fuel conditions. Radiation that will produce a second-degree burn on exposed skin in 5 s takes more than 27 min to ignite wood (Cohen 2000). In contrast, sparks from a campfire do not look or feel dangerous compared to the flames, but they are often far more effective at causing wildfire spread and igniting nearby homes.

A second reason for the difficulty of answering basic questions about wildfire is that the wildland environment differs considerably from other fire contexts, such as buildings and industrial settings. In these environments, the principles of fire spread are reasonably well understood and substantially different from those in wildlands. In buildings and industrial environments, unlike wildlands, thermal radiation *is* critically important in transferring heat because they have large, continuous surfaces of solid materials. By contrast, because many of the fuels in wildland fires are fine materials like grass and pine needles separated by large gaps of air, they heat and cool more efficiently by convection than by radiation (**Chapter 3 and Chapter 6**). Wildland fuels are also composed of both living and dead vegetation, with heating and cooling properties that are influenced by metabolism and decay – processes that rarely influence fire in other kinds of environments. This book focuses on understanding wildfire behaviour as a dynamical physical process and on the distinctive features of combustion, heat transfer, and ignition in the wildland context. We consider the influences of wildland fuels in their infinite variability, local and synoptic scale weather, and topography on physical processes as they determine wildfire spread and behaviour.

Fire science and the need for experiments

Fire is studied in many scientific disciplines, but the research has been conducted in two main contexts: the wildland environment and engineered environments, especially buildings and industrial settings. Although related, these settings differ in some important ways, particularly the control over and certainty of the materials and conditions of burning, the size and range of scales of the fire phenomenon, the diversity of relevant disciplines involved in research, and the uses of scientific knowledge.

Wildland fire, meaning both free-burning wildfires and intentional burning, is best studied by integrating knowledge from multiple sciences: physics, chemistry, biology, landscape ecology, combustion, fluid mechanics, and atmospherics. Because so many scientific disciplines are needed to understand wildland fire, it is difficult for any individual or research group to acquire comprehensive expertise. In addition to being highly interdisciplinary, the study of wildland fire spans a huge range in space and time scales, from millimetres and milliseconds for combustion and ignition, to tens of kilometres and centuries for large wildfires and their ecological effects (Simard 1991).

Wildfire science should help practitioners manage fire appropriately as an essential ecological disturbance while also making the human presence in and near wildlands more sustainable and safer. However, many land management agencies and the public have used the science principally to support suppression technology and attempt to exclude fire from wildlands.

Fire science in engineered environments also integrates knowledge from several disciplines, but the environments are well characterised and well controlled (Williams 1992), and the scales of space and time are much more limited. Fire research in built environments is generally conceived with broader applications in mind than fire research in wildland environments: fire protection engineering aims to anticipate, prevent and mitigate fire in built environments, and seeks improvements in materials and construction practices. The research focuses on improving utility and safety of habitations and industrial settings.

The importance of experiments in understanding fire in both engineered and wildland environments cannot be overstated. Experience-based use of wildland fire, essentially practical and qualitative experiments in applying fire, for millennia enabled indigenous peoples to understand fire, use it to meet objectives, and reliably anticipate its consequences. In recent times, experiments had been the mainstay of fire research worldwide through the 1970s. In the past century, research programs at Borhamwood Station in the United Kingdom (Read 1994), and the USDA Forest Service in Berkeley, California; Macon, Georgia (see USDAFS 1991; Weise and Fons 2014; Smith 2017); and Missoula, Montana (Smith 2012, 2017), used experiments to investigate many critical questions about fire behaviour. These research efforts led to the development of correlations that described burning rate, heat transfer, ignition and spread. The results applied to both fire protection engineering and wildland fire. Subsequently, fire protection engineering maintained an emphasis on experimental research at the National Institute of Standards and

Technology (NIST, formerly National Bureau of Standards) in Maryland and at the Factory Mutual Corporation (now FM Global); the resulting knowledge was developed into a solid theoretical foundation for understanding fires in buildings and industrial settings and for training fire investigators (Quintiere 2006; Torero 2013).

Sometime after the 1970s, the emphasis on experiment-based research in wildland fire began to diminish. Notable contributions have continued with laboratory- and field-based experimental research from Australia, Canada, Europe and the United States, and we synthesise these in this book. However, research has increasingly departed from experimental work towards modelling. The departure has been driven in part by dramatic improvements in computing and information technology and in part by the view that wildland fire is a suppression problem that should be approached as an emergency which requires modelling solutions and an instantaneous response. Thus, the operational and predictive needs of wildfire suppression have driven research investment at the expense of interests in proactive fire uses in land management, ecosystem function, and even basic scientific understanding. The emphasis on suppression has ultimately reduced both ecosystem health and human safety, as is now profoundly realised by urban and rural populations around the world.

For a perspective on the consequences of investing mainly in reactive wildfire suppression, consider these comparisons: what would be the consequences of focusing health-related investments mainly in paramedics, ambulances and emergency rooms as opposed to public health policy, testing, preventive care and physician visits? *What would be the consequences of structural-fire investments mainly in fire engines and urban firefighters rather than in the design of buildings and materials?* Modern, recently constructed urban buildings throughout the world have materials and designs to prevent catastrophic fires due to past fire disasters that stimulated scientific research, engineering solutions, and consequent changes to building codes and standards (Arnold 2005; Quintiere 2006). Properly directed wildland fire research could similarly benefit proactive and sustainable management of fire in our wildlands and mitigate risk to nearby human communities.

Wildland fire disasters for human communities and natural resources have been increasing in the past several decades despite greater and greater expense and effort to combat them. Real solutions in the future will almost certainly depend upon sound understanding of fire behaviour, which will indicate ways to modify and use wildland fires rather than waiting for ignitions and then reacting. Our modern laws, regulations and cultural expectations demand great certainty and accountability in wildfire management, but we have little of the understanding of fire itself that would allow us to meet these stringent requirements (Finney 2020). It is interesting to note that, before European settlement of the 'New World', the indigenous peoples of the Americas and Australia used fire routinely to reliably and safely manage their environments. Without electronic communications, motorised transportation and construction technology, indigenous peoples were ostensibly more vulnerable to wildfire, but their comprehensive knowledge and routine use of fire meant they were less threatened by it than we are today (see reviews by Stewart 2002; Gammage 2012). Our science must advance to new levels in order to support skills in proactive and beneficial uses of fire that could once again match the experiential competence of ancient peoples.

Wildland fire science since 1900

Scientific investigation of wildfire in the early 1900s was driven by the need for predictive tools to support planning efforts and operational decisions that were both tactical and strategic. Initial efforts focused on what is now called *fire danger rating*, a process that provides current indices and forecasts for rating the potential for ignition and fire spread, and the difficulty of fire control (Hardy and Hardy 2007). Fire danger rating is used to determine staffing requirements for fire suppression forces, dispatch firefighters, set fire restrictions for camping and logging, and pre-position firefighters in anticipation of upcoming weather conditions. In the United States in the 1920s and 1930s, Harry Gisborne pioneered methods for measuring fuel moisture and weather and produced a series of meters to indicate fire danger (Hardy 1983). Subsequent efforts to formalise fire planning based on fuels and potential fire behaviour were developed by Hornby (1936) and Barrows (1951). In Canada, Wright (1932) and Beall (1947) produced methods for rating fire danger and fuel moisture conditions based on weather data. Scientists also sought ways to predict the spread rate of fires, because it could be used to estimate changes in fire perimeter and area over time and thus the effort required to contain the expanding fire. Scientists in Australia and Canada developed predictions of fire spread through a field-based empirical approach; they collected fire spread data from wildfires and prescribed fires in various vegetation types and related the data to moisture and wind (McArthur 1966; 1967) or danger rating indices (Van Wagner 1990). However, the ability to generate fire behaviour estimates directly from environmental factors would require more research and development.

A milestone was reached in the United States in 1972, when Richard C. Rothermel at the Northern Forest Fire Laboratory in Missoula, Montana (now the Missoula Fire Sciences Laboratory) developed an equation to predict wildfire spread rate (Rothermel 1972; Andrews 2018). The semi-empirical model was based on laboratory experiments from the previous decade and a theoretical physical framework (Frandsen 1971). It was originally directed to serve as the quantitative foundation of a new fire danger rating system in the United States (see Deeming *et al.* 1972, 1977). However, soon after its development, Frank Albini applied the Rothermel equation to predicting fire behaviour, writing a computer program to calculate a fire's rate of spread and flame length (FIREMODS) (Albini 1976a) and developing nomograms for graphically performing the calculations (Albini 1976b). The distinction compared to danger rating is that fire behaviour is focused at finer scales, using the same data on fuel, topography and weather to make specific calculations of spread and energy release and then predict fire movement and size (see Andrews 1986, 2018). The assumption was, and still is, that coarse descriptors of the environment (fuel, topography, weather) can be used to make fine-scale calculations of fire behaviour. The disparity in scales between inputs, actual physical processes in fire behaviour, and outputs makes the modelling more diagnostic than predictive. Nevertheless, Rothermel's equation has proven for nearly 50 years to be practical and useful for both purposes because it could be supplied with reasonable inputs for describing the environment and was robust to their uncertainties. This model offered fire managers an objective framework to mechanistically link independent environmental factors to quantitative fire characteristics. The rapid rise in computing and information technology that began in the 1970s allowed for the development of more predictive tools and decision support systems based on the Rothermel equation (e.g. the BEHAVE system, Andrews 1986). Soon the term *fire modelling* became largely synonymous with fire behaviour research in general – a trend that continues to this day.

Since the 1980s, the overwhelming operational success and great utility of Rothermel's spread rate equation has inspired more and more modellers to offer physical or semi-physical formulations that could improve upon his semi-empirical one (see Pastor *et al.* 2003; Sullivan 2009). Their objective was to use environmental characteristics to calculate a one-dimensional steady fire spread rate perpendicular to a linear flame zone. Aided by improvements in computing and information technologies, physical sciences were quite rightly seen as the means to overcome limitations and assumptions in empirical wildfire models and to refine the scales of data and processes used to drive fire spread. Empirical models of fire spread rate correlated with coarse environmental variables (e.g. fuel type, moisture, wind) are restricted to the set of observations without offering physical explanation of the phenomenon or addressing of interactions among factors outside the input data. Each fuel or vegetation 'type', for example, requires its own dataset and sometimes different variables and equation forms for statistical fitting of a fire spread model. Ironically, though, the plethora of attempts to replace these with physical fire spread rate models has not been unifying. In fact, the proliferation of models, with their diversity in physical formulations, has succeeded in mystifying rather than enlightening the subject of wildland fire behaviour. We seem to have verified for wildfire the observation by Williams (1992) concerning the difference between technology and science as applied to combustion:

> *It is relevant to distinguish between the science and the technology of the subject. The march of technology has never hesitated. It uses science whenever possible but often, especially in combustion, forges ahead by trial and error, or fortuitously by application of scientific misconception, but without scientific understanding, as it did during the first half-million years.*

The diversity of models has not yielded a convergence of thinking nor has it resulted in advancing development of practical tools. Why is this? Our answer is the rationale for this book and the research approach it advocates. It is worth a bit of discussion as to why our approach is distinct. First, most models had the all-consuming objective of calculating a *fire spread rate* from initial environmental conditions (fuel, topography, weather). Focus on an assumed steady-state spread was seen as simplifying, much as Rothermel had done, because it could ignore time- and space-dependent processes that determined how the fire somehow managed to spread in the first place (i.e. the fire itself). The physical processes have long been known and listed for many fire spread phenomena (Williams 1977), but their configuration or organisation in the special condition of steady spread among discrete wildland fuel particles had to be assumed. This was often conceived as a balance between energy released and energy required for ignition (Frandsen 1971). Dr Don Latham, research physicist at the Missoula Fire Sciences Laboratory humorously asked, 'Do fires spread as fast as they can or as slow as they can?', as a way of illustrating the conundrum posed by the so-called simplifying steady-state assumption. So, the artifice of steady spread meant that any model thus proposed didn't naturally address *how* or *if* the fire was able to spread (i.e. threshold behaviours) and didn't address the means required to achieve spread. *Post hoc* assumptions are required to limit spread, for example, from high moisture or fuel discontinuity or an upper limit for very high winds, whereas an accurate physical formulation would predict these limits from the model. Rothermel's equation introduces an extinction moisture to set the lower bound of spread and scale moisture effects in the calculations, and fuels are assumed continuous and homogenous.

Second, the desired steady-state solution was implicitly based on the notion that spread resulted from steady processes. Anyone who has watched wildfires or even a campfire becomes transfixed by the movement of flames. But most model formulations assumed these motions away as noise – especially if radiation could be justified as the principal and sufficient heat transfer mechanism. Radiation had long been rationalised as sufficient alone (see reviews by Pitts 1991; Baines 1990), and this is evidenced by the schematic drawings in many research papers of a smooth flame profile stiffly tilted forward to ignite fuels by radiation. Likewise, experiments were designed to gather data on *how fast* fires spread in their steady state. Experiments were not designed to find out *how* fires spread and very few to discover *if* fires spread, which of course would mean fire was not steady. The physical processes underpinning fire spread are necessarily at such a fine scale of space and time that they would challenge fuel description and modelling. The preoccupation with model building and predictions of steady spread rate has also confounded attempts to compare predictions with observations, in other words *validation*, but this will be discussed in more detail later concerning the use of field experiments.

Modelling itself and the ability to predict fire spread became the objectives for applied research, rather than understanding the mechanics of fire spread, so no verified basic theory has been available to explain the processes that interact to produce spreading wildfires. Without such theory to anchor model design, the diversity of models for fire spread rate has offered confusion rather than clarity. The absence of highly controlled experiments designed for the purpose of developing and verifying a theory is understandable, however, since this kind of research is expensive and time-consuming, with little certainty as to the cost and timeline of needed but unknown discoveries. Such experiments require a facility staffed with specialists to deal with fabrication, instrumentation, data collection and analysis. Few laboratories are designed with the ability to accommodate spreading fires even at small scales. Even more important, to be useful in explaining the mechanics of wildland fire, experiments must be designed to answer questions that derive from both understanding of physical principles and experience with wildland fires. There are few places in the world where the necessary interdisciplinary expertise, laboratory facilities, time and financial support are available and can be devoted to developing a validated basic theory of how wildland fire behaves. Funding has often been less available for basic research than for modelling, partly because of the expense and uncertainty of delivering results in a short time frame is incompatible with the system of competitive research grants. Many of the models developed in the past 50 years have been extrapolated beyond what their underlying theories and data can support. New models and predictive tools are needed, based on a sound understanding of the physical principles that drive fire behaviour in wildlands. In this book, we survey the science that must underpin this understanding and summarise the research that is now contributing to the understanding needed for new models and new applications.

Modelling and field-scale research

Our emphasis here – on highly controlled, laboratory-based experiments for the purposes of developing a theory of fire spread – by no means obviates the value of modelling or field-scale research. The experiments needed to develop a physical theory of fire spread and behaviour are by necessity small in scale. They are intentionally simplified and isolated to pieces of a phenomenon set in the context of wildland fires (as opposed to building fires). Experiments designed to elucidate sub-processes, such as heat transfer and ignition, must then be integrated with the other sub-processes of the wildfire system to capture larger scale interactions. **Chapters 2–6** introduce the physical processes that drive wildland fire behaviour. **Chapter 7** presents the theoretical modelling that provides this integration and the inductive reasoning needed to understand the emergence of fire spread (or no spread) as an outcome of a dynamical system.

When modelling is used in research (as opposed to application), it often suggests hypotheses that warrant further experimental work. The cycle of modelling and experiments is essential to advancing knowledge and developing a physical theory of fire spread. Such a theory would serve as the foundation for many models and end-user systems for many different applications. For practical uses, models must meet a particular business case defined by intended use and required precision. Practical models or systems must be robust to the uncertainty of user-supplied inputs, target the needed precision and time frame of the predictions, and match the expertise of intended user groups. Highly complex and physically explicit models are of little use if the intended clients cannot learn how to use and interpret them, if the model cannot be fed reliable inputs, or if the model cannot produce usable outputs fast enough to inform decision making.

It may seem incongruous that simple experiments at laboratory scales can be relevant to understanding and modelling larger scale complex free-burning wildfires. The contrasting strengths and limitations of laboratory *v.* field-scale research are the 'two solitudes in forest fire research' addressed by Van Wagner (1971). Laboratory experiments offer measurements and control variation but lack direct applicability to phenomena at field scales. Field-scale fires, whether experimental or wild, are difficult to control and instrument, and experimental burns may also be restricted to milder ranges of environmental conditions than wildfires (Stocks *et al.* 2004; Clements *et al.* 2007; Gould *et al.* 2008; Ottmar *et al.* 2016). But applicability of the measurements from field-scale fires to empirically derived models or predictions of fire behaviours is straightforward. Empirical modelling has shown great utility for management systems in Canada (Forestry Canada Fire Danger Group 1992) and Australia (Cruz *et al.* 2015). Van Wagner (1998) explained:

> *The Canadian approach uses all available field data for the statistical links between weather, fuel moisture, and fire behavior, tying the whole together with a combination of physical principle and mathematical design.*

Although these statistical-empirical models apply directly at the needed time and space scales (e.g. for fire spread rates over kilometres and hours), they are not intended to reflect processes or circumstances at finer or coarser scales than the range of data collected. Fire behaviour response to management actions that change fuel structure, such as prescribed burning and forest thinning, require new sets of data. Fires in stronger winds, or on slopes steeper than the range of original datasets, require extrapolations. Empirical models also aren't attempting to *explain* how any of the processes fit together physically, even though some kind of physical reasoning may be used to organise the modelling approach (as Van Wagner (1971, 1998) states). Thus, the model utility that comes from intuitive correlations often requires that multiple factors and their complex interactions be combined into single variables for model building. For example, vertical and horizontal continuity of fuels and proportions of living and dead materials are implicit properties of generic fuel types (i.e. grasslands, conifer forests). Wide variations in their properties introduces thresholds in fire spread that are not explained by the models and can only be accommodated in fire predictions with considerable judgement on the part of the practitioner.

The challenge of validation

Field-scale research is important to validate scaling of physical processes identified from principles shown at the laboratory scale. In other words, do the same processes occur in small fires as in big fires? This is not commonly the way that validation has been used in wildfire research, but it is the way that we suggest it is most useful. In **Chapter 9** we review methodologies and challenges for obtaining measurements from fires at laboratory and field scales. Most previous attempts at model validation attempt to compare observations with spread rate predictions from wildfire models. These attempts have been difficult and inconclusive. Consider this passage from Albini and Anderson (1982), in which they describe their attempts to compare observations with calculations from the Rothermel model as implemented in the FIREMODS program (Albini 1976a):

> *Attempts to evaluate the theoretical fire behavior models described above through operation or field-level experiment resulted in unclear definitions of the reasons for output inconsistencies. It is impossible for the most part to isolate the cause of output variations to either the accuracy or resolution of the input data, applicability of the model to the real world, or (in some cases) to errors which may have crept into the program during the implementation process. Even in the simplest model, FIREMOD, complex relationships between input data readily available to field personnel and internal model parameters exist. This makes it extremely difficult to evaluate which input variable is suspect*

when deviations are noted between predicted and actual fire behavior characteristics.

The difficulty of validating fire spread rate predictions from field data can be traced to two main factors. One is the ambiguity among sources of error in fire modelling and the second has to do with the metric of fire spread rate itself.

The concept of *validation* involves isolating and quantifying *model error* separate from other sources, such as *data error* and *user error*. Ideally, we want to know how much error and bias is introduced by the model itself. Given perfect input data and observations of the phenomenon, it should be possible to determine model error, but, as Oreskes *et al.* (1994) describe, the data required for input to a model and comparison with model results are always under-represented, so validation is technically impossible. Data errors and under-representation in field data come from unknown and uncontrolled variation in many factors. Examples include overgeneralised fuel descriptions in both research and operational settings, and wind speeds and directions that are never constant in time or stationary in space. Data errors come from measurements and observations of fire itself. For example, predictions and measurements aim to quantify flame length and spread rate as simple numbers, but both of these phenomena actually represent highly unstable processes with huge variability. There is also a possibility that users will introduce error by assuming constant conditions or compensating for conditions that are unknown. Examples include the moisture contents of fuel particles of different sizes, three-dimensional wind variation induced by complex terrain or the fire itself, and the choice of which nearby weather station to obtain data for use in predictions.

The main challenge to validating modelled fire spread rate has to do with how wildfire spread occurs:

- First, fire spread rate is not a physical quantity as is temperature, heat release rate or heat flux (energy received per unit time – see **Appendix A**). **Instead, spread rate is an outcome of sequential ignitions** (Fons 1946; Frandsen 1971). In the wildland context, therefore, fire spread rate is not a continuous function because it emerges from a series of ignitions of discrete fuel particles, which occur over some unit of time and space. Its instantaneous values vary wildly, so it is subject to variation over multiple scales, from centimetres and seconds in pine needle fuel beds to kilometres and hours for large-scale crown fires. The variation and sources of variation are likely to change with those scales. For example, a model's assumption of uniform fuel and stationary weather becomes untenable for estimating spread rate over scales of kilometres and hours.
- Second, fire spread rate is a one-dimensional representation of a two-dimensional and sometimes three-dimensional advancing front. Since fire growth occurs in two or three dimensions, the fire front can meander in spread direction as winds shift or patchy fuels drive uneven movement. However, these directional variations are not captured in measurements of spread rates or predictions from one-dimensional models.
- Third, the physical means of spread often varies over long distances and times. For example, spotting may contribute to moving a fire over barriers or rivers.
- Fourth, there is no unique combination of environmental factors that will produce a given spread rate. Different combinations of fuel characteristics, moisture content, and wind or slope, their sequence in time and space, and their variability can cause the same fire spread rate. Errors in estimating one property may be compensated for by opposite errors in another.

In this book, we do not focus on spread rate as the only metric of validation because it is closely linked with all the other characteristics of the flaming zone. Instead, we address primarily how the physical processes are assembled in a dynamical system, and whether interactions of the components at a fine scale yield reasonable behaviours at the larger scale.

Scale modelling has not seen common use in wildland fire for many years (e.g. Byram 1959; Van Wagner 1971) but is common in fire protection engineering and combustion-related engineering disciplines. Scale modelling seeks to understand the main factors controlling a particular physical behaviour and develop simplified relations among these factors that can be used consistently across scales of, for example, time, distance, velocity, and energy release. With respect to wildfire and fires in general, very useful relations have been developed. They allow for practical application of laboratory-scale experiments at much broader scales. The vast difference in scale between laboratory experiments and field-scale wildfires suggests that great value can be found in scale modelling with proper design of experiments (see summary by Saito and Finney 2014). There are numerous examples of studies that have used this approach. Thomas (1971), for example, concluded in his study of wind-driven fires that 'a judicious mixture of theory and empiricism allows idealized experiments to represent the main features governing this kind of wild fire'.

Outline of the book

This book is organised to develop a theory of wildland fire spread and behaviour as a physical system. The principles of combustion, heat transfer and ignition are explained generally and also specifically for the wildland context. Experimental results are the primary source of information; they are compared where possible against observations at field scale. The strong coupling exhibited by the wildfire system is explored through a simplified one-dimensional model and the model behaviours are compared with observations. Limits to the modelling are then illustrated by discussion of large-scale behaviours of fires in complex terrain and with spatially variable wind and fuel patterns. If the underlying theories are sound, then they will underpin explanations for the behaviours found even in complex circumstances. A passage by Byram (1959) offered the same vision:

If fire can be reduced to its basic-energy processes and physical component parts, and if the interactions and relationships between the parts can be determined, then the resulting fire system model represents the physical system needed to unify the various fire behavior phenomena. A test of the effectiveness of such a model is its ability to anticipate new fire behavior situations and predict new fire behavior phenomena, as well as explain observed fire behavior. In addition, the model should represent a physical system which applies to fires of all sizes and intensities.

This statement implies that, with a sound understanding of wildfire behaviour, we will have the knowledge essential not only for making predictions but also for identifying new fire behaviour situations, recognising new physical phenomena, and realising new management opportunities. This understanding of wildland fire will ultimately be of the greatest value to humanity. This book is an attempt to develop such knowledge through reviewing what is known and what remains unknown about wildland fire behaviour.

Chapter 1: Introduction to wildfire science. Why should we care about wildfire behaviour? This chapter reviews some history of the science of wildfire behaviour and offers some perspective on why we continue to struggle to understand how wildland fires spread. Although readily observed, wildfires are difficult to explain. Decades of modelling and technology development in wildfire prediction systems have done little to develop a firm theory of dynamical wildfire behaviour. This chapter sets the stage for experiment-based research to gain knowledge with the aspiration to ultimately achieve compatibility with wildland fire.

Chapter 2: Fire and wildland fire behaviour. In this chapter, we introduce fire as a system of interacting physical processes. Although these processes take place at very small scales of time and space, we will see that these same processes operate during even the largest of wildland fires. We will see how these processes work together to produce the behaviours of wildfires by interacting within a physical system. The basic characteristics of the fire environment will be introduced, and we will see conceptually how they affect the traditional metrics used to describe wildland fire behaviour: spread, growth and energy release.

Chapter 3: Thermodynamics, fluid mechanics and heat transfer. This is the first of three chapters that describe the physical processes operating in wildland fires. Thermal science is introduced as the framework for thinking about concepts and characteristics of heat, energy, temperature and how properties of matter are affected. Fluid mechanics is introduced, because it deals with the physical properties of gases and liquids; our discussion is applied narrowly to the specific characteristics relevant to wildland fire. Basic principles of fluid flows and heat transfer are also covered. These principles are essential to understanding how heat from a burning fire can affect and ignite adjacent fuel particles in the process of fire spread.

Chapter 4: Combustion. Combustion is a science discipline with applications across diverse fields of engineering and industry, from engines to explosions. Here we focus on combustion science in the wildland fire context. We would not have wildfires without heat released from unconfined combustion of vegetation fuels and oxygen in the atmosphere. This chapter addresses the chemistry of fuel and combustion reactions, flame structure and sizes, combustion limits of gases, burning and heat release rates of fuel, and solid phase combustion.

Chapter 5: Ignition. Wildfires must ignite new fuel to spread and be sustained. Ignition is a discrete change of state in fuel material. Wildfire spread is a non-steady outcome of a complex system that generates a spatial sequence of repeated ignitions. The processes involved in ignition are functions of material properties, the heat transfer processes, and the physical sizes and shapes of the materials. Ignition requirements for flaming and smouldering of wildland fuel are discussed.

Chapter 6: The environment in wildfire dynamics. The wildland fire environment, particularly the fuel, topography and weather conditions, are considered the essential ingredients of fire behaviour. The overview in **Chapter 2** introduced them along with the basic principles of wildland fire behaviour. In this chapter, these environmental factors are described in detail and presented in terms of their effect on the vital physical processes of fire spread and behaviour: heat transfer, combustion and ignition. The state of knowledge of these environmental components is summarised and illustrated using experiment-based examples.

Chapter 7: Wildfire spread. The coupled system of wildfire spread is examined through a simple one-dimensional physical model. Simplified components of combustion, heat transfer and ignition are assembled into a particle-based dynamical model that represents fire spread as a product of feedbacks operating within a system. Fire behaviours of spread and intensity emerge from the model, as do the time- and space-dependent responses of acceleration and steady-state spread, dependencies on initial conditions, extinction from wind and moisture and fuel discontinuity, and spread thresholds. The model elucidates system level behaviours that are not properties of the individual component processes or environmental inputs, and thus it serves to inductively expand our explanation of fire behaviour as a dynamical system.

Chapter 8: Behaviours of large fires. This chapter surveys some wildfire phenomena, including crown fires, spotting, fire shapes, mass fires and fire storms. These phenomena are not well approximated by small-scale studies because they exhibit feedbacks at the scale of the entire fire with strong atmospheric interactions or because they involve large length-scales and high fluxes of thermal radiation. Most fire behaviour research has been focused on relatively small fires and representative segments extracted conceptually from larger fires (line fires). The same basic physical processes operate in all fires, but the largest fires interact with increasingly larger volumes of the earth's atmosphere, essentially becoming mesoscale meteorological events.

Chapter 9: Measurements in fire behaviour. Explanations of wildfire behaviour are founded on quality observations and measurements taken from laboratory and field experiments as well as large wildfires. This chapter summarises traditional and newer methods of obtaining measurements of fuel consumption, physical fire processes and fire behaviours.

Chapter 10: Ignition techniques for experimental burning. We present a discussion of the various ignition configurations used in experimental burning and the appropriate research questions for each technique. Successful use of most ignition techniques in experimental fires relies on understanding fire interactions and dynamical behaviours that well exceed the assumptions of steady-state spread. Experimental fires are vital to testing hypotheses of fire behaviour and testing the scaling of physical processes by application to the field level. This chapter examines basic ignition techniques and presents qualitative differences in spread and behaviour resulting from ignition geometry and spatial patterns of ignitions.

Chapter 11. We conclude with a review of key principles and insights as they pertain to researchers and managers.

References

Albini FA (1976a) *Computer-based Models of Wildland Fire Behavior: A User's Manual*. USDA Forest Service, Intermountain Forest and Range Experiment Station, Ogden, UT.

Albini FA (1976b) 'Estimating wildfire behavior and effects'. General Technical Report INT-30. USDA Forest Service, Intermountain Forest and Range Experiment Station, Ogden, UT.

Albini FA, Anderson EB (1982) 'Predicting fire behavior in U.S. Mediterranean ecosystems'. In *Proceedings of the Symposium on Dynamics and Management of Mediterranean-type Ecosystems*. (Eds EC Conrad and WC Oechel). General Technical Report PSW-58. USDA Forest Service, Pacific Southwest Forest and Range Experiment Station, Albany, CA.

Andrews PL (1986) 'BEHAVE: fire behavior prediction and fuel modeling system-BURN subsystem, Part 1'. General Technical Report INT-194. USDA Forest Service, Intermountain Forest and Range Experiment Station, Ogden, UT.

Andrews PL (2018) 'The Rothermel surface fire spread model and associated developments: a comprehensive explanation'. General Technical Report RMRS-GTR-371. USDA Forest Service, Rocky Mountain Research Station, Fort Collins, CO.

Arnold J (2005) 'Large building fires and subsequent code changes'. Department of Development Services Building Division, Las Vegas, NV, <http://ddwei.info/pdf/subsequent/0.pdf>.

Baines PG (1990) Physical mechanisms for the propagation of surface fires. *Mathematical and Computer Modelling* 13(12), 83–94. doi:10.1016/0895-7177(90)90102-S

Barrows JS (1951) *Fire Behavior in the Northern Rocky Mountain Forests*. Station Paper No. 29. USDA Forest Service, Northern Rocky Mountain Forest and Range Experiment Station, Missoula, MT.

Beall HW (1947) 'Research in the measurement of forest fire danger'. Paper Inf. Rep. FF-X-8. Forest Fire Research Institute, Canadaian Forestry Service, Ottawa, Ontario.

Bowman DM, Balch JK, Artaxo P, Bond WJ, Carlson JM, Cochrane MA, D'Antonio CM, DeFries RS, Doyle JC, Harrison SP, Johnston FH (2009) Fire in the Earth system. *Science* 324(5926), 481–484. doi:10.1126/science.1163886

Brown JK, Arno SF (1991) The paradox of wildland fire. *Western Wildlands* 17, 40–46.

Byram GM (1959) Forest fire behavior. In *Forest Fire: Control and Use*. (Ed. KP Davis) pp. 90–123. McGraw Hill, New York, NY.

Calkin DE, Thompson MP, Finney MA (2015) Negative consequences of positive feedbacks in US wildfire management. *Forest Ecosystems* 2(1), 9. doi:10.1186/s40663-015-0033-8

Clements CB, Zhong S, Goodrick S, Li J, Potter BE, Bian X, Heilman WE, Charney JJ, Perna R, Jang M, Lee D (2007) Observing the dynamics of wildland grass fires: FireFlux – a field validation experiment. *Bulletin of the American Meteorological Society* 88(9), 1369–1382. doi:10.1175/BAMS-88-9-1369

Cohen JD (2000) Preventing disaster: home ignitability in the wildland–urban interface. *Journal of Forestry* 98(3), 15–21.

Cruz MG, Gould JS, Alexander ME, Sullivan AL, McCaw WL, Matthews S (2015) Empirical-based models for predicting head-fire rate of spread in Australian fuel types. *Australian Forestry* 78(3), 118–158. doi:10.1080/00049158.2015.1055063

Deeming JE, Lancaster JW, Fosberg MA, Furman RW, Schroeder P (1972) 'National fire-danger-rating system'. Research Paper RM-84. USDA Forest Service, Rocky Mountain Forest and Range Experiment Station, Fort Collins, CO.

Deeming JE, Burgan RE, Cohen JD (1977) 'The national fire-danger rating system – 1978'. General Technical Report

INT-39. USDA Forest Service, Intermountain Forest and Range Experiment Station, Ogden, UT.

Finney MA (2020) The wildland fire system and challenges for engineering. *Fire Safety Journal*. doi:10.1016/j.firesaf.2020.103085

Fons WL (1946) Analysis of fire spread in light forest fuels. *Journal of Agricultural Research* **72**(3), 93–122.

Forestry Canada Fire Danger Group (1992) 'Development and structure of the Canadian Forest Fire Behavior Prediction System'., Inf. Rep. ST-X-3. Forestry Canada, Ottawa, Ontario.

Frandsen WH (1971) Fire spread through porous fuels from the conservation of energy. *Combustion and Flame* **16**(1), 9–16. doi:10.1016/S0010-2180(71)80005-6

Gammage B (2012) *The Biggest Estate on Earth*. Allen and Unwin, Sydney.

Gould JS, McCaw WL, Cheney NP, Ellis PF, Knight IK, Sullivan AL (2008) *Project Vesta – Fire in Dry Eucalypt Forest: Fuel Structure, Fuel Dynamics, and Fire Behaviour*. Ensis-CSIRO and Department of Environment and Conservation, Canberra and Perth.

Hardy CE (1983) 'The Gisborne era of forest fire research: legacy of a pioneer'. Report FS-367. USDA Forest Service, Northern Rocky Mountain Forest and Range Experiment Station, Missoula, MT.

Hardy CC, Hardy CE (2007) Fire danger rating in the United States of America: an evolution since 1916. *International Journal of Wildland Fire* **16**(2), 217–231. doi:10.1071/WF06076

Hornby LG (1936) 'Fire control planning in the northern Rocky Mountain Region'. Progress Report No. 1. USDA Northern Rocky Mountain Forest and Range Experiment Station, Missoula, MT.

McArthur AG (1966) *Weather and Grassland Fire Behaviour*. Forestry and Timber Bureau Australia, Canberra.

McArthur AG (1967) *Fire Behaviour in Eucalypt Fuels*. Forestry and Timber Bureau Australia, Canberra.

Oreskes N, Shrader-Frechette K, Belitz K (1994) Verification, validation and confirmation of numerical models in the Earth sciences. *Science* **263**, 641–646. doi:10.1126/science.263.5147.641

Ottmar RD, Hiers JK, Butler BW, Clements CB, Dickinson MB, Hudak AT, O'Brien JJ, Potter BE, Rowell EM, Strand TM, Zajkowski TJ (2016) Measurements, datasets and preliminary results from the RxCADRE project–2008, 2011 and 2012. *International Journal of Wildland Fire* **25**(1), 1–9. doi:10.1071/WF14161

Pastor E, Zárate L, Planas E, Arnaldos J (2003) Mathematical models and calculation systems for the study of wildland fire behaviour. *Progress in Energy and Combustion Science* **29**(2), 139–153. doi:10.1016/S0360-1285(03)00017-0

Pitts WM (1991) Wind effects on fires. *Progress in Energy and Combustion Science* **17**(2), 83–134. doi:10.1016/0360-1285(91)90017-H

Pyne SL (2010) The ecology of fire. *Nature Education Knowledge* **3**(10), 30.

Quintiere JG (2006) *Fundamentals of Fire Phenomena*. John Wiley & Sons, Chichester.

Read REH (1994) 'A short history of the fire research station, Borhamwood'. Building Research Establishment Report. IHS BRE Press, Garston.

Rothermel RC (1972) 'A mathematical model for predicting fire spread in wildland fuels'. Research Paper INT-115. USDA Forest Service, Intermountain Forest and Range Experiment Station, Ogden, UT.

Saito K, Finney MA (2014) Scale modeling in combustion and fire research. *Nihon Nenshou Gakkaishi* **56**(177), 194–204.

Simard AJ (1991) Fire severity, changing scales, and how things hang together. *International Journal of Wildland Fire* **1**(1), 23–34. doi:10.1071/WF9910023

Smith DM (2012) 'The Missoula Fire Sciences Laboratory: A 50-year dedication to understanding wildlands and fire'. General Technical Report RMRS GTR-270. USDA Forest Service, Rocky Mountain Research Station, Fort Collins, CO.

Smith DM (2017) Sustainability and the origins of wildland fire research. USDA Forest Service, Washington Office Publication FS-1085, https://www.fs.fed.us/rm/pubs_series/wo/wo_fs1085.pdf.

Stewart OC (2002) *Forgotten Fires: Native Americans and the Transient Wilderness*. University of Oklahoma Press, Norman, OK.

Stocks BJ, Alexander ME, Lanoville RA (2004) Overview of the International Crown Fire Modelling Experiment (ICFME). *Canadian Journal of Forest Research* **34**(8), 1543–1547. doi:10.1139/x04-905

Sullivan AL (2009) Wildland surface fire spread modeling, 1990–2007. 1. Physical and quasi-physical models. *International Journal of Wildland Fire* **18**, 349–368. doi:10.1071/WF06143

Thomas PH (1971) Rates of spread of some wind-driven fires. *Forestry: An International Journal of Forest Research* **44**(2), 155–175. doi:10.1093/forestry/44.2.155

Torero JL (2013) Scaling-up Fire. *Proceedings of the Combustion Institute* **34**, 99–124. doi:10.1016/j.proci.2012.09.007

United States Department of Agriculture Forest Service (USDAFS) (1991) 'Thirty-two years of forest service research at the Southern Forest Fire Laboratory in Macon, GA'. General Technical Report SE-77. USDA Forest Service, Southeastern Forest Experiment Station, Asheville, NC.

Van Wagner CE (1971) 'Two solitudes in forest fire research'. Inf. Rep. PS-X-29. Canadian Forestry Service, Petawawa Forest Experiment Station, Chalk River, Ontario.

Van Wagner CE (1990) Six decades of forest fire science in Canada. *Forestry Chronicle* **66**(2), 133–137. doi:10.5558/tfc66133-2

Van Wagner CE (1998) Modelling logic and the Canadian Forest Fire Behavior Prediction System. *Forestry Chronicle* **74**, 50–52. doi:10.5558/tfc74050-1

Weise DR, Fons TR (2014) Wallace L. Fons: fire research pioneer. *Forest History Today Spring/Fall,* 57–59.

Williams FA (1977) Mechanisms of fire spread. *Symposium (International) on Combustion* **16**(1), 1281–1294. doi:10.1016/S0082-0784(77)80415-3

Williams FA (1992) The role of theory in combustion science (Hottel Plenary Lecture). *Symposium (International) on Combustion* **24**, 1–17.

Wright JG (1932) Forest fire hazard research. *Forestry Chronicle* **8**(3), 133–151. doi:10.5558/tfc8133-3

2

Fire and wildland fire behaviour

Fire is easy to recognise but difficult to describe. In its most basic sense, fire is a high-temperature reaction of fuel and oxygen that releases energy. Much of the energy is released as heat, so we can say that *fire is exothermic oxidation of fuel*. Exothermic means giving off heat, and oxidation is a chemical reaction between fuel and oxygen. While these definitions imply fire is a process rather than a 'thing', they do not give us a clear picture of how the process operates or how we experience it. Our experience is that most fires are composed of flames that give off light as well as heat, just as a candle does and just as campfires do. So that is how we introduce fire here – as a physical process – using our familiar experiences with burning candles and campfires.

The ubiquitous burning candle has been used for lighting by people for centuries. As such it was chosen as the subject of the now-famous Christmas lecture series by renowned British physicist Michael Faraday in 1848 (Faraday 1860). He used the candle (**Figure 2.1**) to explain to public audiences how physical and chemical processes produce the familiar colours and shapes of a candle flame. As Faraday recognised, we can learn a lot from a candle and, amazingly, most of it can be applied directly to wildland fires. In this chapter, we survey the processes and principles that describe the candle flame. The remaining chapters of this book describe the physical processes in greater detail.

The burning candle as a fire process

When we look at a burning candle, we see some familiar features that are also important to wildland fire (**Figure 2.2**). The flame is the most obvious part of the burning candle, but it is not solely 'the fire'. For the flame to be ignited and self-sustaining, physical processes must interact with the wax and the wick. To understand the connections among these parts in a fire, we need some definitions:

Fuel: stored chemical energy that can be released by combustion (i.e. wax)
Combustion: exothermic chemical and physical processes that convert fuel in the presence of oxygen to heat
Heat: thermal energy that can be transferred from high-temperature materials to colder materials
Flame: Gaseous fuels are so hot during combustion that glowing soot particles emit visible light.

A burning candle, like fire in general, is a *coupled system* – that is, a sequence of processes that all depend upon each other. A burning candle is also a remarkably *self-sustaining* system. A candle flame is sustained by the upward flow of liquid wax fuel within the wick, which comes from the melting of solid wax by heat from the flame itself. These

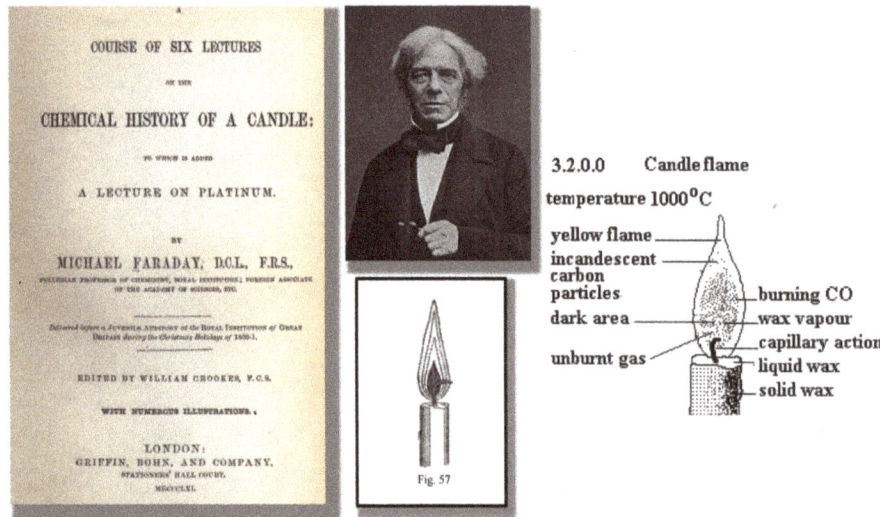

Figure 2.1: The nature of fire as a system was described by Michael Faraday in his 1848 Christmas lecture series on candles (published in 1860). Wildfires have many of the same physical processes as do burning candles.

interdependent processes are sustained until the wax in the candle is consumed (or you blow out the flame).

Interdependencies in a coupled system are called *feedbacks*. Feedbacks can be positive or negative. *Positive feedbacks* drive the system to increase the rates of change within the system, such as energy release rate or fire spread rate. *Negative feedbacks* do the opposite; they reduce the rates of change within the system. Both kinds of feedback are involved in fire at different stages and times. The feedbacks within coupled systems can be

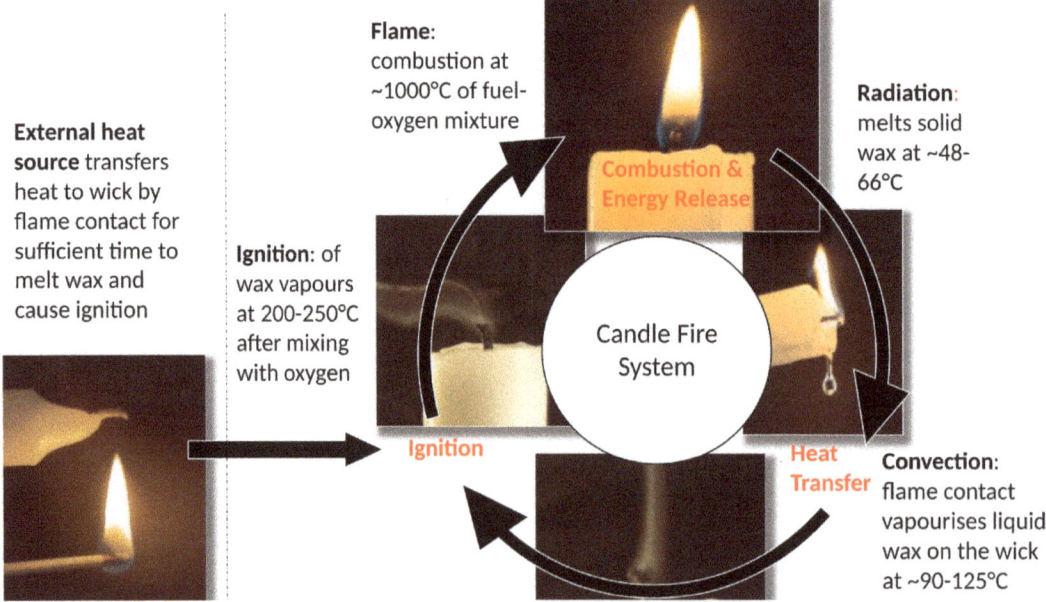

Figure 2.2: Illustration of the coupled system for a candle shows the same components as in wildfires: fuel combustion and energy release, heat transfer by radiation and convection, and ignition (images by Trevor Finney).

highly *nonlinear*, meaning that the system does not respond proportionally to stimuli or changed conditions. In regard to fire, for example, doubling one factor, such as the wind speed or the amount of fuel available, may more than double a system response such as the rate of combustion and energy released.

Unlike candles, wildland fires seldom have a stationary flame zone for long periods of time. To be self-sustaining – that is, to spread – a wildland fire must transfer its heat to adjacent fuels. This makes its flame zone move. It has been difficult to determine how fast a wildfire must move in order to sustain its flame zone because of the tightly coupled nature of the processes in the system. This coupling means that there is no logical place to start describing the system – no beginning and no end. Thus, what goes around comes around; that is, each process affects one or more of the others through space and time. Wildfires, like burning candles, are tightly coupled systems that cannot be understood based on the individual components in isolation. This concept is so important that we will return to it several times throughout the book.

Igniting and burning a candle

The burning of a candle is a product of a coupled system that only begins to function when supplied with ignition from outside (**Figure 2.2**). That is, we light the candle with a heat source like a match or a lighter that is not part of the candle itself. Wildfires are like this too: ignition typically is caused by lightning or human sources external to the wildland fire system, and then the fire itself ignites adjacent fuels if the fire is to be sustained.

Most of us light candles without thinking about two critical physical processes: heat transfer and ignition of fuel. Lighting a candle requires contact between the wick and a heat source like the flame on a match (**Figure 2.2**). The easiest way to light the candle is by positioning the candle wick above the match flame so the flame rises to touch or envelope the wick. By doing this we rely on the fact that a flame is hot and, since heat rises by a property called *buoyancy*, flames are naturally elongated upward. Contact with hot flame easily ignites the wick. This heat transfer between the flame and the wick occurs primarily by *convection*, which is the exchange of heat caused by contact of a fluid (like flame gases) with an object.

How long do you hold the match flame in contact with the wick? This matters a great deal to successful ignition, and you can prove it by pulling the match away in ~1 s. If the heating is insufficient for ignition, you will see smoke or wax vapours coming from the wick, but you will not see flame. As Faraday demonstrated, these vapours are volatilised wax. They are flammable, which you can demonstrate by inserting a lit match into the vapour stream and watching the 'train of fire' descend and attach to the wick. Ignition of solid, gas or liquid fuel requires a certain rate of heating over a certain period of time. Ignition of wildland fuels is similar, in that wildland fires can persist and spread only if heat from the combustion zone can repeatedly ignite new fuels. Here is a detailed description of the steps involved in sustaining a burning candle:

- *Initial ignition.* Heat from the match increases the temperature of the fuel so combustion can begin. In the candle, wax is the fuel that ignites most easily at low temperature, ~200–250 °C. **Chapter 5** covers ignition in more detail.
- *Fuel vaporisation.* Paraffin waxes boil at 90–125 °C. As the wax vapours heat up, they expand in volume, which causes them to rise in the surrounding denser, cooler air. This buoyancy is critical to combustion because the rising vapours mix with oxygen in the atmosphere. When fuel vapours are hot enough and properly mixed with oxygen, they can combust.
- *Combustion.* A series of chemical reactions occur as the vaporised wax mixes with oxygen. These reactions produce a rapid increase in temperature to around 1000 °C. The flame temperature in the burning candle is approximately the same as that in wildland fires, and thus the buoyancy for each is similar.

- *Visible flaming.* Combustion of both candles and wildfires forms carbon (soot) particles in significant quantities. You can see flames because the soot is glowing – that is, releasing heat in the yellow and orange wavelengths, which we can see.
- *Melting wax.* The heat released by the glowing soot melts the wax below the flame into a small liquid pool. Paraffin waxes melt at temperatures between 48 and 66 °C.
- *Liquid wax moving into flame zone.* The melted wax moves up the wick – against gravity – by capillary action. This supplies the flame zone with fresh fuel, which boils and then burns, and the processes continue as a coupled system.

Flame shape

Everyone knows that the candle flame extends vertically to form a teardrop shape (**Figure 2.3a**). Without the earth's gravity, however, the candle flame takes on the appearance of a feeble blue sphere (**Figure 2.3b**). Three ingredients are necessary for combustion – fuel, oxygen, and heat; these comprise the *combustion triangle.* All three are present in both cases, but the flame shapes differ because of gravity. Gravity has a pervasive role in fire behaviour through its effect on buoyancy.

Buoyancy is caused by differences in density between fluids (either gases or liquids) in the presence of gravity. The high-temperature gases in a candle flame are much less dense than ambient air, so they float upward in the presence of gravity (**Figure 2.3**). Their buoyancy removes combustion products from the vicinity of the fuel (the candle's wick). This allows for continuously replenished contact of oxygen with fresh fuel vapours. Without buoyancy, the movement of fresh air into the flame zone and the removal of combustion products from it is severely limited, and the flame structure is very different.

When flames first originate at the fuel source in a candle, they have little upward velocity. In the small area near the wick, there is very little oxygen, as most of the volume is vaporised wax. This warm vapour begins to rise and accelerate upward. As the rising wax vapour mixes with oxygen and combustion begins, their temperature rises dramatically to around 1000 °C. The flame gases in a candle achieve a very smooth appearance, which is termed *laminar.* When a candle flame is examined using a technique called shadowgraphy (**Figure 2.4**), we can see that the laminar flow of the flame gases transitions to *turbulent* flow as the gases move longer distances and at higher velocities. The

Figure 2.3: Photographs of flames (a) with gravity on earth in still air; and (b) in the absence of gravity in space. At sea level, a flame (0.28 kg^{-3} at ~1000 °C) is ~1/4 the density of air (1.15 kg m^{-3} at 30 °C); therefore, the flame expands and rises to take the teardrop shape familiar to us (photographs courtesy of NASA).

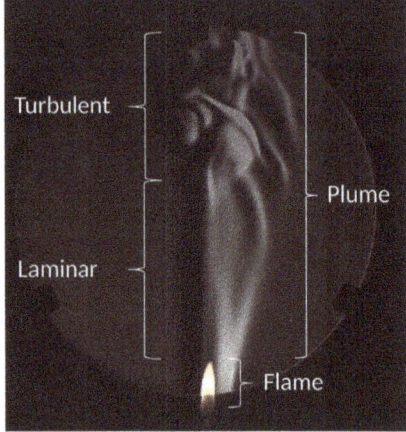

Figure 2.4: Visualisation of flame and plume structure using shadowgraphy reveals the laminar region in calm conditions, which transitions at some point to a turbulent regime. A slight breeze accelerates disruption of the laminar flow (image by Ian Grob).

visible flame occupies a small portion of the plume base; above it, the invisible flow of the plume eventually becomes wrinkled as it transitions to a turbulent flow regime. The distinction between laminar and turbulent flow is important for understanding fire behaviour and convection heat transfer from flames to fuels and from fuels to air, and is covered in **Chapter 3**.

Flame size

Why are candle flames always about the same size or height? Flame height and other dimensions are certainly not consistent among all fires, such as candles, campfires or wildfires. Why? When fires are burning in the open air, the answer lies with the rate of energy release. The more gaseous fuel that is fed into a flame from the solid fuel source, the longer the gases take to mix with oxygen and the higher they rise before mixing and combustion is accomplished – meaning larger or taller flames. In a candle, the diameter of the wick limits how quickly wax can be fed into the base of the flame. The amount of wax vapour produced at the surface of the wick then determines how far the fuel vapours must rise above the wick before they can mix with enough oxygen to burn. If we put two or more candles together so the flames merge, the resulting flame will become taller (**Figure 2.5**) because oxygen cannot penetrate the larger volume of wax gas and react until it rises farther. Effects of fuel production and oxygen mixing on flame size are similar in many kinds of fires. For example, if you are standing around a campfire but still feeling cold, you probably know that you can increase the amount of heat released and the flame size by piling more wood on the fire. The rate at which air mixes with gaseous fuel, whether in a turbulent campfire or a laminar candle, is quite constant and very slow compared to rates of combustion.

What controls the amount of gaseous fuel entering the flame zone in wildland fires? Like campfires, wildfires do not have a constant amount of fuel entering the flame zone. The rate of fuel entry depends upon other factors, including the rates of heat transferred to the fuel and the rates of

Figure 2.5: Flame size is a function of energy release rate in candles (a), where larger flames are produced when flames from several candles merge, and in campfires (b), when more wood is added to the fire (photographs by Trevor Finney).

ignition; these, in turn, are influenced by fuel particle sizes and how densely they are packed within the fuel bed. The faster a fire moves, the more fuel is consumed, the more energy is released, and the larger the flames. This is typical of a tightly coupled system, and fire is a perfect example! All of the parts and processes affect all of the others. It is difficult to distinguish the role of each individual piece of the system because changes in one cause changes in the others. To understand wildfire spread and behaviour, it is critical to understand heat release from the fuel and how it is, in turn, dependent upon specific heat transfer processes that ignite those fuels. In candles, most heat transfer is by radiation that melts the wax. As we will see throughout this book, the process is different and much more complex in wildland fires.

Candles and wildfires as coupled systems

Candles and wildfires are coupled systems that involve similar components and processes (**Figure 2.6**). For small wildfires, the system begins with an externally supplied heat source that causes ignition, whether that is a lightning strike or a human agent. The system begins to function when combustion of solid fuel particles releases energy, some of that energy evolves as heat, and some of that heat is transferred through radiation and convection to adjacent fuel particles. For the wildfire system to continue to function, the heat must pyrolyse the solid fuel particles (not melt liquid wax) to produce enough flammable gases for flaming ignition. *Pyrolysis* is the thermal degradation of a solid material, which releases flammable vapours from the fuel and leaves behind carbon as charcoal, and is discussed in detail in **Chapter 4, Fuels**.

Similarities between small wildfires and candles disappear when we consider that wildfires must spread to be sustained. Only if new fuel ignites and burns will the 'circle of life' for a wildfire continue by spreading. The physical processes in wildfires play different roles depending on the fuels involved and the size and shape of the flame zones; these will be examined in **Chapter 6**. System behaviours of small wildfires in fine fuel materials are explored in **Chapter 7** using a physical model to explain the dynamics of fire spread and energy release over a range of environmental factors. Wildfires at the largest scales have also been characterised as a system driven by feedbacks among energy release, the atmosphere, the ground surface and gravity (Byram 1959a). Some of these large-fire system behaviours will be described in **Chapter 8**.

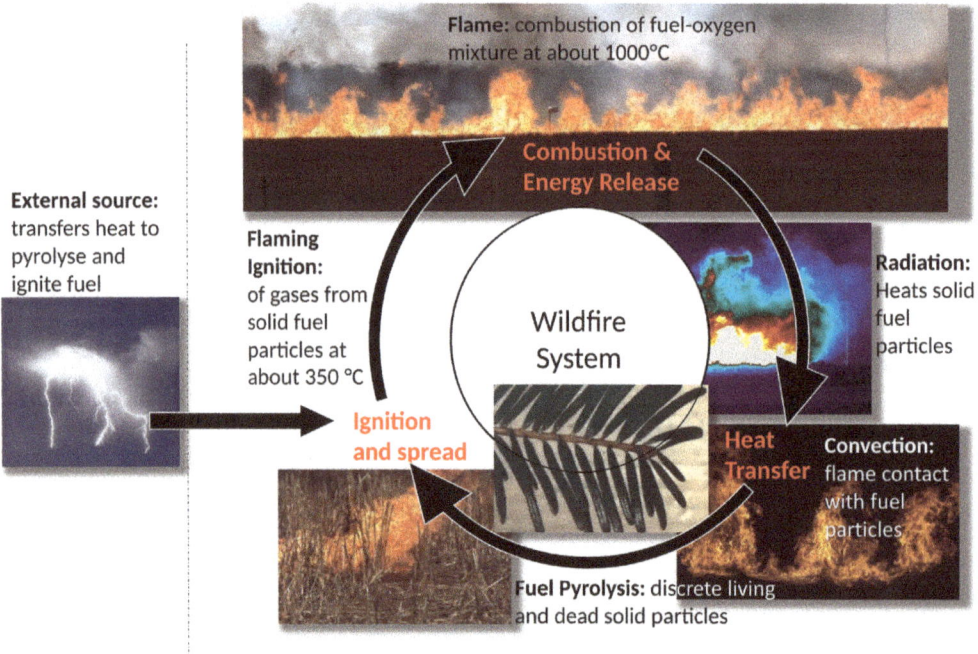

Figure 2.6: Wildfire spread is a self-sustaining coupled system like that in a burning candle (Figure 2.2) (images by Ian Grob, Andrew Gorris and Don Latham). The two systems differ in that the flaming ignition of solid fuel particles in wildfires can occur only after they have been pyrolysed by radiant and convection heating to produce gases enough for combustion.

Wildfire behaviour triangle: fuels, weather, topography

Introductions to wildland fire behaviour must address the role of three factors: *fuels*, *weather* and *topography*. This conceptual model is called the *fire behaviour triangle* (**Figure 2.7**). Each factor exerts strong influences on wildfire characteristics. The fire behaviour triangle alone, however, is insufficient to explain how fires respond to their environment because it omits the state of the fire itself (Countryman 1972). In other words, it does not explain fire behaviour as a *dynamical system* – a system that is time dependent. For example, if we knew the length of a candle, its orientation (vertical or angled), and the strength of a breeze in the room, we could estimate how long it would burn, the flame angle relative to the wick, and how much wax would drip onto the floor. But we could not understand how the physical candle system produces these behaviours, and thus we could not predict how the candle would burn if those factors and others changed over time. Similarly, the fire behaviour triangle does not inform us how wildfire behaviour will be affected by innumerable permutations of fuels, weather and topography or by changes in its own behaviour as it grows in area, morphs to different shapes, or interacts with nearby ignitions. If we seek to improve the science of fire behaviour, if we seek to explain it rather than just describe it, we must consider the fire triangle within the context of a dynamical system. For now, we use the fire behaviour triangle in the conventional manner to introduce some basic concepts in wildfire behaviour.

The fire behaviour triangle provides a useful framework for discussing how environmental factors affect wildfire behaviour. Wildland fuels comprise living and dead biomass distributed across large land areas and have two natural levels of organisation: fuel particles and fuel beds. *Fuel particles* are discrete objects. They come in all sizes and shapes, but many are less than a few millimetres in diameter. Fuel particles derive from grasses and herbaceous vegetation, fresh and fallen foliage from shrubs and trees, lichens and moss, and dead woody logs and sticks that may be rotten or sound. The life cycle of plant growth, maturation, senescence, death, collapse and decay affects all fuel

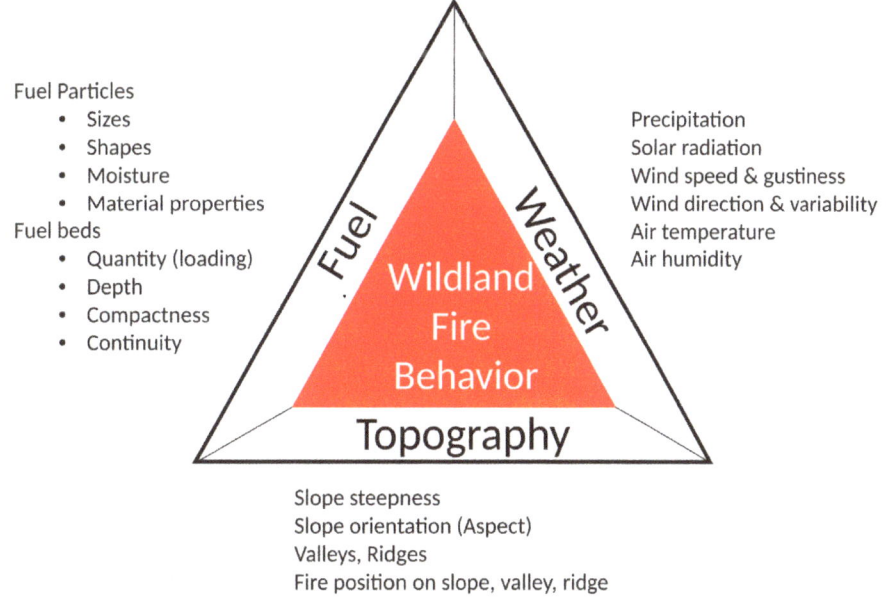

Figure 2.7: The wildland fire behaviour triangle comprises three environmental components: fuel, weather and topography. Each component has numerous characteristics that affect how fires behave – that is, how they spread and release energy.

properties and consequent burning and ignition behaviours. A *fuel bed* is a collection of fuel particles. Fuel beds can be simple or highly complex combinations of particles with an infinite number of possible arrangements. Fuel beds are traditionally classified by quantity (or *loading*, mass per unit area (kg m^{-2}), particle sizes (m), compactness (*bulk density*) (kg m^{-3}), depth (m), moisture content (%), spatial arrangement (*continuity*), and whether living or dead.

Fuels for wildfires provide a unique context among fire phenomena because they are discrete materials, some of them living vegetation, separated by large air gaps and are typically fine in size (a few millimetres in diameter). Structure fires – homes, factories or other buildings – burn large pieces of non-living materials. Buildings can be made of organic products such as wood, but they are typically joined into wide surfaces (e.g. walls, roofing) and combined with non-organic and synthetic materials. Industrial fires can consume liquids or gases as well. The properties of wildland fuels affect wildfire behaviours by the way that they burn and ignite.

Unlike our candle, wildland fuels are solid materials that burn in both *flaming* and *glowing* phases as well as *smouldering* (see **Chapter 4**). Flaming is the combustion of gaseous vapours produced from the pyrolysis of solid fuel. For natural woody fuels, a maximum of ~60–70% of the dry fuel mass is typically burned in flaming. Non-combustible mineral ash typically comprises ~5% of the mass, leaving ~25–35% of the total fuel mass as carbonaceous char (charcoal) that may or may not burn in glowing combustion. Smouldering is similar to glowing combustion and with some fuels (duff, peat, rotten logs) may occur instead of flaming. Note that the heat released in the flaming phase (per unit mass of fuel consumed) is less than that released purely in glowing combustion of char. Smouldering of natural fuel material releases considerably less heat per unit mass than flaming because gaseous pyrolysis products remain unburned and are driven off as smoke. Flaming continues until the *burnout* or *flameout* time, which can vary from ~5–30 s for grass fuels, 20–60 s for brush and tree foliage, and 5–60 min (or longer) for woody fuel. After flaming ceases, combustion may continue as glowing char for a considerable period. Other material may also smoulder. For example, very dense beds of leaves or needles and moist large rotten logs may mostly smoulder, while tall grass will burn rapidly in flaming combustion with very little smouldering.

Fires frequently do not consume all the fuel present. The amount of different fuels consumed (kg m^{-2}) in flaming and non-flaming phases depends on the properties of the fuel particles and fuel bed and also on the moisture and weather conditions during the fire. The fuel that is consumed is called the *available* fuel because it was available for combustion. This means it released heat during the fire, whether that heat was released during flaming or non-flaming combustion. The available fuel is typically not known before the fire occurs, but it must be measured or estimated if we want to know how much heat is released (see **Chapter 9, Fuel consumption**). Available fuel can be partitioned into fuel available in the flaming phase and fuel available afterward in the smouldering or glowing phases. As we'll see shortly, these phases are important for describing energy release and explaining how fires spread.

Weather influences wildfire behaviour indirectly, by changing the moisture content of dead fuels and directly, through the impact of wind. The moisture content of dead fuel particles depends on particle size, contact with the ground, the season of the year, and short-term effects of weather (e.g. solar radiation, precipitation, humidity and air temperature). Fuel moistures – both living and dead – affect fire behaviour by changing the amount of heat released and its rate of release, and by increasing the amount of heat required for ignition. Wind speeds range from calm to more than 100 kph (28 m s^{-1}) and consequently produce wide-ranging responses in fire behaviour. These include rate of spread, size of area burned, and the shape of fire growth patterns. The effects of wind on fire spread are modified by the fire and the fuel structure themselves. In forests, for example, winds near the ground are greatly reduced by the overstorey canopy. Wind speeds and directions are not steady, and short-term gustiness and oscillations produce considerable variability in fire behaviour.

Topography affects fire behaviour directly through slope steepness and indirectly through interactions with other environmental factors. Fires generally spread faster upslope than downslope, so the direction of fire spread is determined partly by the fire's position on a ridge, midslope or valley bottom. Slope orientation and wind direction can combine to influence fire spread. For example, fires spread upslope more quickly if the wind is also blowing upslope (aligned with the slope) than when wind is blowing across the slope or downslope. Topographic interactions with weather are often more complex, especially where wind flow across mountainous terrain is channelled by multiple valleys and ridges and varies with the daily cycle of solar heating.

Wildfire classification

The vegetation and fuel that wildfires burn is often used to classify types of wildfires (**Figure 2.8**). A *ground fire* burns organic material at or underneath the ground surface (**see Figures 4.28** and **9.2**); it spreads primarily by smouldering through rotten or partially decomposed porous material (duff). Ground fires often spread slowly without exhibiting flames. *Surface fires* burn the fuels above the ground surface, including litter (dropped needles and leaves), dead and downed woody material

Figure 2.8: Photographs of wildland fire behaviour in different fuel types: (a) backing surface fire in litter understorey (photo by Scott Stephens); (b) heading surface fire in litter, brush and timber understorey (photo by Scott Stephens); (c) heading surface fire in grass (photo courtesy of USDA Forest Service); (d) heading fire in brush (photo by Scott Stephens); (e) torching tree (photo by Mark Finney); (f) group torching in pine forest (photo by Kelly Close); and (g) crown fire in spruce–fir forest (photo by Roger Ottmar, USFS).

(e.g. sticks, logs), grass, shrubs and small trees. Flames are visible in surface fires during spread, although both flaming and smouldering may alternate throughout the life of a fire as the weather changes. In some kinds of forest vegetation, particularly conifer forests, *crown fires* occur if the live and dead foliage and branches in the tree canopy ignite and fuel fire spread. The ignition of a single tree crown is called *torching*; if the fire does not spread continuously from crown to crown, it is not considered a spreading crown fire. Crown fires often exist in conjunction with surface fires and produce very tall flames because of the elevated nature of the fuel bed and the large amount of burning surface and canopy fuel. S*pot fires* refer to ignitions taking place at distances beyond the main fire that are caused by transport of burning embers, which ignite separate new fires. This book will consider all of these kinds of fires. We build the foundation for understanding them here by explaining the simplest kinds of fire behaviour.

Initial fire growth

We began introducing fire in terms of a candle, which focused on a small, symmetrical, stationary flame source – essentially a point (also called an *axisymmetric fire* in engineering to contrast with a *line fire* and an *area fire* as discussed later) – burning a liquid wax fuel. Like a candle, a wildfire is typically first ignited at a point. Unlike a candle, it can only sustain itself by growing larger. To grow, it must ignite and burn adjacent solid fuel materials.

When we observe the life history of a small flaming surface fire in very uniform fuel without effects of wind or topography, we see that – in the time shortly after ignition – the fire spreads in all directions, increasing the amount of burning area uniformly. For perhaps a few minutes, the fuels first ignited remain flaming in the centre while fuels at the periphery are just igniting and begin flaming. Across the burning area, the hot flames rise and merge into a single volume that is drawn towards the centre (**Figure 2.9a**). This phase continues until flaming combustion of the fuels in the centre is complete (*burnout*). The ring of smaller flames around the periphery of the fire remain tilted towards the middle, since indrafts to the small area of the buoyant flame zone can only come from outside the fire.

A similar initial phase occurs in fires that begin in windy conditions (**Figure 2.9b**). However, fires generally spread faster downwind than upwind. The downwind flames tilt away from the centre both before and after flaming ceases in the middle because the wind overwhelms any indrafts created by the buoyancy in the centre. Flames along the downwind side of the fire are larger because the fire is spreading faster and releasing more energy than the fire along the upwind side. As winds get stronger, the flame profiles extend downwind from the burning zone, making direct contact with unburned fuels.

Steep slopes also produce asymmetric fire growth with faster uphill than downhill spread (**Figure 2.9c**). Flames often become *attached* to the ground surface on steep slopes for a considerable distance ahead of the burning region, so they may be oriented nearly parallel to the slope.

As little fires grow larger, we can illustrate their growth as a series of time contours. That is, lines to represent the instantaneous position of the leading edge at different times. For the no-wind, no-slope case, fire spreads at the same rate in all directions away from a single ignition point and the fire shape is roughly a circle (**Figure 2.10a**). More typically, wildfires grow faster in one primary direction because of wind or slope and take on characteristics associated with each spread direction (**Figure 2.10b**). Terminology describes these directions, referring to the *head* of the fire as the direction where the fastest spread occurs (i.e. *heading* spread). The *back* of the fire (also known as the *tail* or *heel*) spreads the slowest (*backing* spread) and the flanks along each side spread at rates somewhere in between (*flanking*). These same fire spread directions can be identified on fires growing from any ignition geometry – from points or line ignitions, for example (**Figure 2.10c, d, e**). To really understand wildfires, we need to know *how* wind or

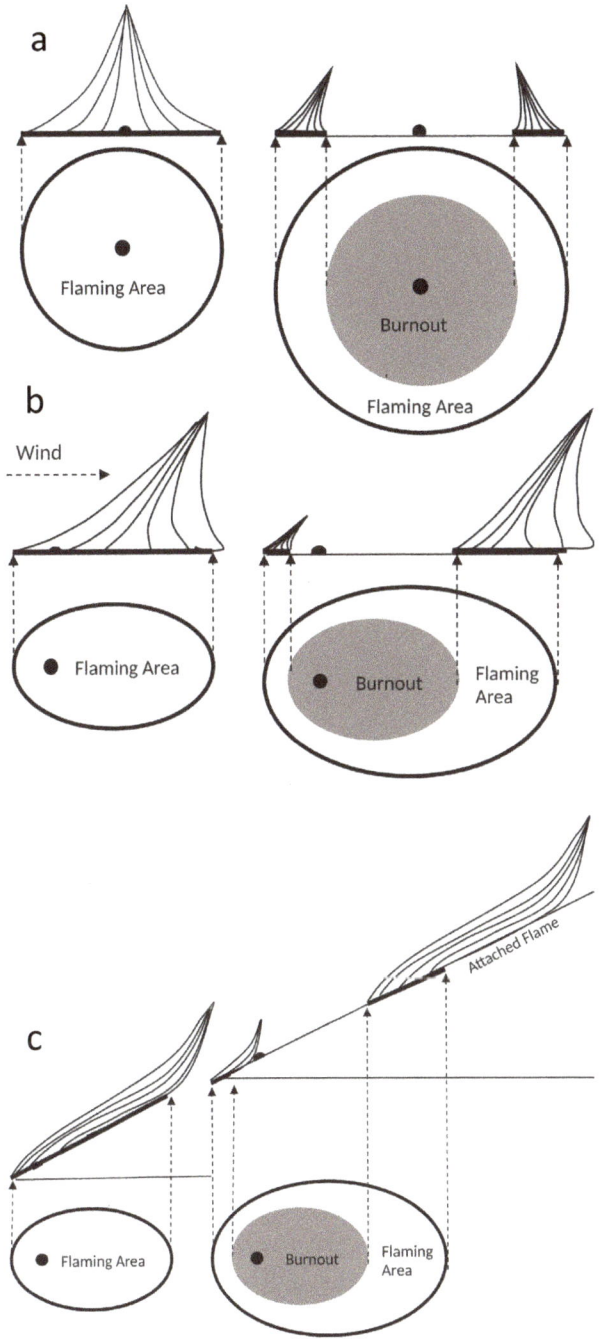

Figure 2.9: Initial growth of a wildfire from point ignition (black dot) is depicted (a) for no wind and no slope conditions, where the flames are drawn towards the middle of the burning area even as fuels burn out in the centre; (b) for wind-driven conditions, where the wind tilts the flames and elongates the fire shape by increasing fire spread rate and flame size in the downwind direction; and (c) for slope-driven conditions, where flames attach upslope, become larger, and increase fire spread rate upslope (adapted from Kerr *et al.* 1971).

slope makes the fire spread faster in a particular direction, as well as how much faster.

Line fire concept

After fires reach an area of a few hectares or more, they become large enough that the local curvature of the fire edge is not very noticeable. A given segment along the front appears nearly linear. This linear segment is called a *line fire* which, in concept, applies to any part of the fire (e.g. heading or backing). Often an *infinite line fire* is assumed, as if the flame zone could be considered infinitely long and the line fire segment behaves as it would without any lateral edge effects. The line fire concept generally assumes that fire spread can be considered independently of the size and geometry of the overall burning area and is not influenced by interactions with the fire's interior or other segments of the perimeter. A very powerful simplification that is frequently used for modelling and prediction is to approximate the growth of a large fire as a series of independent spreading line-source segments (e.g. see Anderson *et al.* 1982 and **Chapter 8, Fire shapes and growth patterns**). The assumption of an infinite line fire is violated in many cases, but the simplicity of the concept is so appealing that it has historically dominated most wildfire modelling and research, especially in studying heading fires because they spread at the fastest rates and burn the greatest areas. We will use the concept later in this book to examine heading fire spread, partly because there are many challenges to understanding even how this rudimentary behaviour occurs.

Like the small line fires discussed above, each imaginary line fire segment along a fire front represents the leading edge of a burning zone, which extends back behind it. If we take a slice perpendicular to the fire edge, we can visualise the burning zone in profile (**Figure 2.11**). This profile shows the burning zone in two dimensions (length and height). The burning zone profile can be depicted for fire spread in any direction, including heading, backing and flanking. Notice that the flames in

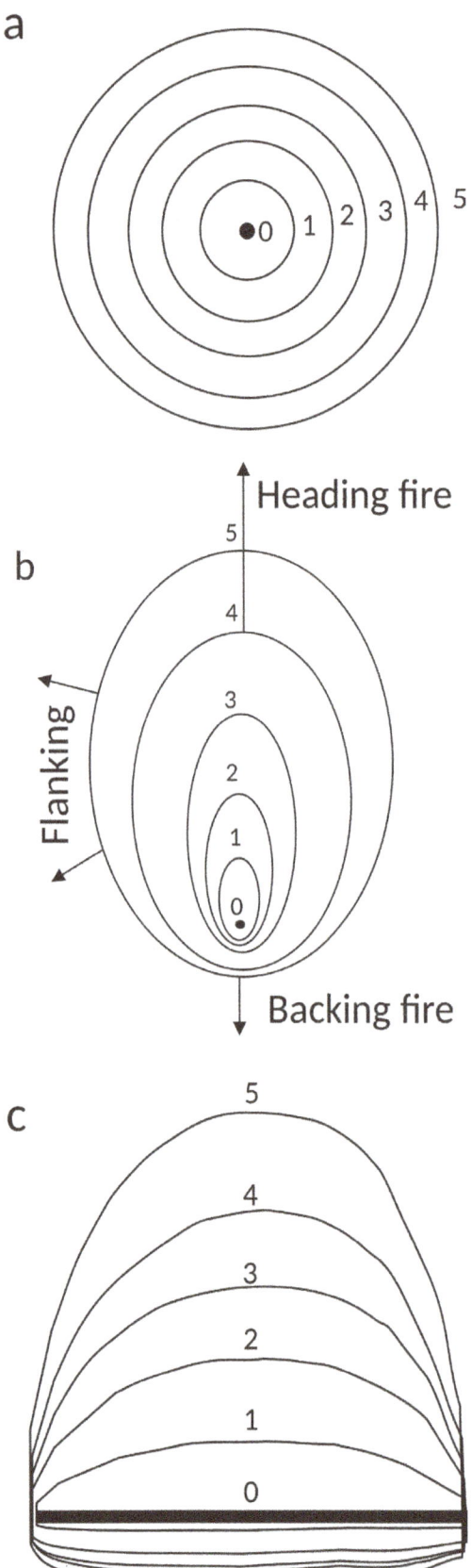

Figure 2.10: Time contours show fire progression from (a) point source ignition with no wind or slope; (b) point source ignition with wind or slope, illustrating heading, flanking and backing fire spread; and (c) line source ignition. Aerial photographs (University of Canterbury, NZ) of fire growth in wheat fields showing (d) the different rates of spread and fire shapes from point and line source ignition and (e) from a wide line source ignition. For scale, note the size of people in these photographs.

backing fires are tilted towards the burned interior of the fire (away from unburned fuel), but in heading fires the flames are tilted towards the unburned fuels. This dramatically changes the rate of heat transfer to adjacent fuel particles. The photograph in **Figure 2.11** shows a profile of a heading fire in a laboratory wind tunnel. The wind tilts the flames forward and ignites the upper tips of the fuel particles (point x_a). As the particles burn downward, flaming ceases first at the top of the fuel bed (point x_b, where ignitions first occurred) and finally at the bottom of the bed (point x_c). Thus, in **Figure 2.11**, the *flame zone depth D* for both backing and heading fires is defined as the linear distance x_a to x_c from the leading edge of ignition to the rear edge of visible flaming. In many wildfires, however, it is difficult to identify and measure points x_a and x_c; we will discuss these challenges in **Chapter 9, Flame zone properties**.

A closely related characteristic of the flame zone depth is the *flame residence time t_r* (s), defined as the time during which the fuel bed at a given point on the ground is flaming. The flame residence time is calculated as the difference between ignition time at point x_a and flame burnout time at point x_c in **Figure 2.11**. Thus, flame residence time for a spreading fire traveling at rate r (m s^{-1}) is defined by Rothermel (1972) as the time required to spread the distance equal to the flame depth D:

$$t_r = D/r \quad [2.1]$$

Flame residence time is also referred to as the *flame reaction time* (Rothermel 1972; Nelson 2003) to distinguish it from the *particle flaming time*, which could be defined as the time difference between ignition at point x_a and burnout at x_b. Particle flaming time has been the subject of laboratory studies

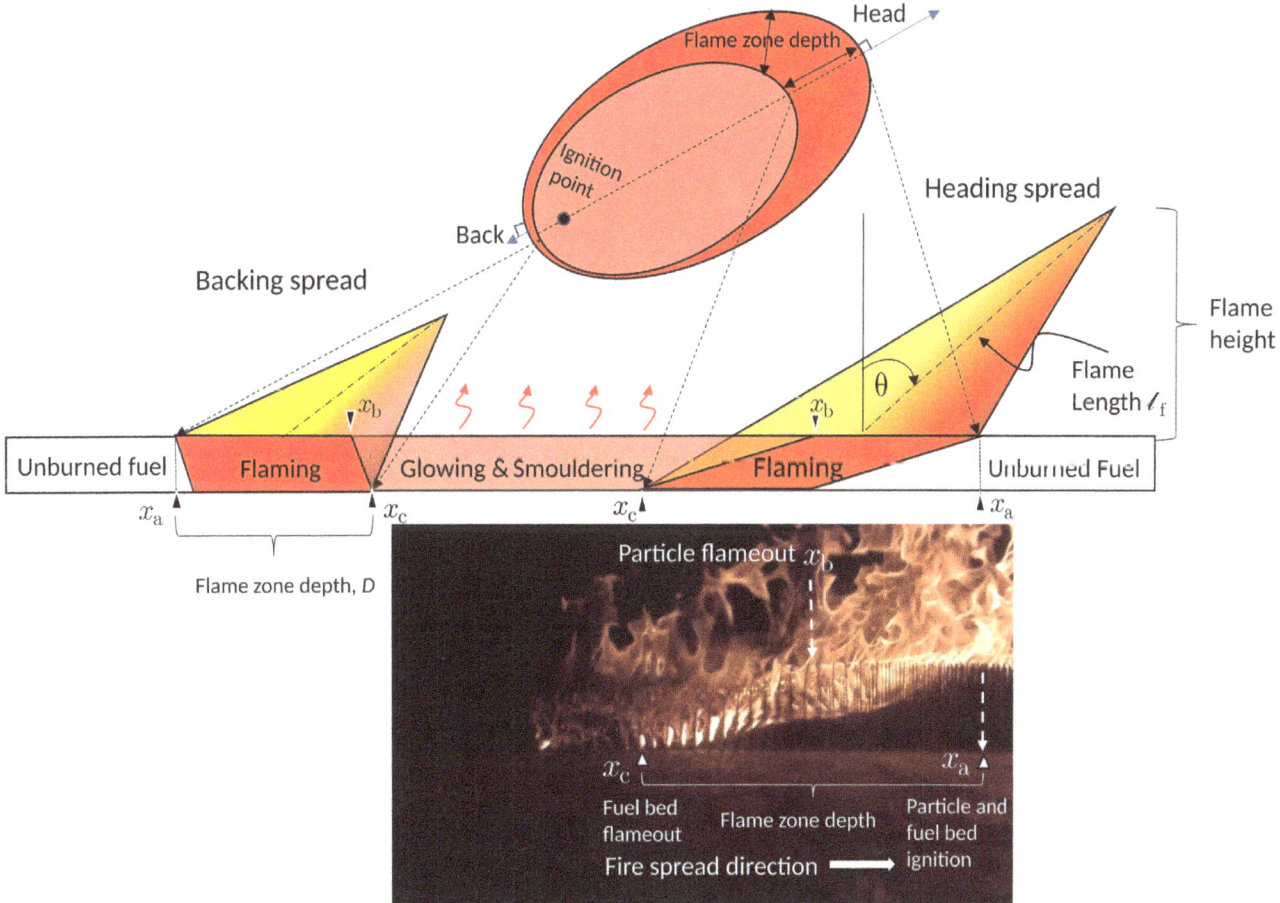

Figure 2.11: Illustrations of the burning zone for an elliptical fire shape that includes flaming, smouldering and the flame profile. They show both heading and backing spread with spread rate measured in perpendicular to any place on the front (blue arrows). The leading edge of the flame zone in a spreading fire is defined by the forward ignition point x_a; the trailing edge is defined as the fuel bed flameout location x_c. Thus, the *flame zone depth* extends from point x_a to point x_c in both the heading and backing spread directions. Individual fuel particles on the upper surface of the fuel bed burnout at point x_b. Flame height is the vertical extent of the flames. If flames are tilted at some angle θ, flame length is defined as the distance from midway through the flame zone to the upper flame tip.

where individual particles are burned and closely observed. When particles are combined into fuel beds in wildland fire settings, however, there is little means of distinguishing the particle behaviour from that of the fuel bed. Furthermore, in many wildland fuel beds, the particle flaming time strongly depends upon the properties of the fuel bed (e.g. mass per unit volume, the fraction of fuel bed filled with fuel, and particle size distributions); particles tend to burn longer in dense beds than individually because oxygen is too limited inside the fuel bed to support flaming combustion and high temperatures around the fuel particles. The flame residence time, rather than the particle flaming time, is important to fire spread in wildland fires because it delimits the maximum time during which the leading edge of the flame zone can transfer enough heat to ignite new fuels (see **Chapter 6, Fuel burning rates and consumption**).

As discussed for candle flames, the size of flames in wildfires relates to the energy release rate (which is also the rate of fuel consumption). The line fire profile illustrated in **Figure 2.11** defines the *flame height* (m) as the vertical extent of flames, whether in calm or windy conditions. *Flame length* is the distance from the flame tip to a point midway through the flame zone. In the presence of wind or on sloping topography, flames are tilted. In this diagram, the flame length is greater than the flame height. In wind-driven fires, the angle of flame tilt is a function of both wind speed and flame length. Flame dimensions apply to any portion of a flame zone; the figure depicts it for both heading and backing fires.

As fires get larger, they often exhibit behaviours that are very different from those of a linear flame zone. In general, these behaviours arise when the assumptions regarding the line fire, such as the simple elliptical shape of the front and the independence of multiple fronts, are no longer valid and the behaviours are determined by influences at the scale of the entire fire. These complex behaviours are seen on small as well as large fires. For example, we see in **Figure 2.9a** that the flame tilts around the centre of the small fire rather than tilting outward into unburned fuels. This case illustrates the fact that a different set of factors may shape fire behaviour at the *fire scale* than at the *flame scale*. In other words, in these cases, we cannot ignore the energy release and geometry of the whole burning area and instead focus on an arbitrary linear segment of the flaming front.

Wildfire behaviour

Wildfire behaviour is a general term that encompasses (a) how fast fires spread, (b) how much heat they release, and (c) how long burning lasts. The science of wildfire behaviour is concerned with how these metrics respond to a range of environmental conditions (i.e. the fire behaviour triangle) and to different fire configurations (e.g. small fires, large fires, points, lines). The rest of this chapter will introduce these common fire behaviour characteristics – fire spread rate, fire growth, fire intensity and flame length – and the relationships of these to environmental factors. Note that these fire behaviour characteristics are introduced in the conventional fashion without the perspective of fire as a dynamical system. We will address the dynamical aspects of the fire system throughout this book, particularly in **Chapters 6** and **7**.

Most of the fire behaviour characteristics have already been presented in the context of the line fire (i.e. a segment of the flaming front) and thus are measured in one dimension. For example, the flame zone depth and the residence time of fires described above are examples of 'how much' and 'how long' flaming exists in a line fire. Flame length is another important characteristic of fires; it reflects the rate of energy release. Fire behaviour concepts for a spreading line fire are typically applied to a condition called a *steady state*. We saw this in **Eqn [2.1]**, in which the flame zone depth and residence time are related to a steady rate of fire spread (Rothermel 1972). But fire spread is generally not steady or constant, so the steady-state assumption requires that we explicitly define some distance and time interval over which the properties of fire behaviour are summarised. The variability of fire in time and space, caused by heterogeneity of many of the factors in the fire behaviour triangle, means that some behaviours are more properly characterised as a distribution of values for which we can quantify the mean and other descriptors of the distribution such as mode, median and variance.

Although the assumption of a steady state is convenient, it may be more illusion than reality in wildland fires. For example, a fire in patchy or discontinuous fuels will sometimes spread quickly or not at all depending on small changes in wind or moisture. Such *threshold behaviour* or *step-changes* in fire spread rate cannot be quantified by an average in time or in space; it cannot be described as a steady state. Even with uniform and relatively continuous fuel, fire acceleration and deceleration after ignition vary and thus demonstrate nonlinear, non-steady behaviour.

Fire spread rate

The average or mean *spread rate* of a fire is perhaps the most frequently described and modelled fire behaviour characteristic of line fires. Fire spread rate is defined as the one-dimensional rate of fire movement in a specified direction (i.e. heading, flanking, backing). In this book we describe fire spread rate using units of m min^{-1}, which is widely used in practice and easy to relate to. Heading spread has an upper range of ~200–300 m min^{-1}. By comparison, backing fire spread rates are much slower, often less than 1–2 m min^{-1}. On sloping topography, the fire spread rate is measured *parallel to the ground* and then transformed to horizontal units for mapping to a two-dimensional plane (**Figure 2.9c**). Because laboratory studies from engineering often complement our field-scale research, we should mention that heading spread in wind or upslope is often defined in the fire engineering literature as *concurrent flow conditions*, meaning that the air is flowing in the direction of fire movement or up a slope. By contrast, backing fire corresponds to the *opposed flow conditions* because it is spreading against the direction of blowing air or downward against gravity. This terminology is important because we will be referring to a mixture of wildland and engineering studies of fire spread processes throughout the book.

Heading spread rate has traditionally been the single most sought-after fire behaviour quantity because it is the fastest moving portion of a fire and is most difficult to control. Nevertheless, the ability to know or estimate fire spread rates in all directions is essential for analysis and decision making by fire management. The difficulty of estimating or predicting the spread rate for any portion of a fire for specific environmental conditions has remained a persistent challenge since scientific investigation of wildland fire began in the early 1900s. Part of the difficulty derives from the fire configuration (i.e. the width and curvature of the flame front). Wide line fire segments tend to spread faster than narrower segments, whether ignited intentionally as lines or because they naturally broaden as a fire grows larger. Other difficulties derive from the inability to quantify fuel characteristics along a line fire segment and to know or measure the wind affecting the fire as it changes in space and time. Sullivan (2009a, b) reviews many attempts to develop predictive models for fire spread, some by empirical means and some by physical approaches. We will cover more details of the physical modelling in **Chapter 7** and methods for measuring spread rate in **Chapter 9**. Even a model with high-fidelity representations of the physical processes cannot produce accurate results if it has uncertain and erroneous inputs. Other difficulties, already explained in **Chapter 1**, are that fire spread is an outcome of sequential discrete particle ignitions and its rate is therefore inherently not a steady-state quantity, even if it could arise from steady physical processes. Spread rates can vary considerably over a specified time or distance even under constant and uniform conditions of fuel, topography, weather and the fire configuration.

Although fire spread rate is affected by all environmental conditions in the fire behaviour triangle, it is not a unique function of any of them. In other words, if we tried to graph spread rates over the range of each of the dozens of environmental characteristics implied by the fire behaviour triangle, we would need a hyperdimensional response surface. On this surface, the same spread rate would appear in many places as a product of different combinations of factors. For example, higher moisture and faster winds can produce the same spread rate as lower moisture and slower winds. While trends in wildfire spread rate are not cleanly delineated along the axis of a single variable, some general tendencies help illustrate typical ranges of spread rates (**Figure 2.12**).

Fuel type tends to be associated with a characteristic range of wildfire spread rates for a given set of environmental conditions. **Figure 2.12a** illustrates the general pattern of the fire spread rate as a function of fuel type and wind speed. The fastest fire spread often occurs in grasses (up to 200–300 m min^{-1}), followed by brush or shrubs (up to 50–150 m min^{-1}). It is highly variable for forest fuels depending on which fuel stratum is available to burn (1–100 m min^{-1}). Many natural fuel complexes are mixtures of fuel material from grass, brush and timber, so they exhibit wide-ranging

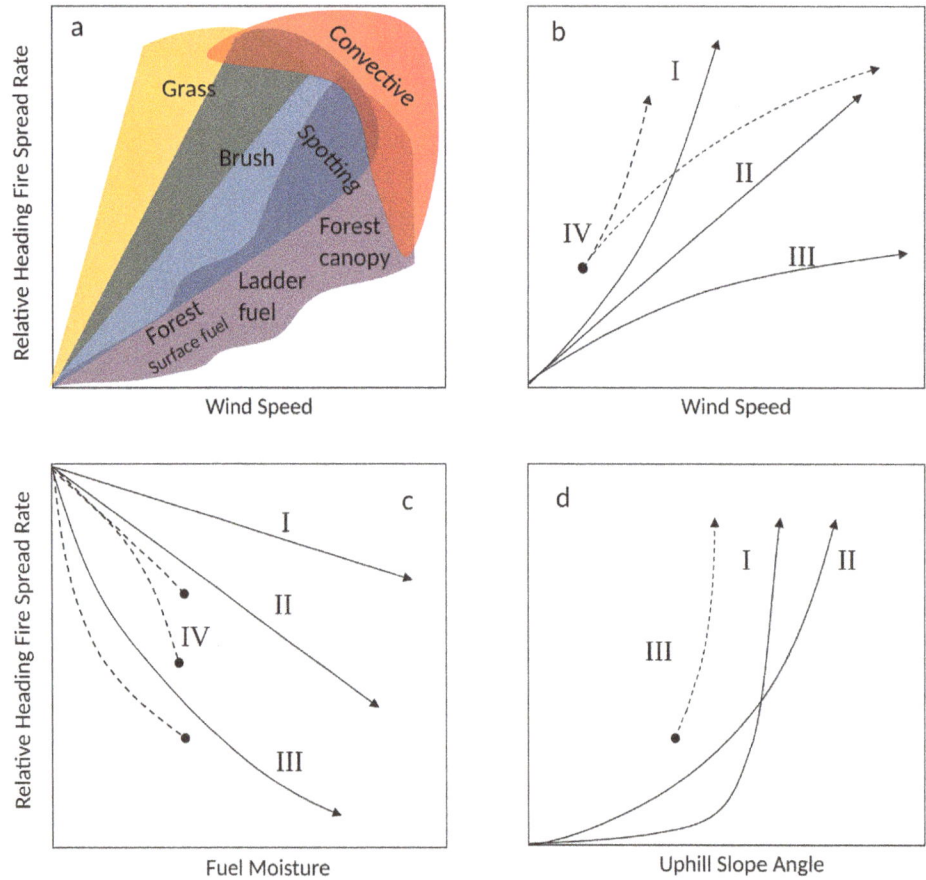

Figure 2.12: Heading fire spread rate as a function of some environmental variables in the fire behaviour triangle. (a) Grass fuel types typically produce the fastest spread relative to brush and forest fuels, but very fast rates of spread are often driven by large-scale interactions involving the heat plume and perhaps upper atmosphere (convection). (b) Wind increases spread rate according to different possible functions (e.g. I, II, III). Some fuel types or conditions exhibit thresholds (IV), in which spread cannot occur below a critical wind speed. (c) Fuel moisture diminishes spread rate as described by various possible functions (I, II, III). Some moisture conditions exhibit thresholds (IV), in which spread only occurs in drier fuels. (d) Slope increases spread rate with a response that may be gradual (e.g. II) or thresholding above a specific angle (I). In some fuel types or conditions, slope increases spread rate only above a threshold angle (III).

and overlapping variations in fire spread rates. Multistoreyed fuel structures often support distinct transitions in fire behaviour from burning in surface litter and woody fuel, igniting elevated or ladder fuels, and then spreading as crown fire through the canopy. These transitions are depicted by the wavy contours at the edges of fire spread rate for brush and forest. At higher wind speeds, spotting becomes an important means of spreading fire as well, further complicating simple trends between spread rate and fuel type. For the largest fires in many fuel types, the fastest spread rates are caused by fire-induced convective storms and associated surface winds triggered by strong atmospheric interactions that influence fire spread and energy release (**Chapter 8, Plumes and pyroconvective storms**).

Wind affects fire spread rates in several possible ways (**Figure 2.12b**). A review of the literature demonstrates remarkably little agreement as to the effect of wind on fire spread rate despite an abundance of studies (Sullivan 2009b). Some trends have positively curved functions (I), some are nearly linear (II), and some are negatively curved (III). Uncertainty in these trends is partly related to the types of fuel and range of wind speeds over which the spread rates have been observed, so the functional form may depict extrapolations either below or above the range of wind speed data. For example, functions with solid lines all imply fires will spread slowly with minimal wind. While this may be mostly true for relatively continuous types (i.e. dry leaf litter), fuels with spatial discontinuities or gaps (dashed lines), as found in some shrubs and tufted grasses, may allow no fire spread at all until winds exceed a threshold sufficient for heat transfer across the gaps (IV). This threshold behaviour can also occur in fuels with high moisture content. It is difficult to obtain fire observations in the highest range of wind speeds because of logistical difficulties in gaining access to a free-spreading wildfire driven by strong winds. At the highest winds, fires may spread primarily by other means, such as spotting (**Chapter 8**). All trends are also dependent on the size or width of the flaming front, which is not consistent among studies.

Empirical studies suggest that fire spread rate slows in response to higher moisture fuel content (**Figure 2.12c**), but with a variety of possible trajectories (Sullivan 2009b). Moisture content in these studies may refer only to dead fuels (litter, woody branches or logs, or grass), but fire spread rate is also strongly affected by living material in the fuel bed and the seasonal dynamics of growth and senescence. The complexities of live and dead fuel moistures add great uncertainty to any trends in fire spread rates. Studies have revealed linear decreases in fire spread rates that descend both slowly (I) and steeply (II) as fuel moisture increases – following both linear (I, II) and nonlinear (III) trajectories. As with wind (**Figure 2.12b**), some of the data in these studies probably come from fuel types with large spatial gaps, which would make the fire spread rates very sensitive to moisture content. These conditions produce threshold behaviour (IV), in which no spread exists at all beyond some critical lower limit of fuel moisture. Of course, all of these trends interact with the effects of wind and other variables.

Slope effects on heading fire spread rate are poorly understood, although there are a few small-scale observations and experiments that suggest the functional forms shown in **Figure 2.12d**. Part of the challenge is caused by a phenomenon associated with a threshold angle. As slope nears and exceeds a critical angle of around 18–20°, fires can exhibit dramatic increases in uphill spread rate even with minor changes in slope (I). Wider fires may transition at lower slope angles than narrow ones, but this is uncertain, and an important research question discussed in **Chapter 6, Flame front width and shape**. Trend II characterises an increase in spread with slope angle that is more typical of many predictive models and rules of thumb (Sullivan *et al.* 2014). Trend III depicts a threshold condition where fire spread is not possible below a critical angle. No wildfire observations have been collected for slopes greater than ~45° or for long distances, so potential spread rates at these upper limits are unknown.

Fire shapes

Although fire spread rate in the heading direction has received the greatest interest, spread rates in all directions around a fire (i.e. flanking, backing) are essential to estimating the area burned and the length of the active fire perimeter. Ideally, we would understand fire spread well enough to be able to calculate fire movement in any direction relative to wind or slope. We would then know the shape of a fire growing under any environmental condition. But this is not presently the case and given the difficulty of explaining the physical processes for heading fire spread alone, there has been little effort to understand the general role of direction in fire spread. Thus, in order to calculate fire growth in all directions, as well as area and perimeter, we must rely upon empirical models of fire shape.

Observations of wildfires have long suggested that their two-dimensional growth could be characterised by a simple shape. The fine-scale contours of the fire edges will vary with environmental heterogeneity and change over time, but the overall shapes of small and large fires often appear remarkably well approximated by simple shapes (**Figure 2.13**).

The most commonly used two-dimensional shape in modelling fires is the single mathematical ellipse (**Figure 2.14**), defined traditionally in fire science as the ratio (l_y/l_x) of the lengths of the major axis l_y to the minor axis l_x. References on fire shapes (Van Wagner 1969; Anderson 1983; Alexander 1985) provide details on the historical development and uses of ellipses. An elliptical shape assumes fires are spreading under constant winds and across uniform fuel and topography (see **Chapter 8, Fire shapes and growth patterns**). Although it is rare for any environmental factor to remain constant and uniform for very long, these assumptions have proven surprisingly useful. Many point-source fires evolve shapes close to perfect ellipses (**Figure 2.10d**), but the physical rationale for how wildfire shapes emerge has not been established. Laboratory and field data on fire growth have not been reliable enough to rigorously test the assumption that fires grow elliptically, because fires are almost constantly accelerating or decelerating, and the environment is rarely uniform and constant enough for obtaining the necessary observations.

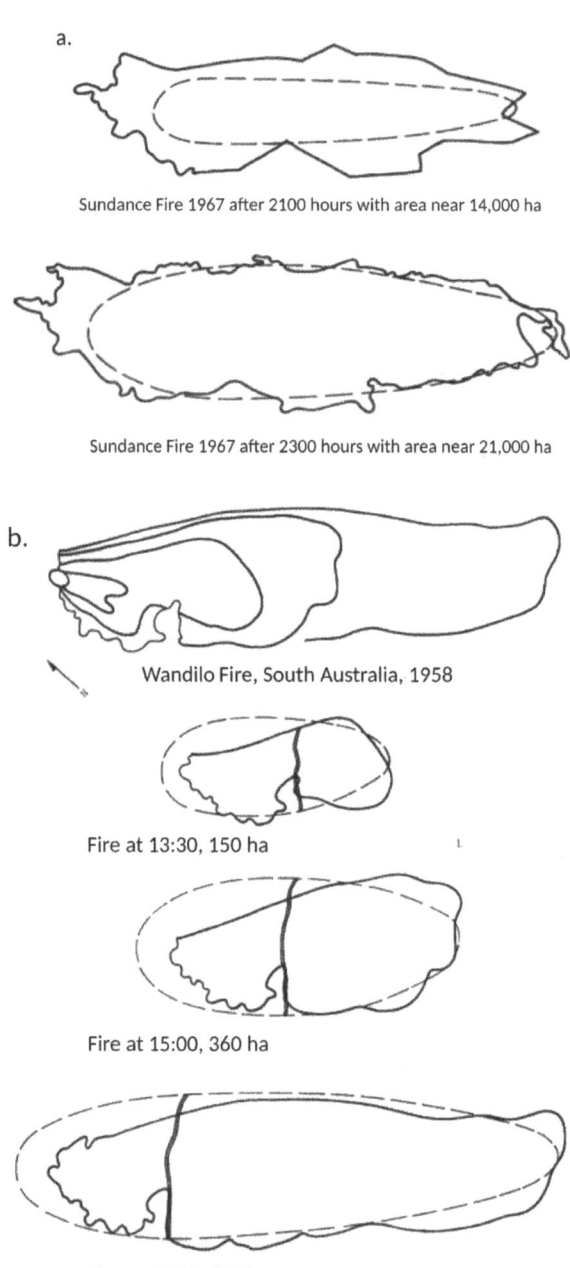

Figure 2.13: The growth patterns of large fires over a range of sizes for (a) the Sundance Fire in northern Idaho, USA; and (b) the Wandilo Fire, South Australia, are often approximated by simple geometric shapes such as the ellipse or double ellipse shown here from Anderson (1983).

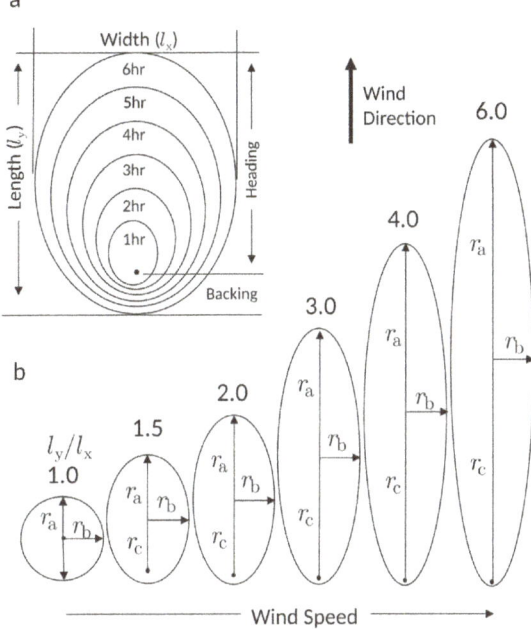

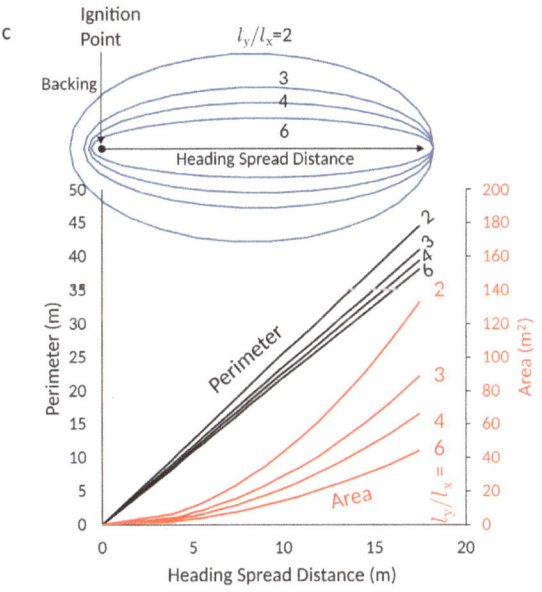

Figure 2.14: The simple ellipse is the most common fire shape assumed for wildfires growing under uniform and constant environmental conditions. Ellipses are illustrated for (a) fire growth over time showing dimensions for heading, backing, length l_y and width l_x; (b) fire shapes becoming more eccentric (defined by length to width ratio (l_y/l_x) or by the elliptical dimensions r_a, r_b, and r_c) with higher winds (adapted from Alexander 1985); and (c) fire perimeter and area trends associated with forward spread rate for different elliptical shapes (l_y/l_x = 2, 3, 4, 6). Each figure assumes the ignition point to be located at the rear focus of the ellipse.

Ideal elliptical fire shapes tend to become elongated by stronger winds. **Figure 2.14b** illustrates l_y/l_x ratios from 1.0 (for the no-wind, no-slope condition) to 6.0. More elongated fire shapes are certainly possible and are likely to occur if fire spread is driven by different processes along different portions of the fire perimeter. This can happen even under uniform and constant environmental conditions. For example, spotting from the head of a fire in strong winds can primarily advance forward spread because most embers are driven downwind, effectively increasing the ratio of forward to lateral expansion (and l_y/l_x ratio). Maps of fire progression do not often reflect evidence of spotting as a means of fire growth, particularly if the interval between observations is longer than a few hours. Long intervals between observations depict only the outer contour of the fire rather than the processes of spot fire ignition and merging. Discontinuous fuels often create spread thresholds where fire may only spread in heading directions and not in flanking and backing directions. This can also lead to inflated l_y/l_x ratios.

Steeper slopes are assumed to produce elliptical fires similar to the shapes produced by wind. As explained above, however, there are many unresolved questions concerning the effect of sloping surfaces on fire spread rates. Science has not resolved how physical processes of fire spread may differ with wind versus slope. Furthermore, there has been little research on the combined influences of wind and slope on fire shape or spread rates, particularly when wind is not aligned with the upslope direction but instead flows across slope or downslope. Practical modelling has thus far assumed that wind and slope effects can be mathematically combined by vector algebra, but this is an active area of research. Experiments on fire growth with wind and slope are discussed in **Chapter 6, Winds and Topography**.

Fire area and perimeter

Fire growth is used to describe the change in area or perimeter with time. Area growth has units of $m^2\ min^{-1}$ (or $km^2\ h^{-1}$), and perimeter growth has units of m min^{-1} (or km h^{-1}). The ability to estimate fire area and its changes over time are important for fire management purposes, and thus has been the subject of considerable attention. Even without scientific certainty of an ideal fire shape, the ellipse's simple mathematics makes it a powerful tool for predicting two-dimensional fire growth.

For an elliptical fire shape with specified l_y/l_x ratio (**Figure 2.14**), the area and perimeter can be calculated at any time if we know the forward spread rate and the elapsed time, t. Referring to **Figure 2.14b**, we see that for each fire shape, there are elliptical dimensions of r_a, r_b, and r_c. These are relative dimensions and need no inherent units, but if we assume their units are rates (m min^{-1}), then these become *elliptical expansion rates*, and we can account for elapsed time only when area and perimeter are to be calculated. The r_a dimension of the ellipse is half the major axis from the centre point to the fire edge at the head ($r_a = l_y/2t$). The r_b dimension is the semi-minor axis, or half-width measured from the centre to the edge at the widest part of the ellipse ($r_b = l_x/2t$). The r_c dimension is the offset along the major axis between the assumed ignition point and the centre. Thus, the steady heading fire spread rate r can be expressed as:

$$r = r_a + r_c \quad [2.2]$$

Backing spread rate r_d is calculated as:

$$r_d = l_y/t - r = r_a - r_c \quad [2.3]$$

Because we cannot directly calculate the backing spread rate, you might wonder where the ignition point is located on an elliptical fire. In other words, what is the r_c dimension of the ellipse? As with fire shape in general (and fire spread for that matter), there are no physical answers yet to this question. We will spend some time discussing this in **Chapter 8, Fire shapes and growth**. The mathematically convenient location for the ignition location, and the most commonly assumed, is the *rear focus* of the ellipse. Assuming the focus as the ignition point allows us to estimate the other dimensions of the ellipse using the heading to backing ratio (r/r_d) (from Alexander 1985):

$$r/r_d = \frac{l_y/l_x + \sqrt{(l_y/l_x)^2 - 1}}{l_y/l_x - \sqrt{(l_y/l_x)^2 - 1}} \quad [2.4]$$

The other dimensions of the ellipse are as follows:

$$r_a = (r + r_d)/2 \quad [2.5]$$

$$r_b = r_a/(l_y/l_x) \quad [2.6]$$

$$r_c = r_a - r_d = r_a\sqrt{1 - (l_x/l_y)^2} \quad [2.7]$$

For an example calculation of fire area, let us assume that the fire has a shape with l_y/l_x of 2.0 and spreads at a constant rate of $r = 8.0$ m min^{-1}. Using **Eqn [2.4]** we calculate $r/r_d = 13.92$. Thus, the backing spread rate is $r_d = \frac{r}{r/r_d} = 8.0/13.92 = 0.57$ m min^{-1}, $l_y/t = r + r_d = 8.57$ m min^{-1}, $r_a = l_y/2t = 4.28$ m min^{-1}, $r_b = r_a/(l_y/l_x) = 2.14$ m min^{-1}, and $r_c = r_a - r_d = 4.28 - 0.57 = 3.71$ m min^{-1}. To calculate elliptical fire area, we must now include the elapsed time t:

$$\text{Area} = A_e = \pi r_a r_b t^2 \quad [2.8]$$

where the symbol π is the number pi with approximate value of 3.141592. In this example, the burned area after 60 min (1 h) of fire growth is $A_e = \pi * 4.28 * 2.14 * 60^2 = 103\ 936$ m$^2 \cong 10.4$ ha. After 2 h it would be ~41.6 ha, and after 3 h it would be 93.5 ha. The fire area grows as a function of both length and width dimensions and thus increases nonlinearly with time as depicted in **Figure 2.14c**. The fire area at any time depends both upon the forward spread rate and the shape of the fire.

The perimeter (circumference) of the fire is useful because it represents the distance around the fire to be extinguished if firefighters were to contain it. Unlike fire area, which can be calculated with an exact solution, fire perimeter must be estimated with mathematical approximation. An exact solution for an elliptical circumference does not exist, but numerous formulas are available to estimate it. The simpler ones routinely over- or underestimate the true value. However, for practical purposes in describing wildland fires, the mathematical errors are probably minor compared to those involved in estimating the heading or backing fire spread rates or the fire shape. A simple formula for approximating fire perimeter is:

$$\text{Perimeter} = P \approx \pi t \left(3(r_a + r_b) - \sqrt{(3r_a + r_b)(r_a + 3r_b)} \right) \quad [2.9]$$

For the example given above, this formula yields $P \approx 1246$ m (1.2 km) at 1 h, 2492 m (2.5 km) at 2 h and 3738 m (3.7 km) at 3 h. Notice that the perimeter increases linearly with time as shown in **Figure 2.14c**. Like fire area, the fire perimeter at any time depends on heading fire spread rate and the shape of the fire.

The assumption that fires are elliptically shaped is also useful in characterising the variation of fire spread rate as a function of spread direction (Catchpole et al. 1982). With the heading fire spread rate r and the elliptical expansion rates r_a, r_b and r_c from the fire shape derived above, the spread rate of a fire front $r(\theta_h)$ as a function of the angle θ_h relative to the heading direction (see Supplementary calculations at the end of this chapter), is:

$$r(\theta_h) = r_c \cos(\theta_h) + \sqrt{r_a^2 \cos^2(\theta_h) + r_b^2 \sin^2(\theta_h)} \quad [2.10]$$

The calculations for our previous example with $r(0°) = 8.0$ m min^{-1} and fire shape of $l_y/l_x = 2.0$ produce the identical values of spread rates for flanking ($r_b = r(90°) = 2.14$ m min^{-1}) and backing ($r_d = r(180°) = 0.57$ m min^{-1}).

We can see in **Figure 2.15a** how the fire shape affects the spread rate distribution around the flame front. The relative spread rate of the fire compared to head fire spread rate decreases with increasing angle away from the heading direction, more steeply with increasing ellipse eccentricity. The relative spread rate around the edge of an elliptical fire can also be illustrated by the fractional length of fire perimeter from the heading direction (**Figure 2.15b**). This procedure described by Catchpole et al. (1992) shows the fraction of head fire spread decreases with increasing eccentricity or l_y/l_x ratio.

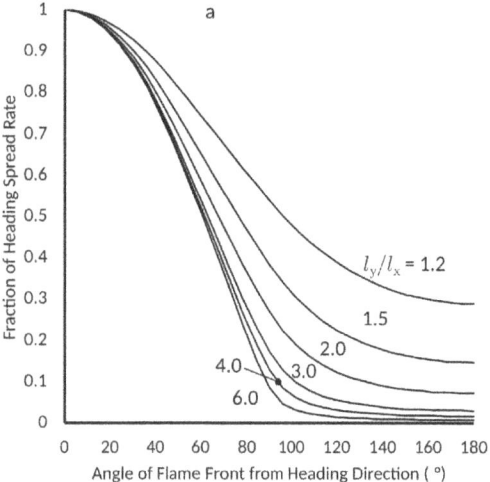

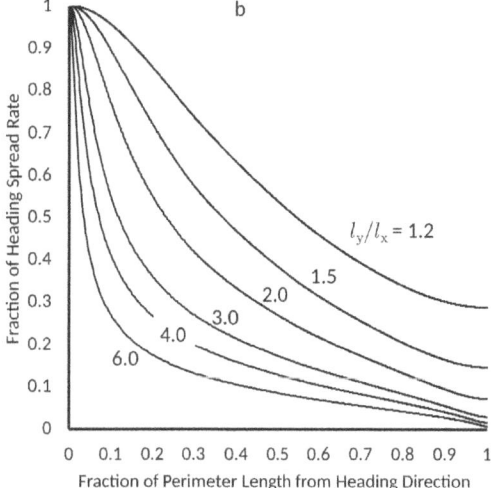

Figure 2.15: Graphs of relative fire spread rate for different elliptical shapes as it varies with (a) orientation of the flame front relative to the heading direction and (b) the fraction of fire perimeter length measured from the head of the fire (adapted from Catchpole et al. 1992).

Heat release and fireline intensity

Up to this point, we have talked about mainly the geometric characteristics of fire behaviour – its flame zone dimensions, how fast it spreads, and the sizes and shapes of fires. But the heat release characteristics of fires are extremely important because these determine what the fire does to vegetation, houses and the heat transfer to other fuels; determine how close firefighters can approach it; and actually begin to explain how fires spread. Fires release heat that can be quantified with different units, depending on the possible use of the measurement (refer to **Appendix A**). **Table 2.1** provides some examples. Heat release calculations require converting the mass of fuel consumed per unit area (kg m^{-2}) to heat (kJ); for this calculation, we need to know the heat of combustion (kJ kg^{-1}) (see **Chapter 4**). Depending on the purpose, we must decide on the dimensions or scale of the quantity desired:

1. Do we need to know the energy released per unit area (i.e. m^{-2}) or summed across the entire fire?
2. Do we need to partition the heat release between flaming and non-flaming phases or simply use the total?
3. Do we want to express the heat release as a rate (i.e. per unit time)?

At the largest spatial scale – the entire fire – total heat release and heat release rate may be valuable for understanding energy input to the atmosphere or comparing large wildfires. These applications would require other calculations or assumptions, such as how much heat from the fire actually heats the air, the ground and other fuels, and how much is required to vaporise water. Large-scale quantities would require spatial data on fuel mass and its distribution across the fire area, and also periodic or continuous estimates of spread progression. Partitioning fuel combustion into flaming and non-flaming phases remains a persistent challenge and possible ways to do this are discussed in **Chapter 9, Fuel consumption and Heat release**. Knowing the duration of heat release allows us to quantify the rate, which is often called fire *intensity* or *power* (i.e. kJ s^{-1} or kW).

The most commonly used metric for wildfire heat release rate is called *fireline intensity* (kJ m^{-1} s). It describes the energy release rate per unit length of a linear flaming front (length perpendicular to the direction of spread). Fireline intensity has many uses in fire management and for estimating effects on vegetation (Van Wagner 1973; Alexander 1982). Fireline intensity is an important property of line fires that varies around the entire fire, just like the spread rate described previously.

For a spreading line fire, Byram (1959b) defined fireline intensity as I_B (kW m^{-1} or kJ s^{-1} m^{-1}) as the product of the *heat yield* (H_c, kJ kg^{-1}) (see **Chapter 4, Flame height** for further discussion of this term), dry fuel mass per unit area consumed in flaming m_c'' (kg m^{-2}) and steady rate of spread r (m s^{-1}):

$$I_B = H_c m_c'' r \qquad [2.11]$$

Table 2.1. Some measures of heat release in wildland fires.

Each variable can be computed for the flaming phase, non-flaming (glowing and smouldering) phase, and/or total.

Variable	Spatial scale, units	Quantities involved	Possible use
Heat released	Entire fire, MJ	Product of fire area (m^2, ha, or km^2), fuel mass consumed (kg m^{-2}) and fuel heats of combustion (kJ kg^{-1})	Emissions, heat output
Heat per unit area	Unit area, kJ m^{-2}	Product of fuel mass consumed (kg m^{-2}) and heats of combustion (kJ kg^{-1})	Fire danger rating, fire characteristics chart
Heat release rate (i.e. overall intensity)	Entire fire, MJ h^{-1}	Product of fire area (m^2, ha, or km^2), fuel mass consumed (kg m^{-2}) and heats of combustion (kJ kg^{-1}), divided by time interval, such as duration of fire or combustion phase or total (min, h)	Gross input to atmospheric and wind models
Unit area heat release rate (i.e. unit area intensity)	Unit area, kJ m^{-2} s^{-1} or kW m^{-2}	Product of fuel mass consumed (kg m^{-2}) and respective heats of combustion (kJ kg^{-1}), divided by time interval (s, min, h)	Modelling of fire and atmospheric interactions
Fireline intensity	Line fire, kJ m^{-1} s^{-1} or kW m^{-1}	Product of unit area fuel consumption in flame zone (kg m^{-2}), heat of flaming combustion (kJ kg^{-1}) and fire spread rate (m s^{-1})	Fire suppression, prescribed fire, fire effects

The variables H_c and m_c'' vary with the phase of fuel combustion. The burning of wildland fuels is partitioned into flaming and smouldering phases, and the heat of combustion differs for each phase. Byram's (1959b) original definition of fireline intensity used a general value for heat yield that did not differentiate between the flaming and glowing phases. *Heat yield* was approximated by subtracting heat for fuel moisture evaporation, incomplete combustion, ash fraction and radiation from the total heat of combustion for oven-dry material. The original intention was that H_c would reflect only the heat contributing to fire spread (fuel heating and ignition) and the energy in the convective plume above the fire. Byram offered various estimates of heat yield that varied from ~14 000–15 000 kJ kg^{-1}. These values are consistent with more recent estimates of the 'effective' heat of combustion of gaseous pyrolysates from dry woody fuel (**Chapter 4**), which are appropriate for approximating heat released only from fuel burned in the flaming front.

The component m_c'' in **Eqn [2.11]** is the *available fuel* – that is, only the fuel mass consumed by the fire. Although not specified by Byram, m_c'' is now generally interpreted as the fuel mass consumed in the flaming phase only, because he related fireline intensity directly to the size of flames (discussed below). Thus, strictly speaking, H_c and m_c'' only describe the heat from the fuel combusted in flaming. Both of these can be challenging to estimate. Values for m_c'' vary considerably across the range of wildland fuel types. For example, the mass of fine material that typically burns in flaming ranges from ~0.1 to 1.0 kg m^{-2} for grasses, 0.5–2.0 kg m^{-2} for shrubs, and 1–3 kg m^{-2} for forest litter and woody debris. If a crown fire occurs, another 1–2 kg m^{-2} would be added from the contribution of canopy fuel mass consumed in the flame zone.

Studies are still being conducted on the effect of wind on fuel consumption during flaming, but the effect is likely to be minor compared to the effect of moisture content. The amount of available fuel m_c'' is highly sensitive to moisture content, and higher fuel moisture leaves a higher fraction of char and uncombusted fuel remaining after the flame front passes. Fuel moisture decreases the H_c component of intensity to a minor extent for dead fuels but considerably more for live vegetation. The impact of fuel moisture on fire behaviour is expressed as the product of $H_c m_c''$, which is called *heat per unit area*, *heat density* or *unit energy* (kJ m^{-2}), and some of the uses for this are described later.

Heat per unit area is not known to be strongly affected by wind, so it is valuable as an indicator of the effects of other weather variables (temperature, humidity, precipitation) on potential fire behaviour. Fireline intensity is influenced by environmental factors through their effects on both spread rate r and heat per unit area ($H_c m_c''$). Fire spread rate is sensitive to wind and slope (**Figure 2.12**). Therefore, for a constant heat per unit area ($H_c m_c''$), higher winds and steeper slopes will increase both spread rate and fireline intensity.

Flame length

Because we cannot see fireline intensity, it is difficult to develop intuition for it and directly judge the intensity from observations of a fire. However, flame dimensions are readily observed and, fortunately, they can be used as a general proxy. Longer flames in a spreading line fire mean greater energy release rates. Just as with candles and campfires (**Figure 2.5**), longer flames result as more fuel is consumed per unit time. The geometry of *flame length*, ℓ_f (m), for a line fire profile was illustrated in **Figure 2.11**. Many empirical functions relating flame length to fireline intensity have been developed from experimental fire data. There is considerable variation among the various functions, however – partly because different data sources were used and partly because of the uncertainty associated with measuring important variables (e.g. length of flames, fuel consumption). As an example of this variation among empirical functions, the relationship presented by Byram (1959a) is commonly used for surface fires:

$$\ell_f = 0.0775 I_B^{0.46} \qquad [2.12]$$

Another relation developed by Thomas (1963) has been used to estimate the flame length in crown fires (Rothermel 1991):

$$\ell_f = 0.0276 I_B^{2/3} \quad [2.13]$$

Differences in the estimates of flame length resulting from these equations are illustrated in **Figure 2.16**. **Figure 2.16a** shows the relationship between fireline intensity and fire spread rate for different loadings of available fuels (m_c''). Fireline intensity is a linear function of spread rate for a given fuel bed, but its relationship to flame length is not linear. The labels along the secondary y-axis (right-hand side) of **Figure 2.16a** show both Byram's and Thomas's scaling of flame length to fireline intensity; neither scale is linear, and the results from the two calculations **Eqns [2.12]** and **[2.13]** differ greatly. **Figure 2.16b** illustrates the nonlinear relation of fireline intensity to flame length. The y-axis is the same as that in **Figure 2.16a**; the x-axis shows flame length on a linear scale. The pair of graphs can be used to estimate flame length using both Byram's and Thomas's equations. For example, a fire spreading at 83 m min^{-1} and burning 2 kg m^{-2} fuel in the flaming front would have a fireline intensity of ~37 000 kW m^{-1} (37 MW m^{-1}). At this intensity, Byram's formula suggests a flame length of 10 m, while Thomas's formula suggests a flame length of ~30 m. This example illustrates some of the uncertainty with using flame length as a direct measure of fireline intensity.

Fireline intensity applies to an arbitrary segment of the fire's leading edge with the heat generated from the entire depth of the flame zone (**Figure 2.11**), irrespective of how deep the flame zone (D) is or where on the fire front the line fire segment is located. Because fireline intensity varies with spread rate, it also varies with the spread direction around the edge of the fire front (**Figure 2.11**), which is measured by the angle of orientation relative to the heading direction (**Eqn [2.10]**, **Figure 2.17**). Thus, fireline intensity can be calculated for any spread direction using **Eqn [2.11]**.

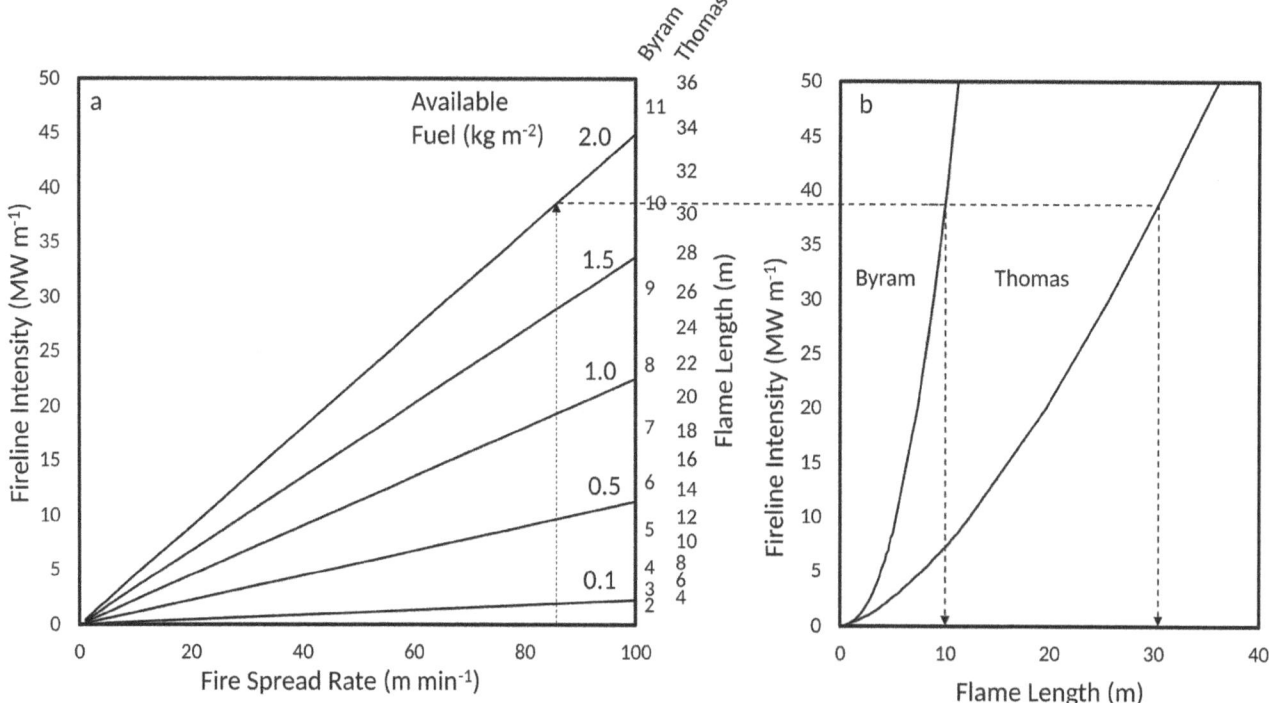

Figure 2.16: Fireline intensity (a) as a function of spread rate and available fuel (adapted from Byram 1959b) and (b) as it relates to flame length calculated by Byram's (1959b) and Thomas's (1963) equations.

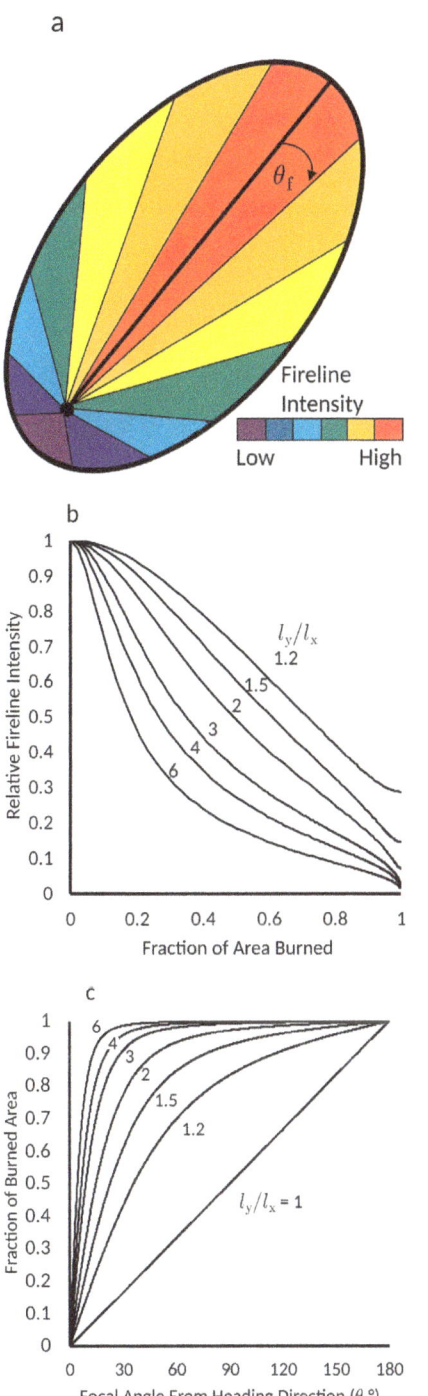

Figure 2.17: Fireline intensity variation within elliptical fire shapes (l_y/l_x ratios) show (a) the radial pattern of intensity within an elliptical fire; (b) the distribution of fireline intensity for elliptical areas burned within different l_y/l_x ratios (Catchpole *et al.* 1992); and (c) the fraction of burned area as a function of focal angle $\theta_f°$ (i.e. the radial angle measured from the ignition point away from the heading direction shown in (a)).

As a wildfire grows across the landscape in different directions, it produces an intensity distribution within the area burned. These intensity patterns are difficult to generalise in heterogenous environmental conditions, where the elements of the fire behaviour triangle vary in time and space. But to illustrate overall characteristics of intensity distributions within burned areas, we can turn once again to an idealised elliptically shaped fire growing from a point ignition under uniform and constant conditions. Within the footprint of an elliptical fire, the calculations described by Catchpole *et al.* (1982, 1992) show the intensity pattern radiating in straight lines outward from the ignition point in all directions (**Figure 2.17a**). (The formulas required for generating these area distributions are contained in the Supplementary calculations at the end of this chapter.) Graphs of the results (**Figure 2.17b, c**) show that the majority of the area is burned by the higher intensities, increasingly so as the ellipses take on more eccentric shapes (higher l_y/l_x ratios).

The fireline intensity patterns and the proportions shown here will be greatly altered if a fire experiences a change in wind speed or direction or encounters complex topography, varying fuel conditions or heterogeneous landscape fuel patterns. The resulting distributions for these conditions must be generated by simulation models. Some examples are discussed in **Chapter 8, Fire shapes and growth patterns** (**Figures 8.11** and **8.12**) and presented in a review of simulation approaches by Sullivan (2009c).

Fire characteristics chart

Fireline intensity and its dependency on the components of **Eqn [2.11]** can be graphically depicted on a *fire characteristic chart* (**Figure 2.18a**) (Andrews and Rothermel 1982; Andrews *et al.* 2011). The fire characteristics chart can be used to depict fire spread and intensity at any location around a fire perimeter. The x-axis of the chart is the heat per unit area ($H_c m_c''$), and the y-axis is the fire spread rate r. The curved contours within the chart are lines of equal intensity and flame length. Pictograms within each zone of the

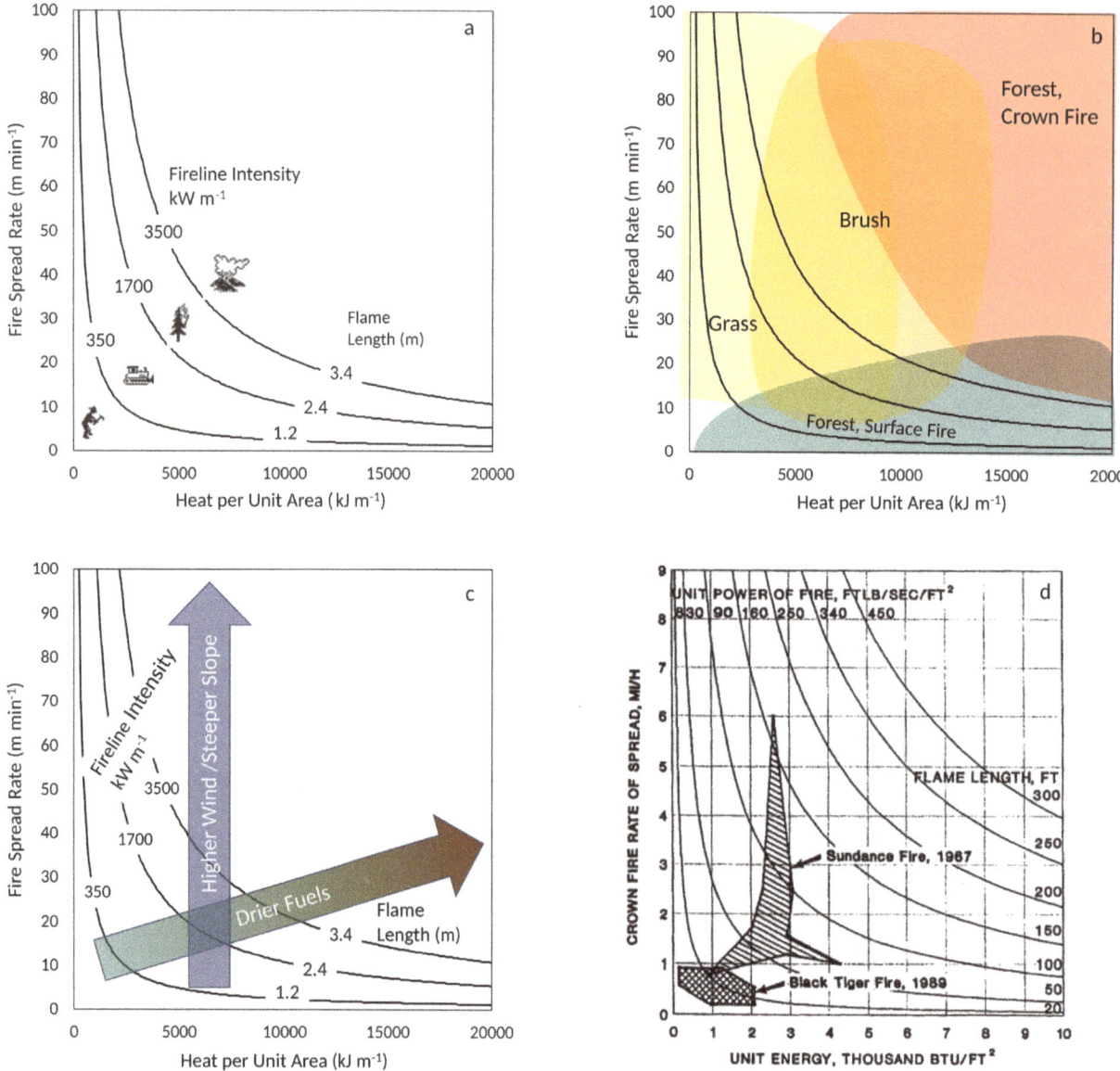

Figure 2.18: Fire behaviour characteristics charts show (a) fireline intensity (kW m⁻¹) and flame length (m) in relation to the separate terms of heat per unit area (kJ m⁻²) and spread rate (m s⁻¹) and the traditional limits of firefighting techniques (pictograms); (b) general ranges of fire behaviour by different fuel types; (c) the general effects of changes in wind, slope, and moisture on fire behaviour metrics; and (d) fire behaviour resulting from changes in moisture and wind throughout the day as displayed by an envelope of points (hatched polygons), shown here for two fires (note English units on *x*- and *y*-axes, from Rothermel 1991).

chart indicate that fire suppression by hand tools can be safely conducted at low intensities (< 1.2 m flame length), but heavy equipment such as a bulldozer is required at high intensities. At the highest intensities (> 3.4 m flame length), firefighting becomes increasingly dangerous and is largely ineffective.

The fire characteristics chart is useful for understanding the role of different environmental factors in producing fire behaviour. Different fuel types, for example, are associated with overlapping but distinctive regions on the chart (**Figure 2.18b**). Grass fires are concentrated along the left side of the chart because grass fuels typically have low loadings of fine fuels and hence produce less heat per unit area. Grass fires also exhibit the highest ranges of spread rate. Note that they exceed the maximum value shown on the *y*-axis. Fires in shrubs, in the middle of the chart, show higher fuel loadings and thus greater heat per unit area but slightly lower ranges of spread rate. Forest fires occupy two general regions of the chart, which correspond to surface fires and crown fires. Surface fires, which occur in the litter and downed woody material, spread more slowly than grass and brush fires and display a wide range of heats per unit area because of the fuel variability. If a crown fire develops, the chart depicts a region to the right produced by the additional fuel mass consumed in the canopy (higher m_c'') and the increased spread rates. Note that natural fuel complexes commonly consist of mixed vegetation (e.g. grass and shrub mixtures, timber and brush mixtures) that blur the distinctions between fire behaviours shown on the fire characteristics chart.

The fire characteristics chart is also useful for graphically depicting changes in fire behaviour caused by changes in wind and moisture (**Figure 2.18c**). For a given fuel type and moisture condition (indicated in the chart by heat per unit area), higher wind speeds and steeper slopes increase the fire spread rate and thus the fireline intensity. Moisture content affects fireline intensity through both axes. On the *x*-axis, lower moisture increases the heat per unit area by increasing available fuel mass (m_c''). On the *y*-axis, lower fuel moisture increases fire spread rate. Thus, a wildfire's behaviour may be characterised as it changes throughout the day or among days, as illustrated for two fires in **Figure 2.18d** (Rothermel 1991). The units on the axes are different from those in **Figure 2.18c**, but the relationships among variables follow the same patterns. The figure shows the ranges of fire behaviour experienced by each fire as humidity and temperature changed from morning to afternoon and as winds increased. Spotting behaviour of fires often increases the effective spread rate but is not represented on these axes.

Fire acceleration

Acceleration is the rate of change of a rate. The rate of change in a fire spread rate is thus called *fire acceleration* (m s^{-2}). In a mathematical sense, all fires that increase their spread rates over time are accelerating. Accelerating spread rate also implies according to **Eqn [2.11]** accelerating fireline intensity and associated flame lengths. Under uniform and constant environmental conditions, fire acceleration occurs immediately after ignition – caused by the positive feedbacks in the fire system that increase the spread rate in response to its larger size and intensity. Acceleration can also describe a fire's response to a rapid change in an external stimulus such as a gust of wind or switch in wind direction (Cheney and Gould 1997). In both cases – a gust of wind or change in wind direction – acceleration describes a temporary phase of adjustment in fire spread rate over a time scale longer than the time taken for the external conditions to change. 'Acceleration' is not commonly used for changes in fire spread rate that occur over time scales similar to those in the environmental conditions. For instance, a fire spread rate that gradually increases through the hours-long drying of fuels from morning to afternoon is called a daily trend in fire behaviour, not acceleration. In contrast, a sudden increase in intensity and spread rate following fire transitioning into canopy fuels or exceeding a threshold for spread imposed by fuel gaps could be described as acceleration.

Fires can also decelerate, gradually as they grow more pointed (increasing curvature of the fire front) (Viegas 2004) or rapidly if they encounter an abrupt fuel discontinuity. Fires naturally slow down if relative humidity increases or winds decline. Rates of deceleration seem to be much greater than rates of acceleration – that is, they tend to occur over shorter time frames because the ignition processes that keep fire spreading depend primarily on short-range and short-term heat exchange, not on processes with long-term memory of past burning conditions.

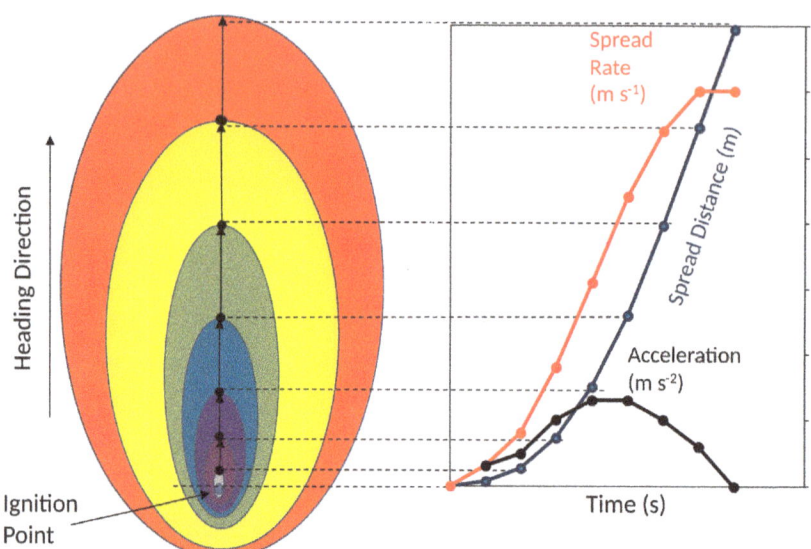

Figure 2.19: Fire acceleration, heading spread rate and heading spread distance for a hypothetical point source ignition in the head fire direction. After ignition, heading spread rate increases, with acceleration being fastest soon after ignition. As the fire grows and reaches a steady spread condition, acceleration declines to zero.

Fire growth and acceleration are illustrated in **Figure 2.19** for the wind-driven heading fire spreading from a point-source ignition. Although the heading fire spread rate increases slowly after ignition, the acceleration is the greatest in this early stage. Notice that acceleration ceases (drops to zero) as the fire reaches a steady state. It has long been assumed that fires will achieve a quasi-steady state in uniform fuels under constant weather and topographical conditions. Laboratory experiments in settings where the fire front is confined to a constant width and curvature seem to confirm this. Longer burning and larger fires, however, are more likely to experience varying environmental conditions (e.g. winds, fuels, topography); these introduce some ambiguity in determining the cause of accelerating or decelerating fire spread rates. We discuss these trends and the physical processes responsible in **Chapters 6** and **7**.

In wildland fire, a common way of thinking about acceleration is to identify the time or distance required to achieve a steady spread rate or some fraction of a steady state (see McAlpine and Wakimoto 1991). The time and distance required for this acceleration are useful metrics because they indicate how long or how far fire spread and intensity will remain below their presumed maxima. The acceleration period spans the time interval when the acceleration curve in **Figure 2.19** is above zero. The distance of fire spread during acceleration spans the same time interval.

Evidence suggests that acceleration times are longer with increasing wind speed, for point ignitions compared to line ignitions (Johansen 1987), and for heavier fuel loads. Point-source ignitions in slash fuels, for example, may have acceleration times of tens of minutes (McRae 1999). Similar acceleration times were reported for grass fires with high winds (Cheney and Gould 1995) but were shorter with widely alternating wind directions, perhaps because wind oscillations effectively widened the fire front. It is not completely clear how fuel type influences the acceleration period, but analysis by Viegas (2006) suggests rapid acceleration on steep slopes is delayed for fuel types with higher fuel loadings. Slower acceleration of point-source ignitions is routinely used in prescribed burning ignition patterns to limit the intensity of the fire (discussed in **Chapter 10**), whereas line ignitions are used to accelerate over shorter times and distances.

Fire acceleration occurs when ignition of fuel particles occurs faster than they are consumed by combustion, meaning that the flaming zone grows deeper. Put another way, the leading edge of the fire front spreads faster than the trailing edge (where flaming ceases). The physical processes producing acceleration involve the positive feedback of increasing rates of heat transfer from the increasing intensities of faster fires. In wind-driven fires, negative feedbacks related to buoyancy and other factors eventually become strong enough to compensate, so acceleration of spread rate slows or stops. On steep slopes, however, fire acceleration appears to respond considerably differently from acceleration in strong winds. This may suggest some physical reasons why fires in steep terrain are thought to be less predictable than wind-driven fires. Fire acceleration is thus a good example of a fire behaviour that provokes important questions about how physical processes operate in spreading fires.

Summary

This chapter has introduced the idea of fire as a general system of interacting physical processes. We began with a burning candle as an example of a stationary self-sustaining fire system and then showed how its behaviour applies to larger scale wildland fires. Wildfires, however, must ignite new fuel and spread in order to be sustained. For spreading wildfires, we introduced traditional descriptions of wildfire behaviour and their dependency upon the role of the fire environment embodied by the fire behaviour triangle (fuels, topography and weather). These environmental factors generally produce asymmetric wildfire spread that differentiates fast-spreading heading fires from slower-spreading backing fires. Conventional metrics of fire behaviour were examined, including spread rate, fire growth, fireline intensity and flame length, and we showed how these relate to the principal environmental drivers. To now extend our understanding of fire behaviour to consider wildfire as a product of a dynamical system, we must turn our attention to the physical processes of heat transfer, combustion and ignition. These processes and their functioning within the wildfire system determine if fire can spread or not, how fast it will spread, and at what intensity it will burn.

Supplementary calculations

The calculation of fireline intensity distributions within an elliptical fire area requires some additional definitions and formulas. The idea was put forth by Catchpole *et al.* (1992) to calculate both the perimeter and area distributions of fireline intensity I_B and inspired some of the content of **Figures** **2.15** and **2.17**. Using the r_a, r_b and r_c

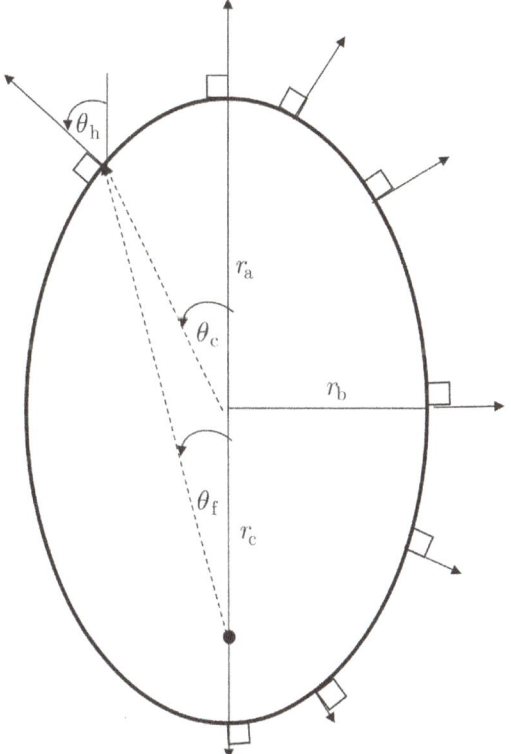

Figure 2.20: Angles and dimensions required to determine the fireline intensity distribution within elliptical fires. All angles are measured relative to the heading direction. The angle θ_h is measured normal to the fire front, θ_c is measured from the centre of the ellipse at the intersection of the major and minor axes, and θ_f is the focal angle measured from the ignition point.

parameters of the ellipse, which are assumed to represent expansion rates, the calculations of area and associated intensity rely on several different angles within an ellipse (**Figure 2.20**). For any fire ellipse, the centre angle θ_c is measured from the heading spread direction to any radius extending between the centre of the ellipse to any point on the perimeter. The focal angle θ_f is similar but for radii originating at the ignition point (often, but not necessarily, the rear focus). The angle θ_h is measured relative to the heading direction normal to the elliptical front (**Eqn [2.10]**).

Because the spatial I_B pattern in an elliptical fire shape radiates from the ignition point (**Figure 2.17a**), we must calculate the area distribution for $I_B(\theta_f)$ using the focal angle θ_f to obtain the focal sector area $A_f(\theta_f)$. The method given by Catchpole et al. (1992) calculates both I_B and area $A_f(\theta_f)$ using θ_c for a specified angle θ_f:

$$\cos\theta_c = \left(\frac{r_b\cos\theta_f\left(r_b^2\cos^2\theta_f + (r_a^2 - r_c^2)\sin^2\theta_f\right)^{0.5} - r_a r_c \sin^2\theta_f}{r_b^2\cos^2\theta_f + r_a^2\sin^2\theta_f}\right) \quad [2.14]$$

which is used to obtain the focal sector $A(\theta_f)$ in terms of angle θ_c:

$$A_f(\theta_f) = \frac{1}{2}r_b(r_a\theta_c + r_c\sin\theta_c) \quad [2.15]$$

Variation in θ_f changes I_B through its effect on the local spread direction θ_h which can also be calculated as a function of θ_c:

$$r(\theta_c|\theta_f) = \frac{r_b(\cos\theta_c + r_a)}{\sqrt{r_b^2\cos^2\theta_c + r_a^2\sin^2\theta_c}} \quad [2.16]$$

Thus, the values of $I_B(\theta_f) = H_c m_c'' r(\theta_c|\theta_f)$ can be obtained for any angle θ_f and the area $A_f(\theta_f)$. Since θ_c and θ_f are angles from half an ellipse, the area fraction burned by an entire elliptical fire for a given intensity $I_B(\theta_f)$ requires $A_f(\theta_f)$ be divided by the nominal area of half an elliptical fire expansion $\left(\frac{1}{2}\pi r_a r_b\right)$.

Note that $A_f(\theta_f)$ is not equal to an elliptical sector from the centre of the ellipse $A_f(\theta_c) = \frac{1}{2}ab\tan^{-1}\left(\frac{a}{b}\tan(\theta_c)\right)$. With possible confusion over the use of θ_c in calculating $A_f(\theta_f)$, a formula credited to David Cantrell[i] uses θ_f directly:

$$A_f(\theta_f) = \frac{1}{2}r_a r_b \left(\theta_f + \left(\sqrt{1-\psi^2}\,\frac{\sin\theta_f}{1-\psi\cos\theta_f}\right) + \left(2\tan^{-1}\left(\frac{\psi\sin\theta_f}{1-\psi\cos\theta_f}\right) + \sqrt{1+\psi^2}\right)\right) \quad [2.17]$$

where ψ is the eccentricity of the ellipse:

$$\psi = r_c/r_a \quad [2.18]$$

Even with **Eqn [2.17]**, the calculation of $I_B(\theta_f)$ requires use of θ_c in **Eqns [2.14]** and **[2.16]**.

Endnote

[i] https://groups.google.com/forum/#!topic/sci.math/E3qABMMnWkc

References

Alexander ME (1982) Calculating and interpreting forest fire intensities. *Canadian Journal of Botany* **60**(4), 349–357. doi:10.1139/b82-048

Alexander ME (1985) Estimating the length-to-breadth ratio of elliptical forest fire patterns. In *Proceedings of the 8th Conference on Fire and Forest Meteorology*, 29 April 29 – 2 May 1985, Detroit, Michigan. pp. 287–304. Society of American Foresters, Bethesda, Maryland, MA.

Anderson DH, Catchpole EA, De Mestre NJ, Parkes T (1982) Modelling the spread of grass fires. *The ANZIAM Journal* **23**(4), 451–466.

Anderson HE (1983) 'Predicting wind-driven wildland fire size and shape'. Research Paper INT-305. USDA Forest Service, Intermountain Forest and Range Experiment Station, Odgen, UT.

Andrews PL, Rothermel RC (1982) 'Charts for interpreting wildland fire behaviour characteristics'. General Technical Report INT-131. USDA Forest Service, Intermountain Forest and Range Experiment Station, Ogden, UT.

Andrews PL, Heinsch FA, Schelvan L (2011) 'How to generate and interpret fire characteristics charts for surface and crown fire behavior'. General Technical Report RMRS GTR-253. USDA Forest Service, Rocky Mountain Research Station, Fort Collins, CO.

Byram GM (1959a) Forest fire behavior. In *Forest Fire: Control and Use*. (Ed. KP Davis) pp. 90–123. McGraw Hill, New York, NY.

Byram GM (1959b) Combustion of forest fuels. In *Forest Fire: Control and Use*. (Ed. KP Davis) pp. 61–89. McGraw Hill, New York, NY.

Catchpole EA, de Mestre NJ, Gill AM (1982) Intensity of fire at its perimeter. *Australian Forest Research* 12, 47–54.

Catchpole EA, Alexander ME, Gill AM (1992) Elliptical-fire perimeter- and area-intensity distributions. *Canadian Journal of Forest Research* 22(7), 968–972. doi:10.1139/x92-129

Cheney NP, Gould JS (1995) Fire growth in grassland fuels. *International Journal of Wildland Fire* 5(4), 237–247. doi:10.1071/WF9950237

Cheney NP, Gould JS (1997) Letter to the editor fire growth and acceleration. *International Journal of Wildland Fire* 7(1), 1–5. doi:10.1071/WF9970001

Countryman CM (1972) *The Fire Environment Concept*. Pacific Southwest Forest and Range Experiment Station, Berkeley, CA.

Faraday M (1860) The chemical history of a candle. Christmas lecture series, <https://www.princeton.edu/ssp/joseph-henry-project/electric-light/The_chemical_History_of_a_Candle.pdf>.

Johansen RW (1987) Ignition patterns & prescribed fire behavior in southern pine stands. *Georgia Forestry Commission – Georgia Forest Research Paper* 72, 1–8.

Kerr JW, Buck CC, Cline WE, Martin S, Nelson WD (1971) *Nuclear Weapons Effects in a Forest Environment*. Thermal and Fire (No. DASIAC-SR-112). General Electric Company, Santa Barbara, CA.

McAlpine RS, Wakimoto RH (1991) The acceleration of fire from point source to equilibrium spread. *Forest Science* 37(5), 1314–1337.

McRae DJ (1999) Point-source fire growth in jack pine slash. *International Journal of Wildland Fire* 9(1), 65–77. doi:10.1071/WF99006

Nelson RM, Jr (2003) Reaction times and burning rates for wind tunnel headfires. *International Journal of Wildland Fire* 12(2), 195–211. doi:10.1071/WF02041

Rothermel RC (1972) 'A mathematical model for predicting fire spread in wildland fuels'. Research Paper INT-115. USDA Forest Service, Intermountain Forest and Range Experiment Station, Ogden, UT.

Rothermel RC (1991) 'Predicting behavior and size of crown fires in the northern Rocky Mountains'. Research Paper INT-438. USDA Forest Service, Intermountain Forest and Range Research Station, Ogden, UT.

Sullivan AL (2009a) Wildland surface fire spread modelling, 1990–2007. 1: Physical and quasi-physical models. *International Journal of Wildland Fire* 18(4), 349–368. doi:10.1071/WF06143

Sullivan AL (2009b) Wildland surface fire spread modelling, 1990–2007. 2: Empirical and quasi-empirical models. *International Journal of Wildland Fire* 18(4), 369–386. doi:10.1071/WF06142

Sullivan AL (2009c) Wildland surface fire spread modelling, 1990–2007. 3: Simulation and mathematical analogue models. *International Journal of Wildland Fire* 18(4), 387–403. doi:10.1071/WF06144

Sullivan AL, Sharples JJ, Matthews S, Plucinski MP (2014) A downslope fire spread correction factor based on landscape-scale fire behaviour. *Environmental Modelling & Software* 62, 153–163. doi:10.1016/j.envsoft.2014.08.024

Thomas PH (1963) The size of flames from natural fires. *Symposium (International) on Combustion* 9(1), 844–859.

Van Wagner CE (1969) A simple fire-growth model. *Forestry Chronicle* 45(2), 103–104. doi:10.5558/tfc45103-2

Van Wagner CE (1973) Height of crown scorch in forest fires. *Canadian Journal of Forest Research* 3(3), 373–378. doi:10.1139/x73-055

Viegas DX (2004) On the existence of a steady state regime for slope and wind driven fires. *International Journal of Wildland Fire* 13, 101–117. doi:10.1071/WF03008

Viegas DX (2006) Parametric study of an eruptive fire behaviour model. *International Journal of Wildland Fire* 15, 169–177. doi:10.1071/WF05050

3

Thermodynamics, fluid mechanics and heat transfer

In this chapter we introduce concepts related to the state of gases, liquids, and solids and how they exchange energy (i.e. heat) with their surroundings. In **Chapter 2**, we discussed how heat is required to ignite fuel, whether it is the wax of a candle or a blade of grass, and that heat must be transferred to and from substances for fire to spread. Many phenomena are involved in heat transfer, and it is one of the essential processes in the system of wildfire behaviour.

The sciences of thermodynamics, fluid mechanics and heat transfer are fundamental to wildland fire behaviour. We learned in **Chapter 2** that, at the simplest level, wildfires are chemical reactions between a fuel and an oxidiser (which is the oxygen in air). As discussed in **Chapter 4**, the chemical reactions that comprise fire are extremely complex. However, only a basic understanding of chemistry is needed for understanding fire behaviour. In contrast, a broad, thorough understanding of thermodynamics, fluid mechanics and heat transfer is essential. Firefighters, fuels specialists, fire ecologists and managers need practical knowledge about questions like how a fire will spread over variable terrain and how wind will change the heat release rate of a fire. The path to understanding these aspects of fire behaviour leads through the sciences of thermodynamics, heat transfer and fluid mechanics. This knowledge will enable you to explain the fire behaviour you see in the field and, to a certain extent, to predict how a fire will behave in the future.

Thermodynamics is a science that deals with quantities of heat and how heat changes its surroundings. It is very important in many fields, including electrical power generation, refrigeration and automobile engines. In the context of wildland fire, we are most interested in questions like, 'How much heat is produced by burning 3 kg of wood?' or 'How much heat is needed to evaporate 50 mL of water?' Here we provide the background needed to answer questions like these and understand why they are important in fire behaviour.

Fluid mechanics is a science that concerns the movement of liquids and gases. The principles of fluid mechanics apply at all scales, from the largest hurricanes to the movement of blood through the tiny capillaries of our lungs. The wind that drives a fire is a fluid, and so are the hot combustion gases produced by the flames. We need to understand fluid mechanics because fluid flow is at the heart of convection heat transfer, the most important mode of heat transfer for explaining how and why wildfires spread.

Heat transfer is closely related to thermodynamics. Instead of examining absolute heat quantities, it deals with how long it takes for quantities of heat to move from one location to another. A question about heat transfer in fire behaviour might be, 'Why does a wildfire spread mostly through fine fuels like grass and leaves rather than through bigger branches and logs?' There are three modes of heat transfer. Conduction heat transfer occurs when thermal energy is transferred through matter by interactions of molecules. Radiation heat transfer occurs when heat moves from one surface to another as electromagnetic energy. Convection heat transfer occurs when heat moves into or out of a surface by the flow of a fluid (such as air or combustion gas) over that surface.

Basic concepts, material properties and terminology

To understand how the principles of thermodynamics, fluid mechanics and heat transfer apply to wildfire behaviour, we need to examine the concepts of energy, temperature, and heat and how they are related.

Energy is the capacity to change the configuration of matter. Such changes include the increase or decrease of temperature of matter, the change in phase of matter (e.g. evaporating water into steam), and moving matter from one place to another (e.g. airborne embers). The air you breathe, the water you drink and the clothes you wear are all matter made of atoms. Any time atoms are reconfigured, there is an accompanying change in energy. Sometimes energy is released or absorbed, and at other times energy changes forms. Energy can take many forms. For our purposes, the most important are chemical energy, electromagnetic radiation energy and thermal energy (heat). The unit of energy is the joule (J). Burning a kilogram of wood releases ~20 million joules of thermal energy or heat, which, for the purposes of fire behaviour, will bring about changes in temperature, changes in phase of other materials (i.e. fuels, water), and motion of the hot buoyant plume. A detailed explanation of how heat is released by combustion is found in **Chapter 4, Thermodynamics of combustion**.

Chemical energy is energy stored in chemical bonds between atoms and molecules. Wildland fuels like grass, leaves and wood all contain chemical energy. Chemical bonds are broken or formed in chemical reactions and, depending on the circumstances, energy is either released or absorbed. The chemical reactions of combustion release the chemical energy bound up in wildland fuels as heat and light (**Figure 3.1**). The interplay between chemical energy, heat and the thousands of chemical reactions that underlie fire are extremely complex. However, for the purpose of understanding fire behaviour and the physical processes of fire, we can distill the intricacies down to one important concept: *Fire produces heat, which drives all the processes that, combined, contribute to fire behaviour seen in the field and the laboratory.*

Electromagnetic radiation is a stream of energy that passes through transparent media like air in the atmosphere, the vacuum of space, or glass. It moves at or very near the speed of light: 300 000 km s^{-1}. The photon is the fundamental physical unit of electromagnetic radiation. It has a property called *wavelength*, which is related to the amount of energy it represents. In this book, we use units of micrometres (μm) to describe wavelength quantitatively.

Thermal energy (or *heat*) is a type of energy that, when gained or lost by matter, causes a change in temperature, phase (e.g. changing a liquid to a gas) or chemical composition. We deal with heat on a daily basis. When cooking food, we add heat to the ingredients in a controlled manner, using a stove or microwave oven. When the weather is cold outside, we wear a jacket to minimise loss of body heat to the environment. When a firefighter sprays water on a fire, heat that would otherwise sustain burning goes instead to heating and boiling the water, thereby inhibiting the chemical reactions of combustion.

Power is a concept that describes a change in energy over time. In the context of heat transfer, we most often use power to describe how fast energy is transferred from one location to another. The unit of power is the watt, which is equivalent to energy divided by time.

$$1 \text{ watt} = 1 \text{ W} = 1 \text{ J s}^{-1}$$

For example, thermal energy could be transmitted from a flame to a log at a rate of 10 kW. Another

Figure 3.1: Wildland fire burning in timber with moderate dead and down fuels. Photograph by Ian Grob, USDA Forest Service.

common use of the concept of power is in describing the rate at which energy is released by the burning of fuel. Many factors affect this rate, including the size of the fuel element, fuel density, and moisture content. This will be covered in **Chapter 4**.

Heat flux describes the amount of thermal energy transmitted to an area over an amount of time. We frequently use heat flux when we discuss heat transfer and ignition. Heat flux has units of watts per meter squared:

$$1 \text{ W m}^{-2} = 1 \text{ J s}^{-1} \text{ m}^{-2}$$

For example, a crown fire will routinely produce a radiant heat flux of 100 kW m^{-2}. Heat flux is essential for describing how much heat from fire is impinging on fuel particles. In this book, heat flux is represented by the variable, $\dot{q}''$, which reads, 'q dot double prime'. The letter q represents energy, the dot over the q indicates energy per unit time, and the double prime means the quantity is also expressed per unit area.

In everyday life, we think of *temperature* as a measure of how hot or cold an object is. The most common way to quantify temperature is by using the *Celsius temperature scale*, in which 0 °C is the temperature at which pure water freezes and 100 °C is the temperature at which it boils (at atmospheric pressure at sea level). While the Celsius scale is familiar and convenient, its scaling relative to the properties of water is arbitrary, making it inappropriate for some heat transfer and thermodynamics calculations. For these problems, we use the *Kelvin temperature scale*, an absolute temperature scale in which zero Kelvin (abbreviated 0 K) refers to the coldest an object can possibly be and 273.15 K is the freezing temperature of pure water. The numerical increment of the Kelvin scale is identical to that of the Celsius scale, so a change in temperature of 1 K is the same as a change in temperature of 1 °C. Zero Kelvin is extremely cold, equal to approximately –273.15 °C. This is a temperature we would never encounter in fire scenarios, much less in daily life. For the majority of equations and calculations in this book, the Celsius and Kelvin temperature scales are interchangeable. In those instances where the Kelvin scale is mandatory – such as thermal radiation and ideal gas calculations – it will be stated so explicitly.

Temperatures commonly encountered in wildland fire range widely from 35 °C (308 K) (the air temperature on a hot summer day) to around 350 °C (623 K) (the approximate ignition temperature of wood) to roughly 1000 °C (1273 K) (a typical flame temperature). Even though we rely on our ability to feel hot or cold to survive and protect ourselves, we cannot rely on this physical sensation to measure temperature, because it is imprecise and can be misleading. For example, because different substances transfer heat differently, a piece of wood will feel warmer to us than a piece of metal even when both objects have the same temperature. Instead of relying on our skin's sensitivity to heat to measure temperature, we use instruments that take advantage of the way materials change predictably with changes in temperature. A glass thermometer, for example, takes advantage of the way the red-coloured alcohol inside expands and rises when it is heated.

Fuels in wildland fires are heated and cooled simultaneously, often by different mechanisms. We use the concept of *energy balance* to account for the heat going into an object, the heat going out, and the heat produced by the object as it burns. The following 'leaky-bucket analogy' is useful for understanding energy balance: imagine you have a bucket full of water with a small hole in the bottom, as illustrated in **Figure 3.2**. As water flows through the hole, the amount of water in the bucket decreases (a). Now suppose you pour water into the bucket faster than the water is leaking out through the hole (b). In this case, the level of the water goes up. Next you reduce the flow of the hose so you are pouring water in more slowly than it is leaking out (c). This causes the level water in the bucket to go down. Finally, by carefully adjusting the flow of water, you exactly match the amount pouring into the bucket to the amount leaking out (d), causing the water level to remain the same. The logic of the leaky bucket extends to the energy balance of a fuel element. The fuel element may be gaining or losing thermal energy through convection, radiation heat transfer, or some combination of the two. If the *net heat transfer* is positive, then the fuel element will increase in temperature. If the net heat transfer is negative, then the fuel element will decrease in temperature. (Note that the leaky-bucket analogy does not extend to heat produced by an object as it burns.)

Eqn [3.1] formally expresses the concept of energy balance. If the energy balance $\dot{q}$ of an object accounts for the rate of energy going into the object $\dot{q}_{in}$, the rate of energy leaving the object $\dot{q}_{out}$, and the rate of energy released by combustion on the object surface $\dot{q}_b$, then:

$$\dot{q} = \dot{q}_{in} - \dot{q}_{out} + \dot{q}_b \qquad [3.1]$$

Subscript b represents 'burning'. Suppose you want to determine the energy balance of a small twig near a fire. For example, say the adjacent flames

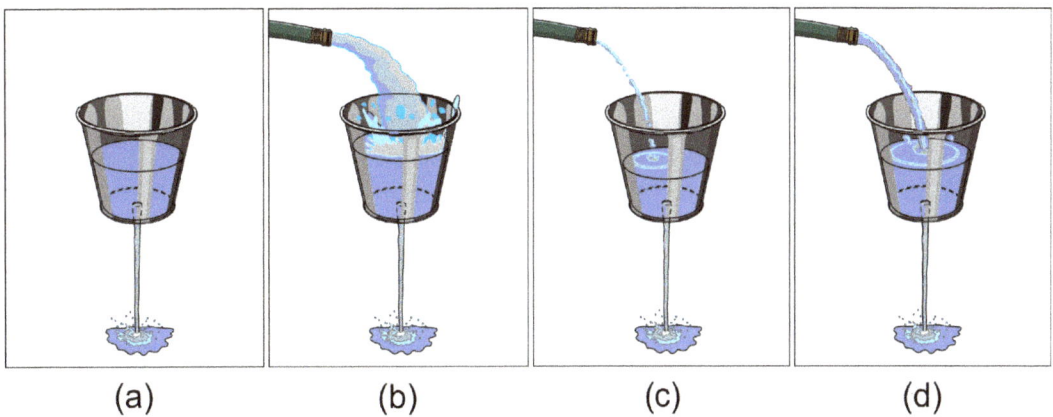

(a)   (b)   (c)   (d)

Figure 3.2: The 'leaky-bucket analogy' uses the flow of water to illustrate the concepts of energy balance and net heat transfer. Illustration by Brian Elling.

transmit 130 W of thermal energy to the twig via radiation heat transfer, where it is completely absorbed: $\dot{q}_{in}$ = 130 W. An ambient breeze is cooling the twig, carrying away heat at a rate of 50 W: $\dot{q}_{out}$ = 50 W. The twig is undergoing smouldering combustion, generating thermal energy at a rate of 120 W: $\dot{q}_b$ = 120 W. The energy balance for the twig is: $\dot{q}$ = 130 − 50 + 120 = 200 W. This means that the twig gains energy at a rate of 200 W causing its temperature to increase. A negative $\dot{q}$ would mean the twig was losing thermal energy, causing its temperature to decrease.

The *ideal gas law* is an equation that relates the pressure, volume, quantity, and temperature of a gas:

$$pV = NRT \qquad [3.2]$$

In Eqn [3.2], p is pressure (Pa), V is volume (m³), N is the number of moles of gas (mol), R is the universal gas constant (R = 8.314 J mol⁻¹ K⁻¹), and T is temperature (K). Note that temperature must be expressed in Kelvin when using the ideal gas law, not Celsius. The ideal gas law is useful because if you know any three of these variables (pressure, volume, moles and temperature), you can use it to solve for the fourth. Virtually all applications of the ideal gas law in wildland fire deal only with air and combustion gas. That being the case, a simplification of **Eqn [3.2]** can be made. The ideal gas law for air is written as:

$$p = \rho_a R_{air} T \qquad [3.3]$$

where ρ_a is air density (kg m⁻³) and R_{air} is the individual gas constant for air (R_{air} = 287.05 J kg⁻¹ K⁻¹). Combustion gas is sufficiently similar to air in composition that it can be treated as the same with minimal loss of accuracy. Therefore, we can use **Eqn [3.3]** for air, combustion gas and any mixture of the two. The specific gas constant for any particular gas is related to the universal gas constant by:

$$R_{gas} = RM_{gas}^{-1} \qquad [3.4]$$

where M_{gas} is the molecular mass of the gas. Air is a mixture of several gases (nitrogen, oxygen, argon and numerous trace gases), but has an apparent molecular mass of 29.0 g mol⁻¹.

One final concept pertaining to heat transfer and fluid mechanics is *diffusion*, the net movement of something from a region of high concentration to low concentration. The 'something' that moves is often a fluid – milk diffusing into coffee or perfume diffusing throughout a room – but it can also refer to energy and heat. For example, conduction heat transfer is a form of *thermal diffusion*, since thermal energy diffuses through a material from the hot side to the cold side. The rate of diffusion through a given medium is quantified with a diffusivity coefficient. Regardless of what is diffusing, diffusivities always have units of length-squared divided by time (m² s⁻¹).

Thermodynamics

Thermodynamics is an accounting of quantities of energy within a system. In the context of wildfire, thermodynamics describes the conversion of energy from one form to another, changes in the temperature of a substance, or changes in the phase of a substance. For example, using the principles of thermodynamics, we can track the energy released by burning of plant materials, the amount of heat required to heat fuel to a certain temperature, or the heat required to vaporise sap and pitch in live fuels.

All chemical bonds – including those contained within lignin and cellulose, the structural material of plants – are forms of stored chemical energy. Remarkably, plants derive energy from sunlight and, along with carbon dioxide gas from earth's atmosphere, water and trace minerals from the soil, convert it into lignin and cellulose. These strong, lightweight materials are made up of complex assemblies of sugar molecules. Lignin and cellulose make up the structural parts of wood, leaves, grass, and other plant biomass. They enable a plant to have rigid structure rather than being formless sacks of fluid. When plants burn, the stored chemical energy is released when lignin and cellulose break down into smaller molecules and react with oxygen (i.e. burn) to produce energy in the form of heat and light. Just how much stored energy does wood contain?

Heat of combustion, H_c, refers to the release of stored chemical energy per unit mass of fuel (J kg^{-1}) when it is completely consumed by fire. We can use heat of combustion to calculate the total heat and light generated by fire with the following equation:

$$q_b = m H_c \qquad [3.5]$$

where q_b (J) is the heat released by the burning of fuel of mass m (kg). The term *heat release* refers to the heat and light converted from stored chemical energy by fire. Suppose you have made some measurements and found that the litter loading in a 1 ha stand of timber is 80×10^3 kg. Let us further suppose you have determined the heat of combustion of the litter in this stand is 20×10^6 J kg^{-1}. A fire spreading through the stand and completely burning all litter would produce the following heat release:

$$\begin{aligned} q_b &= (80 \times 10^3 \text{ kg})(20 \times 10^6 \text{ J kg}^{-1}) \\ &= 1.6 \times 10^{12} \text{ J} = 1.6 \text{ TJ} \end{aligned}$$

No fire completely burns all fuel, but a calculation like this one provides the maximum possible heat release for a given situation.

How much thermal energy is needed to raise the temperature of a material by a certain amount? Answering this question requires a material property called *specific heat capacity* or simply *specific heat*. It is expressed with units of J kg^{-1} °C^{-1} and is a unique property for any given material. Specific heat for wood and other plant biomass varies somewhat depending on species, but a reasonable estimate is 2000 J kg^{-1} °C^{-1}. Woody biomass tends to have relatively high specific heat capacities, meaning that more thermal energy is required to raise the temperature compared to other materials, such as metals.

Suppose we have a 100 g (0.1 kg) block of dry wood at a temperature of 25 °C. How much thermal energy is needed to raise the block of wood to 275 °C? We can use the following equation:

$$\Delta q_{th} = m c_s (T_x - T_o) \qquad [3.6]$$

where Δq_{th} is the thermal energy absorbed (J), m is mass, c_s is specific heat capacity of the solid, T_x is the final temperature, and T_o is the initial temperature. Thus we can calculate:

$$\Delta q_{th} = 0.1(2000)(275 - 25) = 50 \times 10^3 \text{ J} = 50 \text{ kJ}$$

This calculation is based on some assumptions that are worth noting. First, it is assumed that the specific heat capacity of wood does not change with temperature. This is not strictly true for wood or any other material, but the difference in specific heat capacity of wood at 25 °C and 275 °C is small enough that we can ignore it in this simple example. Second, it is assumed that the thermal energy is introduced uniformly throughout the volume of the wood block. This never occurs in wildland fire, and more discussion on this topic follows in the **Heat transfer** section below.

A third crucial assumption in the calculation of Δq_{th} is that no *phase change* occurs as the temperature of a substance rises. *Phase change* is the transition of matter from one phase (solid, liquid, or gas) to another, a process driven by the absorption or release of thermal energy. In the context of wildland fire, the no-phase-change assumption is especially problematic when we try to account for the thermal energy absorbed by water moisture contained in fuel. Suppose we have 2 L of water in a container and we want to find out how much heat is needed to raise its temperature from 25 °C to 120 °C. We could look up the specific heat of water and use **Eqn [3.6]** to calculate an answer, just as we did for the wood block. However, the answer would be incorrect because the water will undergo a phase change that is not addressed by the equation. We know that water vaporises – that is, changes from the liquid phase to the gas phase – at around 100 °C (at atmospheric pressure at sea level). When liquid water just reaches 100 °C, additional thermal energy does not increase the water's temperature; instead, it vaporises (boils) the liquid water into water vapour (gas). The opposite effect occurs when water vapour condenses into liquid water; during that phase change, thermal energy is released – the same amount of energy as the amount absorbed when the water transitioned from the liquid to the gas phase.

Eqn [3.6] applies only when there is no change in phase, but in our example (raising the water temperature from 25 °C to 120 °C), water undergoes a phase change, so we need a new approach to calculate Δq_{th}. First let us describe the process of boiling water in more detail, using the example in **Figure 3.3**. In the figure, the circled numbers in the plot indicate key transition points in the heating process. The liquid water starts at state ① with an initial temperature of 25 °C. We add heat, causing the temperature to rise until the liquid water just reaches the boiling temperature of 100 °C (state ②). If we add more heat at this point, it will not increase the water temperature; instead, it will cause the water to boil – that is, to vaporise. If we keep adding heat until all of the water has vaporised, we reach state ③. During boiling (from state ② to state ③), both liquid and water vapour remain at 100 °C. In fact, phase changes in all matter occur without a change in temperature as long as the pressure remains unchanged. Suppose we capture all of the water vapour at state ③ in a perfectly insulated bag. If we continue to add heat at this point, the temperature of the water vapour will increase. **Figure 3.3** shows the addition of heat until the temperature reaches 120 °C (state ④).

The total heat absorbed by the water as it is heated, boiled, and heated further is given by:

$$q_{tot} = q_{1 \to 2} + q_{2 \to 3} + q_{3 \to 4} \qquad [3.7]$$

where $q_{1 \to 2}$ is the heat absorbed over ①→②, $q_{2 \to 3}$ is the heat absorbed over ②→③, and $q_{3 \to 4}$ is the heat absorbed over ③→④. No phase change occurs during the two heating processes, ①→② and ③→④, so **Eqn [3.6]** applies. The specific heat capacities for liquid water and water vapour are ~4185 J kg^{-1} °C^{-1} and 1893 J kg^{-1} °C^{-1}, respectively. Two litres of liquid water has a mass of ~2 kg. Given this, we can calculate:

$$q_{1 \to 2} = 2(4185)(100 - 25) = 627\,750 \approx 628 \text{ kJ}$$

$$q_{3 \to 4} = 2(1893)(120 - 100) = 75\,720 \approx 76 \text{ kJ}$$

We need a new equation to calculate $q_{2 \to 3}$ because ②→③ is a phase-change process:

$$q_v = mH_v \qquad [3.8]$$

where q_v is the heat required to completely vaporise a substance (J), m is the substance mass (kg), and H_v is the *heat of vaporisation* for the substance (J kg^{-1}). The heat of vaporisation is the amount of energy required to vaporise a substance per unit mass. The heat of vaporisation for water is approximately 2.26×10^6 J kg^{-1}. Given this, we can calculate:

$$q_{2 \to 3} = (2)(2.26 \times 10^6) = 4.52 \times 10^6 \text{ J} = 4520 \text{ kJ}$$

We now have all the terms needed to evaluate **Eqn [3.7]**, allowing us to find $q_{tot} = 5224$ kJ. Notice how much larger $q_{2 \to 3}$ is than $q_{1 \to 2}$ and $q_{3 \to 4}$. This means that the majority of the heat absorbed by

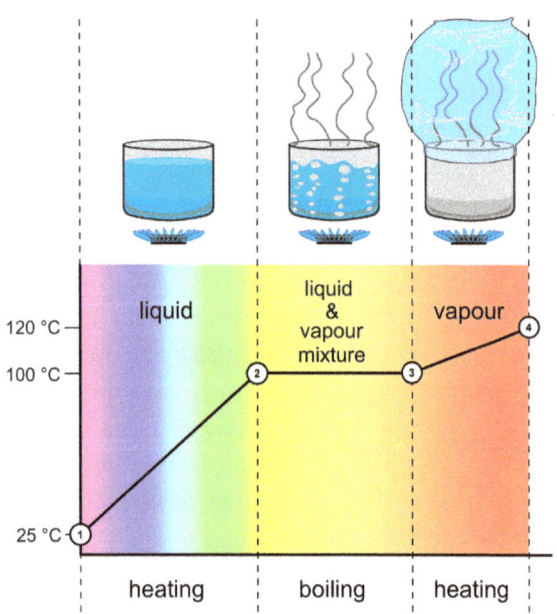

Figure 3.3: Phase change of water from liquid to gas. Water is heated until it begins to vaporise. It stays at the same temperature until all water has evaporated. The temperature of the resulting water vapour, which has been captured in a bag here, will rise if more heat is added. Illustration by Brian Elling.

the water while going from states ① to ④ was absorbed during the phase change. The importance of this fact for wildland fire cannot be overstated because the vaporisation of fuel moisture has a profound effect on the rate that fuel burns.

Any given wildland fuel element will ignite and burn only after it has been heated sufficiently to thermally decompose lignin and cellulose into flammable vapours (see **Chapter 5**). The temperature at which this process begins is called the *pyrolysis temperature*. Wildland fuels always contain some moisture, even in the driest desert landscapes. The boiling temperature of water is well below the pyrolysis temperature. Therefore, at least some of the fuel water moisture must be heated and vaporised before a fuel particle can ignite and burn. The example calculation above illustrates that a substantial amount of heat is needed to vaporise water – much more heat than that required for raising its temperature. This means that the thermal energy required to heat up and burn a fuel element is very sensitive to its fuel moisture. By extension, fire behaviour is very sensitive to relative humidity, because it drives the moisture content of the fine fuels that carry a spreading fire. The energy intense nature of water vaporisation largely explains dramatic changes in wildland fire behaviour observed following seemingly small changes in relative humidity.

We have examined three key concepts from a thermodynamics perspective: quantities of stored chemical energy in wildland fuels, the energy needed to change the temperature of a material, and the energy needed to cause a phase change, especially in water. All three of these areas are profoundly important for understanding the physics of fire and fire behaviour. We have focused on thermodynamics in this section, but we have not yet discussed heat transfer, the physical mechanisms of moving heat from one area to another. Heat transfer is introduced later in this chapter, but first we address the science of fluid mechanics, which is required to understand the mechanism of convection heat transfer.

Fluid mechanics

There are three common phases of matter: solid, liquid and gas. The science that describes the flow of liquids and gases in particular is called *fluid mechanics* (or *fluid dynamics*). Fluid mechanics includes a broad array of subtopics, including weather forecasting, canal design, and aerodynamics. In this book, we focus on buoyancy and how fluids flow around solid objects, because these topics are important for understanding convection heat transfer in wildland fire.

Perhaps the most basic fluid mechanics question is, 'What is a fluid?' In everyday use, the term 'fluid' normally refers to liquids like water or petrol. However, in science and engineering, *fluid* is a technical term that refers to *both* liquids and gases. In the science of wildland fire, we are concerned mostly with flow of gases – that is, air and the combustion gases produced by fire. The air that we breathe is a mixture of nitrogen, oxygen, argon, and trace amounts of other gases. The relative proportions of gases in air are ~78% nitrogen, 21% oxygen, and 1% argon and other gases.

In wildland fires, flowing fluids (e.g. wind) often interact with stationary solid surfaces, including fuels and the ground. The boundary between a flowing fluid and a solid surface is called the *fluid–solid interface*. For example, on a cold blustery day, you are very aware of the fluid–solid interface between your body and the rushing air because you have to add layers of clothing to prevent the wind from chilling you to the bone. Moreover, you have to lean into the wind to stay on your feet as your body is buffeted by gusts. This example illustrates two interactions between a fluid and a solid object that are key to understanding wildland fire.

First, the transfer of heat: the wind causes you to wear more clothing to reduce the loss of body heat to the flowing air. Similarly in wildland fires, the flow of cool ambient air over a fuel surface draws away heat applied by thermal radiation (see next section) from adjacent flames. Second, the wind is applying a force to your body, which will propel you in the direction it is flowing unless you oppose it by leaning into it. Similarly in wildland fires, the wind (whether ambient or fire-induced) is applying a force to embers, which can carry them through the air to ignite spot fires away from the main fire.

There are two kinds of pressure in fluid mechanics that affect the behaviour of air flow and flames in wildfires. The first is called *hydrostatic pressure*. We experience the hydrostatic pressure of the atmosphere every day from the hundreds of kilometres of air that extend above us into space. Although we don't perceive it, standard atmospheric pressure is 101 325 Pa at sea level and lower at higher altitudes (see **Appendix B**). Another example of hydrostatic pressure is what scuba divers experience as they descend deeper and deeper into water. The increased pressure on the diver is caused by the weight of the water above them. We can calculate hydrostatic pressure for the scuba diver using the following equation:

$$p_h = \rho g \ell \quad [3.9]$$

where p_h is hydrostatic pressure (Pa), ρ is the density of the fluid (kg m^{-3}), g is gravitational acceleration (9.81 m s^{-2}), and ℓ is the height of the fluid layer (m). Let's use **Eqn [3.9]** to calculate the pressure on a scuba diver who is 5 m underwater. If we assume the density of water is 1000 kg m^{-3}, then:

$$p_h = (1000)(9.81)(5) = 49\,050 \text{ Pa}$$

This is the pressure due solely to the weight of the water above the diver. The total pressure experienced by the diver is the sum of hydrostatic pressure from the water and the atmospheric pressure at the water surface: 49 050 + 101 325 = 150 375 Pa ≈ 150 kPa. We do not often think about hydrostatic pressure in fire unless we are pumping water uphill through a hose. However, it plays an important role in fire whirls and other manifestations of vorticity, which are discussed in **Chapter 8, Vorticity**.

The second type of pressure important in fluid mechanics is *dynamic pressure*, which is produced when a moving fluid presses against an object. As in the wind example above, you perceive dynamic pressure when you feel wind pressing against your body. Dynamic pressure is also responsible for tilting flames and lofting embers. The equation that describes the magnitude of dynamic pressure is given by:

$$p_d = 0.5 \rho U^2 \quad [3.10]$$

where p_d is dynamic pressure (Pa), ρ is fluid density (kg m^{-3}), and U is fluid flow speed (m s^{-1}). Let us calculate an example. Suppose you are riding in a vehicle with a speed of 30 m s^{-1} with your hand out the window. What is the dynamic pressure applied to your hand by the air? We will assume air density is ρ = 1.16 kg m^{-3}. Then, dynamic pressure is: p_d = (0.5)(1.16)(30)^2 = 522 Pa. We can calculate the force applied to your hand by multiplying the dynamic pressure by the area of your hand, say, 150 cm^2 = 0.015 m^2. Then (522)(0.015) = 7.8 newtons.

Boundary layers

In this chapter and frequently in wildland fire science as a whole, whenever we concern ourselves with the flow of fluids, we do so in the context of its interaction with a solid surface. For example, the exchange of heat between a flow of air and a fuel element occurs through the interface between the two. This exchange of heat is called *convection heat transfer* and is the subject of the next section. In another example, dynamic pressure applied by the wind causes a smouldering ember to become airborne and travel away from the main body of a wildland fire.

When a fluid like air passes around an object like a pine needle or a branch, there is a layer of fluid adjacent to the surface where the flow velocity changes from the ambient wind speed far from the object to zero relative to the object. This layer is called the *velocity boundary layer* (also *momentum boundary layer*), which we shorten to 'boundary layer' in this book.[i]

The thickness of the boundary layer depends on the speed of the flow, size of the object, surface features of the object, fluid properties and several other factors. However, what is common to all boundary layers is that the flow velocity immediately adjacent to the surface is always zero, in what is termed the *no-slip condition*. Throughout the thickness of the boundary layer, the flow velocity increases smoothly from zero at the surface, to the speed of the undisturbed wind, or *freestream velocity*.

We can visualise the effects of a boundary layer in **Figure 3.4**, which shows you working on a fire on a windy day. Let us zoom in very close to the fluid-solid interface on your arm and draw a series of arrows to show the air speed at various distances from your arm. Right at the surface where the flow velocity is zero, the air seems to 'stick' due to the no-slip condition. The further we go away from your fire shirt sleeve, the greater the air speed – until it reaches the *freestream velocity*. The *boundary layer thickness* is the distance over which the air flow's velocity transitions from zero to the freestream velocity. The thickness of the boundary layer on your arm is around 1–2 mm, depending largely on the wind speed and other factors listed above.

Why does a boundary layer form? Why does a fluid like air or water 'stick' to the solid surface instead of just sliding over it? The boundary layer is caused by friction within a fluid due to a fluid property called *viscosity*. The viscosity of a fluid is responsible for friction between adjacent volumes of fluid that are moving relative to one another. In other words, viscosity indicates how thick or thin a fluid is. In this book, we use the term 'viscosity' alone to refer to *dynamic viscosity*, which is represented by the Greek letter mu (μ), and has units of Pa s (or kg m^{-1} s^{-1}). In the science of fluid mechanics particularly, dynamic viscosity is often normalised

Figure 3.4: All flows around solid objects have a boundary layer, including you. At the surface of your shirtsleeve, the flow velocity is zero. Away from your shirtsleeve, the flow velocity increases until it reaches the velocity of the freely flowing air. Illustration by Brian Elling.

by fluid density. This quantity is termed *kinematic viscosity*:

$$v = \frac{\mu}{\rho} \quad [3.11]$$

where Greek letter nu (v) is kinematic viscosity ($m^2\ s^{-1}$) and ρ is fluid density ($kg\ m^{-3}$).

Returning to dynamic viscosity, let us consider some intuitive examples. Air is a very thin fluid and has a viscosity of 18.2×10^{-6} Pa s at 20 °C. Water is more than 55 times thicker than air and has a viscosity of 1.0×10^{-3} Pa s. Honey is 14 000 times thicker than water and has a viscosity of 14.1 Pa s. All three fluids will form a boundary layer as they flow past a solid object, but because they have different viscosities, the boundary layer thickness will be very different. Honey has a very high viscosity, so the slow fluid near the surface will drag heavily on the fluid further from the surface, creating a thicker boundary layer than that of the other fluids. The opposite is true for air, with its low viscosity. The slow air next to the surface will drag much less on the air further from the surface, creating a much thinner boundary layer than that of honey.

Boundary layer thickness is not constant over the surface of an object, but tends to grow thicker along its length. Let us consider the thickness of the boundary layer in the simple case of air flow over a smooth, flat surface (**Figure 3.5**). The dashed line shows the evolution of the boundary layer thickness (represented by the Greek letter gamma, γ) as a function of distance along the plate. Notice that γ is zero at the edge of the plate, called the *leading edge*, and increases as air moves along the length of the plate. **Figure 3.5** shows the three regions of a boundary layer: the first region, which starts at the leading edge, is the *laminar* zone of the boundary layer. Laminar flow is smooth and layer-like. In the second region, the boundary layer begins to spontaneously generate rotating parcels of air called *vortices* in the otherwise smooth laminar flow. This is called the *transition* zone. As you move through the transition zone, the vortex disruptions become more frequent and larger, and the boundary layer thickens rapidly. Finally, the

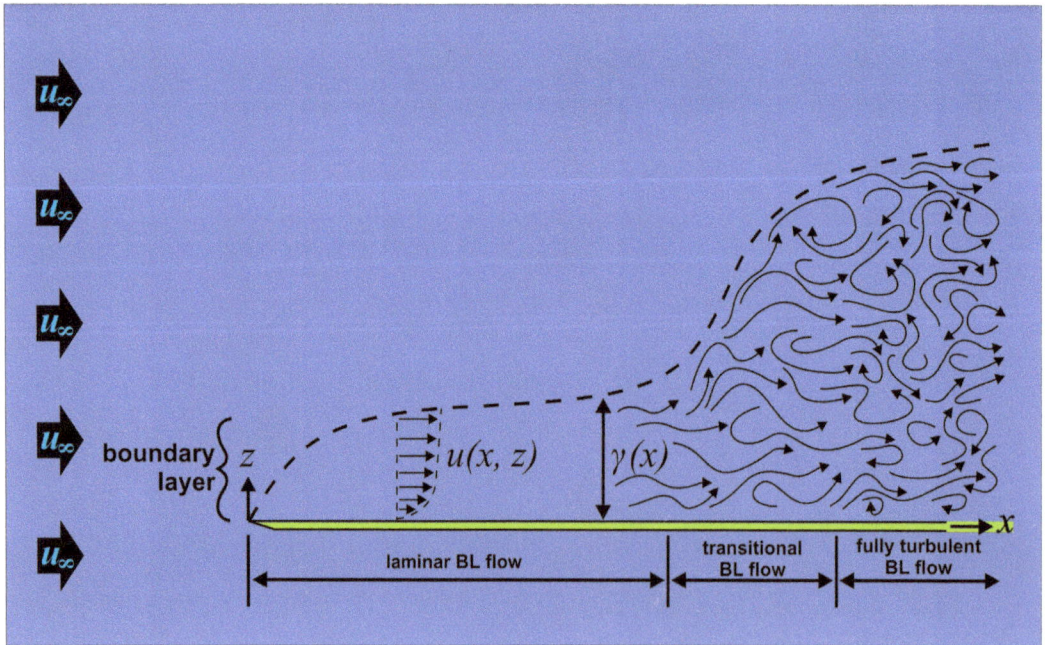

Figure 3.5: A developing boundary layer (BL) on a flat plate shows the laminar, transitional, and turbulent regions. Illustration by Brian Elling.

vortex disruptions become pervasive; this marks the beginning of the third zone, the *turbulent* zone. *Turbulence* is fluid flow that is made up of highly disordered patterns of vortices. The vortices have a range of sizes, with the biggest similar to the size of γ and the smallest less than 1 mm across. These vortices churn the air of the boundary layer, mixing fast-moving air of the freestream with slow air of the lower boundary layer. Note that for illustration purposes, the horizontal scale has been greatly compressed and the vertical scale likewise exaggerated in **Figure 3.5**.

The vortices of turbulent flow form due to the presence of *shear*, which occurs when two adjacent fluid streams have different velocities. The slower fluid stream exerts drag on the faster stream, slowing it down, while the faster stream exerts pull on the slower stream, speeding it up. However, fluid layers do not simply glide past one another as we learned when discussing viscosity above. Instead, shear at the interface of two fluid streams produces vortices. In fact, shear is responsible for the spontaneous generation of vortices in transitional flow and the continuous vortex formation in fully turbulent flow (**Figure 3.5**).

One kind of shear flow in wildland fires occurs within boundary layers. For example, slower air flow near the ground and fuel surfaces interacts with faster air flow in the freestream. Another common kind of shear flow in wildland fire occurs at the edges of a *plume* – the column of hot gas, air, and smoke that rises above a fire. The plume has a higher vertical velocity than the surrounding ambient air, so shear occurs at the interface between the two. If the plume is relatively symmetrical, shear will produce doughnut-shaped vortices called *toroidal* vortices; if the plume is not symmetrical, shear will produce less structured vortices. The upward velocity of a plume is caused by *buoyancy*, the force applied to gases of a density that is different from that of the surrounding fluid. The ideal gas law (**Eqn [3.2]**) explains plume buoyancy: when a fire warms the gases and air in the plume, it causes them to have lower density than the surrounding air. In the presence of gravity, the higher-density ambient air sinks and the lower-density hot gases in the plume rise. Buoyancy is critical to understanding fire behaviour, even the behaviour of candle flames.

The flow of smoke from an incense stick (**Figure 3.6**) shows the transition from laminar to turbulent flow (see also **Figure 2.4**). The smoke released from the tip of the burning object acts like a flow tracer, which means it makes the flow of otherwise invisible gases visible to our eyes (and to a camera). Notice the three regions of boundary layer transition that were illustrated in **Figure 3.5**: the smoke rises smoothly upward in the laminar region.

Figure 3.6: Smoke plume from a smouldering incense stick shows the laminar, transitional and turbulent regimes. Photograph by Ian Grob, USDA Forest Service.

In the transition region, spontaneously generated flow fluctuations are visible as small waves and a distinct broadening of the smoke plume. As the smoke rises further, the plume becomes fully turbulent; vortices of all sizes dominate the flow and cause the smoke plume to appear segmented. Transition between flow regions is seen often in wildland fires, especially in flames and plumes. Flames generally transition to a turbulent regime when they become longer than 10–20 cm. This affects their appearance but, more importantly, it affects how they heat solid objects (namely fuel) and their behaviour when contacting solid objects. These effects are discussed more fully later in this chapter.

Turbulence is a description of a fluid flow regime – a property, not an object. In that sense, it is similar to temperature. Temperature describes a property of matter – the amount of thermal energy in an object. We cannot see temperature, but we can certainly experience cold air on a winter day and hot air in the summer. Similarly, turbulence describes a property of fluid flow. We cannot see turbulence, but we can certainly hear and feel turbulent air when it rushes past our ears and buffets our bodies in a strong wind. The nature of turbulence is worth emphasising because it is ubiquitous in wildland fire: most winds and flows of combustion-related gases are turbulent.

Figure 3.5 illustrates the development of a boundary layer over a smooth, flat plate. But this is an idealised case and much simpler than what happens in wildland fires where we are concerned with the flow of air and combustion gases over fuel elements like needles, twigs, branches and logs (**Figure 3.7**). For many of these objects, we can estimate the thickness and properties of the boundary layers by assuming that they are cylinders. (Flat leaves and bark pieces would obviously require a different approach.) The flow of air over any object is complicated because small changes in flow velocity can have strong, seemingly disproportionate effects on the nature of the flow. Later in this chapter, we will examine how such effects change the rate of convection heat transfer.

We have learned that friction between a fluid and a solid surface causes a boundary layer to form, in which the fluid flow velocity changes from zero at the surface to the freestream velocity at a distance γ away from the surface. If the fluid and the surface are at different temperatures, an analogous *thermal boundary layer* also forms. This is a fluid layer adjacent to a solid object through which temperature varies from the surface temperature to the temperature of the bulk fluid – referred to as the *freestream temperature* (T_∞). Suppose a fluid at temperature T_∞ flows over a thin, flat plate that has a fixed surface temperature, T_s. **Figure 3.8** illustrates how the thermal boundary layer thickness (γ_t) is zero at the leading edge of the plate, then increases with increasing distance from the leading edge. This is analogous to the growth of the velocity boundary layer as shown in **Figure 3.5**. In fact, whenever a surface and an adjacent flowing fluid have different temperatures, the velocity and thermal boundary layers will develop simultaneously. Both boundary layers are important in determining the rate of convection heat transfer, so it is useful to quantify the relative thickness of the two using the Prandtl number. The *Prandtl number* (abbreviated Pr), which relates to the diffusion of momentum and thermal energy in a fluid, will be covered in more detail in the section, **Convection heat transfer,** below. However, in the context of boundary layer growth, it is worth noting that fluids with Pr < 1 have a thermal boundary layer thicker than the velocity boundary layer, while fluids with Pr > 1 have a velocity boundary layer thicker than the thermal boundary layer. Fluids with Pr = 1 have velocity and thermal boundary layers that are the same thickness. See **Appendix B** for the Prandtl number of air and combustion gas as a function of temperature.

Vortex flows

We have learned that turbulent flows are comprised of disordered patterns of vortices. Let us take a closer look at vortex flows to prepare for the discussion on fire whirls and other fire-related vortices (**Chapter 8, Vorticity**). *Vortex flows* are fluid

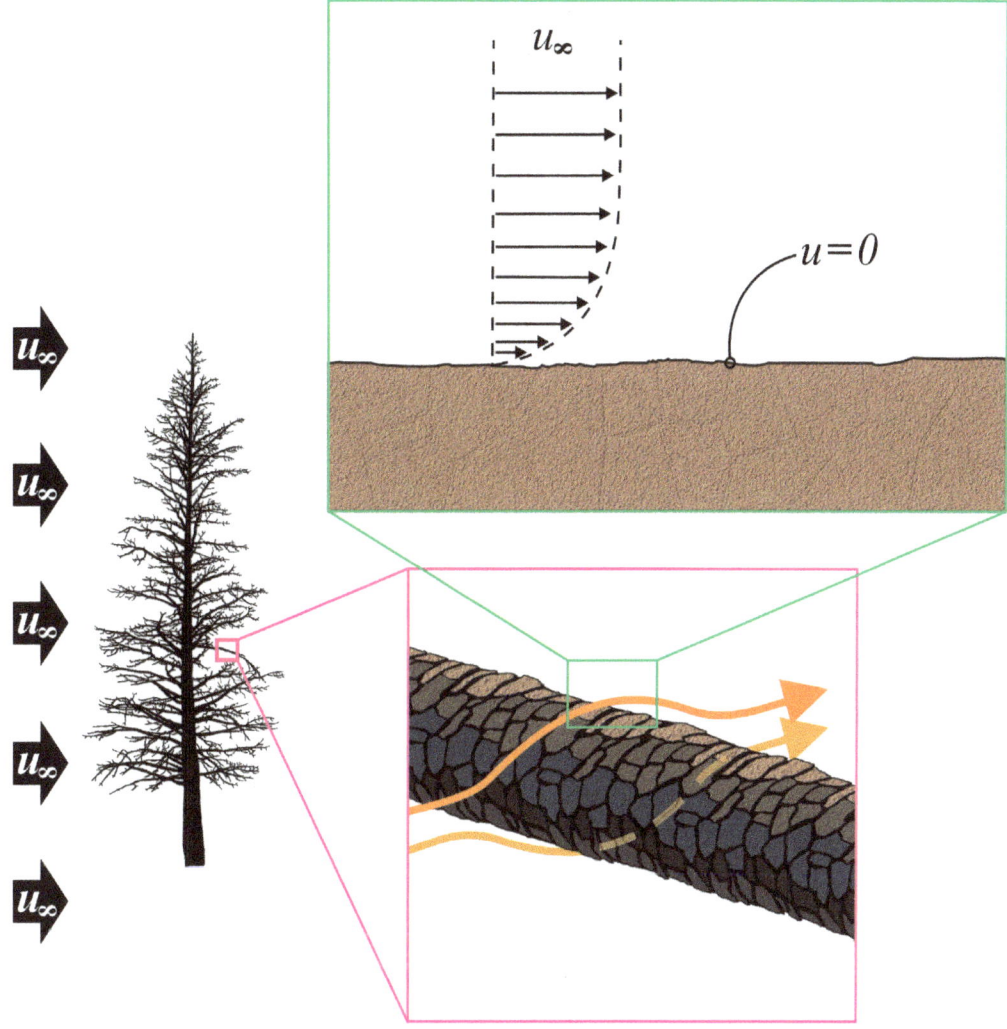

Figure 3.7: Flow of air and gas around a branch produces a boundary layer. Illustration by Brian Elling.

paths that go around in a circle. Fire whirls, tornadoes, and a stirred cup of coffee are all examples of vortex flows. Two basic types of vortices are characterised by the rotational velocity of the fluid elements that comprise them.

Imagine a vortex in a large body of water. To make it easier to see the motion of the fluid, we position boats on the surface of the water within the vortex (**Figure 3.9**). In the left-hand illustration, the boats rotate as they orbit around the axis of the vortex so that they are always pointed outward. This is a called a *rotational vortex* because, just like

the boats, the fluid that comprises the vortex rotates as it orbits around the axis. In the right-hand illustration, the boats do not rotate but remain pointed in the same direction as they move around the axis of the vortex. This is called an *irrotational vortex* because the fluid elements, like the boats, stay oriented in the same direction. We used water as the fluid in this example, but any fluid, including air and combustion gases, exhibit the same behaviour.

Angular velocity describes how fast something spins around in a circle. The angular velocity of a

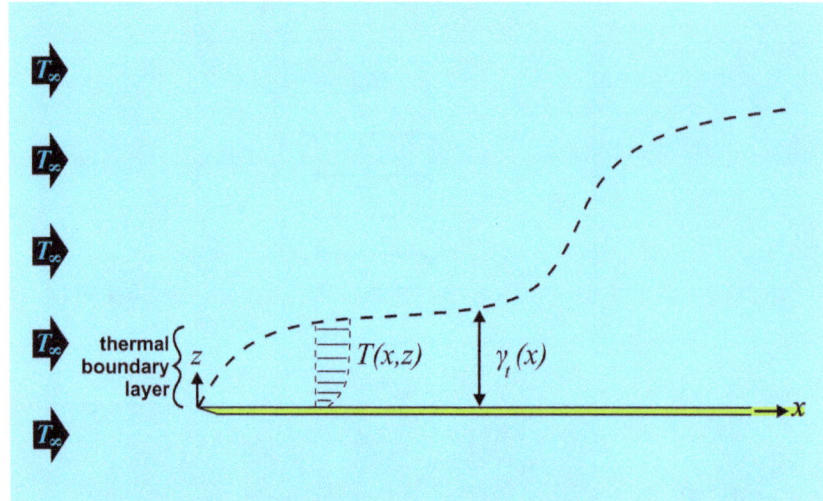

Figure 3.8: Just like the velocity boundary layer, the thermal boundary layer has zero thickness at the leading edge, then grows with distance along the plate surface. Illustration by Brian Elling.

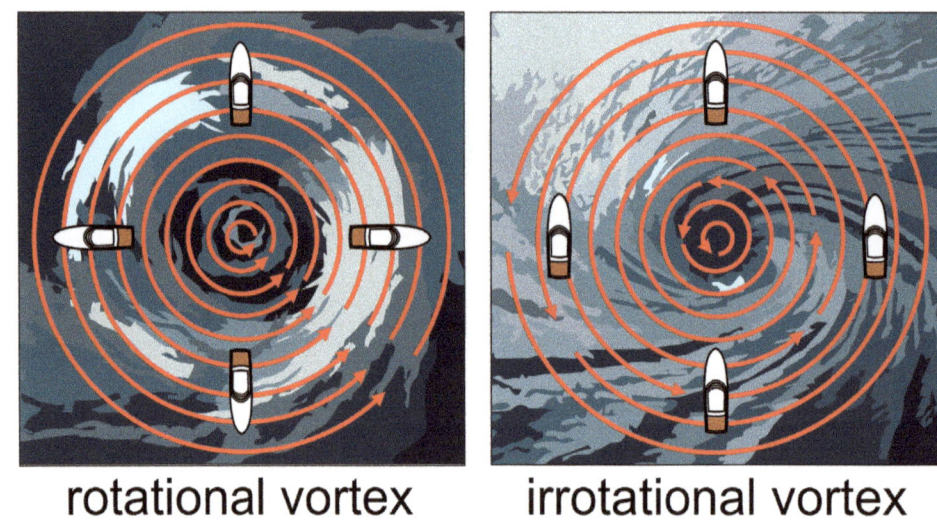

Figure 3.9: Rotational and irrotational vortices are illustrated by observing the orientation of boats as they travel around a vortex flow in water. Illustration by Brian Elling.

solid object spinning about an axis is the same at any distance from the axis. For example, any point on a radius of a spinning frisbee has the same angular velocity because no part of the disk makes a full revolution about the axis any faster or more slowly than any other part. This is also true for fluids in a rotational vortex. For this reason, rotational vortices in fluids are sometimes said to exhibit *solid body rotation*. A different principle applies to irrotational vortices. In an irrotational vortex, the angular velocity of the fluid is proportional to ξ^{-1}, where ξ is the radial distance from the axis of the vortex to any fluid element. Therefore, as one moves closer to the axis, the angular velocity gets faster and faster. At the axis of a theoretical irrotational vortex, where ξ = 0, the angular velocity is infinite. Of course, this is physically impossible in reality because friction from fluid viscosity will limit the maximum angular velocity. Instead, real vortices like fire whirls have a central core that acts like a rotational vortex, which transitions to irrotational behaviour further from the axis.

Note that the direction of rotation of a vortex, its *sense*, can be either clockwise or anticlockwise. The flow features and obstacles that contribute to the formation of a vortex dictate its rotational direction. The Coriolis 'force' does not influence vortices in wildland fire; it only affects vastly larger fluid flows, like earth's atmosphere and ocean currents. Whether a vortex is in the Northern or Southern Hemisphere has no influence on its sense.

Ember lofting

Spot fires are the result of ignition and ensuing combustion of fuels separated from the main body of a wildland fire. *Spotting* occurs when a chunk of burning fuel, called an *ember* (or *firebrand*), is carried by the wind through the air and deposited in unburned fuel outside of a control line. Spot fires are an important means of fire spread, particularly for fires in strong winds. Wildland firefighters receive substantial training on the importance of being vigilant for spotting because spot fires can lead to loss of control, or even entrapment and burnover.

How far will an ember travel before landing? We can address this question using the principles of fluid mechanics. The size, shape and mass of the ember all affect the way it gets lifted by the hot plume and carried away from the fire by the wind. Of course, in order to ignite a spot fire, the ember must still be burning when it lands in receptive fuel. Consider **Figure 8.5** where it is shown a small, light ember can be lofted long distances. However, if it is too small the ember will lose heat to the air faster than it is produced by combustion, causing it to extinguish before it lands. In contrast, a large ember may burn for a relatively long time, but it may be too heavy to be carried far by wind. The optimal ember size to ignite a spot fire far from the burning front is a balance of heat loss with ember mass. Embers are often comprised of segments of wood or bark and are usually a few centimetres in size. Maximum spotting distance varies with the intensity of the fire, but hundreds of metres is not unusual. However, fires in 'stringybark' eucalyptus timber are an exception. In these fuels, burning ribbons of bark in excess of 30 cm long are regularly observed flying through the air well ahead – even several kilometres – of the main fire front. Spotting is discussed in detail in **Chapter 8, Spotting and spot fires**.

Heat transfer

Conduction heat transfer

Conduction is a type of heat transfer in which thermal energy is transmitted from one molecule to adjacent molecules. Conduction is especially important in transferring thermal energy through solids. In the context of wildland fire, instead of concerning ourselves with activity at the molecular scale, we simply need to know that conduction is the diffusion of heat from warm to cold regions in a substance. How fast heat moves by conduction depends on the temperature difference between the warm and cold regions, and the thermal conductivity of the substance. *Thermal conductivity*, which has units of W m^{-1} °C^{-1}, is a material-specific property that quantifies the rate of conduction through a given material thickness for a given difference in temperature.

To introduce the idea of conduction heat transfer, we will focus on one-dimensional conduction – that is, heat being moved in only one direction. Imagine a wooden board (e.g. plywood) that is 1 m tall, 1 m wide and 20 mm thick (**Figure 3.10**). Suppose we place the board outside so the sun is heating one side and the wind is cooling the other side. Energy from the sun will raise the temperature of the sunlit side, and the wind will reduce the temperature of the other side. Thermal energy will flow from the sunlit side through the wood towards

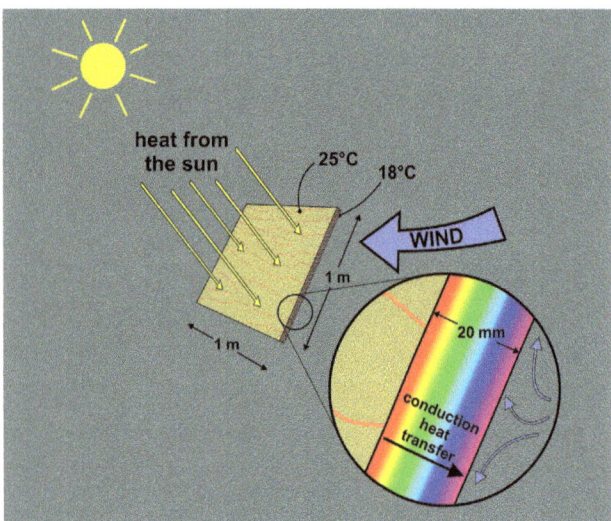

Figure 3.10: A wooden board is heated by the sun on one side and cooled by the wind on the opposite side. Heat is transmitted through the 20 mm thickness of the board by conduction heat transfer. Illustration by Brian Elling.

Table 3.1 shows thermal conductivity for selected materials. Substances with a low thermal conductivity, like the fibreglass insulation used in home construction, are poor conductors. This is, of course, exactly why they make good insulation. Materials with a high thermal conductivity, like metals, are good conductors but poor insulators. The materials that concern us in the context of wildland fire behaviour are wood and the moisture within wood. The thermal conductivity of wood increases as its moisture content increases.

Let us consider how conduction heat transfer participates in wildland fire, as it heats and dries dead woody fuel particles. As flames draw near, the surface of the unburned fuel is heated by *convection* and *radiation heat transfer*. The heat is then transmitted by conduction heat transfer from the surface of the fuel particle to the interior. As described in **Chapters 4** and **5**, the surface of the fuel must be heated to the ignition temperature before it can ignite and burn. However, this cannot occur until the moisture is driven out, because any water inside the fuel will absorb thermal energy instead of heating the fuel. Thus moisture can delay or even prevent the onset of ignition of dead fuels. Burning of live fuels, which can occur for moisture content drastically higher than dead fuels, is a special case that deserves further discussion (see **Chapter 5, Live fuels**).

the wind-cooled side by way of conduction heat transfer. Suppose the temperature on the hot side is 25 °C, and it is 17 °C on the cool side. We can calculate the rate of conduction heat transfer using Fourier's Law of one-dimensional heat conduction:

$$\dot{q}_k = -kA_s \frac{T_L - T_U}{\ell} \quad [3.12]$$

where $\dot{q}_k$ is the rate of conduction heat transfer (W), k is thermal conductivity of the material (W m^{-1} °C^{-1}), A_s is the cross-sectional area through which the heat transfer occurs (m²), T_L is the lower temperature, T_U is the upper temperature (K or °C), and ℓ is the distance over which the heat transfer occurs (m). Let us apply **Eqn [3.12]** to the board example and calculate the rate of heat transfer. The area A_s of the board is easy to calculate: 1 m × 1 m = 1 m². T_L and T_U are 17 °C and 25 °C, respectively. ℓ is 20 mm, which is equivalent to 0.020 m. Finally, the thermal conductivity of plywood is 0.13 W m^{-1} °C^{-1}. Now we can calculate the rate of heat transfer through the board:

$$\dot{q} = -(0.13)(1)\left(\frac{17-25}{0.020}\right) = 52 \text{ W} \quad [3.13]$$

Table 3.1. Thermal conductivity of some common materials.

Material	K (W m^{-1} K^{-1})
Air	0.026
Soil	0.3–0.7
Wood (oak)	0.17
Wood (pine)	0.14
Plywood	0.13
Water	0.61
Fibreglass insulation	0.04
Copper	390
Steel	46
Diamond	2300

Conduction heat transfer acts relatively slowly in comparison to convection and radiation, especially in the woody fuels of wildland fires. This can cause a fuel element to have a much different temperature at the surface than the interior. Anyone who has broiled a turkey or a large roast knows the challenge of thoroughly cooking the interior without burning the surface of the meat. Likewise, when large fuels like branches, limbs, and logs are heated by convection and radiation heat transfer (or even burning), the surface may be at a much higher temperature than the centre. Such fuels are often referred to as *thermally thick*. On the other hand, relatively thin fuels like needles and grass will more likely have a similar temperature throughout their thickness as they heat up. These fuels are often called *thermally thin*. These two examples concern fuel heating, but the same concept applies to the cooling of fuel as well.

The distribution of temperature throughout a fuel element as it heats or cools is clearly dependent on its thickness or diameter. It is also a function of the fuel's thermal conductivity and the combined rate of surface heating due to convection and radiation heat transfer. A rough guideline is provided by the following expressions from **Quintiere (2016)**. Thermally thin fuels have a diameter or thickness ℓ (m) for which:

$$\ell \leq \frac{2k\left(T_{ig} - T_\infty\right)}{\dot{q}_s''} \qquad [3.14]$$

Thermally thick fuels have a diameter or thickness ℓ for which:

$$\ell > \frac{2k\left(T_{ig} - T_\infty\right)}{\dot{q}_s''} \qquad [3.15]$$

In **Eqns [3.14]** and **[3.15]**, T_{ig} is ignition temperature (°C), a concept covered in **Chapter 5**. T_∞ is the freestream temperature (°C), and $\dot{q}_s''$ is the combined surface heat flux (W m^{-2}) at the surface due to convection and radiation heat transfer. Let us calculate an example.

Suppose we wish to determine if a 10 mm diameter oak twig is thermally thick or thin for an ignition temperature of 350 °C, ambient environment temperature of 35 °C, and total surface heat flux of 12 000 W m^{-2}:

$$\frac{2(0.17)(350-35)}{12\,000} \approx 0.009\,\text{m} = 9\,\text{mm}$$

Since the oak twig is larger than the 9 mm calculated above, **Eqn [3.15]** applies and we conclude it is thermally thick. Here we use T_{ig} = 350 °C, which is a reasonable value but not one that is supported by an unassailable theoretical or experimental basis (see **Chapter 5, Flaming ignition criteria**). Keep in mind that ignition is an extremely complex process, so using a quasi-arbitrarily determined value for ignition temperature is a drastic simplification. As such, it is worth re-emphasising that the criteria embodied by **Eqns [3.14]** and **[3.15]** are rough guidelines.

Now let us change the example above so that the heat flux is much larger: $\dot{q}_s''$ = **200 000 W m^{-2}**. This is a very high heat flux, but one that is typical of that experienced by a fuel element completely immersed in thick flames. The new result is:

$$\frac{2(0.17)(350-35)}{200\,000} \approx 500 \times 10^{-6}\,\text{m} = 0.5\,\text{mm}$$

What this means is that all but the thinnest wisps of fuel (e.g. seed fluff from cottonwood trees [*Populus balsamifera* ssp. *trichocarpa*]) are thermally thick, even pine needles and grass – fuels we would normally characterise as 'fine'. However, this result makes sense when you consider that an especially large heat flux would heat the surface of fuel very rapidly, and conduction, being a relatively slow mode of heat transfer, is simply inadequate for rapidly distributing the heat throughout the interior of even a pine needle.

Now let us consider the opposite case in which heat flux is particularly small: $\dot{q}_s''$ = 900 W m^{-2}, which is approximately equivalent to the heat of the sun on a clear day. The calculation now yields:

$$\frac{2(0.17)(350-35)}{900} \approx 0.120 \text{ m} = 120 \text{ mm}$$

In this case, limbs and logs as large as 120 mm in diameter will behave thermally thin. Despite the fact we would normally consider fuels in this size class as 'heavy,' they would nonetheless heat evenly throughout because the heat flux is so small. These last two calculations show that while fuel size is important in how heat is conducted into the interior, extreme heat fluxes can result in fuels behaving in non-intuitive ways.

Thus far we have focused largely on how conduction participates in fuel heating, but its role in fuel cooling is equally important. The heating and cooling of fuel near a flame front is intermittent because of the turbulent nature of flames in wildland fire. For example, in one moment, a pine needle might be engulfed in and heated by hot combustion gases but, a tenth of a second later, it might be bathed in and its surface cooled by ambient air (e.g. **Figure 6.24**). In this latter case, the needle surface is cooler than the interior, so heat is conducted from the inside to the outside of the needle. The alternating pattern of heating and cooling is important for how fuel particles heat and ignite, and thus how fire spreads.

Conduction heat transfer is also important for explaining how heat from flames can damage living tissues of trees and other plants above and below ground. Some tree species have evolved thick, insulating bark, which protects the living cambium cells beneath from the heat of fires. Two common North American examples are ponderosa pine (*Pinus ponderosa*), with bark sometimes exceeding 8 cm thick on mature trees, and giant sequoia (*Sequoiadendron giganteum*) with bark up to 60 cm thick. Having low thermal conductivity, thick bark limits the rate of conduction heat transfer into the interior of trees when they are bathed in flames and hot gas.

Conduction is responsible for transmitting heat into the soil over which a wildland fire burns. This can kill microorganisms and small fauna that are important for soil health and viability, and damage shallow root systems of trees and other plants. Soil heating can be particularly intense when long-burning fuels like downed trees or heavy logging slash are present. Even fuels that merely smoulder for several days like thick duff and wood chips can cause significant soil heating.

Some lodgepole pine (*Pinus contorta*) and other tree species evolved to produce *serotinous* cones. Serotiny means that seeds are dispersed only after an environmental trigger, like fire. The cone scales of lodgepole pine are sealed with resin, which when melted by the heat of a fire, causes the seeds to be released. Hot gas and flames heat the cone surface and conduction heat transfer carries heat deeper into the cone scales, leading to the melting of the resin. Trees bearing such cones are able to store vast quantities of seed over many years of cone production that is ready to disperse immediately after the passing of a fire. Abundant seed enables these tree species to dominate regions experiencing infrequent, but relatively high-intensity fires that kill adult tree populations.

Radiation heat transfer

Radiation heat transfer is important to wildland fire behaviour because hot flames and glowing coals radiate thermal energy to nearby objects, including fuel particles and plants (and firefighters). These objects are also radiating heat away to cooler surroundings. When you stand outside on a sunny day, you can feel the heat of the sun on your skin. How does the heat reach your skin even though the sun is ~150 million km away? It cannot be conduction, which requires continuous molecule-to-molecule interaction, because the vacuum of space between the earth and the sun contains virtually no matter through which heat can be transmitted. Nor can convection (see below) be responsible because there is no fluid present in outer space to carry the heat. Instead, heat from the sun is transmitted from the sun's surface, through the vacuum of space, through earth's atmosphere, and onto your skin through radiation (or 'radiative' or 'radiant') heat transfer.

Radiation heat transfer is the transmission of thermal energy by electromagnetic radiation. The *electromagnetic spectrum* (see **Figure 3.11a**) includes all electromagnetic radiation over the entire range of possible wavelengths, from one-trillionth of a metre (gamma rays) to thousands of metres (radio waves). Visible light (the *visible spectrum*) (**Figure 3.11c**) is the narrow range of wavelengths within the electromagnetic spectrum that we can see with our eyes. We perceive the visible spectrum as the colours of the rainbow: violet light has a wavelength around 0.4 µm, and visible red light has a wavelength of around 0.7 µm. Electromagnetic radiation with wavelengths longer than visible red light is called *infrared* (0.7–1000 µm). We cannot see infrared light, but we can feel its effects – for example, when your skin is warmed by sunlight or nearby flames. Electromagnetic radiation with wavelengths shorter than those of visible violet light is called ultraviolet (0.01–0.4 µm).

Sunlight contains a small component of ultraviolet radiation, which is responsible for suntans and sunburns. Fire produces radiation in the *thermal spectrum* (**Figure 3.11b**) that includes an extremely small amount of ultraviolet radiation (much too weak to cause harm to human skin) and a tremendous amount of visible and infrared radiation. Light in this range of wavelengths, for sufficiently high heat flux, can burn your skin by raising its temperature to a dangerous level – essentially, by cooking it.

The *thermal radiation spectrum* contains wavelengths ranging from ~0.1 to 100 µm (see **Figure 3.11b**). All objects emit a range of wavelengths within the thermal radiation spectrum as a function of temperature, as shown in **Figure 3.12**. The horizontal axis of **Figure 3.12** is wavelength and the vertical axis is *spectral emissive flux*, which can be thought of as the emission rate of thermal radiation for a given area and a given wavelength. Notice that because the range of values is so large on both axes, the curves are drawn on logarithmic

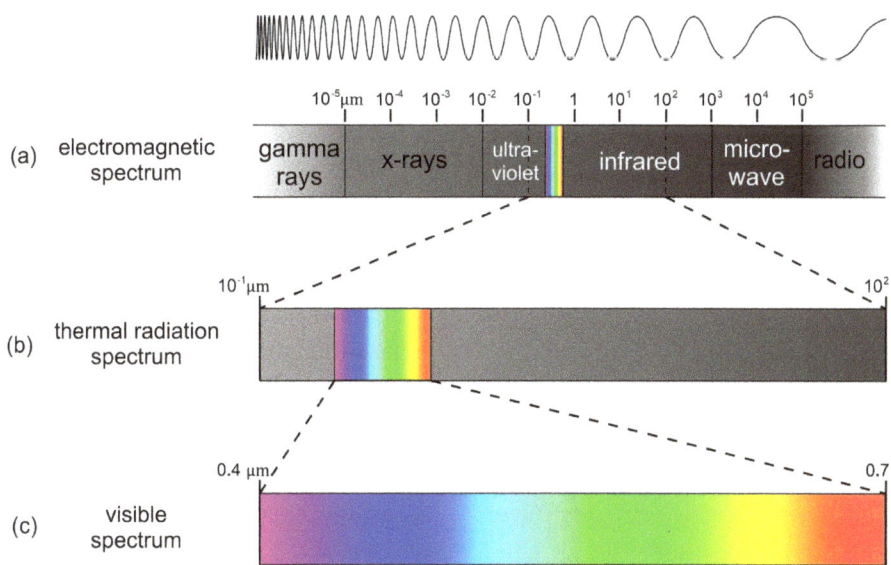

Figure 3.11: The electromagnetic spectrum (a) is the continuum of all wavelengths of electromagnetic radiation. Thermal radiation (b) includes some ultraviolet, all visible, and some infrared wavelengths. The visible spectrum (c) is a subset of the thermal radiation spectrum. Illustration by Brian Elling.

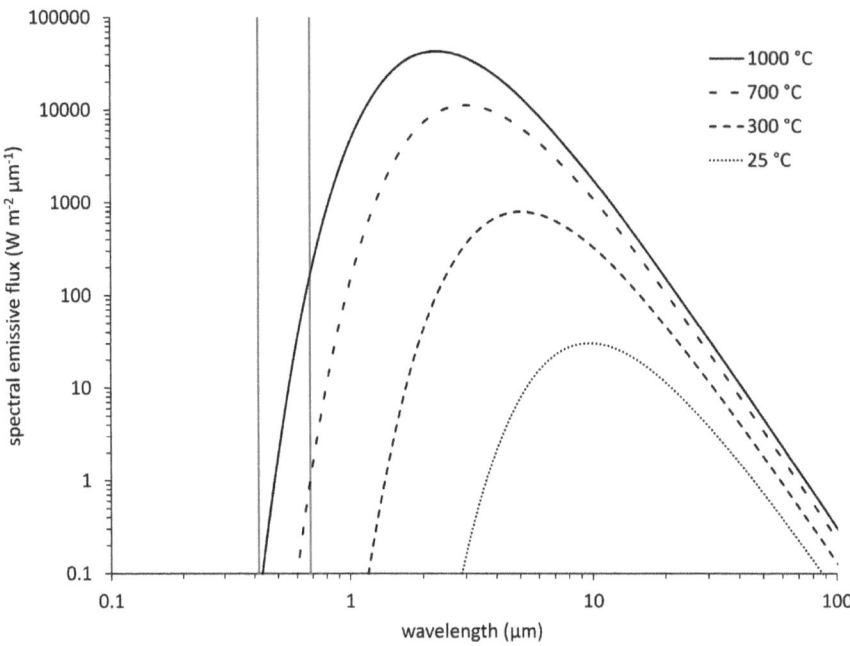

Figure 3.12: Wavelength-dependent thermal radiation flux for a blackbody at selected temperatures. The violet and red vertical lines indicate the extent of the visible spectrum.

scales. The violet and red vertical lines show the range of the visible light spectrum. The four curves are thermal radiation emission spectra for blackbodies (described below) at the indicated temperatures. Objects at room temperature (~25 °C, shown by the dotted line) and temperatures near ignition (~300 °C, the inner dashed line) only emit infrared thermal radiation outside of the visible spectrum. Thermal radiation becomes just visible with a dark red glow around 700 °C, as shown by the outer dashed curve in **Figure 3.12**. The curve for 1000 °C – a typical flame temperature in wildland fire – shows that thermal radiation emissions have significant red, orange, and yellow components, which is borne out by visual observations of flames. Radiometric temperature of a surface can be measured using this physical principle (see **Chapter 9, Heat transfer**). The intensities of certain wavelengths of an object's thermal radiation emission spectrum are dependent on its surface temperature. Radiometric temperature is determined by comparing the relative magnitudes of these emissions.

Let us define some terms that are used to discuss heat transfer via thermal radiation:

- An *emitter* is any object that is a source of thermal radiation.
- A *target* is an object onto which thermal radiation is falling.
- A *body* is a general term for an object that can be an emitter or target. Flames can be considered bodies for the purposes of radiation heat transfer calculations because most of their thermal radiation originates from the cloud of glowing carbon particles that form them.

We use 'emitter' and 'target' to refer to the role of two bodies in our calculations. All bodies pertinent to wildland fire are constantly emitting and receiving thermal radiation simultaneously, irrespective of temperature differences between them. *Net radiation heat transfer*, however, is *always* from a hot body to a cold one. Here we will simplify our calculations by assuming that radiation is only flowing in one direction, from an emitter (hot body) to a target (a colder body). Later in this section, we will introduce more realistic calculations, which will account for net radiation from all objects.

There is a special kind of body, called a *blackbody*, which is important to the theory of radiation heat transfer. A blackbody is a theoretical object that emits and absorbs thermal radiation at the maximum possible rate for a given temperature. A blackbody can be either an emitter or a target, depending on whether it is the hottest or coolest object in the environment. The rate of emission of thermal radiation $\dot{q}_r$ (W) for a blackbody is described by the Stefan-Boltzmann equation:

$$\dot{q}_r = \sigma_B A_s T^4 \qquad [3.16]$$

where σ_B is the Stefan-Boltzmann constant (W m^{-2} K^{-4}), A_s is the surface area of the emitter (m^2), and T is temperature in Kelvin (K). Note the temperature *must* be in Kelvin; the Celsius scale cannot be used with **Eqn [3.16]**. The Stefan-Boltzmann constant has the value:

$$\sigma_B = 5.67 \times 10^{-8} \text{ W m}^{-2} \text{ K}^{-4} \qquad [3.17]$$

Let us calculate an example. Suppose you are walking through a stand of timber shortly after the fire front has passed. You find a 1 m × 2 m bed of smouldering coals, and you want to know the rate at which thermal radiation is being emitted (see **Figure 3.13**). You use an infrared thermometer to measure the surface temperature of the coals; it is 620 °C. You convert the temperature to the Kelvin scale: 620 °C + 273.15 ≈ 893 K. While a blackbody is a theoretical object, smouldering coals and charred woody material behave so much like a blackbody that the difference between the real object and the theoretical blackbody can be ignored. Therefore, you use **Eqn [3.16]** to estimate the rate of thermal radiation emission from the bed of coals:

$$\dot{q}_r = \left(5.67 \times 10^{-8}\right)(2)\left(893^4\right) = 72\,114 \approx 72 \text{ kW} \cdot$$

Notice that temperature T in **Eqn [3.16]** is raised to the fourth power. This means that a small change in temperature will result in a larger change in the rate of thermal radiation emission. For example, if the temperature of the coals were 50 K lower (a

Figure 3.13: A bed of coals from a wildfire emits thermal radiation in all directions. Illustration by Brian Elling.

5.6% reduction in absolute temperature), then the rate of thermal radiation emission would decrease to 57 310 W (a 21% reduction).

A *greybody* refers to a real-world object that emits less thermal radiation than a blackbody would at a given temperature. Essentially everything in the fire environment is a greybody, although coals and charred materials act very much like blackbodies. Since a blackbody would emit and absorb thermal energy at the maximum possible rate for a given temperature, it is 100% effective at transmitting thermal energy, and we say its *emissivity coefficient*, ε, is 1.0. The emissivity coefficient (often shortened to simply *emissivity*) of a greybody is the ratio of thermal radiation emission from a greybody to that of a blackbody at the same temperature. For example, a typical emissivity for bare wood is 0.90, which is to say that wood emits 90% of the thermal radiation that a blackbody would emit at the same temperature. A typical soil emissivity is slightly higher, around 0.92; the emissivity of human skin is even higher, ~0.97. When we calculate the rate of thermal radiation emission by a greybody, we use the following equation:

$$\dot{q}_r = \varepsilon \sigma_B A_s T^4 \qquad [3.18]$$

where ε accounts for the reduction in the rate of thermal radiation emission relative to that of a blackbody.

Flames are an interesting radiation heat transfer problem because they are made up of clouds of microscopic carbon particles and hot gas, instead of solid objects like a piece of wood or a bed of coals. The thermal radiation emitted and absorbed by flames is the *net radiation heat transfer* of all the carbon particles (and, to a much lesser extent, the hot gases) that comprise flames. While individual carbon particles alone behave very nearly like true blackbodies, their diffuse distribution throughout a flame (~1 g of carbon per cubic metre of gas) makes the apparent flame emissivity a function of flame thickness. Consider **Figure 3.14**, which shows how apparent emissivity changes with flame thickness for typical wildfire conditions. The curve was produced using the emissivity model presented in Taylor and Foster (1975). It shows that small, thin flames have very low emissivity, while flames beyond ~3 m thick radiate nearly like a black body. The change in emissivity with respect to flame thickness is important for understanding how radiation heat transfer differs between small and large flames in wildland fires. Although the apparent flame emissivity is not a true emissivity in a technical sense, it is convenient to treat it as such when making radiation calculations.

So far, we have discussed radiation produced by the emitter. Now let us consider the target. The thermal radiation that falls on a target is called the *incident thermal radiation* (or simply incident radiation). Emitters like flames send thermal radiation in all directions, so only a fraction of the total thermal radiation produced by the emitter will land on a target as incident radiation; the rest will go to other targets in the environment. The amount of incident radiation varies, depending on the target's shape and orientation, and on the distance between the emitter and the target. Consider the logs in **Figure 3.15**. The advancing flame front is emitting thermal radiation in all directions, but only a fraction of it lands on the logs. Moreover, the orientation of the logs relative to the flame front affects the amount of incident radiation they receive: the upper log is lying broadside to the fire, so it receives the most incident radiation. The lower log is lying mostly end-on to the fire, so it receives much less incident radiation. The distance between the emitter and the target matters because thermal radiation decreases as distance from the emitter increases.

A rule of thumb often introduced in wildland firefighter training is that thermal radiation decreases with the square of the distance between the emitter and the target. Where this is true, it means that incident radiation declines rapidly as the distance from the emitter increases. However, this rule of thumb only applies when the emitter is an infinitesimally small point or the distance

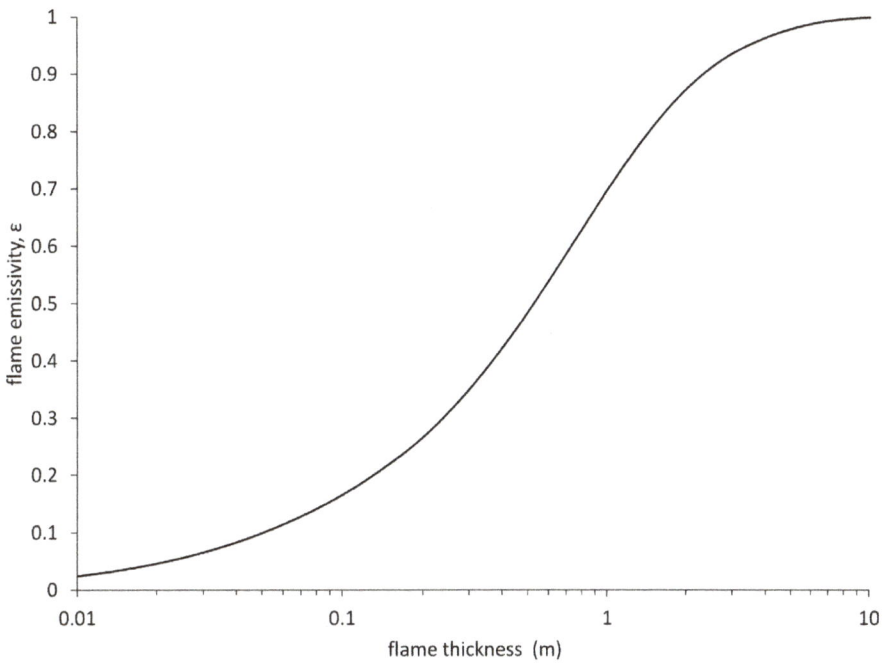

Figure 3.14: Apparent flame emissivity varies with flame thickness for typical wildland fire conditions. Flames with thickness greater than 3 m can be safely approximated as blackbody emitters ($\varepsilon = 1.0$). The curve was produced using the flame emissivity model presented in Taylor and Foster (1975), wherein $T = 1400$ K, $c_p = 0.001$ kg m^{-3}, and $p_w p_c^{-1} = 2$.

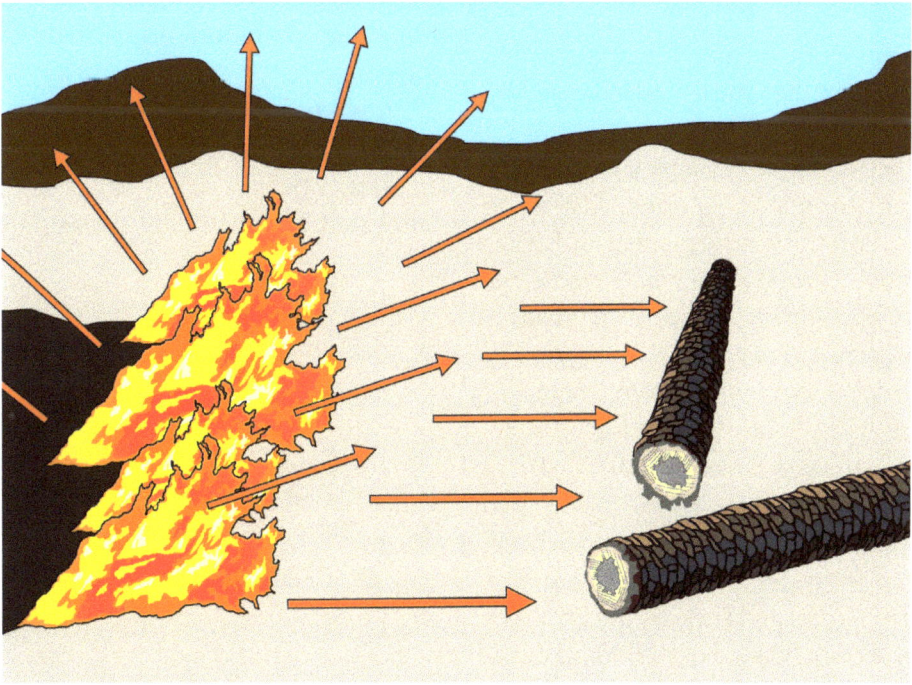

Figure 3.15: The amount of thermal radiation emitted by the flames that falls on the logs depends on their shape and orientation, and their distance from the flames. Illustration by Brian Elling.

between the target and emitter is much larger than the size of the emitter and target. Neither of these assumptions applies to practical wildland fire problems because flames and other emitters are never points and the pertinent radiation problems are those where the target is relatively close to the emitter. Therefore, thermal radiation from emitters in wildland fire will decrease with distance, but never as rapidly as predicted by the conventional rule of thumb, the square of the distance.

When calculating incident radiation, we combine considerations of target shape, target orientation, and distance between the emitter and the target into a non-dimensional quantity called the view factor. *View factor* (or *configuration factor*) is the ratio of incident thermal radiation received by a target to the total thermal radiation produced by the emitter. View factor ranges from 0 to 1 and is normally indicated by the symbol F_{12}, which means the fraction of thermal radiation leaving surface 1 that lands on surface 2. View factor is employed in a third version of the Stefan-Boltzmann equation:

$$\dot{q}_r = \sigma_B \varepsilon_1 A_1 F_{12} T_1^4 \qquad [3.19]$$

Notice that emissivity, area and temperature in **Eqn [3.19]** are those of the emitter, as indicated by the subscript 1.

View factor, because of the complex geometrical relationships that underlie it, is often difficult to determine for an arbitrary arrangement of bodies that are exchanging thermal radiation. Nonetheless, the book *Thermal Radiation Heat Transfer* (Howell *et al.* 2016) contains technical details on how to do this. An extensive catalogue of view factors for characteristic arrangements is available online (Howell 2010). For example, see configuration factor 'C-10' of Howell (2010) to determine the view factors for two perpendicular planes with a common edge. One could use this to estimate thermal radiation flux between a sheet of flames and fuels on the ground.

So far in this section, we have derived an equation (**Eqn [3.19]**) for calculating the rate at which radiation falls on a target for a given emitter emissivity and geometrical arrangement. A blackbody will absorb all of the incident radiation, while a real target object – a greybody – will absorb only a fraction of it. The remaining radiation will be reflected or, in the case of transparent materials like glass, transmitted (see **Figure 3.16**). In the context of wildland fire, we are almost entirely concerned with opaque objects, so transmission of radiation can be disregarded, with the possible exception of translucent leaves and windows of firetrucks.

We use two terms to quantify reflection and absorption of incident radiation. *Absorptivity* ζ is the fraction of total incident radiation that is absorbed by a target object. *Reflectivity* η is the fraction of total incident radiation that is reflected by a target object. As we would expect in the absence of transmission, the sum of absorptivity and reflectivity is 1 (see page 83 of Howell *et al.* 2016):

$$\zeta + \eta = 1 \qquad [3.20]$$

Notice that the origin of *reflected* thermal radiation is different from that of *emitted* thermal radiation. Both kinds of radiation leave the surface of a greybody object. However, emitted radiation is produced by the object itself, whereas reflected radiation originated from a separate object.

It is easy to use absorptivity and reflectivity in calculations. Simply multiply the rate of incident radiation by one or the other to find the fraction allocated to that mode. For example, suppose we

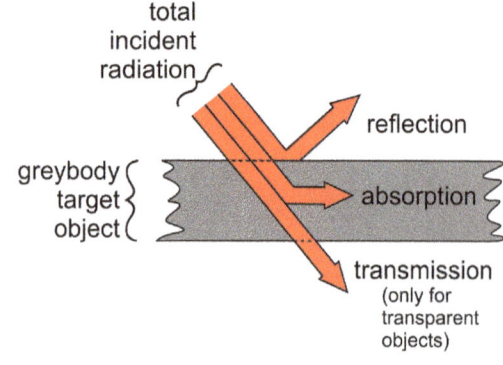

Figure 3.16: Thermal radiation falling on an object can be reflected, absorbed, and (for transparent media) transmitted. Illustration by Brian Elling.

use **Eqn [3.19]** to calculate $\dot{q}_{12}$, the rate of radiation release from emitter 1 that falls on target 2. We use the absorptivity of target 2, ζ_2, to calculate the fraction of that incident energy absorbed by target 2:

$$\dot{q}_{abs} = \zeta_2 \dot{q}_{12} \qquad [3.21]$$

Likewise, the rate of radiation produced by emitter 1 that is reflected by target 2 is:

$$\dot{q}_{ref} = \eta_2 \dot{q}_{12} \qquad [3.22]$$

Absorptivity and reflectivity of materials can be found in textbooks (e.g. Howell *et al.* 2016; Çengel *et al.* 2017) or online.

In the preceding, we have treated objects as *either* emitters or targets; however, any object undergoing radiation heat transfer with one or more other objects will absorb, emit, and – in the case of greybodies – reflect radiation simultaneously. We can write an equation for the rate of *net thermal radiation heat transfer* for a body:

$$\dot{q}_r = \begin{pmatrix} \text{absorbed} \\ \text{radiation} \end{pmatrix} - \begin{pmatrix} \text{emitted} \\ \text{radiation} \end{pmatrix} - \begin{pmatrix} \text{reflected} \\ \text{radiation} \end{pmatrix} \qquad [3.23]$$

$$= \dot{q}_{abs} - \dot{q}_{emit} - \dot{q}_{ref} \qquad [3.24]$$

where $\dot{q}_{emit}$ is calculated using **Eqn [3.18]**. If an object is absorbing energy faster than it emits and reflects, it experiences a net gain of energy ($\dot{q}_r > 0$). Conversely, if an object is absorbing energy more slowly than it emits and reflects, it experiences a net loss of energy ($\dot{q}_r < 0$).

At this point, we have all the mathematical tools we need to calculate the rates of emission, absorption, and reflection from both blackbodies and greybodies. Now we can calculate heat transfer between two bodies for some specified difference in temperature. Doing so for greybodies is rather complex because thermal radiation can be exchanged among bodies many times before it is fully absorbed. However, the same calculation for blackbodies is much simpler because thermal radiation is never reflected, and we use that

simplification here. It is not overly compromising to assume blackbody behaviour for most wildland fire problems because thick flames, woody material, soil and skin all behave at least roughly similar to a blackbody ($\varepsilon \geq 0.9$). Moreover, thermal radiation calculations are rough approximations anyway, since it is impossible to fully characterise every aspect of a real fire and its environment. So, given that we can assume blackbody properties within a fire, let us calculate the net rate of radiation heat transfer between body 1 and body 2:

$$\dot{q}_{12} = \begin{pmatrix} \text{thermal radiation} \\ \text{emitted by 1 that} \\ \text{is incident on 2} \end{pmatrix} - \begin{pmatrix} \text{thermal radiation} \\ \text{emitted by 2 that} \\ \text{is incident on 1} \end{pmatrix} = \sigma_B A_1 F_{12} T_1^4 - \sigma_B A_2 F_{21} T_2^4$$

[3.25]

Eqn [3.25] can be simplified by using the shape factor *reciprocity relation*, which states that $A_1 F_{12} = A_2 F_{21}$ (see p. 166 of Howell *et al.* 2016). The resulting equation is:

$$\dot{q}_{12} = \sigma_B A_1 F_{12} \left(T_1^4 - T_2^4 \right) \qquad [3.26]$$

Notice that if $\dot{q}_{12}$ in **Eqn [3.26]** is positive, blackbody 2 has a net energy gain while blackbody 1 has a net energy loss. Conversely, if $\dot{q}_{12}$ is negative, blackbody 1 has a net energy gain while blackbody 2 has a net energy loss. Let us calculate an example:

Consider again the bed of coals shown in **Figure 3.13**. Let us calculate the net rate of radiation heat transfer between the bed of smouldering coals (surface 1) and the trunk of the tree nearby (surface 2). Remember from the earlier example that the bed of coals is at a temperature of 893 K and has area 2 m². Assume that the tree has a temperature of 30 °C (~303 K) and $F_{12} = 0.02$. Now we can calculate the net rate of radiation heat transfer using **Eqn [3.26]**:

$$\dot{q} = (2)(0.02)(5.67 \times 10^{-8})(893^4 - 303^4)$$

$$= 1423 \approx 1.4 \, \text{kW} \qquad [3.27]$$

Note that the solution is a momentary heat transfer rate, not a fixed value over a long period of time (say, 30 min). The magnitude of $\dot{q}$ will change over time as the surface of the tree trunk and the smouldering coals change temperature.

In discussing rates of radiation heat transfer in this subsection, we have quantified heat transfer between bodies in terms of an energy rate $\dot{q}$, which is energy per unit time – that is, power, which we have expressed in kilowatts (kW). Suppose someone told you the heat transfer rate between two bodies was $\dot{q} = 10$ kW. This quantity would be meaningful only if you knew the shape and surface area of the two bodies, their orientation, and the distance between them (in other words, all of the parameters needed to determine the view factor). A more general term for describing heat transfer – and one with more practical application in wildland fire – is *heat flux*, expressed as $\dot{q}''$, which is the energy transfer rate over a specific surface area ($\dot{q}'' = \dot{q}A_s^{-1}$). We can use heat flux to communicate the incident radiation on any arbitrary object. Now suppose you learned that the rate of heat transfer between the same two bodies was a heat flux of 8 kW m^{-2}. This is much more useful for practical application than the rate of energy release, because you do not need further information about the two bodies to understand the quantity. Therefore, in wildland fire science, heat transfer is often described in terms of heat flux. For example, the thermal radiation flux emitted by a robust crown fire can be ~100 kW m^{-2}.

We have been discussing the release of thermal radiation from flames in wildland fires, but *blocking* of thermal radiation is also important. Just as the foliage of trees shades the ground from the sun's rays, vegetation and other opaque matter can block thermal radiation from reaching a target. Blocking limits the distance over which radiation from a fire can play a significant role in heating fuels where the vegetation or the fuel bed is particularly dense. Blocking also occurs when fires produce a thick, dark smoke column that completely obscures the interior flames (**Figure 3.17**). This is called *soot blocking*, because the particulates comprising the smoke are preventing thermal radiation emitted by the flames from escaping into the

Figure 3.17: Soot can block emissions of thermal radiation from flames to the surroundings. Illustration by Brian Elling.

environment. Soot blocking can limit the rate of adjacent fuel heating through thermal radiation. Nevertheless, heavy black smoke indicates extremely high rates of fuel combustion.

Convection heat transfer

Convection heat transfer is critical to wildland fire spread because it is responsible for heating, drying and igniting fuel elements by contact with hot gas and flames. It is unique among the three heat transfer modes because it relies on the motion of fluids to transmit heat. Therefore, understanding convection heat transfer requires some knowledge of fluid mechanics. The flow of fluids is very complex, especially in a turbulent fire environment. However, empirical correlations provide a practical, but sufficiently accurate, means to predict rates of convection heat transfer. Calculations of this kind will be presented later in this section.

Convection heat transfer (often shortened to 'convection') is separated into two categories based on how the flow of fluid is driven. *Natural convection* refers to heat transfer enabled by buoyancy-driven flows – for example, the rising of hot air and the sinking of cold air. *Forced convection* refers to heat transfer occurring due to flows that are driven by external forces like wind. Both natural and forced convection often play a combined role in wildland fire; this is called *mixed convection*.

The rate of heat transfer between a fluid and a solid surface for any category of convection is given by Newton's law of cooling:

$$\dot{q} = hA_s(T_s - T_\infty) \qquad [3.28]$$

where $\dot{q}$ is the heating rate (W), h is the *convection heat transfer coefficient* (W m^{-2} K^{-1}), A_s is the surface area (m^2), T_s is the surface temperature and T_∞ is the freestream fluid temperature (K or °C). The convection heat transfer coefficient (normally shortened to 'heat transfer coefficient' or HTC) describes the rate of heat transfer – whether it be heating or cooling – over a given surface, for some specified temperature difference between the surface and fluid.

The magnitude of the heat transfer coefficient is dependent on a significant number of input variables: thermal and geometric properties of the fuel; characteristics of the flow; and physical properties of the air. *Dimensionless numbers* are algebraic combinations of these inputs that simplify convection problems by reducing the number of parameters and facilitating insight into the relative effects of the various inputs. As implied by the term, the definitions of dimensionless numbers are formulated so that all units cancel out. There are several ways to derive dimensionless numbers, including their appearance in non-dimensional formulations of the governing equations of fluid mechanics and heat transfer. The magnitude of a given dimensionless number indicates the relative influence of one input variable over the other in the associated physical process. The range of possible values and the meaning of a specific numerical magnitude is unique to a given dimensionless number. In wildland fire, we use four dimensionless numbers that pertain to fluid flow and heat transfer: Nusselt number, Prandtl number, Reynolds number and Grashof number.

The *Nusselt number* (Nu) is given by:

$$\mathrm{Nu} = \frac{h\ell}{k} \qquad [3.29]$$

where ℓ is a characteristic length (e.g. the diameter of a twig), h is the convection heat transfer coefficient, and k is the thermal conductivity of the fluid (e.g. air or combustion gases). Nu quantifies the extent to which heat transfer is enhanced by fluid motion relative to a condition in which there is no fluid motion. Nu of around 1 means that there is virtually no heat transfer due to convection; instead heat moves through the fluid solely through conduction heat transfer. Larger values of Nu indicate the degree to which convection is responsible for heat transfer relative to pure conduction. For example, consider a flow of hot combustion-gas-and-air mixture at 500 °C and speed 10 m s^{-1}. Nu for fuel elements of diameter 1 mm (a pine needle) and 100 mm (a tree branch) is 5.9 and 62, respectively.

The *Prandtl number* (Pr) is given by:

$$\Pr = \frac{\nu}{\alpha} \quad [3.30]$$

The Prandtl (pronounced PRAN-dl) number is the ratio of the rate of diffusion of momentum to the rate of diffusion of heat in a given fluid. Unlike the Nusselt number, the Prandtl number is a physical property of a given fluid and can simply be looked up in a reference or online. The numerator ν is *kinematic viscosity* (m² s⁻¹) and is discussed in the **Boundary layers** subsection above. The denominator of the Prandtl number is *thermal diffusivity* (α):

$$\alpha = \frac{k}{\rho c_p} \quad [3.31]$$

where k is thermal conductivity (W m⁻¹ K⁻¹), ρ is fluid density (kg m⁻³), and c_p is the specific heat of the fluid for a constant pressure process (J kg⁻¹ K⁻¹). Thermal diffusivity indicates how fast heat diffuses through a substance. It has units of m² s⁻¹ and is tabulated for most common fluids (see **Appendix B** for thermal diffusivity of air and combustion gas). Prandtl number represents a comparison between diffusivity of momentum in a fluid flow and the diffusion of heat. Liquid metals like mercury and molten sodium diffuse heat readily but have low viscosity, so they have very small Prandtl numbers (Pr ≈ 0.004 to 0.030). In contrast, oils have large Prandtl numbers because they diffuse heat very slowly (Pr ≈ 50 to 100 000) and have high viscosity. A fluid with Pr = 1 will diffuse momentum and heat equally. The fluids we are interested in, air and combustion gases (see **Appendix B**), have Prandtl numbers that near this intermediate case: Pr ≈ 0.7.

The *Reynolds number* is important for understanding forced convection heat transfer in fire because it indicates the flow regime (laminar, turbulent or transitional) exhibited by the fluid flow. Reynolds number (Re) is:

$$\text{Re} = \frac{U \ell}{\nu} \quad [3.32]$$

where U is the flow velocity (m s⁻¹), ℓ (m) is a characteristic length (usually diameter), and ν is kinematic viscosity (m² s⁻¹). The Reynolds number is the ratio of the magnitude of inertial forces of fluid flow (numerator) to the magnitude of viscous forces (denominator). In heat transfer problems related to wildland fire, the Reynolds number for fuels can range from around 30 for the smallest fuel elements (e.g. a 1 mm pine needle) to more than 1.2×10^6 for large (~1 m) logs. If the Reynolds number is small, the viscous forces in a fluid are more important than the inertial forces of flow. Viscous forces prevent the growth of spontaneous perturbations in the flow, keeping the flow laminar. If the Reynolds number is large (greater than some critical value), the inertial forces of fluid flow are large relative to the viscous forces. The inertial forces overwhelm the viscous forces, so flow perturbations grow in size and number, and the flow becomes turbulent. The critical Reynolds number denotes the transition between laminar and turbulent flow and for flow around a cylinder is generally agreed to be around 200 000:

$$\text{Re}_{crit} \approx 200\,000 \quad [3.33]$$

However, this is a rough guideline rather than an absolute criterion, because it is possible to have turbulent flow well below Re = 200 000 in some circumstances and laminar flow well above Re = 200 000 in others. Moreover, fluid flowing past the cylinder may already have some level of turbulence generated by interactions with objects upstream.

The mixing action of turbulent flow enhances convection heat transfer by bringing freestream fluid closer to the surface over which it is flowing. This is why the Reynolds number so strongly affects the rate of convection heat transfer and why Re always appears in correlations (see below) for the convection heat transfer coefficient (e.g. **Eqn [3.36]**).

The *Grashof number* (Gr) indicates the flow regime (laminar, turbulent, and transitional) for flows that are driven by buoyancy. Therefore, it plays a key role in understanding natural convection heat transfer. The Grashof number represents the ratio of buoyant forces (numerator) to viscous forces (denominator). The Grashof number is defined as:

$$\mathrm{Gr} = \frac{g\tau |T_s - T_\infty| \ell^3}{v^2} \quad [3.34]$$

where g is acceleration due to gravity (m s^{-2}), T_s is the surface temperature (K or °C), T_∞ is the freestream temperature (K or °C), and τ is the volumetric thermal expansion coefficient (K^{-1}). The Grashof number plays the same role in natural convection as the Reynolds number does in forced convection. If the Grashof number is smaller than a critical value, the flow is typically laminar; if the Grashof number is greater than that number, the flow is typically turbulent. For natural convection on a heated (or cooled) vertical plate, the critical Grashof number is:

$$\mathrm{Gr}_{\mathrm{crit}} \approx 1 \times 10^9 \quad [3.35]$$

Just as with Reynolds number, the critical Grashof number should be thought of as a very rough guideline. Typical values of Grashof number in wildland fire range from around 1 to 1×10^9. The upper range being the same as Gr$_{\mathrm{crit}}$ implies that all purely natural convection problems are laminar. However, purely buoyant flows rarely occur in wildland fire since some ambient or fire-induced wind is nearly always present.

Determining the convection heat transfer coefficient represents the main challenge of making convection heat transfer calculations. There are two methods for evaluating it for a given situation. First, the convection heat transfer coefficient can be calculated from first principles. This involves simultaneously solving the coupled conservation equations for mass, momentum and energy in the fluid flowing over the surface. Exact analytical solutions for any practical circumstances are presently unknown due to the profound complexity of the problem. Thus calculating heat transfer coefficient from first principles is not useful for wildland fire.

The second way to evaluate the convection heat transfer coefficient is to find it using *empirical* methods – that is, by measuring it in carefully controlled experiments and then finding an equation that fits the resulting data. Such an equation, called a *correlation*, will provide a heat transfer coefficient for a given set of input parameters that describe a particular application. It is preferable to calculate heat transfer coefficients from first principles. However, due to the complex nature of all practical fluid flows and particularly those of wildland fire, a correlation is always used to evaluate convection heat transfer coefficient.

Correlations for convection heat transfer coefficient for various configurations relevant to wildland fire can be found in any engineering textbook on heat transfer (e.g. Çengel *et al.* 2017). For example, consider the following correlation for calculating the forced convection heat transfer coefficient h (W m^{-2} K^{-1}) for fluid flow over a cylinder:

$$h = \frac{k}{d}\left(0.3 + \frac{0.62\,\mathrm{Re}^{1/2}\,\mathrm{Pr}^{1/3}}{\left[1 + \left(\frac{0.4}{\mathrm{Pr}}\right)^{2/3}\right]^{1/4}} \cdot \left[1 + \left(\frac{\mathrm{Re}}{282000}\right)^{5/8}\right]^{4/5}\right) \quad [3.36]$$

for $0.5 \leq \mathrm{Re} \leq 4 \times 10^6$ and $\mathrm{Re} \cdot \mathrm{Pr} > 0.2$.

This correlation is found in Çengel *et al.* (2017) and other modern heat transfer textbooks, but it appeared for the first time in Churchill and Bernstein (1977). Many fuels in wildland fire are cylindrical in shape: grass stalks, needles, twigs, branches, limbs and logs. So, we can use **Eqn [3.36]** to, for example, estimate the heat transfer coefficient for a round, 30 mm diameter branch in an 8 m s^{-1} wind at temperature 200 °C (h = 55.3 W m^{-2} K^{-1}). With the heat transfer coefficient known, we can then estimate the rate of heat transfer to the branch using **Eqn [3.28]**.

Despite the complex appearance, **Eqn [3.36]** is relatively easy to apply, because the four variables are numbers that we either look up or calculate using simple formulas: the variable k (W m^{-1} K^{-1}) is the thermal conductivity of air (see **Appendix B**), and d (m) is the diameter of the cylinder. Re is easy to calculate (see **Eqn [3.32]**) and Pr for air is tabulated in **Appendix B**.

The heat transfer coefficient calculated from **Eqn [3.36]** varies with the diameter d of the cylinder. Cylinder diameter affects the calculation twice – first in the $k\,d^{-1}$ factor and second in the calculation of Re. In general, the convection heat transfer coefficient becomes smaller as the cylinder diameter gets bigger. Thus convection heat transfer is less effective at heating large fuels. **Figure 3.18** shows a plot of heat transfer coefficient (h) from **Eqn [3.36]** for cylindrical fuel elements with diameters (d) ranging from 1 mm (a pine needle or a stalk of grass) to 1000 mm (a large log) at four wind speeds and temperature 500 °C. The log–log plot is used to better show the large range of both heat transfer coefficient and diameter. Consider, for example, the 2.0 m s^{-1} wind speed curve, where a 1 mm fuel element has $h = 155$ W m^{-2} K^{-1} and a 1000 mm element has $h = 5.1$ W m^{-2} K^{-1}. Since the 1 mm fuel element has a much larger (×30) heat transfer coefficient, it will heat up and ignite much faster than a 1000 mm fuel element. It is for this reason that fine fuels are said to 'carry' the spread of fire through wildland fuel beds.

The second line of **Eqn [3.36]** indicates that the equation should be used only when the Reynolds number is between 0.5 and 4×10^6, and when the product of Pr and Re is greater than 2.0. Restrictions such as these are common to all correlations for convection heat transfer coefficient. Recall that correlations are equations fitted to experimental data. The restrictions represent the range of conditions over which the experiments were carried out. One can calculate a heat transfer coefficient for conditions outside of these parameters, but it is not recommended because some unforeseen behaviour may cause the actual heat transfer coefficient to be very different from that extrapolated from the correlation. At any rate, the restrictions shown for **Eqn [3.36]** are never a problem for our purposes because the Reynolds number and the Prandtl number for relevant wildland fire calculations will always fall within the permitted ranges.

Perhaps you have noticed that some of the material properties needed to calculate the

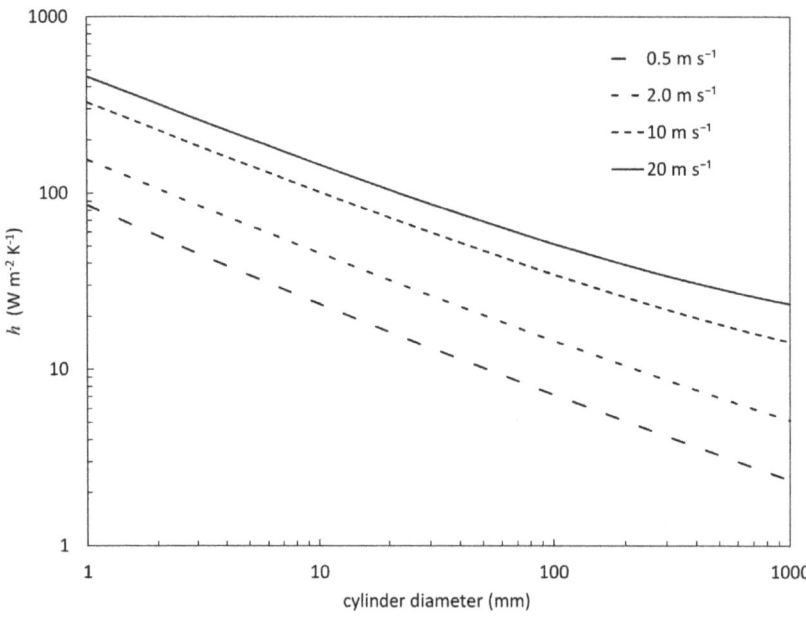

Figure 3.18: Average heat transfer coefficient for cylinders of various diameters for the indicated wind speeds and temperature 500 °C. Calculations based on **Eqn [3.36]**.

dimensionless numbers above are temperature-dependent, meaning that the properties take different values depending on the temperature. For example, two temperature-dependent properties of air are important for calculating dimensionless numbers: The density (ρ) of air, which is used to calculate Reynolds number (**Eqn [3.32]**), is about four times smaller at 1000 °C than at room temperature. The specific heat (c_p) of air, which is used to compute thermal diffusivity (**Eqn [3.31]**), is ~18% greater at 1000 °C than it is at room temperature. Other temperature-dependent properties include thermal conductivity and kinematic viscosity. Both Prandtl number and Grashof number vary with temperature as well.

To simplify matters, heat transfer coefficient correlations will normally specify the temperature at which to evaluate the temperature-dependent properties. For example, the fluid properties required to calculate convection heat transfer coefficient using **Eqn [3.36]** are calculated at the *film temperature*, which is the average of the freestream temperature and the surface temperature:

$$T_{\text{film}} = \frac{1}{2}(T_\infty + T_s) \qquad [3.37]$$

In other correlations, fluid properties may be evaluated at the freestream temperature T_∞ or the surface temperature T_s alone.

Correlations like **Eqn [3.36]** yield an average heat transfer coefficient for the whole object (e.g. a pine needle or a tree branch). In reality, the heat transfer coefficient of an object can vary wildly depending on the location on the surface and its orientation to the wind. The heat transfer coefficient at a specific point on a solid object is called the *local heat transfer coefficient*. For example, consider **Figure 3.19**, which shows how local heat transfer coefficient changes dramatically around the circumference of a 25 mm diameter cylinder in a wind (Giedt 1949). The variation of the curves results from changing boundary layer thickness, transition from laminar to turbulent

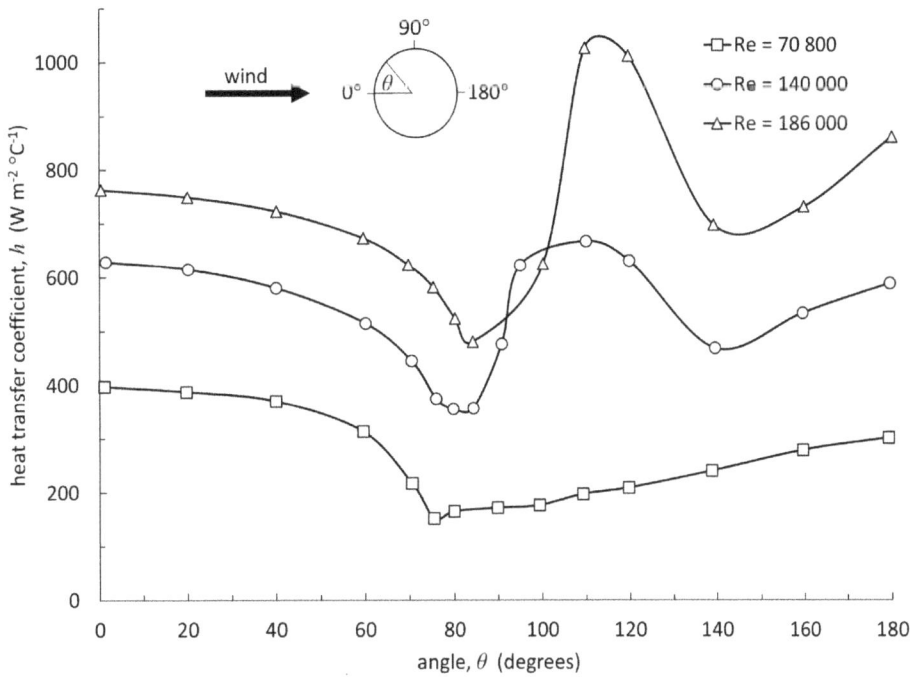

Figure 3.19: Local heat transfer coefficient for a 25 mm diameter cylinder in air for selected Reynolds numbers. Adapted from Giedt (1949).

flow, boundary layer detachment from the surface, and onset of wake flow on the lee side of the cylinder. All of these changes result from the shape of the object (a cylinder in this case, but true for any other shape as well), its orientation relative to the flow, the speed and turbulence level of the incoming flow, and the surface roughness. Using correlations like **Eqn [3.36]** for an average heat transfer coefficient is perfectly suitable for calculating overall heat transfer to an object. However, to understand the process of heating fuel to ignition requires that we understand the circumstances that give rise to particularly high local heat transfer coefficients. It is there that maximum heating occurs and, therefore, that the earliest ignition will first occur.

Recall **Figure 3.18**, which shows convection heat transfer for a range of cylinder diameters. As noted previously, the log–log scaling is used because the relationship is nonlinear and covers a wide range of values on both axes. If we view a small portion of the values from this figure with linear scaling (**Figure 3.20**), we can see how dramatically nonlinear the relationship actually is. Notice how the heat transfer coefficient grows drastically larger only for the smallest cylinder diameters (conifer needles and grass stalks). This effect can be explained by the thickness of the boundary layers that form in flows around fine fuels.

Convection heat transfer coefficient is proportional to the temperature gradient at the surface. So, as the temperature gradient gets larger the local heat transfer coefficient also increases. Elevated local heat transfer coefficients for a given object occur in regions where the boundary layer is particularly thin. This is true for both natural and forced convection. Thin boundary layers enhance convection heat transfer coefficient because the separation between the freestream air and the surface is small, resulting in a larger temperature gradient. Flows of air and combustion gas around a conifer needle and other fine fuels produce thin boundary layers simply because there is so little surface over which they can grow thicker.

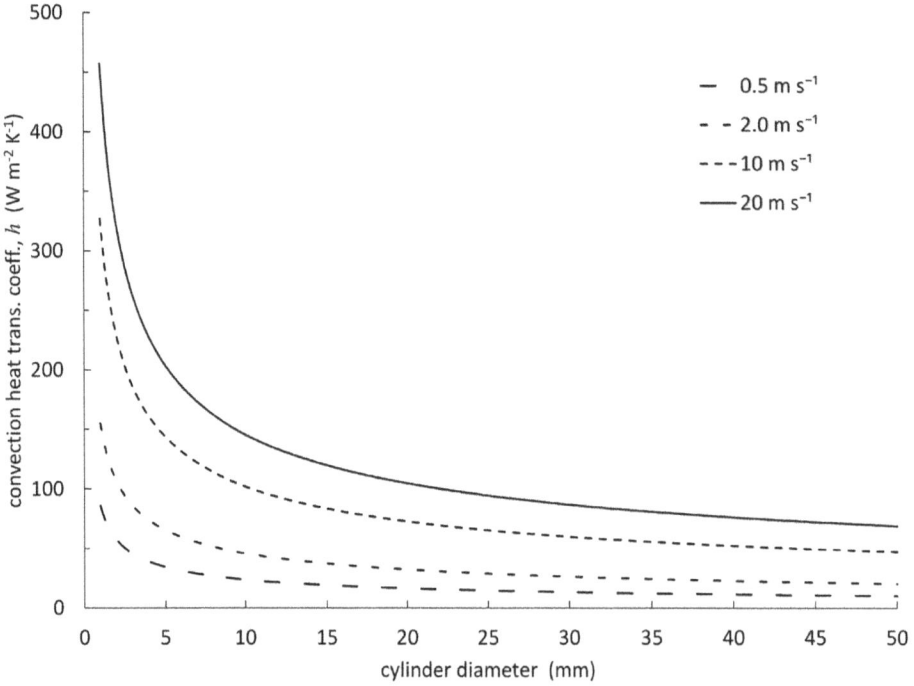

Figure 3.20: A linear plot of the left half of **Figure 3.18**, which shows how convection heat transfer coefficient increases dramatically for cylindrical fuel elements smaller than around 10 mm.

Boundary layers also have small thickness at the edges of thin flat objects like deciduous leaves or blades of grass. Even larger fuels can have locally thin boundary layers at corners and sharp protrusions. The scorching one often sees in such fuels is evidence of enhanced heating due to thin boundary layers. For example, **Figure 3.21** shows a wooden stick with a 12 mm × 12 mm square cross section that was subjected to an air flow at 600 °C. The corners were scorched preferentially because the thin boundary layer enhanced heating there.

Fine fuels like grass and conifer needles usually have Reynolds numbers less than 100, which puts the flow field squarely within the laminar regime. On the other hand, since large fuel elements like logs and high wind speeds lead to high Reynolds numbers, the boundary layers will quickly transition to turbulence in such instances. Even smaller diameter fuels may have boundary layers that transition to turbulence if they are particularly rough and/or the incoming flow is already intensely turbulent from upstream obstacles (like other pine needles). This is one reason why we emphasised earlier that the critical Reynolds number is a very rough guideline for the onset of turbulent flow.

Combined heat transfer

All three modes of heat transfer described in this chapter play important roles in wildland fire. *Combined heat transfer* refers to instances where more than one mode is actively participating in heat transfer. Consider **Figure 3.22**, which shows a tree branch being heated, then cooled, by convection. On the left, the flames are approaching the tree, so hot gas and air are convectively heating the branch as they flow around it. Simultaneously, conduction heat transfer is transmitting heat at the surface to the interior of the branch (lower left). Now suppose there is a rapid wind direction change (**Figure 3.22**, right), which causes cool ambient air to flow around the branch. Moreover, heat is carried away from the surface by convection heat transfer, reducing its temperature. Now conduction heat transfer reverses direction, drawing heat from the warm interior of the branch to the surface.

In the preceding section, we learned that convection heating is particularly effective for fine fuels because the thin boundary layer that is formed leads to large temperature gradients. The same is also true for convection cooling, where a hot fuel element is cooled by ambient air. Let us consider the implications of that fact on ignition of a fuel element. Decades-old conventional wisdom indicates that the heat of flames approaching any fuel – even the finest particles, such as conifer needles – is transmitted largely by radiation heat transfer. As the distance between approaching flames and the pine needle decreases, the thermal radiation flux likewise increases, and hence, it has been assumed, the temperature of the fuel increases until it ignites. This scenario is certainly plausible, but recent experiments (Finney *et al.* 2015) have shown that it does not apply universally, and especially not to fine fuels. Radiation alone is nearly always insufficient to heat fine fuels to a high enough temperature for ignition, because convection is so effective at cooling such fine

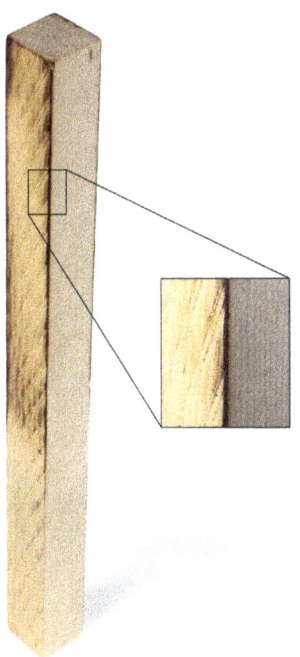

Figure 3.21: A 12 mm × 12 mm square wooden rod subjected to a 600 °C air flow shows scorching at the corners. Photograph by Ian Grob, USDA Forest Service.

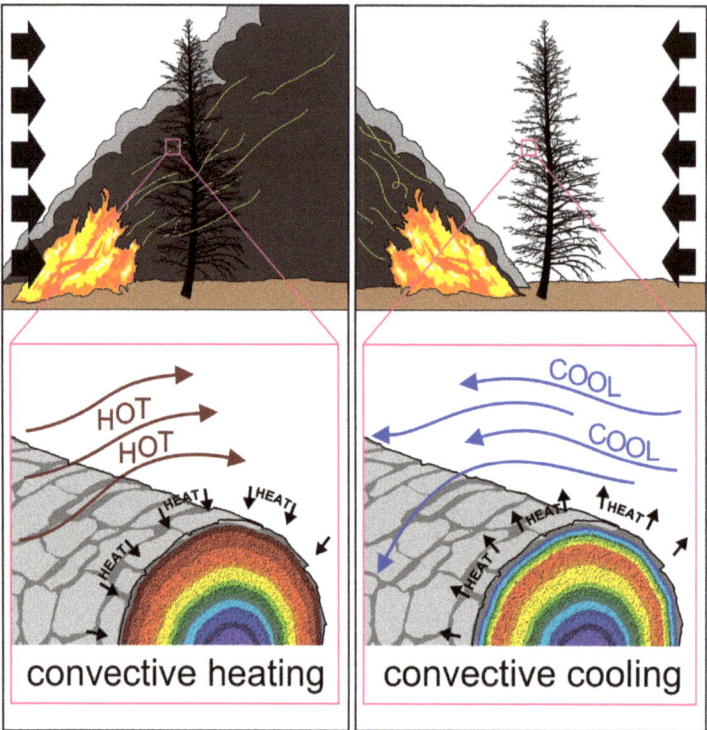

Figure 3.22: Convection heat transfer can enable both heating and cooling of fuel. Illustrations by Brian Elling.

particles with low-temperature ambient air. Even with thermal radiation from large, thick flames, ambient wind flowing around a fine fuel element is sufficient to prevent ignition. In fact, not until the flames have drawn near enough that the needle is being convectively heated by direct contact with flames and hot gas, will it ignite. In contrast, large logs, tree trunks and densely packed litter and duff cannot be cooled as effectively by convection, so they are more sensitive to heating and ignition by absorbing thermal radiation. This topic is discussed further in **Chapters 5** and **6**.

Summary

This chapter has presented many physical processes and concepts that describe how heat is transferred to and from different materials in the wildland fire environment. We emphasised the concepts of heat, temperature and energy, and then we introduced the processes of conduction, radiation, and convection heat transfer. These are fundamental to the spread and behaviour of all fires, especially wildland fires. In the next chapters, we will discuss in detail how energy is released by combustion. Then, using our knowledge of heat transfer, we will see how ignition of wildland fuels occurs.

Endnote

[i] The shortened 'boundary layer' term is common in fluid mechanics and heat transfer sciences generally. However, it should not be confused with the *thermal* boundary layer, a distinct but related phenomena introduced in the next section. In this book, the 'thermal boundary layer' term will never be shortened.

References

Çengel YA, Cimbala JM, Turner RH (2017) *Fundamentals of Thermal-fluid Sciences*. 5th edn. McGraw-Hill Education, New York.

Churchill S, Bernstein M (1977) A correlating equation for forced convection from gases and liquids to a circular cylinder in crossflow. *Journal of Heat Transfer* **99**, 300–306. doi:10.1115/1.3450685

Finney MA, Cohen JD, Forthofer JM, McAllister SS, Gollner MJ, Gorham DJ, Saito K, Akafuah NK, Adam BA, English JD (2015) Role of buoyant flame dynamics in wildfire spread. *Proceedings of the National Academy of Sciences of the United States of America* **112**, 9833–9838. doi:10.1073/pnas.1504498112

Giedt W (1949) Investigation of variation of point unit heat transfer coefficient around a cylinder normal to an air stream. *Transactions of the ASME* 71, 375–381.

Howell JR (2010) *A catalog of radiation heat transfer configuration factors*. http://www.thermalradiation.net/indexCat.html.

Howell JR, Mengüç MP, Siegel R (2016) *Thermal Radiation Heat Transfer*. 6th edn. CRC Press, Boca Raton, FL.

Quintiere JG (2016) *Principles of Fire Behavior*. CRC Press, Boca Raton, FL.

Taylor PB, Foster PJ (1975) Some gray gas weighting coefficients for CO_2-H_2O-soot mixtures. *International Journal of Heat and Mass Transfer* **18**, 1331–1332. doi:10.1016/0017-9310(75)90244-6

4

Combustion

The goal of this chapter is to become familiar with the physical processes of combustion. At its core, combustion is a rapid chemical reaction between a fuel and oxygen which generates heat and light. We will first discuss what fuel is, starting with simple household examples and progressing to wildland fuels. Then we will discuss the important thermodynamic aspects of combustion – that is, the chemical reactions involved, the heat released and the temperatures attained. We will then learn about the speed of reactions (i.e. *chemical kinetics*) and the different types of flames and their properties. These first sections are focused on *flaming* combustion, but we will wrap up this chapter by discussing another vital form of combustion in wildland fires, smouldering, which produces no flames at all.

Fuels

In flaming combustion, fuels can start as any phase of matter. They can be gases, liquids or solids. Regardless of their initial phase, all fuels must first be converted to a gas in order to produce a flame. The chemical reactions we see as flames only occur between *gaseous* fuel and the oxygen in air. This seems obvious for many of the combustion devices that we use in daily life. In your home, for example, you may have a gas stove or furnace that runs on natural gas, or you may have a barbeque that runs on propane. These fuels both burn as gases. Natural gas is primarily methane, delivered into your home in gaseous form. Propane is most often compressed and delivered as a liquid in a cylinder, but it evaporates when released from the tank and is therefore a gas when it enters the burner. Liquid fuels are common in other parts of daily life as well. Examples include the gasoline in a car engine or lawn mower. These too must evaporate before they can burn. In an engine, the evaporation of the fuel is sped up by a carburettor or fuel injectors that spray it as tiny droplets into the hot engine chamber. In these examples, the chemical composition of these fuels is the same whether they are in the liquid or the gas phase, and they do not require further modification before they can burn.

Combustion is more complicated for solid fuels, which are more relevant to wildland fires. Wildland fuels are mostly woody. They are characterised by a large *cellulose* component (40–50%), but they also contain significant amounts of hemicellulose (25–35%) and lignin (18–35%) (Pettersen 1984). Cellulose is a *polymer* – that is, it consists of long, repeating chains of carbohydrate molecules. Carbohydrate molecules consist of chains of carbon (C) atoms with hydrogen (H) and hydroxyl groups (OH) attached. Their general chemical formula is $C_x(H_2O)_y$, where x denotes the number of carbon atoms and y denotes the numbers of hydrogen and hydroxyl molecules. (H and OH are almost always in equal proportions in a carbohydrate.) The main carbohydrate in cellulose is *glucose*. Cellulose consists of long, straight chains of hundreds or even thousands of glucose molecules ($C_6H_{12}O_6$), so it is represented by the formula $(C_6H_{10}O_5)_j$, where j is the number of glucose molecules in the chain. (One H and one OH from each glucose molecule are given up to form the chain.)

Hemicellulose is more general than cellulose in its composition. It is similar to cellulose in that it contains chains of glucose, but it can contain several other simple sugars as well, and it branches rather than forms straight chains. The other sugars within hemicellulose vary, depending on the plant species and even the tissue within individual plants (e.g. bark *v.* root *v.* branch). Hemicellulose polymers typically consist of only a couple of hundred blocks of sugars and are thus much shorter than cellulose. Plants produce cellulose and hemicellulose within their cell walls to generate their own structure; hence they are often called *structural carbohydrates*.

Lignin is another polymer in the cell walls. It acts like glue to further strengthen the structural integrity of the plant. Lignin's water-repelling (hydrophobic) nature is what allows wood to transport water. Lignin is a very complex molecule with no clearly defined composition or structure.

The proportions of the structural components of wood vary, depending on the soundness of the wood and its species. Certain wood-rotting organisms target cellulose only ('brown rot'), leaving behind elevated levels of lignin; others ('white rot') do the opposite. Cellulose, hemicellulose and lignin comprise the majority of the dry mass of dead wood. However, they may comprise as little as half of the dry mass of living plant cells, especially those of live foliage. The other half consists of various quantities of other sugars and starches produced during photosynthesis, as well as fats and proteins. All of these components have different combustion properties, so the chemical composition of the fuel can have important, and sometimes unknown, effects on the overall burning behaviour.

Solid fuels, including wildland fuels, must be converted into gas before they can burn. Woody fuels are solid because their molecules are large and heavy and bonded together in large chains. To be converted to a gas, these molecules must be broken down into much smaller pieces. This process is much more complicated than the simple phase change that takes place in liquid fuels such as gasoline. The process is a chemical reaction called *pyrolysis*, and the resulting gaseous fuels are sometimes called *pyrolysates*. Pyrolysates are chemically different from the original carbohydrates. In fact, they vary dramatically and their composition depends on the temperature of pyrolysis. Pyrolysis at low temperatures produces large molecules that are sticky liquids like tar, while pyrolysis at higher temperatures produces smaller, gaseous molecules that are more ready to react chemically. In most cases, pyrolysis forms a huge variety of individual products, and their chemical composition is largely unknown.[i] In other words, we do not ever really know what the gaseous fuel is in wildland fires. This makes detailed calculations of the combustion chemistry of wood nearly impossible. Because of this, we often take a 'global' approach: we simplify the hundreds of thousands of individual chemical reactions that are likely to be occurring to just a few that describe the overall behaviour well enough for practical purposes.

Part of a simplified approach to combustion in wildland fires is estimating the temperature of the solid fuel when pyrolysis starts to become significant. Known as *pyrolysis temperature*, it depends on the molecular structure of the solid. Some structures are very stable, while others are more likely to break apart. For example, hemicellulose begins to significantly pyrolyse at ~260 °C (Yang *et al.* 2007), cellulose at ~355 °C (Yang *et al.* 2007), and lignin intermittently between temperatures of 350 and 600 °C (Kawamoto 2017). Note that these temperatures only indicate when significant *thermal breakdown begins*, not when *combustion* occurs. As we discuss in **Chapter 5**, environmental conditions and the amount of gaseous fuel present play critical roles in whether and when ignition occurs.

Not all of the wood is consumed during a fire. There is always ash left over, and most often there is char as well. The ash is from mineral compounds in wood that contain elements such as calcium, phosphorus and magnesium. These compounds are not burnable, and they can retard flaming combustion. Mineral compounds typically constitute ~5% of the dry mass of wood (Dimitrakopoulos and Panov 2001), although they may constitute as much as 25% in the leaves of some plants (Philpot 1970). Mineral compounds do not decay as readily as do other fuel constituents, so their relative concentration increases as plant material decomposes (Hough 1969). Char is primarily carbon from the fuel that was never fully pyrolysed. It has been stripped of the majority of the hydrogen and oxygen atoms that were once attached, but it has not been broken down into gas molecules. Char can, however, be consumed in smouldering combustion, as we discuss below.

Thermodynamics of combustion

Combustion reactions

The simplest way to describe the combustion reaction of a hydrocarbon fuel – that is, one with both hydrogen and carbon – in air is:

$$\text{Fuel} + \text{air} \rightarrow H_2O + CO_2 + \ldots \quad [4.1]$$

In this reaction, fuel and air are the *reactants* in combustion, while water and carbon dioxide are the primary *products*. Actual combustion reactions are far more complicated than this, and they are often incomplete (discussed below). The combustion of even the simplest hydrocarbon, methane (CH_4), is complex: the methane molecule itself almost never directly reacts with an 'air molecule', but instead breaks down and reacts in ~500 different 'mini' intermediate reactions. Complete modelling of methane combustion requires calculating these individual reactions. The combustion of pyrolysed fuels in wildland fires is even more complex, consisting of hundreds of thousands, if not millions, of intermediate reactions. However, just as in our study of pyrolysis reactions, we can learn much about combustion by simplifying the many intermediate reactions to a single 'global' combustion reaction. We focus here on combustion of cellulose because of its importance in wildland fuels.

From the previous section, we know that cellulose consists of long chains of glucose and can be written with the chemical formula $(C_6H_{10}O_5)_j$, where j is any whole number. Air is ~21% oxygen (O_2) and 79% nitrogen (N_2). When we write a chemical reaction with air, we express these proportions relative to 1 mole[ii] (mol) of oxygen, so air is described in the chemical reaction as ($O_2 + 3.76N_2$). This indicates that air contains 3.76 moles of nitrogen relative to every 1 mole of oxygen, and air contains a total of 4.76 moles of molecules relative to each mole of oxygen. This satisfies the known proportions of O_2 and N_2 in air (1/4.76 = 21% and 3.76/4.76 = 79%, respectively). A 'global' combustion reaction equation for cellulose[iii] is therefore:

$$C_6H_{10}O_5 + 6\,(O_2 + 3.76N_2) \rightarrow 6CO_2 + 5H_2O + 22.56N_2 \quad [4.2]$$

The number of molecules (or moles of molecules) of each compound is found by making sure both sides have equal numbers of atoms of each element. The number of atoms of carbon, hydrogen, oxygen and nitrogen must be the same on both sides of **expression [4.2]**, since atoms themselves are neither created nor destroyed in a chemical reaction. They are only rearranged.

Looking at the reaction this way, we can see that for every molecule (or mole) of fuel, 6 * 4.76 = 28.56 molecules (or moles) of air are required. In other words, much more air is required than fuel. When the amount of fuel is perfectly balanced with the amount of air, the reaction is said to be *stoichiometric*. In this case, all of the fuel and oxygen is consumed by combustion, and the nitrogen is unchanged as in **expression [4.2]**. If more air (or less fuel) is present than the stoichiometric amount, not all of the air can react, so the combustion products will include the extra oxygen. The reactants in this case are said to be *fuel lean*. Alternatively, if less air (or more fuel) is present than this stoichiometric amount, the reactants are said to be *fuel rich*, and unburnt fuel will be present in the combustion products. Because actual combustion is more complicated than our global reaction implies, the unburnt fuel will take a variety of forms. There will be some intact fuel molecules, parts of fuel molecules, and carbon monoxide (CO). In reality, even if the quantities of fuel and air are perfectly balanced, combustion will be incomplete, and small quantities of unburnt fuel and unreacted oxygen will exist in the combustion products.

Heat of combustion

How much heat is released when a particular fuel is consumed? This is the fuel's *heat of combustion*. In theory, this can be calculated if we know the energies in the chemical bonds of the reactants and products. To know this, however, we would need to know what the reactants and products actually are. As discussed above, the actual gaseous fuel consumed in wildland fires is produced by pyrolysis of the solid material, and its composition is complex and unknown. Furthermore, combustion in wildland fires rarely if ever takes place under perfect stoichiometric conditions or with complete combustion, so details about the various products are also unknown. Because of this, it is far more practical to *measure* the heat of combustion rather than *calculating* a theoretical one.

The standard method for measuring heat of combustion is to use an apparatus called a bomb calorimeter. As dictated by a widely used, standardised procedure, a fixed amount of fuel is inserted into a closed chamber with a surplus of pure oxygen to react with the fuel (see **Figure 4.1**). This chamber is submerged in a water bath of known volume that is well insulated. The fuel is ignited and burned until completely consumed. The heat of combustion is found by measuring the temperature rise in the water surrounding the combustion chamber. Because the properties of water are well known, the heat can be calculated

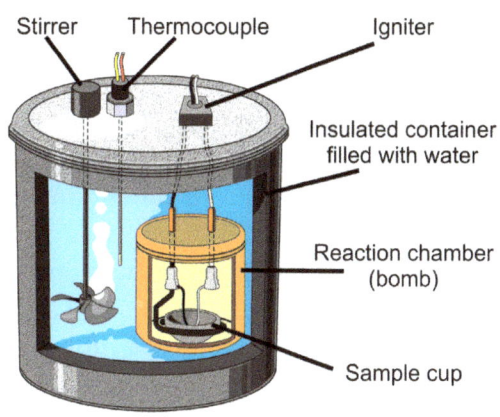

Figure 4.1: General schematic of a bomb calorimeter. Illustration by Brian Elling.

from its final temperature. If the products of combustion are cooled until the water vapour resulting from combustion condenses inside the chamber, the result is the *higher heat of combustion*. If the combustion products still include water vapour, the result is the *lower heat of combustion*. The lower heat of combustion is typically used to estimate heat release and fireline intensity in wildland fire because the combustion products usually contain water vapour.

Because the bomb calorimeter method provides ideal conditions for complete consumption of the fuel, this method produces 'textbook' values for heat of combustion, which may be adequate for use and accurate in many cases, such as industrial applications. Unfortunately, wildland fire is not one of them. In reality, fuels in wildland fires are never completely consumed as they are in bomb calorimeter tests, so the heats of combustion calculated with this method are often much higher than those that actually occur in the flaming zone of wildland fires. This is mainly due to incomplete combustion of both the pyrolysates and solid fuel. Because wildland fires are often very chaotic and turbulent, not all of the gaseous fuel has access to adequate air or the time to combust before it cools to the point where burning cannot continue. The resulting incomplete combustion limits the amount of heat released and also increases emissions of carbon monoxide, particulate matter and other toxic combustion products. In addition, incomplete combustion of solid fuel results in char left behind. As we will see later, the consumption of char (called *char oxidation*) can release a significant amount of heat. This is important in the smouldering stage of wildland fires, but it typically does not occur in large amounts during the flaming combustion stage. Thus, heat released during char oxidation should not be fully included in the heat of combustion during flaming. Furthermore, the process of charring itself is complex and variable. The amount of char formed, its physical and thermal characteristics, and the amount consumed in the various stages of combustion can vary greatly depending on the burning conditions, such as the heating rate of the fuel (see **Figure 4.2** for an example) and the availability of oxygen.

Despite overestimating heat of combustion for wildland fuels, bomb calorimeter measurements have been frequently used in wildfire applications. More realistic values can be obtained using different standardised testing techniques, such as cone calorimeters or fire propagation apparatuses (FPAs) (see **Figure 4.3**). These testing techniques allow the fuel to burn in a more realistic manner so the measured heat of combustion accounts for incomplete combustion and appropriate amounts of char formation and oxidation. This is achieved by burning the fuel in a more natural state in perforated sample holders. The mass of the fuel can be varied within the sample holder volume to produce different fuel bed bulk densities which, as described below, can alter the burning behaviour. An important feature of the cone calorimeter and FPA is that the burning conditions can be controlled using an adjustable radiant heater, and, in the FPA, the air flow through the sample can be varied to examine the wind effect. The heat released during combustion is calculated by measuring the products of combustion. In particular, oxygen sensors measure

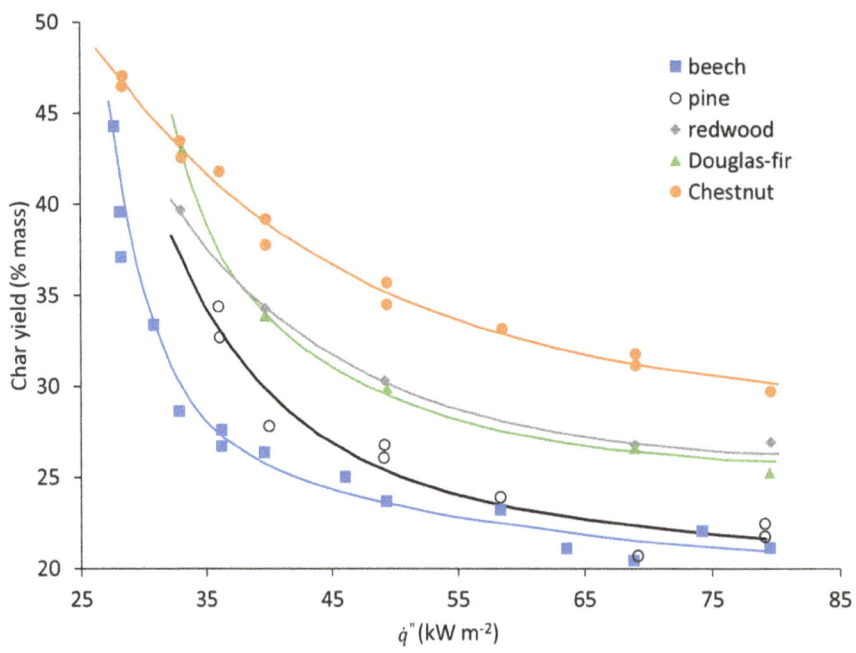

Figure 4.2: The char produced (as percent of initial dry mass) of cylinders of different wood species varies with the heating rate ($\dot{q}''$). Wood species tested include beech (*Fagus sylvatica*), pine (*Pinus pinea*), Douglas-fir (*Pseudotsuga menziesii*), redwood (*Sequoia sempervirens*), and chestnut (*Castanea sativa*). Data replotted with permission from Di Blasi *et al.* (2001).

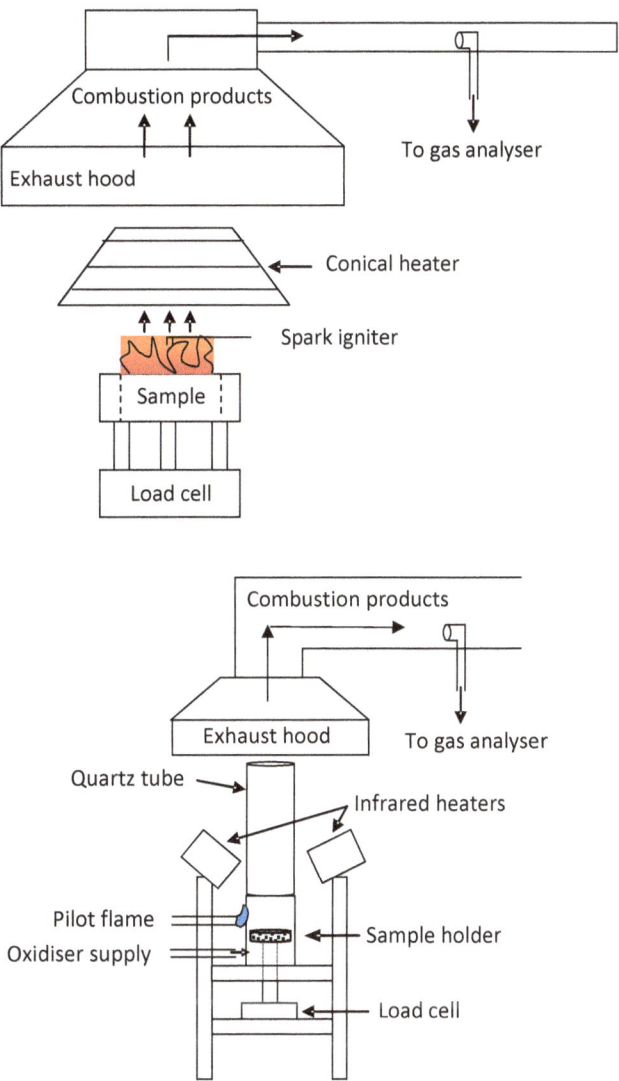

Figure 4.3: General schematic of (a) a cone calorimeter and (b) a fire propagation apparatus (FPA).

the amount of oxygen consumed, which is directly related to the heat released. This is called *oxygen consumption calorimetry*. For details about the cone calorimeter, FPA and oxygen consumption calorimetry, see Janssens (2016). The results from these methods are called the *effective heat of combustion*, and it can be much lower than the value measured by a bomb calorimeter. As a comparison, the lower heat of combustion of Douglas-fir needles measured in a bomb calorimeter is 20.55 MJ kg^{-1} (Williamson and Agee 2002), but the effective heat of combustion measured using oxygen consumption calorimetry, as a function of the moisture content (MC), is calculated as (Babrauskas 2006):

$$H_{c,eff} = 16.52 - 0.057 \, MC \, MJ \, kg^{-1} \qquad [4.3]$$

For example, the effective heat of combustion for completely dry needles is 16.52 MJ kg^{-1}, but for live needles with 100% MC it is 10.82 MJ kg^{-1}. So for even dry needles, measurements from the bomb calorimeter overpredict the heat of combustion by 20%. For live needles, the discrepancy between the methods can be more than 50%.

Let us return to combustion of gaseous fuels to examine a general concept about heat of combustion that can be very useful. Imagine a container with a mixture of some gaseous fuel and air. When the mixture is fuel lean, all of the fuel is consumed. As illustrated in **Figure 4.4**, the heat of combustion for the given amount of fuel (heat per mass or volume = kJ kg^{-1} or kJ mol^{-1}) is therefore constant. In contrast, when the fuel and air mixture is fuel rich, only some of the fuel is consumed because the amount of air is a limiting factor. The more fuel rich the mixture, the less fuel is consumed, so the heat of combustion (per mass or volume of fuel) decreases.

Flame temperatures

Just like the heat of combustion, the temperature of the combustion products, or *flame temperature*, can theoretically be calculated based on the energies of bonds in the molecules and thermodynamic properties of the products and reactants. However, the same issues remain for wildland fires: the compositions and concentrations of the products and reactants aren't actually known. In addition, the transfer of heat from the flame to the surroundings (especially due to the radiation of soot particles within the flame) affects flame temperature substantially. The theoretical, calculated value of flame temperature is thus relevant for only certain types of flames under ideal conditions (when there are no heat losses, i.e. *adiabatic* conditions). Not surprisingly, these conditions do not occur in wildland fires. It is also challenging to actually measure flame temperatures in wildland fires. A traditional method is to place thermocouples (two wires of dissimilar metal welded together) within the flames. However, the environment of a wildfire is turbulent and unsteady, and contains large amounts of radiant heat, which is both absorbed and lost by thermocouples. All of these factors contribute to measurement errors. As we will discuss later, because of the type of flames and fuels involved in wildland fires, a good estimate of the flame temperature is between 1000 and 1200 °C, regardless of whether it's a small prescribed fire or a raging crown fire.

Although adiabatic flame temperatures are not helpful for describing wildland fires, they show general trends that can help us understand how flame temperature varies with the proportions of fuel and air. As shown in **Figure 4.5**, when fuel and air are in stoichiometric proportions (shown by the dashed vertical line), both are completely consumed and the flame temperature is at its maximum. If the mixture is fuel lean, even though the heat released by a given quantity of fuel is the same (as shown in **Figure 4.4**), some of the heat is used to warm the surplus air, so the flame temperature is reduced. If the mixture is fuel rich, only some of the fuel can be consumed and the heat of combustion is less. Additionally, the remaining fuel acts as a heat sink, absorbing some of the heat released, so the flame temperature is reduced under these conditions too. This

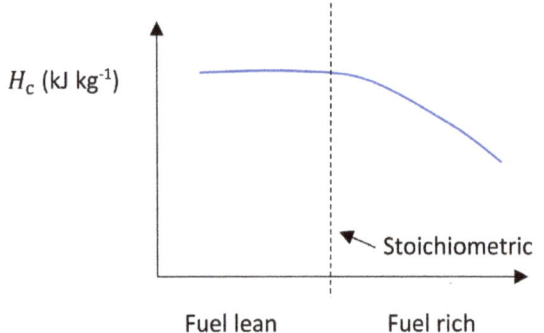

Figure 4.4: Heat of combustion varies with different proportions of gaseous fuel and air.

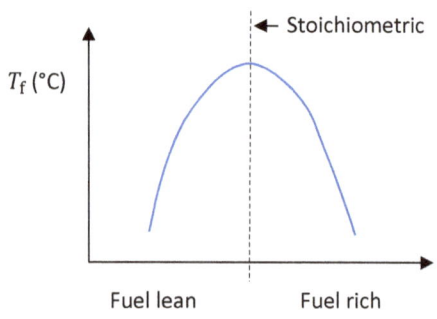

Figure 4.5: Flame temperature varies with different proportions of fuel and air.

concept will be particularly helpful when we discuss the ignition process in **Chapter 5**.

Brief discussion of chemical kinetics

Another important aspect of combustion reactions is how quickly they occur. This is the subject of chemical kinetics. The bottom line is that combustion reactions are fast but not instantaneous. Though the reaction time is usually very short, there are circumstances in which this time may be important, particularly in marginal conditions for burning like ignition and extinction. You may have experienced this if you have ever turned the gas flow up really high on your stove. You may have seen the flames detach from the burner ('lift-off') and hover some distance above it (see **Figure 4.6**). The gas is moving so fast that it travels a noticeable distance before it can combust. Another example is the backfire of a car. If the engine is poorly tuned, the combustion of the fuel–air mixture is slowed just enough for it to occur in the exhaust of the car rather than in the engine itself, causing the loud noise.

The amount of time that chemical reactions take, and the rate they occur, depends primarily on three things: the temperature, the amount (concentration) of gaseous fuel and the amount (concentration) of air. The equation that describes these dependences is called the *Arrhenius reaction rate* equation:

$$\dot{N}''' = A e^{\frac{-E_a}{RT}} [\text{Fuel}]^a [\text{O}_2]^b \quad [4.4]$$

In this equation, $\dot{N}'''$ is the *reaction rate* (mol cm^{-3} s^{-1}), A is the *pre-exponential factor*, e is the base of the natural logarithm, E_a is the activation energy, R is the universal gas constant (8.314 × 10^3 kPa cm^3 mol^{-1} K^{-1}), T is the temperature (K[iv]), [Fuel] is the concentration of fuel (mol cm^{-3}), [O$_2$] is the concentration of oxygen, and a and b are exponents. Although the values of A, E_a, a and b can, in some cases, be calculated, in the wildland fire context they are all experimentally measured. While this equation describes the reaction rate for gaseous fuels, a similar equation can be written for the pyrolysis reactions that generate the gaseous fuels burned in wildland fires.

From **Eqn [4.4]**, we can see how important both the activation energy (E_a) and the temperature (T) are because they are in the exponential term. The greater the activation energy, the slower the reaction rate; the higher the temperature, the faster the reaction rate. As the name implies, the activation energy quantifies how hard it is to get the reaction started. If the bonds holding the molecules together are strong, a lot of energy is required to break them, and therefore the reaction has a high activation energy. It turns out that this is indeed the case for both combustion and pyrolysis reactions. Just putting propane and oxygen together at room temperature isn't enough to start combustion.[v] The fuel–air mixture needs to be heated, giving the molecules more speed and energy so when they collide with each other, the bonds inside them can break. The activation energies for both combustion and pyrolysis are constant properties for a given fuel.[vi] The universal gas constant (R) also appears in the exponent of **Eqn [4.4]**, but it is a physical constant that never varies.

Figure 4.6: Flame lift-off from a Bunsen burner. The gas issuing from the burner is moving too fast for the flame to attach to its surface. Combustion occurs some distance above. Photograph by Ian Grob.

The only true variable in the exponential term of **Eqn [4.4]** is the temperature. As shown schematically in **Figure 4.7**, increasing the temperature can dramatically increase the reaction rate. Due to the large activation energies of both pyrolysis and combustion, they show exponential increases in reaction rate beginning at a distinct point where the curve begins to ascend rapidly. This point on the curve is where 'thermal runaway' of combustion reactions begins (which could lead to ignition). On a curve for pyrolysis reactions, this point is the pyrolysis temperature. It is important to note that, below the ignition or pyrolysis temperatures, those reactions are still occurring, just at a much-reduced rate.

The remaining variables in **Eqn [4.4]** are the concentrations of the fuel and air. The exponents a and b are usually small, so these concentration terms do not have a large effect. For example, for methane combustion, a is usually −0.3 and b is 1.3 (Westbrook and Dryer 1984). However, we learned above that flame temperature is strongly related to the proportions of fuel and air (**Figure 4.5**) so the relative concentrations are still important. Because the reaction rate depends so strongly on temperature, it follows a pattern similar to that of flame temperature (**Figure 4.8**). The reaction rate is therefore at its maximum when the fuel and air are in stoichiometric proportions and the flame temperature is at its maximum. If the reactants are fuel rich

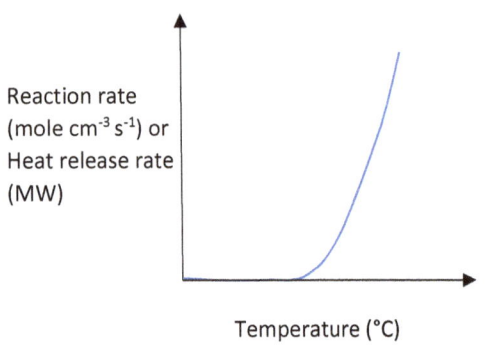

Figure 4.7: Reaction rate and heat release rate of a combustion reaction is strongly dependent on temperature. A distinct bend in the curve indicates where the reaction rate sharply increases.

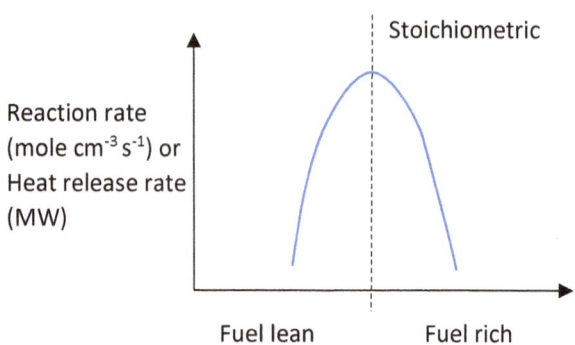

Figure 4.8: Reaction rate and heat release rate depend on the proportions of fuel and air.

or fuel lean, the reaction rate drops dramatically to the point where it *almost* ceases.

The chemical reaction rate of a fire dictates its *heat release rate*. The heat release rate (MJ s^{-1} or MW) is simply the heat of combustion (e.g. MJ kg^{-1}) multiplied by the rate at which fuel is consumed (kg s^{-1}). The heat release rate therefore follows the same qualitative trends as the reaction rate (**Figures 4.7** and **4.8**), and this is what actually drives fire spread. Consider, for example, how the reaction rate and the heat release rate can create an explosion in a mixture of propane and air at room temperature. As discussed above, combustion reactions are actually occurring at room temperature – but very, very slowly. The heat produced from this very slow, dispersed reaction is lost easily and quickly to the surroundings. However, if you were to light a match in the room (please do not!), the temperature of the fuel–air mixture immediately next to the match would be very high. As **Eqn [4.4]** indicates, the reaction rate in that tiny volume of fuel and air would be exponentially higher than the reaction rate in the rest of the room, so the fuel around the match would combust much faster and therefore generate much more heat in that little volume than in the rest of the room. This rapid heat release would not disperse without heating the neighbouring fuel–air volume, raising its temperature significantly and combusting, starting a chain reaction that would sustain and propagate the combustion reaction throughout the room.

Types of flames

There are two basic types of flames. The first is the flame that occurs when the gaseous fuel and air are mixed before burning. Not surprisingly, this type of flame is referred to as a *premixed flame*. Because it is dangerous to keep fuel and air pre-mixed and ready to burn, this type of flame does not occur in many practical situations. It is briefly present inside car engines, however: the fuel and air are injected into the combustion chamber of the engine and compressed before being ignited by a spark plug. Because the engine is running at more than 1000 rpm, the fuel and air are premixed for a very short time. You can see a premixed flame if you open the vents at the bottom of a Bunsen burner (**Figure 4.9a**). Notice the characteristic blue flame in the photograph. As we will discuss in **Chapter 5**, premixed flames are relevant to the ignition process.

The other type of flame, a *diffusion flame* (also called *non-premixed*), is much more relevant to wildland fire behaviour. A diffusion flame occurs when the fuel and air mix at the moment of combustion. This is the type of flame that occurs for all burning solids, be it a candle or wildfire. You can see a diffusion flame if you close the vents of a Bunsen burner (**Figure 4.9b**). The photograph shows its characteristic orange colour. Both types of flames are important in certain contexts of wildland fire.

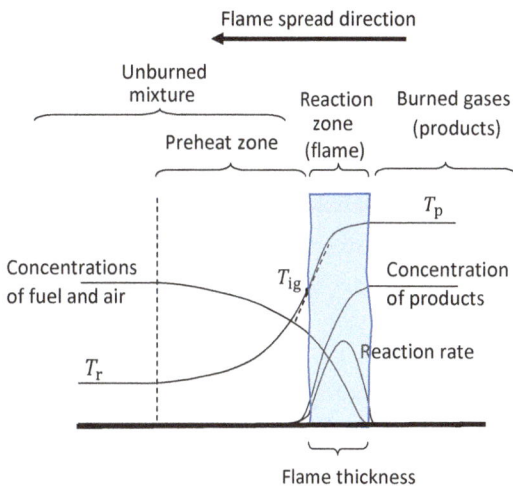

Figure 4.10: Structure of a propagating premixed flame. T_r is the temperature of the reactants, T_{ig} is the ignition temperature of the fuel and air mixture, and T_p is the temperature of the products.

Figure 4.9: A Bunsen burner with vents open (a) produces a premixed flame. With vents closed (b), it produces a diffusion flame. Note the difference in the colour of the flames. Photographs by Ian Grob.

Premixed flames

To understand the combustion dynamics in premixed flames, imagine a container with a mixture of gaseous fuel, say propane, and air. When ignited, a flame front propagates as a thin sheet from the ignition point through the mixture at a particular speed. As shown in **Figure 4.10**, the container holds unburnt mixture on one side of the flame, a preheat zone just ahead of the flame, the flame zone itself, and combustion products behind the flame. The combustion reaction is sustained by the transfer of heat from the flame zone ahead into the unburnt gases, preheating them. The thickness of the flame front is caused by chemical kinetics: reactions take a bit of time, so the flame front has some thickness. The thickness depends on a variety of factors but is typically only a few millimetres. The speed at which the flame front propagates through the mixture is also dictated by chemical kinetics. Both the flame thickness and flame speed are sensitive to the proportion of fuel and air, the temperature of the unburnt fuel and air mixture (reactants), the thermal properties of the gases, and the flow dynamics (laminar versus turbulent). The thermal properties and flow dynamics dictate how the heat from the combustion front gets transferred to the unburnt mixture. The more heat produced by the reaction and the faster that heat can be transferred to the unburnt mixture, the faster the flame can spread. As discussed in **Chapter 3**, convective heat transfer in a turbulent flow is much more efficient than in laminar flow, and flame speeds in turbulent flows are generally much greater than in laminar flows.

When the mixture is either fuel rich or fuel lean, flame temperatures and reaction rates drop dramatically (recall **Figure 4.5**), leading to a drop in flame speed (**Figure 4.11**). At some point in this decline, for both rich and lean mixtures, combustion can no longer be maintained. We call these points *flammability limits*. At the *lean flammability limit*, there is too little fuel for combustion to occur, be sustained and propagate. You have probably experienced conditions outside the lean flammability limit if you have ever smelled a gas leak in your home that did not lead to a fire. At that point, the mixture of fuel and air in the house was below the lean flammability limit and was therefore unable to burn even if a pilot flame was present or a spark occurred. If the leak continued, however, gas could accumulate until there was enough to ignite and sustain flame spread. If the leak continued further and, by some miracle, no ignition occurred, the mixture could pass the *rich flammability limit*, where combustion (once again) could not occur. Oxygen might be present, but beyond the rich flammability limit, there is not enough of it for combustion to occur, be sustained and propagate. Because of the rich flammability limit, you could technically drop a match into a closed container of gasoline on a hot day without causing an explosion (but please don't try it!). Gasoline is very volatile, so it evaporates very easily. When it is in a closed container, the space above the liquid fuel is very fuel rich and thus usually above the rich flammability limit. Because of this limit, urban firefighters use great caution when dealing with gas leaks. If they open a door to a home with a too-rich-to-burn mixture, letting fresh air in, the mixture can become flammable again.

Flammability limits are not fixed points. As shown in **Figure 4.12**, both the lean and rich flammability limits change with temperature. The lean flammability limit tends to decrease as the temperature rises, meaning that less fuel is required to sustain combustion. Alternatively, the rich flammability limit tends to increase as the temperature rises, meaning that less air is required to sustain combustion. This means that a wider range of mixtures can be flammable as the temperature of the mixture increases.

This discussion of flammability is in the context of premixed gaseous fuels. The flammability of solid fuels is more complex. As with premixed gaseous fuels, combustion in solid fuels requires conditions in which they can ignite and sustain a flame.

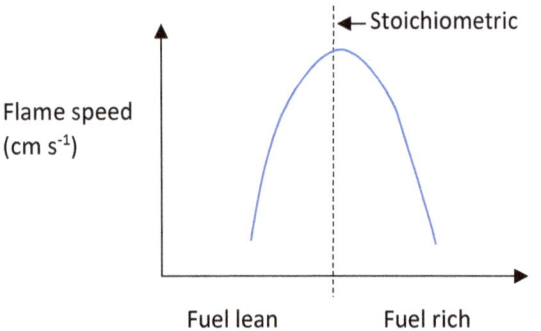

Figure 4.11: Variation in the speed of a flame propagating though a premixed fuel and air mixture as a function of the proportions of fuel and air.

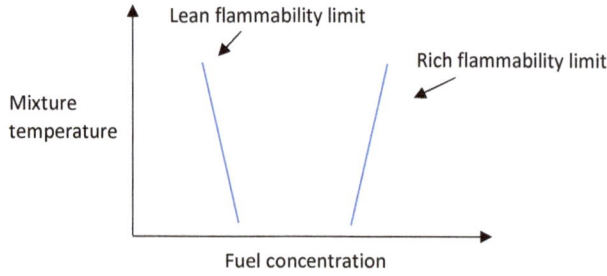

Figure 4.12: Change in flammability limits with the temperature of the fuel–air mixture.

In addition, combustion of solid fuels typically depends in part on the heat release rate and, occasionally, on the proportion of the solid fuel actually consumed. Therefore, the flammability of wildland fuels is said to consist of *ignitability*, *sustainability* and *combustibility* (Anderson 1970), with some adding *consumability* (Martin *et al.* 1993). As discussed earlier, the heat release rate and the proportion of fuel burned in flaming combustion of solid fuels are not as straightforward as they are in combustion of premixed gases and are discussed further in the next section.

Non-premixed or diffusion flames
Basic structure and characteristics

A candle is a classic example of a diffusion (i.e. non-premixed) flame. A candle flame has the same structure as the flames in a wildland fire, although the process of generating gaseous fuel is different. As described in **Chapter 2**, when you light a candle, it takes a few moments to catch fire. That is because the heat from the match or the lighter must first melt the solid wax. Once melted, the wax travels up the wick via capillary action, where it vaporises. Once vaporised, the fuel reacts with the air to produce a flame. It is the vaporised wax, not the solid or liquid wax, that burns in the flame. Once the candle is lit, the heat from the flame continues to melt and vaporise the wax and thus provides the fuel needed to sustain combustion.

Unlike the premixed flames discussed in the previous section, the fuel vapour that burns in a candle is not mixed with the air before burning. As illustrated in **Figure 4.13**, the mixing of fuel and air happens right at the flame front by gas diffusion (hence the name). With abundant fuel on the inside of the flame and abundant air on the outside, the combustion reaction always takes place at stoichiometric conditions. Because of this, the region inside of the flame ideally has no air and the area outside of the flame has no fuel. As discussed in the previous section, combustion cannot occur outside the range of the flammability limits, so the actual flame is a hollow shell with some minimal

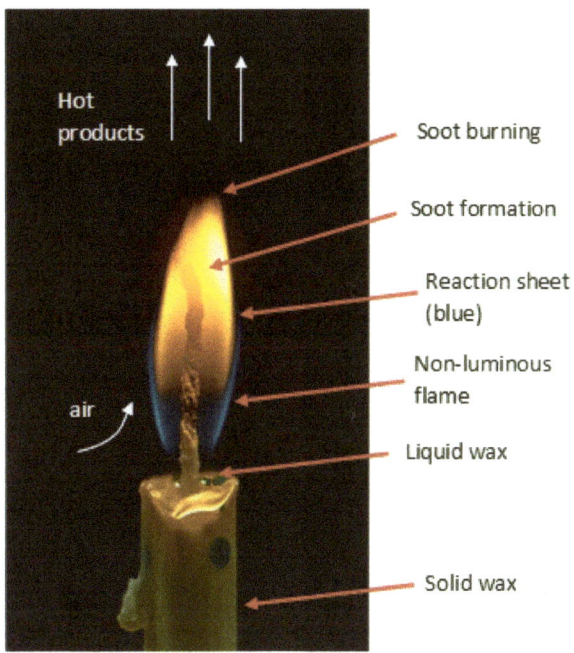

Figure 4.13: Structure of a non-premixed (diffusion) flame. Photograph by Ian Grob.

thickness. This can be demonstrated by using a wire screen or mesh to *quench* the flame (i.e. stop combustion) (**Figure 4.14**).

The source of a diffusion flame's characteristic orange colour is actually carbon (i.e. soot). As the gaseous fuel travels from the source (wick), it is continually heated by the flame and hot combustion products. This breaks apart the fuel molecules into smaller and smaller molecules, stripping off hydrogen and oxygen atoms and leaving more and more pure carbon behind. When the carbon gets hot enough (~600–800 °C), its atoms begin to vibrate and produce orange and yellow light. As the carbon then moves into the flame front itself, most of it is consumed in the combustion reaction. This means that the orange and yellow of the flame originates behind the flame front. The front of the combustion reaction itself is actually faintly blue, but it is much dimmer than the bright orange and yellow emanating from behind it, so it is normally not noticed. Down at the base of the flame, however, little soot has been produced and more air is mixed with the wax vapour, so the blue of the actual combustion reaction can be seen (**Figure 4.13**). If you look

Figure 4.14: By using a wire mesh, you can quench the flame (stop combustion) to observe its structure. Photographs by Ian Grob.

effect is called buoyancy. It stretches the flame upward, giving it the familiar shape. The hot gases rise, drawing fresh air into the flame and transporting the combustion products away.

Temperatures vary dramatically from the centre of a flame to the air outside it (**Figure 4.15**). The temperature at the centre, near the wick, is elevated because combustion products and heat (which are being produced in the outer shell of the flame) are diffusing and being conducted inward. The temperature continues to increase as the edge of the flame is approached. Within the flame shell itself, the temperature rises dramatically due to the combustion reaction. This is the hottest part of the flame. Just outside the flame, the temperature remains elevated as hot combustion products move away and heat is conducted outward through the gases. Because non-premixed flames are always burning at stoichiometric conditions, the temperature in the flame's shell is relatively constant. However, it can be affected by soot production. Heavily sooty, bright orange flames (such as those from burning heavy hydrocarbon fuels like heptane (C_7H_{16})) give off not only orange light but also radiant heat, so they lose heat more effectively

carefully at a candle flame, you may see a tiny wisp of black smoke floating out the top. This is soot that has escaped the flame and cooled off, so it is no longer emitting light. Combustion is never completely perfect (i.e. it never fully oxidises every atom of carbon), so tiny bits of soot manage to escape through the flame unconsumed. In the flames of wildland fire, you can see much more soot, largely due to the highly turbulent nature of the flames. This is discussed further below.

As discussed in **Chapter 2**, the characteristic 'teardrop' shape of a flame is actually due to gravity. The high-temperature gases in a candle flame are much less dense than ambient air, so they float upward in the presence of gravity (**Figure 2.3**). This

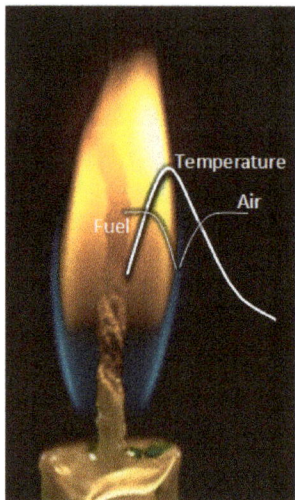

Figure 4.15: Schematic diagram overlaid on diffusion flame of a burning candle. Lines show relative temperature, fuel concentration, and air concentration from the centre to the outside of the flame. Photograph by Ian Grob.

than less sooty, bluish flames (such as those from burning methanol (CH_3OH)). Bright orange flames can therefore have lower temperatures than blue ones. In crown fires, however, the soot produced in the flames creates thick black smoke that has the opposite effect. It can actually help shield the flames from losing heat through radiation.

As flames get larger, they go from being laminar to chaotic and turbulent (as length scales increase, so do the Reynolds and Grashof numbers) (**Figure 4.16** and **Chapter 3 Eqns [3.32] and [3.34]**). Candle flames are typical laminar flames. They do not fluctuate much, and the flame and flow of hot gases are smooth and predictable. The flames in wildland fire, particularly in small prescribed fires, can be laminar at the base, but they often become turbulent towards the tips because the flow inside the flames accelerates due to buoyancy. Generally, we see a transition from laminar to turbulent when flames reach ~10–20 cm in height, but the transition can vary with other conditions too. For instance, if a turbulent wind is blowing. Flames in crown fires are often very turbulent. They are jagged, chaotic and fluctuate rapidly. Because of turbulence, the structure of flames in wildland fires is not like the idealised hollow teardrop of a burning candle.

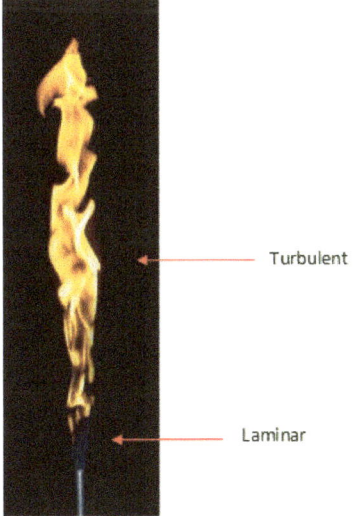

Figure 4.16: Flame from a burner illustrating the transition from a laminar flame at the base to a turbulent flame at the top. Photograph by Ian Grob.

While the candle flame's structure (fuel on the inside, oxygen on the outside with a thin flame sheet in between) remains, turbulence folds everything together, wrinkling the flame sheet considerably. The inside of a wildland flame (or other turbulent flame) is therefore full of 'flamelets' that combine to form what we observe as a large, continuous flame. As we will see, this has important consequences for the temperature and vertical velocity inside the flames of wildland fires.

The turbulence of flames in wildland fires can actually slow or stop combustion. Crown fires, which often produce large columns of very black smoke, provide an example. Although we do not fully understand the causes of this black smoke, the turbulent nature of the flames provides some explanation. When flames are highly turbulent, there can be small pockets where the flow speeds of the gaseous fuel and air are very high as the flame sheet is wrinkled and folded, so there might not be time to fully combust the soot and gaseous fuel. In addition, relatively cool ambient air can be folded into the flames, reducing their temperature and slowing combustion to the point that combustion is incomplete or is entirely stopped. Furthermore, crown fires burn so intensely, consuming such large amounts of fuel, that there may not be enough air deep within the flame to burn the fuels completely. (Recall from **expression [4.2]** that much more air than fuel is required for complete combustion.) Crown fires also release large amounts of water vaporised from the live vegetation, which can further quench combustion. When the flames are locally quenched or combustion is incomplete, there is no chance to consume the soot that is formed. Under these conditions, large amounts of unburnt fuel and soot are lifted upward along with the products of combustion, contributing to the characteristic black smoke produced by crown fires.

Puffing

Most fires exhibit a phenomenon called *puffing*. We can see it in the flickering of a candle flame, but it is difficult to see in wildland fires unless the wind is calm. As illustrated in **Figure 4.17**, the puffing of a fire is observed as a periodic widening of the base of the flame that propagates towards the tip. Puffing is caused by fluid dynamic instability, which arises because there is a large velocity difference between the hot, rising flame gases and the stagnant, cold air surrounding the flame. This velocity difference causes a mushroom-shaped vortex to form, indicated by the widening of the flame (**Figure 4.17**). This vortex is then shed upward as the hot gases rise. As the vortex moves upward, the flame base narrows, or 'necks in'. Another vortex can only form once the previous one is completely out of the way.

Puffing is a periodic phenomenon that occurs at a regular and predictable frequency. It is strictly a fluid dynamic instability due to buoyancy, which means that it is related only to the difference in density between the combustion gases and the ambient air. In fact, this puffing phenomenon can even be seen in helium plumes, with no flame at all (albeit at a different frequency because of differing amounts of buoyancy). Since the temperature of most hydrocarbon diffusion flames is about the same (~1000 °C) and thus the buoyancy of the flames is fairly consistent, the frequency of puffing for these fires is related only to their diameter. Experiments have shown the relation between the frequency of puffing and the diameter of a circular fire is about:

$$f = 0.5(g/d)^{1/2} \qquad [4.5]$$

where f is the pulsation frequency (hertz, or cycles s^{-1}), g is the acceleration due to gravity (9.8 m s^{-2}), and d is the fire diameter (m). This relationship demonstrates that the larger the fire, the smaller the frequency. For example, a fire with a 1 m diameter pulsates ~1.5 times per second. However, a fire with a 100 m diameter only pulsates ~0.15 times per second (or once every 6.4 s). Observations of smoke plumes in large crown fires (see **Chapter 8, Pulsating or puffing**) show around one puff every 20–30 s, indicating that the fires have a diameter of around 1 km (**Figure 8.18**). Notice that the acceleration due to gravity is present in this equation – fires in space do not pulsate!

Flame height

The height of a diffusion flame is largely controlled by how far the fuel has to travel to find sufficient

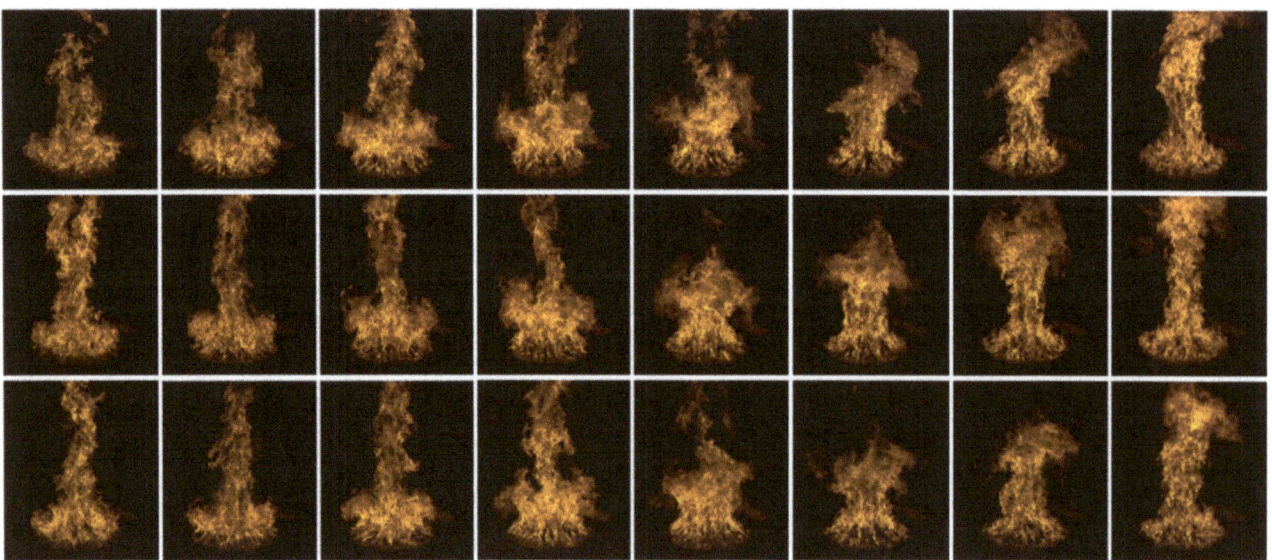

Figure 4.17: Pulsation of an ethanol pool fire (d = 1 m). The time between photographs is 0.1 s. Each row corresponds to a puff cycle, with the puffs originating at the flame base in the left images and moving up to the flame tip in the right images. Photographs by Ian Grob.

Figure 4.18: Gas flames surrounded by different gas mixtures. On the left is air (21% O_2/79% N_2), middle is 50% O_2/50% N_2, and right is 100% O_2. Reproduced with permission from Lee *et al.* (2000).

oxygen to burn. By playing with a Bunsen burner (with the vents closed), one can examine the variables controlling flame height. One parameter that can be changed is the diameter of the burner. If gas velocity is held constant, the wider the burner, the taller the flame because the fuel in the centre of the burner has to go further to encounter enough oxygen to burn. Another important parameter is the composition of the surrounding gases. A flame is much shorter in pure oxygen than in air (which is 21% oxygen), because the fuel does not have to travel as far to encounter enough oxygen to combust (see **Figure 4.18**). A third parameter is the flow velocity of the fuel. The faster the flow velocity of the gas jet, the taller the flame because the fuel is propelled further before it can all combust. This effect is limited, however, by turbulence. If the gas jet and the flame become turbulent, the flow will mix more efficiently with the ambient air than if the flow remains laminar (and thus mixes only by diffusion). Turbulent flames can pull in and mix air at a rate proportional to the gas jet velocity, so the faster the gas jet, the more the air that gets pulled in and mixed. This reduces the flame height compared to the laminar flame, so in turbulent conditions flame height is independent of gas jet velocity (**Figure 4.19**). If the gas jet velocity is too high, however, the flames will lift off the burner (as in **Figure 4.6**), and with further increases in flow rate, the flames will blow off entirely and extinguish.

For flames from solid fuels, such as those in burning candles and wildland fires, the fuel flow rate is not controlled as easily as that of a Bunsen burner. The fuel flow rate is the rate of pyrolysis, which is controlled by heat feedback from the flames. Additionally, the gas velocity is not a fixed constant that we can control, as is the velocity of gas jets in a Bunsen burner. Instead, the gaseous fuel accelerates as it moves upward from the solid fuel. Flames in wildland fires are therefore said to be 'buoyancy driven' as compared to the jet flame from a Bunsen burner, which is 'momentum driven'. The basic concept that the flame height is dictated by how far the gaseous fuel must travel to encounter sufficient oxygen to burn applies to flames from solid fuels as well as gaseous jets. Because the process is so complicated for wildland fires, relationships for predicting the flame height are not derived from first principles but instead are semi-empirical – that is, equations are derived by measuring the parameters important to the process and fitting an equation to the resulting experimental data.

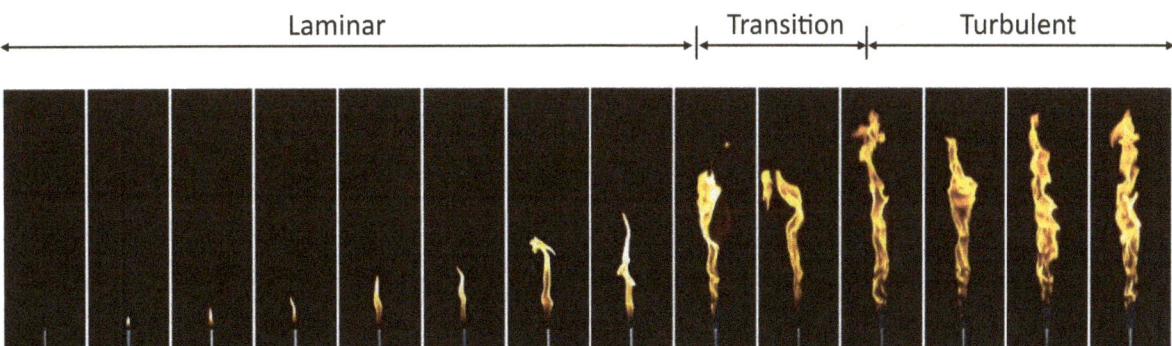

Figure 4.19: Variation of flame height as fuel flow velocity increases. Shows the change in flame behaviour from laminar to turbulent. Photographs by Ian Grob.

Determining flame height in real fires is complicated by the pulsation and flickering of flames. Flames are often said to have a continuous region down at the bottom, an intermittent flame region in the middle, and a smoke plume above, which contains smoke but not actual combustion reactions (**Figure 4.20**). Where exactly is the 'top' of the flame? Accurate measurement of flame height depends on the criteria used to answer this question. This assessment is often done visually (especially in wildland fire), but more robust laboratory methods are available. These use either temperature measurements made using thermocouples or intermittency measurements made using visual video footage. The video footage is processed using a computer algorithm that uses a brightness threshold to convert each frame into a black-and-white image of flame v. no flame. These laboratory methods remove some of the human judgement that complicates measurement of flame height, but they still rely on an arbitrary threshold for deciding where the top of the flame is located (e.g. one criterion might be that the flame temperature drops below 600 °C 50% of the time, or that a threshold brightness is reached 50% of the time). Regardless of how the top of the flame is determined, it is clear that the method used must be clearly stated in any reports, and results from one method cannot be directly compared to those from another method.

There are additional complications in measuring flame height in spreading wildland fires. The variation of height along the span of a line fire is exaggerated due to the peak-and-trough nature of the flow dynamics of a spreading fire (see **Chapter 6, Wind in the presence of fire, Figure 6.20** and **Chapter 9, Flame zone properties**). Because it takes some time for the fuel to burn out, the flaming zone has some depth (**Figure 4.21**), but scientists do not yet agree on whether or not flame zone depth should be included in flame height measurements – or how to do so, if it is needed. A further complication is the fact that flames are usually tilted by the wind (see **Chapter 6**). For this reason, flame length is often used in wildland fires, not flame height. An even further complication is how to determine flame height when the fuel bed itself has significant

Figure 4.20: Flame diagram illustrating the continuous and intermittent flame zones. Above the flaming region is the smoke plume. Photograph by Ian Grob.

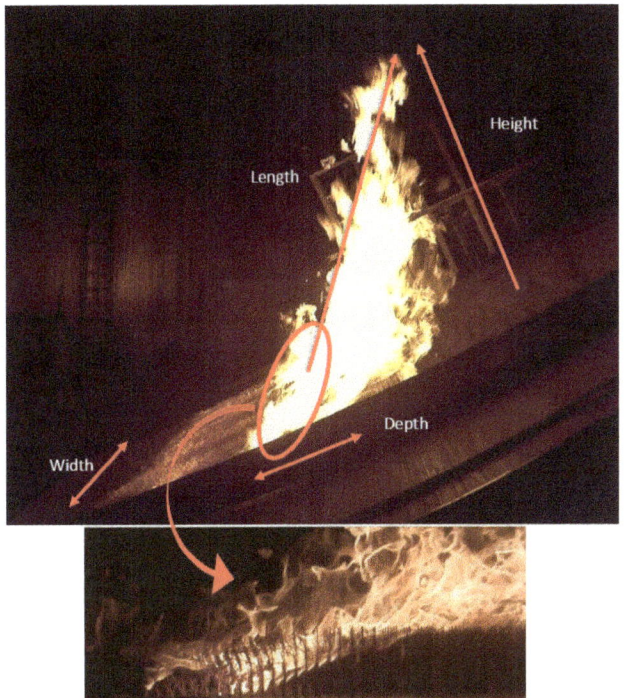

Figure 4.21: Flames in wildland fires have depth, length, and height. As discussed in **Chapter 6**, the width of the fire can also be very important.

height, such as in a crown fire. Does the flame height start at the bottom or the top of the fuel bed? All of these complicating factors make it essential that, in reporting flame height, the measurement methods also be reported, and that results from one method not be directly compared to results from another.

There are several different ways in which to indirectly estimate the height or length of flames in wildland fires. One of the most common methods for surface fires is the equation from Byram (1959). In metric units:

$$\ell_f = 0.0775 \dot{q}'^{0.46} = 0.0775 \left(H_c m_c'' r \right)^{0.46} \quad [4.6]$$

where ℓ_f is the flame length (m), $\dot{q}'$ is the fireline intensity (kW m^{-1}) of a line fire, H_c is the net low heat of combustion (kJ kg^{-1}), m_c'' is the fuel consumed in the active flaming front (kg m^{-2}), and r is the rate of spread (m s^{-1}) (see also **Chapter 2, Eqn [2.12]**). In this equation, the flame length begins at the *top* of the fuel bed and extends to the tip of the visible flame as measured by eye[vii] (**Figure 4.21**). One of the main challenges in using Byram's equation is determining the amount of fuel consumed in the flaming front. Simply using measured pre- and post-fire fuel loadings will grossly overestimate the amount of fuel consumed in some cases because of the variation in the amount of fuel consumed in smouldering combustion behind the flaming front. The amount of fuel consumed in flaming combustion varies with the moisture content, fuel bed properties (spatial arrangement, fuel element size, etc.), and burning environment (e.g. wind, surface fire *v.* crown fire, size of fire). An added challenge in using Byram's equation is the appropriate value for the heat of combustion. The idealised value measured with a bomb calorimeter is often used, but, as described earlier, this is not realistic, mainly because of incomplete combustion. Furthermore, multiple fuel components are usually consumed in wildland fires (duff, litter, woody fuels, live fuels), and each of these may have its own value for heat of combustion. How much each fuel component contributes to the overall heat of combustion depends on the proportion consumed, which depends on the burning conditions. Needless to say, it is very difficult to obtain measurements of either the fuel consumed or the heat of combustion in realistic or field conditions (see **Chapter 9, Combustion and heat release, and Fuel consumption** for further discussion).

The equation commonly used to indirectly estimate flame length in crown fires is that developed by Thomas (1963) for wood cribs (see also **Eqn [2.13]**), which are simply ordered cross-piles of sticks (discussed later, see **Figure 4.26**). In metric units:

$$\ell_f = 0.027 \dot{q}'^{2/3} = 0.027 \left(H_c m_c'' r \right)^{2/3} \quad [4.7]$$

Although the general form of this relationship was derived from theory, the coefficient (0.027) was found by fitting experimental data; hence the equation is semi-empirical. The data were taken in no-wind conditions, in which flame length and flame height are equal, so the equation really predicts the height of the vertical flames from a line fire. Flame height as used in this equation begins at the *bottom* of the fuel bed and extends to the top of the flames as determined by photographs (which, incidentally, differ from measurements made visually). As with Byram's relationship (**Eqn [4.6]**), Thomas's relationship requires that we know the amount of fuel consumed in the flaming front and the appropriate heat of combustion.

The literature reports many more relationships that correlate flame length with fireline intensity using a power-law relationship similar to those mentioned above (see, e.g., Alexander and Cruz 2012). Much of the variation among them is probably due to differences in the technique used to measure the flame length (e.g. by eye, photograph, video, temperature) and what 'length' is actually referred to (e.g. tilted length, vertical height, whether or not the depth of fuel is included). Differences are also due to variation in fuel types and burning conditions (see discussion in **Chapter 9, Flame zone properties**). Note that **Eqns [4.6]** and **[4.7]** are for flames from line fires. Expressions with different exponential dependencies on burning intensity (i.e. heat release rate) also exist for axisymmetric fires, in which the air comes in from all sides, such as would occur for an individual torching tree (see, e.g., Quintiere and Grove 1998). **Eqns [4.6]** and **[4.7]** are for unrestricted, free-burning flames. If air is restricted from entering the flame zone, from one or more sides as described in **Chapter 6, Multiple flame zones and air-flow interactions**, the flame length can increase. Air flow can be restricted by the presence of flames from nearby fires or by solid obstructions from topography or perhaps tree stems. As flame zones from multiple fires begin to merge, the flames in the junction zone tend to grow taller both from the combined effect of increased combustion products within the conjoined flame zone as well as air-flow restrictions at the intersections of the various fires.

Temperature profiles, vertical gas velocity and entrainment

As discussed in **Chapter 3, Convection heat transfer**, there are two important variables about the flow of a fluid that determine the rate of convective heat transfer to a solid surface: the temperature and the velocity of the fluid. In wildland fires, the flowing fluid consists of the flame and the hot combustion products in the smoke plume above it. It is therefore important to understand how temperature and fluid velocity vary in and above a fire. As described in the previous section and illustrated in **Figure 4.20**, there are really three regions of interest: the continuous flame region, the intermittent flame region, and the non-reacting smoke plume above. Both temperature and vertical velocity vary in different ways in each of these regions.

As was described earlier, turbulence folds the sheets of flame together so the continuous flame region of a typical wildland fire is not hollow like a candle flame but filled with flamelets. Because of this, the time-averaged temperature at the centre of the continuous flame region is largely constant from the bottom to the top of the region. As illustrated in **Figure 4.22**, the centreline temperature begins to fall in the intermittent flame region, continuing its decay at an even faster rate in the smoke plume above. **Figure 4.22** demonstrates a common method of examining this *centreline temperature* profile. The x-axis is the height above the base of the flame divided by the heat release rate of the fire (HRR) to the 2/5 power. This x-axis scaling was found empirically by McCaffrey (1979) to collapse the data and aid in developing correlations. The y-axis is the temperature difference between the centreline temperature and the ambient temperature. The graph shows that the centreline temperature profiles for the three regions follow these trends (McCaffrey 1979):

1. $\Delta T \propto z^0$ in the continuous flame region
2. $\Delta T \propto z^{-1}$ in the intermittent flame region [4.8]
3. $\Delta T \propto z^{-5/3}$ in the non-reacting smoke plume.

The symbol $\propto$ means 'proportional to'. These relations confirm that the centreline temperature should remain constant in the continuous flame regime, decline gradually in the intermittent flame regime, and decline more rapidly in the

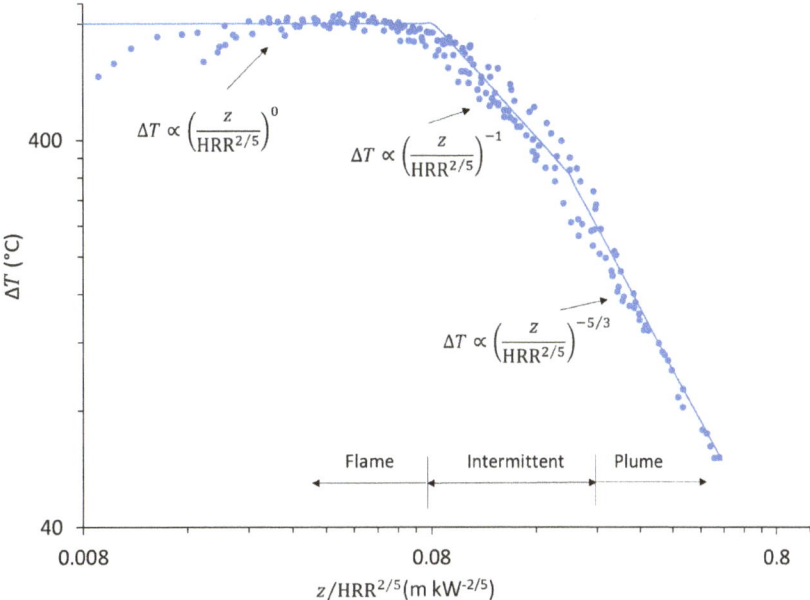

Figure 4.22: Relationship between temperature along the centreline of flames (relative to ambient temperature) and height from the base (scaled) of a small circular fire, showing the gases cooling as they rise in the plume. Height from the base of the fire is scaled (divided) by the heat release rate (HRR) to the 2/5 power. Adapted from McCaffrey (1979).

non-reacting smoke plume. Interestingly, these trends were originally noted for relatively small gas burners, in which the base of the plume is relatively narrow compared to the width of the plume (called 'point source' plumes) (Thomas 1963). It turns out that these trends also hold for much wider fire sources if we treat the base of the plume as if it occurs at a different height than reality, and very similar trends exist for infinitely long line fires (Quintiere and Grove 1998).

The average vertical velocity of the rising gases at the centre of the fire is related to this vertical temperature profile. As pyrolysis gases are produced and emerge from the solid fuel, their velocity is typically very small. As the gases heat up and burn, they become less dense than the surrounding air and accelerate upward due to buoyancy. The maximum vertical velocity is thus right at the top of the continuous flame region. Some very high upward velocities have been reported – up to 15 m s^{-1} in grass fires and 60 m s^{-1} in crown fires (Clark et al. 2005; Coen et al. 2004). As the temperature starts to fall in the intermittent flame zone, the acceleration slows, so the upward velocity remains relatively constant in this region. In the non-reacting smoke plume, the temperature drops more rapidly (**Figure 4.22**), as more and more slow-moving air from outside the plume gets pulled in and slows the rise of the gases. **Figure 4.23** demonstrates a common method for examining the *centre line vertical velocity* profile. The *x*-axis is the same as that in **Figure 4.22**. The *y*-axis shows the centreline vertical velocity divided by the heat release rate (HRR) to the 1/5 power. This scaling was also determined empirically by McCaffrey (1979). As in our examination of the centreline temperature profile above, the graph shows that the centreline vertical velocity (w) for the three regions follows three trends (McCaffrey 1979):

1. $w \propto z^{1/2}$ for the continuous flame region
2. $w \propto z^0$ for the intermittent flame region [4.9]
3. $w \propto z^{-1/3}$ for the non-reactive smoke plume.

These relations confirm that the centreline vertical velocity increases in the continuous flame region, is constant in the intermittent flame region, and declines in the non-reactive smoke plume. As with the temperature profiles, these trends were

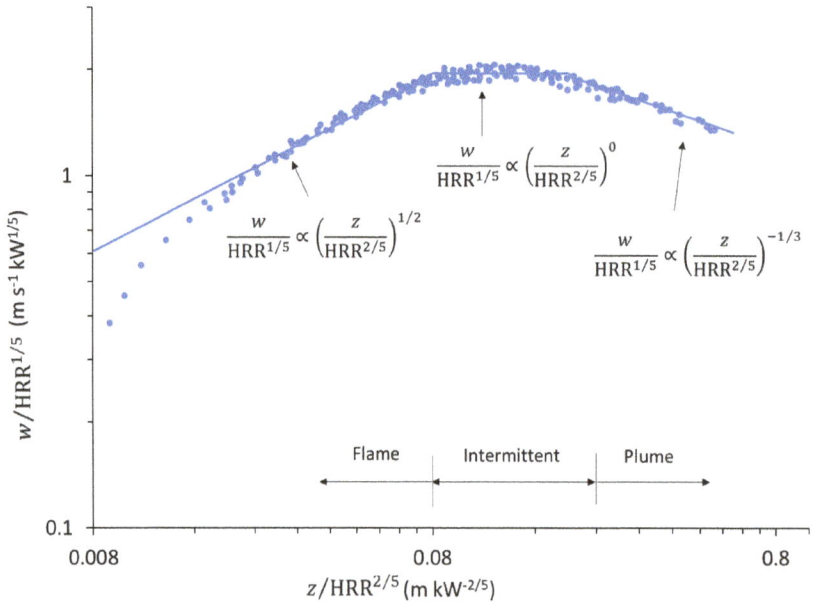

Figure 4.23: Scaled centreline vertical velocity profile with height from the base of the flame (scaled) showing that the gases slow as they rise in the plume. Note that velocity is scaled by the heat release rate (HRR) to the 1/5 power and that height is scaled by the (HRR) to the 2/5 power. Adapted from McCaffrey (1979).

originally observed with small point source gas burners, but they can be adjusted to work with area fires if we treat the base of the plume as if it occurs at a different height than reality, and to infinitely long line fires by using a different exponent in the smoke plume region (Quintiere and Grove 1998).

As we have seen, as gases rise in the flame and smoke plume, temperatures drop and gas velocity declines. This occurs because of a phenomenon called *entrainment*. Entrainment is the process of dragging and mixing fresh air into the plume. Everything has friction, including air. As the flames and hot combustion products rise (because of buoyancy), friction causes the surrounding air to be dragged along with them. Turbulence mixes this dragged-along cooler air into the plume. As the surrounding air is entrained, it cools the plume. Because the temperature (and therefore density) difference between the interior and exterior of the plume drives buoyancy, cooling due to entrainment reduces buoyancy, slows the rise of the plume, and ultimately stops it. Entrainment explains why you can hold your hand above a candle at a reasonable distance above it: cool air has mixed with the hot gases in the plume until they have cooled and are slowed down, and they are no longer hot enough to burn your skin. In wildland fires, several circumstances can restrict entrainment into the plume, so temperatures remain high at a greater distance above the fire and the velocity of rising gases increases. All of these circumstances can have dire consequences for fire fighters. They will be discussed further in **Chapters 6** and **8**.

Note that the relationships shown in **expressions [4.8]** and **[4.9]** and **Figures 4.22** and **4.23** provide an idealised description of the time-averaged temperature and vertical velocity in a flame and smoke plume. However, blobs of hot gas and flames in the intermittent region can introduce substantial variation in these patterns, and this variation can be critical to understanding how wildland fires spread in fine fuels. We will explore this in more detail in **Chapters 6** and **7** with respect to convection heat transfer.

Burning and heat release rate

The *burning rate* is the rate at which the fuel is consumed (kg s^{-1}). For gaseous fuels, it is simply the reaction rate discussed earlier because the gaseous fuels are already in a state where combustion can occur. For solid fuels, the actual consumption of the gas-phase fuels is not the limiting factor; instead, the production of the gaseous fuels is. A solid fuel's burning rate is therefore measured by the change in mass of the solid with time and is intimately related to the amount of heat that is transferred to the solid fuel.

A fire's *heat release rate* is directly related to the burning rate:

$$\text{HRR} = \dot{m}_b H_{c,\text{eff}} \quad [4.10]$$

where HRR is the heat release rate (MJ s^{-1} or MW), $\dot{m}_b$ is the burning rate (kg s^{-1}) and $H_{c,\text{eff}}$ is the *effective* heat of combustion (MJ kg^{-1}). Knowledge of the heat release rate and how much of that heat is transferred to the unburnt fuel ahead allows us to predict a fire's spread rate. As described in **Chapter 2** and above for gaseous fuels (see the **Premixed flames** section), heat release rate is an important consideration for fire safety.

Another important application of the burning rate in understanding wildland fire behaviour is its relationship to the flame residence time. The flame residence time generally means the time the flaming front is present at a particular location. It is related to the burning rate by:

$$t_r = \frac{m_c}{\dot{m}_b} \quad [4.11]$$

where t_r is the flame residence time (s), m_c is the mass of fuel consumed in the flaming front (kg), and $\dot{m}_b$ is the burning rate (kg s^{-1}). Flame residence time is related to other fire behaviour metrics, including flame zone depth (see **Chapter 2, Eqn [2.1]** and **Figure 2.11**) where for a given spread rate, the longer the residence time, the deeper the flame zone. In addition, it is important for predicting fire effects, since it helps to quantify 'dosage' of heat in the flame zone to above ground plant organs such as plant stems and foliage.

The burning rate of a solid fuel depends on many parameters, including the rate of heating, spatial arrangement of fuels, particle size, wind, and moisture content. As discussed above, the fuels consumed by the flame are the gaseous products from the thermal degradation of the solid – that is, from pyrolysis. The pyrolysis rate is thus the burning rate of the solid. Because the pyrolysis rate is driven by temperature, the rate at which the solid is heated drives the burning rate. It follows then that any factors that influence the rate of heating will also influence the burning rate. For this reason, individual fuel particles burn differently than particles of the same size arranged in a fuel bed. An individual particle is heated only by the flame produced by the particle itself. If multiple particles are arranged in a fuel bed, however, they can be heated by the flames around the other particles and radiant heat from the surfaces of the surrounding hot particles. If you have ever built a campfire, you know that this 're-radiation' inside a fuel bed is very effective: you cannot get large logs to burn on their own, but when you arrange them to face each other, they will burn very well.

Because the burning rate depends on the rate of heat transfer into the solid fuel, it varies over time. As **Figure 4.24** illustrates, when the fuel bed is first ignited, the flames are small and the surfaces are still cool, so the burning rate starts off slow. As the flames grow larger and the surfaces heat up, more heat is transferred into the fuel, so the burning rate increases until all of the fuel is fully involved in the fire. At that point, the burning rate peaks. Some time later, the fuel supply begins to run out, the flames shrink, and the burning rate decreases until flaming ceases.

Experience with campfires demonstrates that the spatial arrangement of the fuel elements – how densely they are packed – will also influence the burning rate. Fuel beds that are very densely packed will not burn well because not enough air is available to react with the fuel vapours – that is, the gaseous fuel–air mixture inside the fuel bed is too *fuel rich*. Fuel beds like these are said to be *ventilation limited*. If you start with a ventilation-limited fuel bed and then increase the spacing between fuel elements, you will allow more air in and thus increase the burning rate. If you continue to

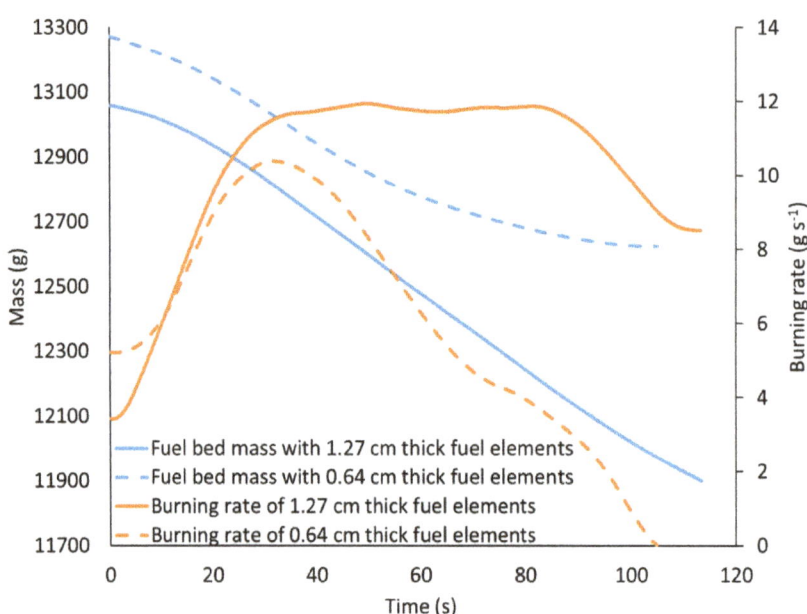

Figure 4.24: Fuel consumption and burning rate over a period of time. Blue lines indicate fuel mass and are associated with the left axis. Orange lines indicate burning rate and are associated with the right axis. Fuels burn slowly at the start of a fire. The burning rate increases as all of the fuel becomes involved, reaches a maximum, then decreases as the fuel begins to burn out. Fuel beds with thicker fuel elements (solid lines, fuels 1.27 cm thick) can maintain the maximum burning rate longer than fuel beds with thinner fuel elements (dotted lines, fuels 0.64 cm thick).

increase the spacing between fuel elements, you will reach the stoichiometric condition, where the air supply is sufficient to burn all of the gaseous fuel. This is called a *loosely packed* fuel bed. From this point, relatively small increases in the spacing of fuel elements will not affect the burning rate. However, as the spacing between particles continues to increase, less and less heat is transferred between fuel elements. Eventually, heat transfer declines to the point that the fuel bed cannot sustain burning. This is illustrated in **Figure 4.25**, where the burning rate in densely packed fuel beds increases with increasing porosity factor until a plateau is reached. The burning rate then remains constant in loosely packed fuel beds until heat transfer among individual elements declines. Then the burning rate declines until the fuels can no longer sustain combustion. **Figure 4.25** presents just one method of characterising effect of fuel bed packing on burning rate. Other methods (such as in Anderson 1990) condense the curve horizontally, giving the appearance of an *optimum packing ratio*, at which the burning rate is maximised.

Figure 4.25 was generated using a research fuel bed called a wood crib (**Figure 4.26**), which consists of ordered cross-piles of sticks. A wood crib

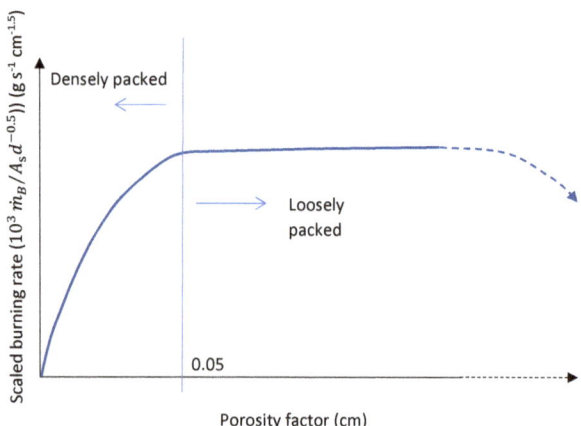

Figure 4.25: The burning rate of a fuel bed depends on the spatial arrangement of the fuel – how densely the particles are packed. Porosity factor is defined in **Eqn 4.12**.

allows for control of many aspects of the fuel bed structure, including fuel element thickness and length, spacing between fuel elements, and height of the bed.

There are several ways to quantify how densely a fuel bed is packed. The fire protection engineering community uses the term *porosity factor* to describe how densely a fuel bed is packed, and that is what we use here and in **Figure 4.25**. For wood cribs, porosity factor is defined as (Heskestad 1973):

$$\phi = s^{1/2} d^{1/2} \left(\frac{A_v}{A_s} \right) \quad [4.12]$$

where ϕ is the porosity factor (cm), s is the horizontal distance between fuel elements (cm), d is the fuel element thickness (cm), A_v is the area of the vertical shafts as seen from above ($s^2(n-1)^2$ where n is the number of sticks per layer) (cm²), and A_s is the exposed surface area of the fuel (cm²). This is the independent variable used for the x-axis in **Figure 4.25**.

Porosity can be defined in other ways. In the wildland fire literature, it is often defined as λ (m), the ratio of fuel bed volume to fuel surface area. Among wildland fire practitioners, however, porosity is often described by *bulk density*:

$$\rho_b = \frac{m_f''}{\delta} \quad [4.13]$$

Figure 4.26: Wood crib used to study how fuel bed parameters affect burning rate.

where ρ_b is the bulk density (kg m⁻³), m_f'' is the dry fuel load (kg m⁻²), and δ is the fuel bed depth (m). However, the total dry fuel load can be inappropriate for describing what happens in the flaming zone because not all of the fuel is consumed in the flaming front. The *effective* fuel load may be more relevant in real fires. Unfortunately, porosity factor cannot be easily linked to bulk density as defined in **Eqn [4.13]**, but both properties can be useful. While the porosity factor includes more detail about the fuel bed and allows for a more thorough understanding of the factors that affect the burning rate, bulk density is much more easily measured in wildland fuels, especially in field conditions.

Heat transfer within the fuel particle itself also plays a role in controlling the particle's burning rate. If the fuel particles are thin and are experiencing a relatively slow rate of heating from the surface, conduction will transfer heat well enough through the solid to keep the temperature inside the fuel particle relatively uniform. That is, the fuel particle surface temperature and centre temperature will be about the same (see discussion about *thermally thin* in **Chapter 3**). In these cases, the larger the diameter of the fuel, the more volume is at a sufficiently high temperature to pyrolyse, and greater production of pyrolysate means that the burning rate will increase with the fuel diameter. However, if the fuel particles are thick or surface heating is very rapid, there will be a temperature gradient within the fuel particle, and the inside of the fuel will be noticeably cooler than the surface (see discussion about *thermally thick* in **Chapter 3**). There will be an outer layer of material that is hot enough to pyrolyse, and this hot region will propagate down into the solid as time goes on. The burning rate will thus be dictated by how fast the hot region moves into the solid. Because there are two heating regimes – thermally thin and thermally thick – the variation in burning rate (and thus the flame residence time via **Eqn 4.11**) with fuel particle diameter should have two different forms.[viii]

For thick fuels, the burning rate is proportional to the square root of the thickness ($\dot{m}_b \propto d^{0.5}$), so burning rate increases with thickness – but not quickly. Let us consider how this pattern differs from burning rate scaled by (divided by) particle surface area. For a square stick, the surface area is four times the stick thickness times the length (L) ($A_s = 4dL$). If burning rate is divided by surface area (($\dot{m}_b/A_s = \dot{m}_b''$), it is proportional to one over the square root of the thickness ($\dot{m}_b'' \propto d^{-0.5}$). This is why the y-axis of **Figure 4.25** scales the burning rate per surface area with $d^{-0.5}$: the burning rate of the loosely packed fuel beds becomes constant. We can use **Eqn [4.11]** and the knowledge that the mass of fuel is the density times the volume (d^2L) to calculate the residence time. Residence time for thick fuels is proportional to the thickness to the 3/2 power ($t_r \propto d^{1.5}$). Thus, for thick particles, residence time increases rapidly as thickness increases.

For thin fuels, the burning rate is likely to be linearly proportional to thickness ($\dot{m}_b \propto d^1$), which means that the burning rate per surface area ($\dot{m}_b''$) does not vary with particle thickness (d). Likewise, the flame residence time (**Eqn [4.11]**) is linearly proportional to the thickness ($t_r \propto d^1$).

Wind can also affect the burning rate. When fuels are densely packed and burning rate is ventilation limited, wind can force air into the fuel bed. This makes the fuel–air mixture leaner and closer to stoichiometric, consuming more fuel gases, increasing the heat transferred to the solid fuel, and thus increasing the burning rate.

Increased wind can also increase the burning rate of the fuel bed even if it is not ventilation limited. Not all combustion during the flaming phase occurs in gaseous fuels; some occurs at the surface of the solid fuels. This kind of combustion is called *smouldering*, which is discussed in detail below. However, even during the flaming phase, some air sneaks through the flame, reaches the fuel surface and reacts directly on the surface. This surface reaction is called *char oxidation* and can give off a large amount of heat. When wind blows on the fuel, more air gets to the fuel surface, increasing the rate of char oxidation and thus fuel consumption. You can see this effect when you blow on the coals of a campfire; the coals burn red, more heat is released, and flaming may increase.

One final variable that affects the burning rate is the moisture content of the fuel. In general, an increase in fuel moisture causes a decrease in the burning rate and the total amount of heat released (**Figure 4.27**). This happens because moisture changes the thermal properties of the fuel (density, conductivity and specific heat). Liquid water requires a significant amount of heat to be vaporised (latent heat of vaporisation). In addition, increased fuel moisture can dilute the pyrolysis vapours in the flame, thus reducing the flame temperature and the reaction rate in the gas phase. All of these factors will change the heating rate of the fuel and thus affect the pyrolysis and burning rate. The dilution of the pyrolysis vapours is mostly noticeable when the fuel is thick, or the fuel bed is very deep because of the temperature gradient within the fuel. If the fuel is very thin, its temperature is uniform throughout its thickness, so by the time it ignites, the majority of the water should have been driven off and should not influence its burning rate. However, if the fuel is very thick (or if the fuel has a very high moisture content, such as in live fuels), it is possible that the entire particle is not above boiling temperature when ignition occurs, so significant amounts of moisture could be vaporising from the fuel as it is burning (see **Chapter 5**). Interestingly, there is some evidence that the reduction in burning and heat release rate begins to taper off for the high moisture contents typically present in live vegetation. As shown in **Figure 4.27**, the change in the peak heat release rate for Douglas-fir Christmas trees burned in a laboratory levelled off for foliar moisture contents higher than ~60% (Babrauskas 2016).

Smouldering and glowing

As alluded to throughout the chapter, not all combustion in wildland fires produces flames. *Smouldering* is flameless combustion that occurs at a lower temperature than flaming combustion. *Glowing combustion* is an intense smouldering reaction that produces visible light. **Figure 4.28** shows examples of both. Not nearly as much is known about smouldering and glowing as is known about flaming combustion, but these are very important processes during wildland fires, especially after the flaming front has passed and even when the fire might be presumed 'out'. For example, large logs and duff can 'holdover', meaning that they smoulder unnoticeably deep inside for a long period, only to transition into flaming mode later on. Smouldering embers or firebrands can be lofted into the air, cross firebreaks and start new fires or ignite houses. Smouldering can also consume large amounts of ground fuel, altering the soil structure and damaging or killing underground organisms and plant parts. Additionally, the ignition of smouldering combustion requires less energy than flaming and with lower ignition temperatures. Because smouldering can abruptly transition to flaming combustion, smouldering can therefore provide a pathway to flaming from sources too weak to cause direct flaming. Smouldering combustion can also be notoriously difficult to fully extinguish.

While the visible flames in wildfires are ~1000 °C, smouldering temperatures are typically closer to 600 °C. Instead of oxygen in the air reacting with gaseous fuel to produce a flame, the oxygen in smouldering is reacting directly with the surface of the solid fuel. Smouldering combustion spreads very slowly, typically at a rate of only 10–30 mm per hour. You can often locate smouldering fuels by the acrid smoke that is emanating from rotten logs or the duff layer. The unpleasant smell of the smoke tells us that the combustion products of smouldering differ somewhat from the products of flaming combustion. Emissions from smouldering include much more carbon monoxide, partially burnt gaseous fuel, and other toxic gases than flaming combustion.

Not all solid fuels are capable of smouldering. Because smouldering combustion occurs throughout the solid material, the fuel must be permeable to oxygen. Solid wood can smoulder due to small cracks and fissures in the wood, but permeability sufficient for smouldering usually occurs when the fuel consists of small individual particles (e.g. litter, rotten or

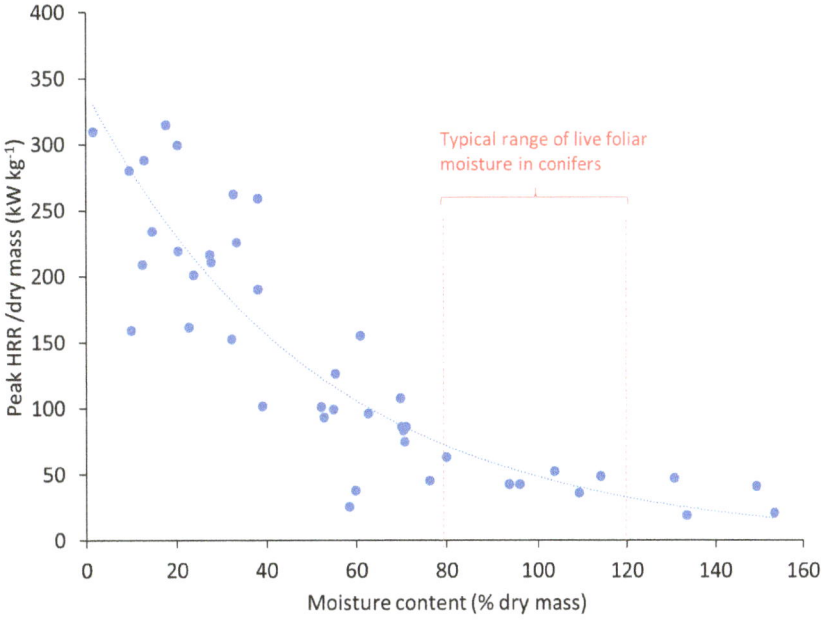

Figure 4.27: The peak heat release rate for Douglas-fir Christmas trees declines for moisture contents up to ~60% of dry weight. For higher moisture contents, such as those typical in live foliage, the peak heat release rate remains nearly constant. Figure adapted with permission from Babrauskas (2016).

Figure 4.28: Left: typical scene of smouldering in wildland fire (photograph from USDA Forest Service, Klamath National Forest). Right: glowing combustion (photograph by Jens Buurgaard Nielsen).

punky wood, duff, sawdust and peat) or is naturally porous (e.g. the foam in household furniture). Fuel permeability not only allows oxygen to access the interior of the fuel but also increases the surface area available for reaction. Furthermore, its complex structure can help insulate against heat loss.

Another requirement for smouldering combustion is that the material must be capable of *charring*. Charring occurs when a material does not pyrolyse completely to form gases but instead leaves behind a solid carbon residue (*char*), which has been stripped of hydrogen and oxygen atoms. All wildland fuels are capable of charring (while some meltable plastics are not). As the char is oxidised, heat is released, and this heat sustains the oxidation reaction. The reaction can be promoted (catalysed) by the alkali metals that are naturally present in vegetation. If the carbon in a fuel is completely oxidised, the alkali metals remain in the ash.

A smouldering combustion front has a structure similar to that of a flame spreading in a premixed gaseous fuel–air mixture. The smouldering front consists of four regions: a preheat zone, an evaporation zone, a pyrolysis and burning zone, and a char and ash zone. The photograph in **Figure 4.29** illustrates these zones. For a smouldering front to spread, it first has to heat the fuel from the ambient temperature up to the moisture evaporation temperature. This occurs in the preheat zone and no gaseous products are formed just yet. All wildland fuels contain some water. This is usually driven off in the evaporation zone before ignition can occur. No other gaseous products are formed in this small region. In the third zone, the pyrolysis and burning zone, fuel is heated sufficiently for pyrolysis, which is roughly above 200 °C. (Because smouldering is such a slow process, pyrolysis becomes important at a lower temperature than in flaming combustion.) Pyrolysis generates gaseous fuels, but they do not combust because the process is so slow. Instead, they diffuse away with other emissions. While pyrolysis is producing gaseous fuels, it is stripping the solid fuel of hydrogen and oxygen atoms to form char. Since char is primarily carbon, it is still significant as a fuel. The porous char can be oxidised to generate heat. Char oxidation occurs at a higher temperature than pyrolysis but at a temperature still lower than that of flaming combustion. The final zone of the smouldering front is where the solid residue cools. The overall smouldering process is often limited by access to oxygen, so a considerable amount of both solid and gaseous fuel is usually unburned. Therefore, the heat released from smouldering combustion can be far lower than estimated using the heat of combustion measured in a bomb calorimeter. The residue

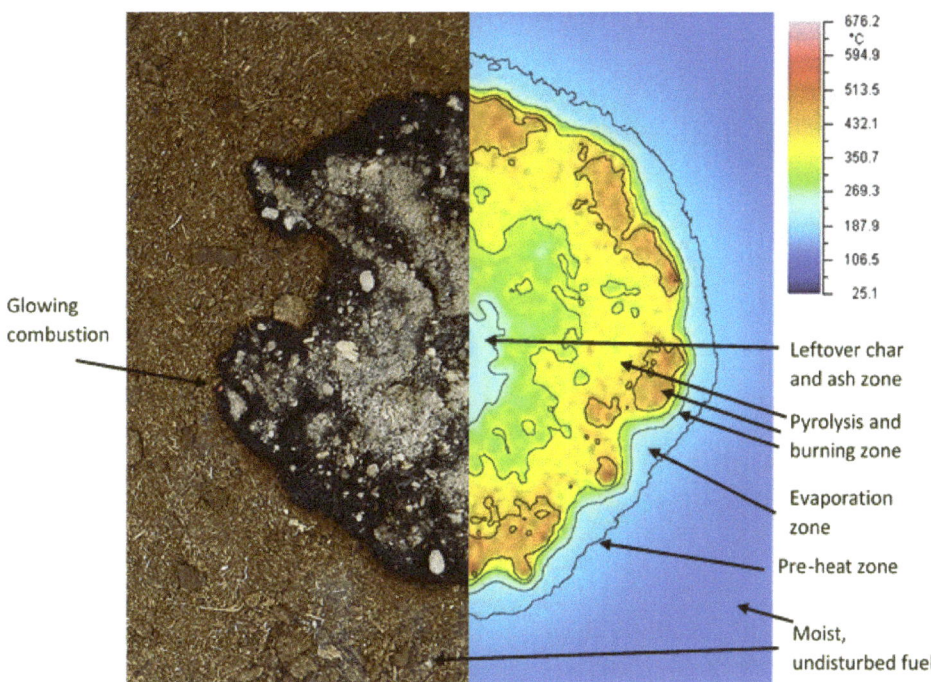

Figure 4.29: Structure of the smouldering front seen both visually (left) and with an infrared (IR) camera (right). The fuel (peat) was ignited in the centre and smouldering spread radially outward. Along the outer edges of the smouldering front, the peat remains moist and undisturbed. However, the light blue band in the infrared image shows that heat from the smouldering reaction extends ahead into the unburned fuel. The pyrolysis and burning zone appears visually as the region where the peat is dark brown (thermally degraded) and black (charred). There is even a spot on the left edge where the reaction is intense enough to glow. From the infrared image, we can see the pyrolysis and burning zone as a ring of elevated temperatures (red, orange, and yellow). The centre of the fuel – and the ignition point for the fire – has begun to cool (returning to a bluish colour), and leftover char and ash remain. Photographs by Ian Grob and Jim Reardon.

from smouldering combustion consists of unburnt char and ash (the minerals and other non-flammable components of the original fuel).

It is interesting to note that oxidation of char actually releases more heat per unit mass of fuel than flaming combustion of pyrolysates. **Figure 4.30** shows the results of bomb calorimeter experiments of a variety of woody fuels (from Sussott et al. 1975). Whole specimens, char and pyrolysates were tested separately. As discussed earlier, the pyrolysis gases tend to be smaller molecules with a lower carbon content than char. As burning progresses, the residual solid material (i.e. char) becomes increasingly composed of carbon. Since the heat of combustion increases linearly with carbon content, the oxidation of char can release more heat per mass than the oxidation of whole specimens or pyrolysates: char can produce up to 30 MJ kg^{-1}, while pyrolysates produce ~15 MJ kg^{-1}. Typical values of the heat of combustion for the entire fuel are thus weighted averages of the values for pyrolysates and char (with proportions varying depending on a variety of burning and fuel conditions).

Because char oxidation results from oxygen reacting directly on the fuel surface, the rate at which it occurs is sensitive to the rate at which oxygen can get to the surface. **Figure 4.31** shows the results of experiments burning charcoal in an air stream (Evans and Emmons 1977). This is almost entirely a char oxidation reaction because, as the name implies, charcoal is primarily char, since most of the pyrolysate gases were driven off

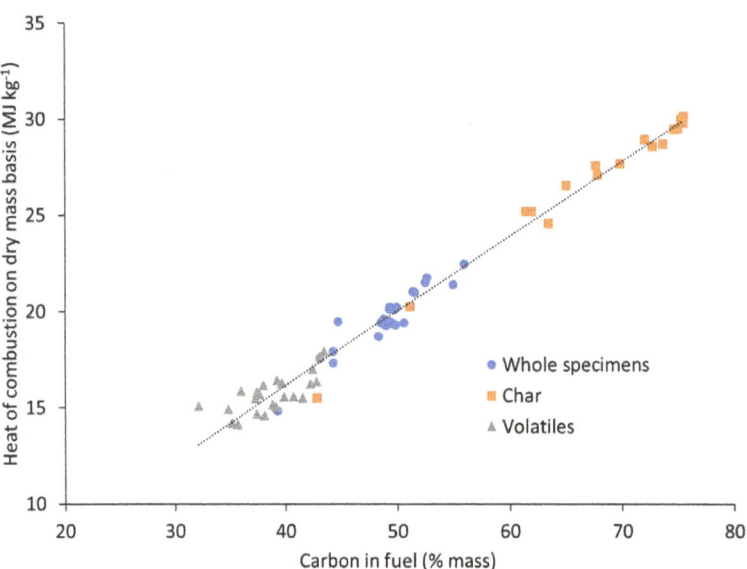

Figure 4.30: Heat of combustion (higher heat of combustion) of a variety of woody fuels as measured in the bomb calorimeter. Whole specimens were tested separately from the pyrolysis gases (i.e. volatiles) and char (adapted from Sussott *et al.* 1975).

when it was created. The graphs in **Figure 4.31** demonstrate that the burning rate and surface temperature of charcoal increase dramatically as the flow of air on the surface increases. This means that the rate of heat release can vary considerably within a fire as the wind changes speed. You have probably noticed this effect when you have blown on the coals in a campfire: the added oxygen causes the coals to glow and occasionally even begin flaming. Rates of char oxidation vary with air velocity in the flaming front as well as behind it, since some air always sneaks through the flame. The effect of wind on smouldering fuels can have important consequences for wildland fires and their effects on soils and vegetation. For example, consumption of large woody fuels and deep duff layers can increase considerably with wind and result in increased soil heating.

As with most combustion processes, sustaining and propagating a smouldering front is largely controlled by the balance between the heat lost and the heat generated. Heat losses include those due to wind and radiation to the cooler surroundings (see **Chapter 3**). These are reduced as the smouldering front propagates deep into the fuel bed, because the layers of fuel and ash surrounding the smouldering reaction serve to insulate the reaction and thus help retain the heat that is generated. However, the same properties of the fuel bed and location of the reaction within it that minimise the heat losses also reduce the availability of oxygen. If heat cannot get out, oxygen cannot get in and the reaction will be choked off. In general, smouldering combustion on the surface is limited by large heat losses, but smouldering combustion deep within a fuel bed is limited by access to oxygen. A gentle breeze will help the oxygen penetrate the fuel bed and thus increase the heat release and spread rate. However, a strong wind will cool the exposed surfaces and thus hinder the smouldering reaction.

As mentioned above, smouldering can transition to flaming combustion. Conversely, as flaming combustion begins to die down, it can transition to smouldering combustion. This transition from smouldering to flaming and back occurs frequently over the course of a wildland fire. Although these transitions are important to fire

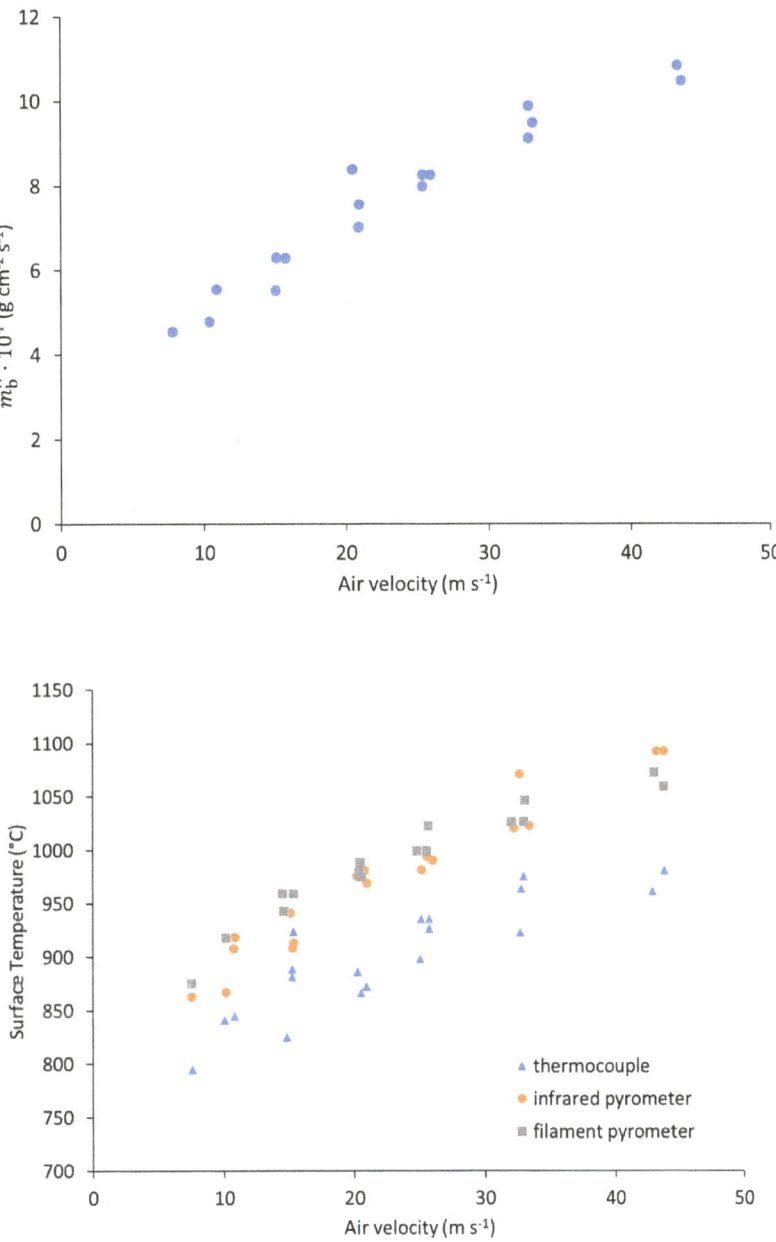

Figure 4.31: Charcoal burning rates (a) and surface temperatures (measured with three techniques) of burning charcoal (b) as functions of air speed. These effects are likely to be found in wildland fires, but have been little investigated for smouldering fire behaviour and fuel consumption. Adapted with permission from Evans and Emmons (1977).

safety and fire effects, they are not well understood and our ability to predict them is poor. It seems logical that the transition from smouldering to flaming would require, first, a flammable mixture of gaseous fuel and air. As discussed above, smouldering usually produces gaseous fuel very slowly. If exposed on an open surface, the gaseous fuel often diffuses away and never builds a high enough concentration to produce a flammable fuel–air mixture. In order to reach a high enough concentration, either the gaseous fuel must be produced faster (i.e. pyrolysis rate must increase) and/or it must be trapped, such as in a void space inside the fuel. To increase the pyrolysis rate, the temperature must be increased either through external heating (e.g. crown fire above smouldering duff) or through increased char oxidation by means of increased access to oxygen (e.g. light breeze). A second requirement for the transition from smouldering to flaming is that the flammable mixture must be located near an ignition source, such as a hot spot of the char oxidation region or a flaming ember. The balance between heat generated and heat lost controls the process. The transition from flaming back to smouldering is perhaps even less predictable, but it seems logical that it should occur when the rate of pyrolysis slows to the point that it cannot sustain a flame – that is, when heat loss from the flame outweighs the heat that it generates.

Summary

In this chapter, we have examined some of the fundamental aspects of combustion that are relevant to wildland fire. We learned that there are two important forms of combustion: flaming and smouldering. For flaming to occur, gaseous fuel must be produced through the thermal degradation process called pyrolysis. The resulting gases burn as non-premixed (diffusion) flames. Smouldering combustion is a flameless form of combustion, in which oxygen directly attacks the solid fuel surface. Smouldering combustion is not as well understood as flaming combustion, even though it can play an important role in wildland fire behaviour and effects. In the next chapter, we will apply this fundamental understanding of combustion to the ignition process. Understanding these aspects of combustion will also serve us well in the other remaining chapters of the book.

Endnotes

[i] A good summary about what is known about pyrolysis and combustion chemistry of cellulose is found in Sullivan and Ball (2012).

[ii] A mole (abbreviated mol) is Avogadro's number of molecules (6.02×10^{23}).

[iii] If we assume that cellulose pyrolyses into individual carbohydrate molecules. This is a big assumption, and it is not correct.

[iv] Kelvin is a temperature unit related to Celsius by $K = C + 273$. See Appendix A.

[v] As discussed below, combustion reactions are still occurring, but not at a rate fast enough to offset heat losses.

[vi] The chemical kinetic parameters for pyrolysis and combustion of wood and other wildland fuels are determined by matching a proposed reaction model to experimental data. The resulting parameters can vary widely depending on the reaction model and the experimental conditions, so there are no universally accepted values for these parameters.

[vii] This is not explicitly stated in the original work, but it is often assumed. Since the fuel bed is shallow compared to the size of the flames, the imprecision due to fuel bed depth is negligible.

[viii] This distinction is not made anywhere in the literature.

References

Alexander ME, Cruz MG (2012) Interdependencies between flame length and fireline intensity in predicting crown fire initiation and crown scorch height. *International Journal of Wildland Fire* **21**, 95–113. doi:10.1071/WF11001

Anderson HE (1970) Forest fuel ignitability. *Fire Technology* **6**(4), 312–319. doi:10.1007/BF02588932

Anderson HE (1990) 'Relationship of fuel size and spacing to combustion characteristics of laboratory fuel cribs'. Research Paper INT-424. USDA Forest Service, Intermountain Research Station, Ogden, UT.

Babrauskas V (2006) Effective heat of combustion for flaming combustion of conifers. *Canadian Journal of Forest Research* **36**, 659–663. doi:10.1139/x05-253

Babrauskas V (2016) Heat release rates. In *SFPE Handbook of Fire Protection Engineering*. (Eds MJ Hurley *et al.*) pp. 799–904. Springer, New York, NY.

Byram GM (1959) Combustion of forest fuels. In *Forest Fire: Control and Use*. (Ed. KP David) pp. 61–89. McGraw Hill, New York, NY.

Clark TL, Reeder MJ, Griffiths M, Packham D, Krusel N (2005) Infrared observations and numerical modeling of grassland fires in the Northern Territory, Australia. *Meteorology and Atmospheric Physics* **88**, 193–201. doi:10.1007/s00703-004-0076-9

Coen J, Mahalingam S, Daily J (2004) Infrared imagery of crown-fire dynamics during FROSTFIRE. *Journal of Applied Meteorology* **43**, 1241–1259. doi:10.1175/1520-0450(2004)043<1241:IIOCDD>2.0.CO;2

Di Blasi C, Branca C, Santoro A, Gonzalez Hernandez E (2001) Pyrolytic behavior and products of some wood varieties.

Combustion and Flame **124**, 165–177. doi:10.1016/S0010-2180(00)00191-7

Dimitrakopoulos AP, Panov PI (2001) Pyric properties of some dominant Mediterranean vegetation species. *International Journal of Wildland Fire* **10**(1), 23–27. doi:10.1071/WF01003

Evans DD, Emmons HW (1977) Combustion of wood charcoal. *Fire Safety Journal* **1**(1), 57–66. doi:10.1016/0379-7112(77)90008-X

Heskestad G (1973) Modeling of enclosure fires. *Symposium (International) on Combustion* **14**, 1021–1030.

Hough WA (1969) 'Caloric value of some forest fuels of the southern United States'. (Vol. 120). Research Note SE-120. USDA Forest Service, Southeastern Forest Experiment Station, Asheville, NC.

Janssens M (2016) Calorimetry. In *SFPE Handbook of Fire Protection Engineering*. (Eds MJ Hurley *et al.*) pp. 905–951. Springer, New York, NY.

Kawamoto H (2017) Lignin pyrolysis reactions. *Journal of Wood Science* **63**, 117–132. doi:10.1007/s10086-016-1606-z

Lee KO, Megaridis CM, Zelepouga S, Saveliev AV, Kennedy LA, Charon O, Ammouri F (2000) Soot formation effects of oxygen concentration in the oxidizer stream of laminar coannular nonpremixed methane/air flames. *Combustion and Flame* **121**, 323–333. doi:10.1016/S0010-2180(99)00131-5

Martin RE, Gorden DA, Gutierrez ME, Lee DS, Molina DM, Schroeder RA, Sapsis DA, Stephens SL, Chambers M (1993) Assessing the flammability of domestic and wildland vegetation. In *Proceedings of the 12th Conference on Fire and Forest Meteorology*. 26–28 October 1993, Jekyll Island. Georgia Society of American Foresters Publication 94-02, pp. 130–137. Society of American Foresters, Bethesda, MD.

McCaffrey BJ (1979) *Purely Buoyant Diffusion Flames: Some Experimental Results*. NBSIR 79-1910. National Bureau of Standards, Center for Fire Research, Washington, DC.

Pettersen RC (1984) The chemical composition of wood. In *The Chemistry of Solid Wood* (Ed. RM Rowell), Advances in Chemistry Series, vol. 207, p. 984. American Chemical Society, Washington, DC.

Philpot CW (1970) Influence of mineral content on the pyrolysis of plant materials. *Forest Science* **16**(4), 461–471.

Quintiere JG, Grove BS (1998) A unified analysis for fire plumes. *Symposium (International) on Combustion* **27**, 2757–2766.

Sullivan AL, Ball R (2012) Thermal decomposition and combustion chemistry of cellulosic biomass. *Atmospheric Environment* **47**, 133–141. doi:10.1016/j.atmosenv.2011.11.022

Sussott RA, DeGroot WF, Shafizadeh F (1975) Heat content of natural fuels. *Journal of Fire and Flammability* **6**, 311–325.

Thomas PH (1963) The size of flames from natural fires. *Symposium (International) on Combustion* **9**, 844–859.

Westbrook CK, Dryer FL (1984) Chemical kinetic modeling of hydrocarbon combustion. *Progress in Energy and Combustion Science* **10**, 1–57. doi:10.1016/0360-1285(84)90118-7

Williamson NM, Agee JK (2002) Heat content variation of interior Pacific Northwest conifer foliage. *International Journal of Wildland Fire* **11**, 91–94. doi:10.1071/WF01046

Yang H, Yan R, Chen H, Lee DO, Zheng C (2007) Characteristics of hemicellulose, cellulose and lignin pyrolysis. *Fuel* **86**, 1781–1788. doi:10.1016/j.fuel.2006.12.013

5

Ignition

Ignition is a vital step in the fire spread process. Wildland fuels are not uniform or continuous, as liquid or gaseous fuels are. Instead, they consist of individual fuel elements. For fire to spread, individual fuel elements must ignite progressively, almost like dominoes falling. We must understand the ignition processes that govern these progressive ignitions to understand wildland fire. This chapter starts with a detailed description of how ignition occurs. From there, we discuss ignition criteria, the types of ignition, simple methods to predict how long ignition will take, and the variables that can affect it. We conclude the chapter by discussing the challenge of understanding ignition in live fuel. An important and pervasive theme in this chapter is the balance between heat generated by the combustion reactions and heat lost by the fuel (solid or gas) to the surroundings. For ignition to succeed, the heat generated must exceed the heat lost.

The ignition process

Ignition is the initiation of combustion. The visible flames that we see during flaming combustion of a solid fuel result from the combustion of flammable[i] vapours that have been produced by thermal degradation (i.e. pyrolysis) of a solid fuel. Let us consider ignition of a wildland fuel particle that is resting at the temperature of the ambient air (**Figure 5.1**). The outside surface and the inside material are all at the same ambient temperature. The first step – and really the main step – in igniting the particle is the *heating of the solid* to the point that gaseous fuel is produced. This heating could be due to thermal radiation from an approaching flame, convection from contact with the flame, or direct contact with a burning ember. As the outside surface of the fuel is heated, its temperature begins to increase and heat is conducted into its interior. As heating continues, some moisture in the fuel is evaporated – though maybe not all of it. A relatively large amount of energy is required to convert water from the liquid to the gaseous phase (i.e. the latent heat of vaporisation), so there may be a short 'stall' in the temperature increase of the fuel while the heat evaporates water rather than raising the fuel temperature. This may or may not happen right at 100 °C, especially in live fuels, depending on how water is stored within the fuel. In dead wood, water can either be stored as 'free water' in voids or as 'bound water', which is chemically bound to the cell walls. Free water evaporates more readily than bound water. As we will learn later, live fuels have yet another mode of water storage that affects ignition.

As fuel heating continues, the temperature continues to increase and eventually pyrolysis starts in earnest. As discussed in **Chapter 4**, pyrolysis is a chemical reaction, not a simple phase change process, so the exact composition of the

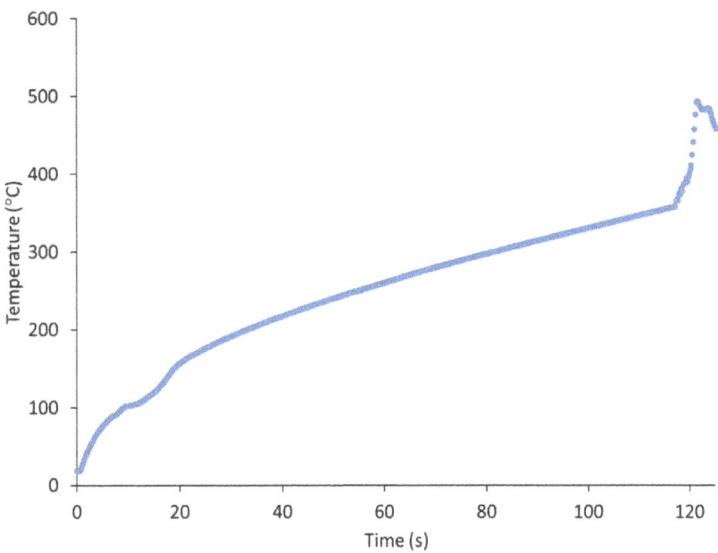

Figure 5.1: Surface temperature of a 10 cm × 10 cm slab of wet wood (18% moisture content) heated with 20 kW m^{-2} radiant heat flux. Ignition occurs at 117 s. Notice the relatively long stall in temperature rise when the surface is ~100 °C.

gaseous fuel is not necessarily known. The rate of pyrolysis is strongly temperature dependent. At room temperatures, pyrolysis of solid fuel materials is almost non-existent, but the rate increases dramatically above some critical temperature. This is often referred to as the *pyrolysis temperature*, and it is typically around 300 °C (see **Chapter 4**). At this point, the fuel gases (*pyrolysates*) may be visible to a keen observer as white or grey smoke like that shown in **Figure 5.2**. Not only is the rate of pyrolysis dependent on the temperature, so is the composition of the gases produced. At low temperatures, the quantity of gases produced is very small. The pyrolysates tend to be large molecules, perhaps even in a liquid phase that resembles tar (a sticky liquid). These pyrolysates cannot burn in their current state. As the temperature continues to rise, the composition of the pyrolysis products changes to smaller, more reactive gases, and their quantity increases.

As pyrolysate gases are released from the solid surface, they begin to mix with the surrounding air, forming a mixture of gaseous fuel and air near the solid surface. At first, this mixture will have too low a concentration of fuel to be flammable because the gas-phase combustion will be too slow to produce enough heat to offset heat losses to the cold surroundings. The mixture will be outside of the flammability limits (**Figure 4.11**), too fuel-lean to burn. As the temperature and pyrolysis rate increase further, this premixed gaseous fuel and air mixture will reach the *lean flammability limit*. At

Figure 5.2: Douglas-fir (*Pseudotsuga menziesii*) needles heating under a 50 kW m^{-2} radiant heat flux with a slight air flow. Notice the abundance of white vapours. These are pyrolysate gases that are not quite flammable, either due to the composition of the gases themselves or to their low concentration.

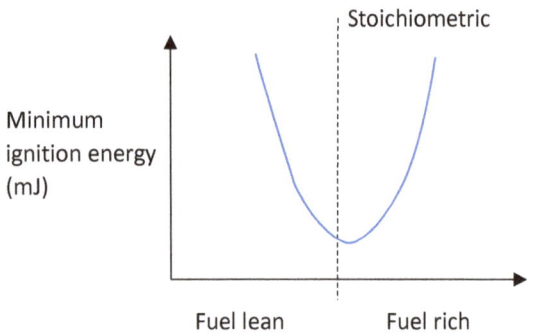

Figure 5.3: Required minimum energy for ignition varies with the proportions of gaseous fuel and air in the mixture.

this point, any spark or hot spot with sufficient energy can ignite the gaseous mixture. If the mixture is close to the lean or rich flammability limit, it needs more energy to ignite than if it is close to the stoichiometric condition (**Figure 5.3**).

We can assume that the minimum energy needed to ignite the gas phase is easily met in an ongoing wildland fire. Once ignited, the flame spreads through the premixed gas-phase mixture away from the source of ignition and towards the solid fuel (**Figure 5.4**). To ignite the actual solid fuel rather than just the gases – that is, to 'anchor' a diffusion flame on the solid fuel – the approaching premixed flame must generate enough heat to offset heat losses both to the surroundings and to the solid itself. When the premixed mixture of gaseous fuel and air is near the lean flammability limit, not enough heat is likely to be generated to offset these losses, so the flame will go out as it approaches the solid surface. This is observed as a quick flash of flame, so it is referred to as *flashing ignition* or the *flash point*. As the solid temperature and pyrolysis rate increase further, the combustion reactions in the premixed mixture become faster and generate more and more heat, eventually offsetting the heat losses and leading to a sustained diffusion flame that is anchored on the surface of the solid fuel. The initiation of sustained ignition is called the *fire point*.

The above discussion applies to flaming ignition, which is the type of ignition generally experienced as a fire *spreads*. Smouldering ignition, however, is probably the process that *ignites* most wildland fires. To initiate smouldering, just as to initiate flaming combustion, the heat-balance requirement must be met; that is, the heat generated must exceed the heat lost. As discussed in **Chapter 4**, smouldering combustion occurs within the solid fuel itself rather than in the pyrolysis gases. The heat generated is due to the chemical reaction of oxygen directly with the fuel surface. This reaction is very slow at room temperature but occurs much faster and releases much more heat as the temperature of the fuel increases. In wildland fires, smouldering ignitions are most common during fire initiation and during spotting, when burning or glowing embers are lofted ahead of large fires and land on receptive fuels. Wildland

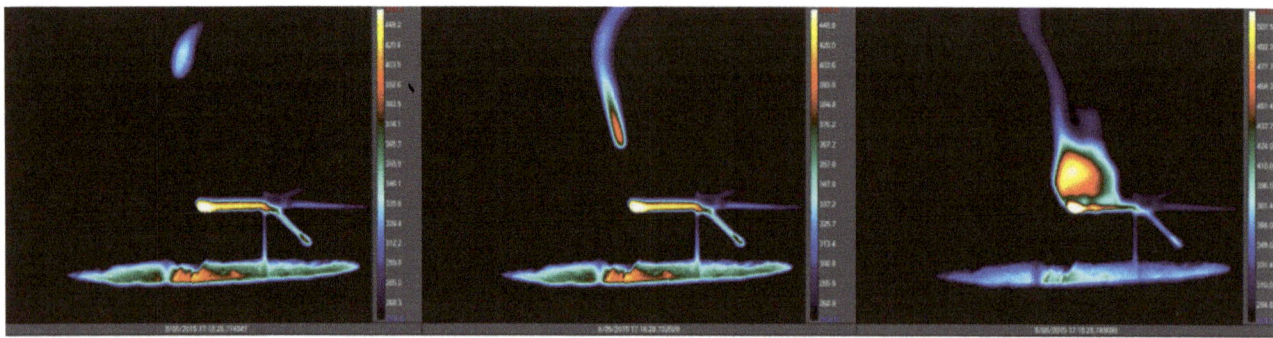

Figure 5.4: Ignition of a wood rod as recorded by high-speed infrared videography. The 0.64 cm diameter rod is heated with air at 600 °C. Ignition first takes place in the gas–air mixture above the solid fuel and then burns back towards the fuel particle itself. Total elapsed time in this series of photographs is 3.5 ms.

fuels can be good insulators – particularly rotten, punky logs and fluffy layers of duff and litter. Their porous structure can prevent embers from losing heat rapidly, so it takes relatively little heating to tip the heat balance so the fuels are producing more heat than they are losing. If a burning ember gets into a punky log, it stays hot and reactive longer than if it were exposed to the air, allowing the heat to build up slowly and eventually ignite the log. The heating of the fuel for smouldering ignition can be at a much slower rate over a much longer time than required for flaming ignition (see **Chapter 4, Smouldering and glowing**).

Flaming ignition criteria

To predict exactly how long it will take to ignite a fuel to flaming combustion (i.e. to predict the *flaming ignition time*) we need a measurable criterion. As we saw above, sustained flaming ignition of a solid fuel requires that enough heat be generated in the gas-phase reactions to offset heat losses to the solid surface and the surroundings. Perhaps the most physically correct criterion for ignition would be a *critical heat release rate* from the pyrolysis gases. However, this is not a practical criterion for use in wildland fire because it requires that we know how the solid is heating up, how heat is transferred between the gas and the solid, and how the reactions are occurring in the gas phase. This would be a coupled model that integrates information on both solid- and gas-phase ignition, which would be especially challenging because the chemical composition of the gases is largely unknown.

Critical mass flux[ii] is a second possible criterion for predicting the flaming ignition time. Critical mass flux is closely related to critical heat release rate, but it is simpler because it only requires knowledge of what is going on in the solid phase. As the solid fuel is heated, its mass decreases. **Figure 5.5** shows how the mass of a block of wood declines as it undergoes ignition. At first, the decline is due only to moisture loss, then to moisture loss and pyrolysis. When the surface temperature passes ~300 °C, the rate of mass loss increases rapidly (**Figure 5.6**). Soon a critical amount of gaseous fuel is produced and ignition occurs. The mass flux required for ignition still depends on achieving enough heat generated in the gas-phase reactions to offset heat losses. As we learned in **Chapter 4**, the temperature and heat release rate of a forming flame are related to the proportions of fuel vapours and air in the mixture (**Figures 4.5** and **4.8**). For the fuel-lean mixtures typical during ignition, the greater the amount of fuel vapours, the greater the amount of heat generated and the higher the temperature of the flame. The critical mass flux is therefore an approximation for the critical heat release rate criterion.

Unfortunately, no single critical mass flux value can be applied in all situations and fuels because it depends on what is happening in the gaseous mixture. If a change in ambient conditions changes the balance between heat generated and heat lost, it will change the production rate of gaseous fuel required for flaming ignition. For example, higher winds increase the rate of available oxygen, so ignition in wind requires a greater production rate of gaseous fuel than ignition in still conditions. Additionally, the chemical composition of pyrolysates changes as temperature increases and, because the ignition characteristics of these compounds vary, the heat release rate and critical mass flux will vary as well. Because the gas phase is not considered when we use a mass flux criterion to estimate the time to flaming ignition, changes such as these cannot be accounted for, so this criterion is rarely used.

Ignition temperature is a third possibility for defining the flaming ignition time. We know that the production rate of pyrolysis gases is highly dependent on temperature. Attaining a critical mass flux for ignition is thus related to attaining a sufficiently high temperature on the surface of the solid fuel. However, this criterion has limitations similar to those for critical mass flux, because ambient and heating conditions can change the temperature needed to achieve ignition. Thus, there really is not a single true ignition temperature. In fact, depending on conditions and the testing method used, the ignition temperature for wood can vary between 296 and 497 °C (Babrauskas 2002)! Despite

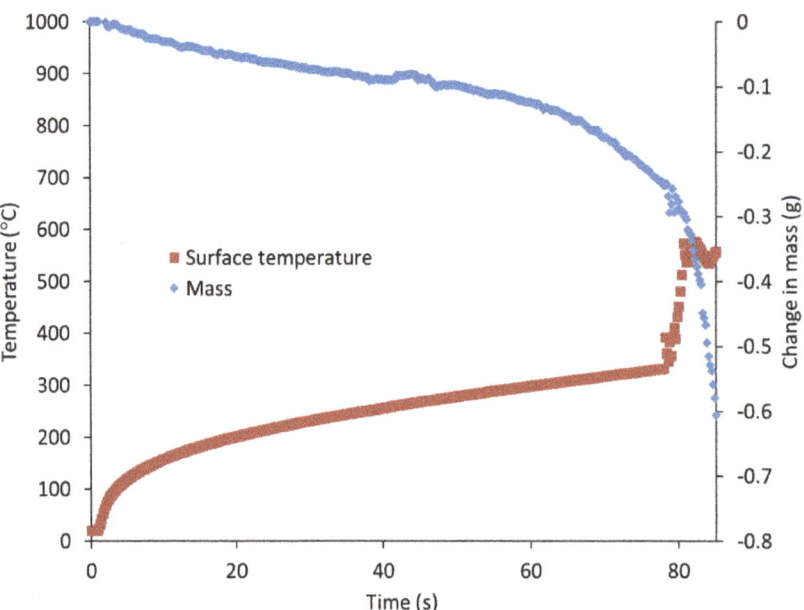

Figure 5.5: Change in mass and surface temperature of a block of mostly dry wood (1–2% moisture content) being heated until ignition by a 20 kW m^{-2} radiative heat flux. Ignition occurs at ~78 s, as indicated by the rapid rise in fuel surface temperature and decline in mass.

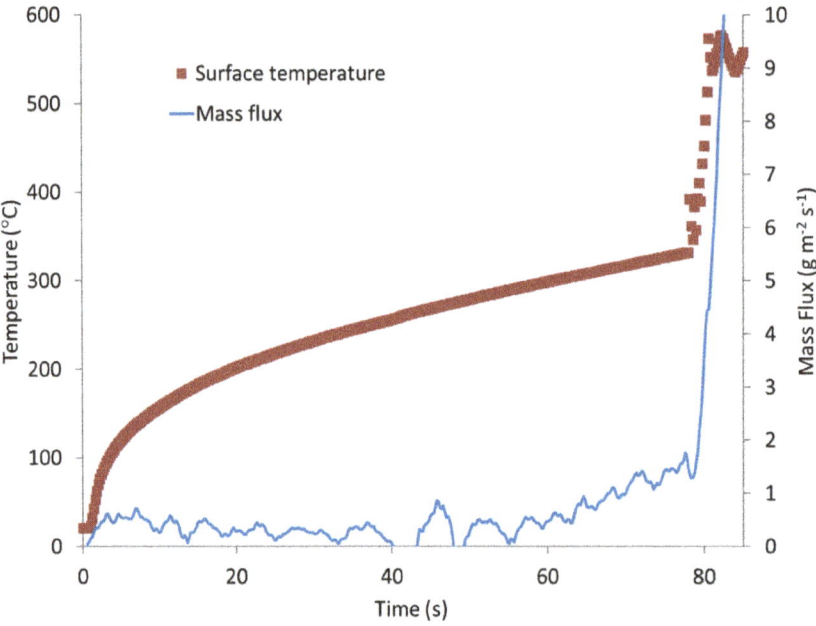

Figure 5.6: Change in fuel mass flux compared to surface temperature for the block of wood shown in **Figure 5.5**. The mass flux is found by taking the slope of the mass curve in **Figure 5.5** and dividing it by the exposed surface area. The critical mass flux at ignition (at ~78 s) in this case is ~1.5 g m^{-2} s^{-1}. Note that the mass flux begins to noticeably increase after ~55 s, when the surface temperature is ~300 °C (pyrolysis temperature).

its limitations, the concept of ignition temperature is intuitive and easy to implement, so it is widely used, including in the remainder of this book. However, it is important to understand that ignition temperature is a rough approximation for the complex ignition process and to be flexible in attaching a numerical value to it.

Types of ignition

Two types of ignition are defined based on how the rapid combustion reaction needed for fire spread is initiated. As described in the description of the ignition process above, there is often a hot spot, spark, or flame that raises the local temperature of the gas-phase mixture (or the solid itself in smouldering ignition), accelerating the local reaction and heat release rate, and finally and triggering self-sustaining combustion. This is called *piloted ignition*, and the hot spot is referred to as a *pilot*. Piloted ignition is probably the most common type of ignition in spreading wildland fires, since the abundant flames and embers in the vicinity of the fire front can easily act as a pilot. Flames from other sources – a drip torch or an escaped campfire, for example – also cause piloted ignitions.

The other type of ignition is *autoignition*, in which there is no localised hot spot. As with any ignition, autoignition is determined by the balance between heat generated and heat lost. Because the gas-phase reaction is not accelerated by a pilot in autoignition, it requires higher temperatures than those needed for piloted ignition. You may be familiar with another term for autoignition: *spontaneous ignition*, which is often used to explain why a pile of oily rags can catch on fire. The chemical reactions responsible for combustion – that is, the reactions that produce and oxidise pyrolysis gases (**Chapter 4**) – occur at all temperatures. At low temperatures, however, they occur at such a slow rate that the small amount of heat generated is easily lost to the surrounding environment. This means that the temperature of the fuels does not increase substantially, so the reaction is not sustained and no chain reaction occurs. Why then is a pile of oily rags susceptible to autoignition?

The oil on the rags is constantly reacting with the oxygen in the surrounding air at all temperatures and constantly releasing heat. A single oily rag at room temperature is producing a minuscule amount of heat and loses it easily. However, when a bunch of oily rags is piled up, they insulate each other and trap the heat being produced by oxidation. Given enough time, the temperature inside the pile increases, increasing the reaction rate. Feedback between increasing temperature and increasing reaction rate accelerates the process until the rags themselves are ignited, initially in smouldering combustion. Autoignition is too slow to spread the active flaming front of a fire, but it can initiate a fire when large piles of hay and manure are heated by microbial decomposition.

Critical heat flux for ignition

As discussed in **Chapter 3, Heat transfer**, there are several forms of heat transfer in fires, and they can occur simultaneously. In many practical situations, fuel particles can both gain and lose heat at the same time. Whether the fuel temperature increases or not depends on the balance between heat gain and heat loss. For example, when you stand outside, your skin receives radiant heat from the sun. At the same time, even on a calm, windless day, your skin loses heat from convection – and the amount of heat lost from convection increases as your skin temperature increases. While you are warming up slightly, the balance between radiant heating and convective cooling usually prevents you from increasing your body temperature to a dangerous level.

This same interplay between heat lost and heat received occurs for fuel particles in wildland fires. Fuel particles receive heat from the fire via radiation and convection, but they simultaneously lose heat through natural convection, forced convection from the wind, and radiation to the cooler surroundings. As we saw in **Chapter 3, Heat transfer**, the size, shape and spatial arrangement of fuel particles strongly affect how efficiently they exchange heat with their surroundings by convection and radiation. For ignition to occur, the temperature of the fuels must increase at least to the ignition temperature. The heating rate that exactly balances the cooling rate and therefore raises the temperature exactly to the ignition temperature is called the *critical heat flux* for ignition. For heating rates below this critical level, ignition will never occur. For a heating rate at exactly the critical level, the time required to ignite the fuel is almost infinitely long. For faster heating rates, the ignition time decreases as the heating rate increases (**Figure 5.7**). Note that the critical heat flux for ignition can vary, because the rate of heat loss depends on the physical characteristics of the fuel particle and the environmental conditions. For any fuel, faster heating is required for ignition on a day with high winds than on a completely still day because the wind is constantly removing heat from the surface of the fuel. This is illustrated in **Figure 5.7**, which shows three different critical heat fluxes are associated with three different wind speeds.

The critical heat flux is one more way to illustrate the interacting roles of ignition, combustion, and heat transfer in wildland fires. A fire can spread only if the heat flux is sufficient to ignite new fuel particles during the time of flaming combustion. As discussed in **Chapter 4, Burning and heat release rate**, the burning rate determines the flame residence time and thus how long flames are present and able to heat new fuel particles. If particles do not ignite within the time of flaming combustion, then the fire will not spread. As we will see in **Chapter 7**, the processes of heat release, heat transfer and ignition are not independent of each other, and the way they come together to produce fire spread is not straightforward.

Predicting ignition times

If we examine the detailed description of the ignition process above, we see that it takes three steps to trigger flaming combustion in a solid fuel:

- First, the solid must be heated to a temperature at which pyrolysis can occur at a sufficient rate.
- Second, the pyrolysis gases must mix with the surrounding air thoroughly enough to form a mixture capable of burning.
- Third, the air–pyrolysate mixture must ignite, propagate a flame towards the solid, and anchor it on the solid surface.

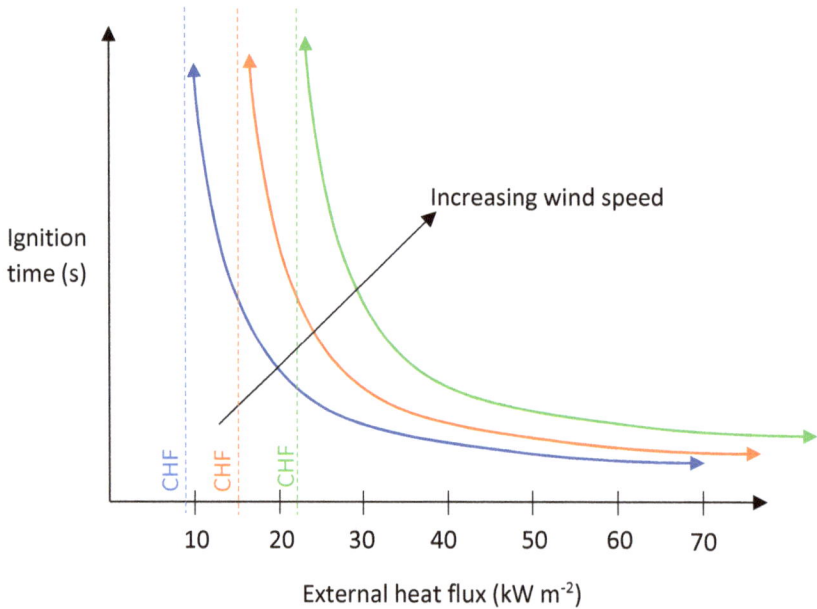

Figure 5.7: Ignition time decreases with external heat flux. Critical heat flux (CHF) for ignition varies with ambient conditions. It increases with wind speed because of convective cooling from the relatively cool surrounding air.

Each of these steps takes a finite amount of time. However, for most wildland fires, the time required to heat the fuel is longer and more significant than the time required for the other two steps. Thus, the time required to heat the fuel to its ignition temperature can provide a good approximation of the time to ignition.

The time required to heat the fuel to ignition can be calculated using the principles learned about heat transfer in **Chapter 3**. Recalling the 'leaky bucket analogy' shown in **Figure 3.2**, energy is being transferred in and around the fuel in three ways: first, radiation and convection from a nearby fire are adding heat to the surface. Second, heat is simultaneously being lost from natural and forced convection and from radiation exchange with the surroundings. Finally, as the surface temperature begins to increase, heat is being conducted towards the centre of the fuel. As discussed in the **Chapter 3, Conduction heat transfer**, two simplifying assumptions can be used to predict how the temperature of the fuel changes:

1. The *thermally thick assumption*[iii] applies to fuel particles that are thick enough to require some time for heat to penetrate to the middle, so a temperature gradient exists, and there could be a large temperature difference between the surface and the centre of the fuel particle.
2. The *thermally thin assumption* applies to very thin fuels. In this case, heat is conducted throughout the entire solid quickly enough to consider the entire interior of the solid to be at the same temperature – that is, there is no temperature gradient within the fuel.

Very few situations in the real world perfectly satisfy these assumptions, but many situations are close enough for the assumptions to be useful. They allow us to write the following equations to predict the ignition time.

For thermally thick fuels:

$$t_{ig} = \frac{\pi}{4} k \rho_s c_s \frac{(T_{ig} - T_o)^2}{\dot{q}_s''^2} \qquad [5.1]$$

where t_{ig} is the ignition time (s), k is the thermal conductivity of the solid (kW m^{-1} °C^{-1}), ρ_s is the density of the solid (kg m^{-3}), c_s is the specific heat of the solid (kJ kg^{-1} °C^{-1}), T_{ig} is the ignition temperature (°C), T_o is the initial temperature of the solid (°C), and $\dot{q}_s''$ is the net heat flux on the surface (kW m^{-2}).

For thermally thin fuels:

$$t_{ig} = \frac{\rho_s c_s d (T_{ig} - T_o)}{\dot{q}_s''} \qquad [5.2]$$

where d is the thickness or diameter of the fuel (m).

For both equations, the net heat flux on the surface $\dot{q}_s''$ is the sum of the heat added and the heat lost via radiation and convection. When the heat added is very high, such as when flames engulf the fuel, the heat losses are minor in comparison and can be ignored. When ignition conditions are marginal, however, both heat additions and losses must be considered. Note that in order to make these relatively simple calculations, we must assume that heat flux is constant with time. This is certainly not realistic for spreading wildland fires, since both the heat added and the heat lost change as the flame front approaches (see **Chapter 6** and **Figure 6.24**). Researchers are only beginning to address the effect that this has on ignition time.

In order to use **Eqn [5.1]** or **[5.2]** appropriately, we must determine which assumption (thick v. thin fuel) is appropriate. In addition to the method provided in **Eqns [3.14]** and **[3.15]**, we can do so with the following equation:

$$\ell_{th} = \sqrt{\frac{k}{\rho_s c_s} t_{ig}} \qquad [5.3]$$

where ℓ_{th} is the *thermal penetration depth*. (If the ignition time is unknown, we calculate the ignition time with both **Eqns [5.1]** and **[5.2]** and use whatever time is longer in **Eqn [5.3]** to get the most conservative estimate for ℓ_{th}.) For fuels heated on only one side, if the fuel thickness is greater than the calculated thermal penetration depth (ℓ_{th}), the thermally thick assumption is appropriate and **Eqn [5.1]** should be used. If the fuel thickness is less than ℓ_{th}, the thermally thin assumption is appropriate and

Eqn [5.2] should be used. For fuels heated on more than one side, half the thickness should be used for comparison. Note that **Eqn [5.3]** contains a time component. This means that a particular fuel particle may not behave consistently as thermally thin or thermally thick. Instead, it could transition between the two conditions over time. For example, a 10 mm diameter fuel particle undergoing a slow rate of heating may initially exhibit thermally thick behaviour, but given enough time for the heat to penetrate the entire particle, it will exhibit thermally thin behaviour.

To obtain a reasonable estimate of ignition time, care is needed in determining whether the fuels are thermally thick or thin. This is particularly true for live fuels, as discussed below. For the rapid heating rates experienced during fire spread, most wildland fuel particles, even pine needles and grasses, do not actually satisfy the thermally thin assumption. For much slower heating, such as from the sun or air temperature changes throughout the day, these fuel particles may indeed be thermally thin.

Factors that affect ignition time

The equations in the previous section can help us picture how various factors affect the time to ignition. As already discussed, the higher the heating rate, the shorter the ignition time. This inverse relationship is apparent in the equations for both thermally thin and thermally thick fuels (**Eqns [5.1]** and **[5.2]**), because the net heat flux ($\dot{q}_s''$) occurs in both denominators. The numerators of both equations indicate that the ignition time is directly proportional to three other fuel properties: the density (ρ_s), specific heat (c_s), and ignition temperature (T_{ig}). These relationships are illustrated in **Figures 5.8–5.10**. One practical application illustrates how ignition time responds to variation in one of the factors in the numerator – fuel density: sound, solid wood, with its high density, takes longer to ignite than punky, rotten wood, with its lower density. This is one reason small firebrands can start spot fires more easily in degraded fuels than in sound ones.

Increased fuel moisture content is another factor that increases the time to ignition (**Figure 5.11**). We

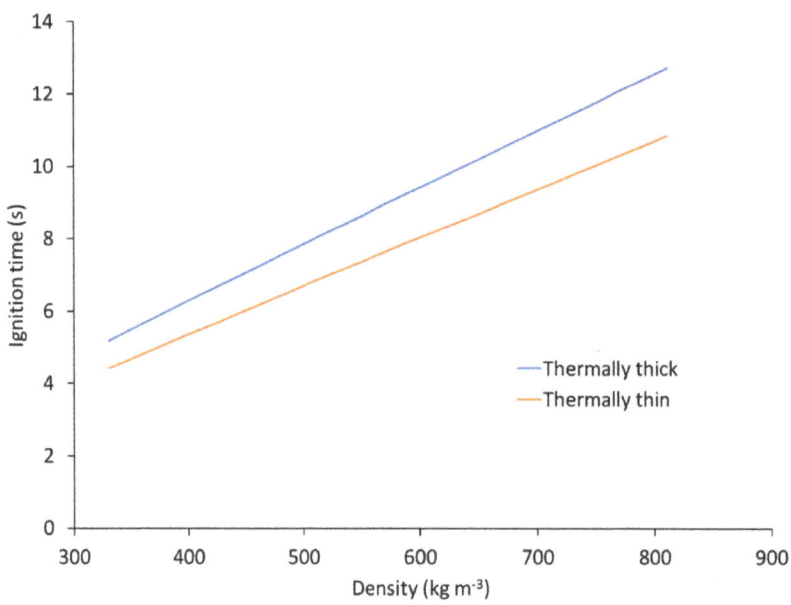

Figure 5.8: Ignition time as a function of the density of the fuel. All other variables are held constant ($k = 1.4 \times 10^{-4}$ kW m^{-1} °C^{-1}, $c_s = 1.255$ kJ kg^{-1} °C^{-1}, $T_{ig} = 350$ °C, $T_o = 30$ °C, $d = 1$ mm, $(\dot{q}_s'') = 30$ kW m^{-2}). The trend is shown for both the thermally thick and thermally thin assumptions as calculated from **Eqns [5.1]** and **[5.2]**.

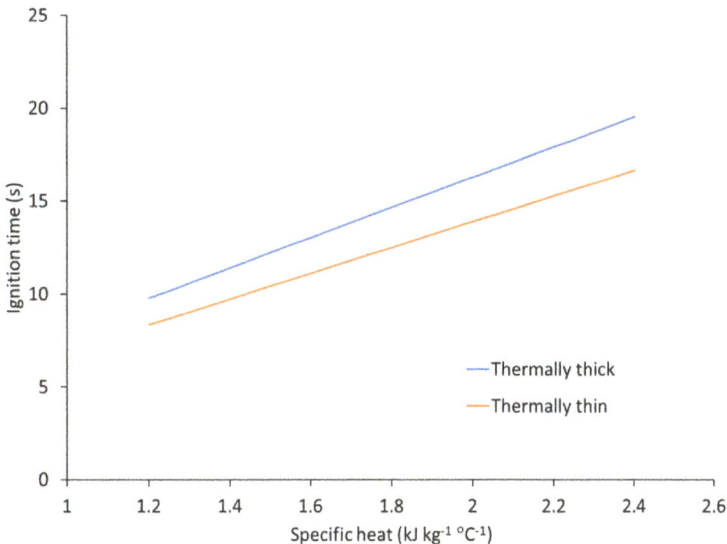

Figure 5.9: Ignition time as a function of the specific heat of the fuel. All other variables are held constant ($k = 1.4 \times 10^{-4}$ kW m^{-1} °C^{-1}, $\rho_s = 650$ kg m^{-3}, $T_{ig} = 350$ °C, $T_o = 30$ °C, $d = 1$ mm, $(\dot{q}_s'') = 30$ kW m^{-2}). The trend is shown for both the thermally thick and thermally thin assumptions as calculated from **Eqns [5.1] and [5.2]**.

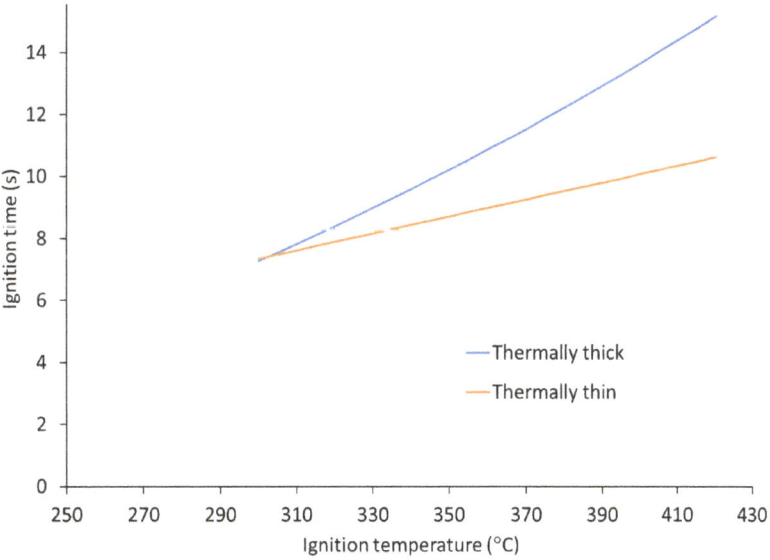

Figure 5.10: Ignition time as a function of the ignition temperature of the fuel. All other variables are held constant ($k = 1.4 \times 10^{-4}$ kW m^{-1} °C^{-1}, $\rho_s = 650$ kg m^{-3}, $c_s = 1.255$ kJ kg^{-1} °C^{-1}, $T_o = 30$ °C, $d = 1$ mm, $(\dot{q}_s'') = 30$ kW m^{-2}). The trend is shown for both the thermally thick and thermally thin assumptions as calculated from **Eqns [5.1] and [5.2]**.

can explain this by looking at the effect of moisture on the heating process. Moisture increases the density, thermal conductivity and specific heat of the solid. For a given heating rate, the temperature of a wet fuel increases more slowly than that of a dry fuel. Additional energy is also needed to evaporate the water, stalling the rise in temperature of the fuel (**Figure 5.1**). Moisture can play an additional role in increasing the time to ignition. If the fuel behaves at all like a thermally thick material, the

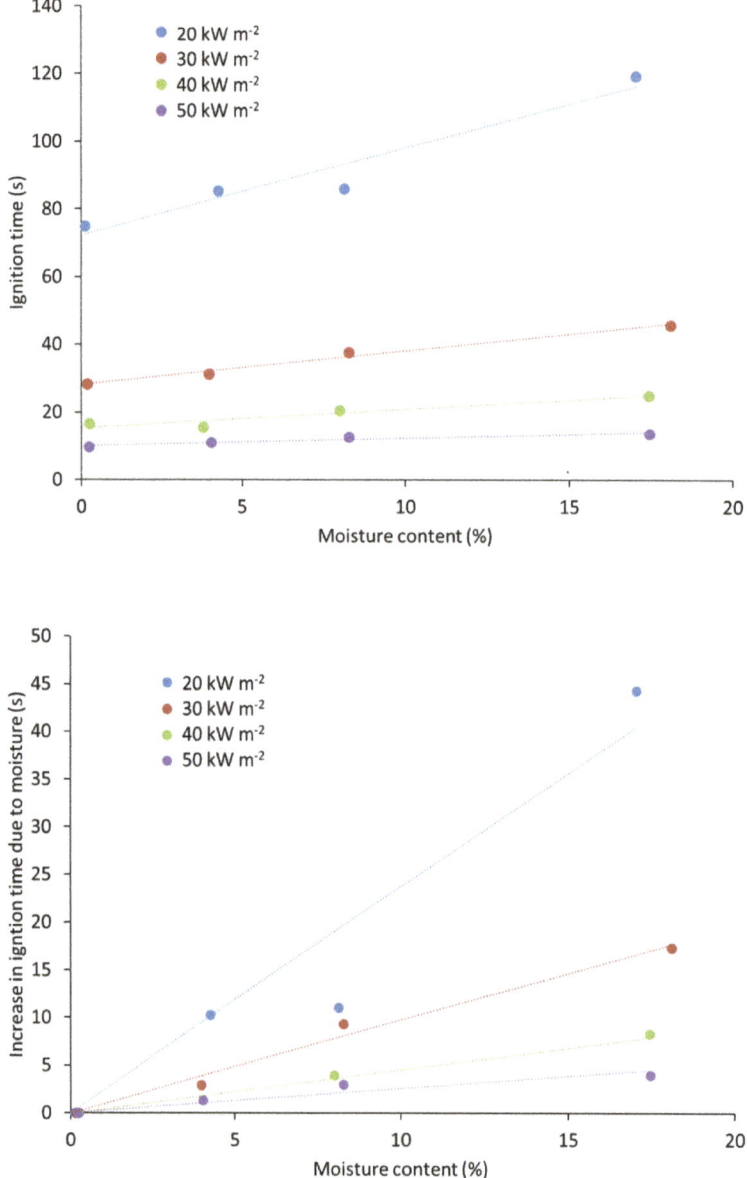

Figure 5.11: Change in ignition time for different moisture contents of wood. Wet wood slabs (10 cm × 10 cm) were exposed to different radiant heat fluxes ranging from 20–50 kW m⁻². (a) The variation in ignition time with moisture content, demonstrating the reduction in ignition time as the radiant heat flux increases. (b) The increase in ignition time relative to the dry case caused by moisture, highlighting that the linear increase in ignition time with moisture is present for all heat flux levels. Data are from McAllister (2013).

temperature gradient inside the fuel causes the water inside to still be evaporating while the surface is hot enough to ignite. The exiting water vapour dilutes the gaseous fuel at the surface, so more gaseous fuel is needed to sustain ignition. Since higher temperatures are needed to increase the rate of pyrolysis (gaseous fuel production), the release of moisture from inside the fuel raises the ignition temperature. In fact, an increase of 1% in fuel moisture raises the ignition temperature ~2 °C (Janssens 1991). Higher ignition temperatures increase the required heating time.

One difference between thermally thick and thin fuels is that the calculated ignition time of thermally thin fuels is sensitive to the actual fuel thickness and the thick fuel is not[iv] (you can see this by comparing **Eqns [5.1]** and **[5.2]**). The ignition time for thermally thin fuels increases as the fuel thickness increases, all else held constant. However, if the net heating rate remains constant as the fuel size increases, at some point this dependence on physical thickness disappears, the fuel gradually transitions from thermally thin to thermally thick behaviour, and the ignition time becomes related to the thermal conductivity of the fuel instead. The transition from thermally thin to thick is not abrupt. For a given set of conditions, there will probably be a range of fuel sizes that do not perfectly follow the definitions of either assumption. In these conditions, either assumption – either **Eqn [5.1]** or **[5.2]** – can produce a reasonably accurate estimate for ignition time.

Ignition time may vary with fuel size even if the fuel is truly thermally thick. As we learned in **Chapter 3, Heat transfer**, the heating rate of a fuel may vary with fuel size. Heating and cooling by convective heat transfer becomes more and more efficient for smaller and smaller fuels. For example, a 1 mm diameter pine needle bathed in flame will ignite more quickly than a 15 mm diameter branch, since the convective heating rate is faster. **Figure 5.12** shows the ignition time calculated using **Eqn [5.1]** for a wood cylinder of varying diameter that is instantaneously heated convectively with 800 °C air. (This is similar to what occurs when flames contact fuels in wildland fires.) Even though the size of the fuel does not appear directly in **Eqn [5.1]**, it does appear in the relationship used for calculating the convection heat transfer coefficient (**Eqns [3.29]** and **[3.36]**), and this coefficient declines with increasing particle size (**Figure 3.20**). The heat transfer coefficient in turn is used to calculate the net heat flux at the fuel surface (**Eqn [3.28]**). As the particle diameter increases, the rate of convective heating declines and therefore the time to ignition increases. Even if the

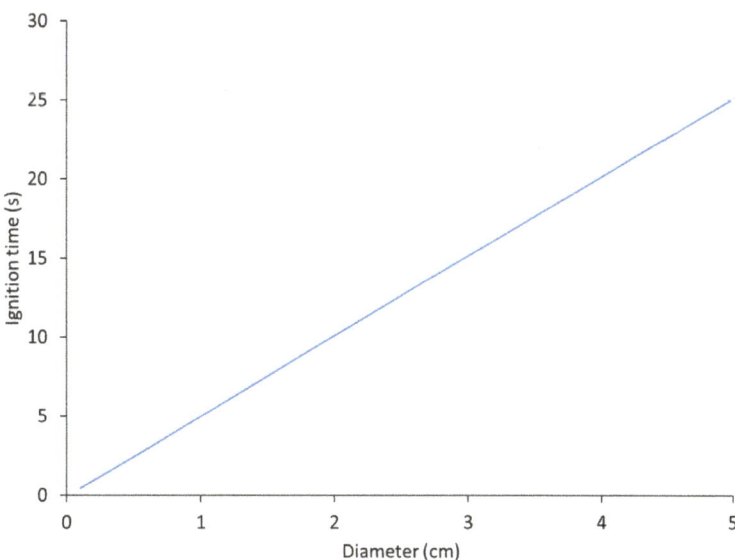

Figure 5.12: Ignition of thermally thick fuels can vary with particle size due to changes in convective heat transfer rate. The relationship shown here is calculated using **Eqn [5.1]** with particles instantly exposed to a convective heat source of 800 °C, which is similar to immersion in flame. The convective heating rate is calculated using **Eqns [3.28]**, **[3.29]** and **[3.36]**, assuming a constant surface temperature of 160 °C (average between initial and ignition temperature). The thermal penetration depth (**Eqn [5.3]**) is less than the radius in all cases shown, so use of the thermally thick assumption is appropriate for this scenario.

heating is caused by *radiation*, the *net* heating rate will probably increase with increasing particle size because heat *losses* due to convection differ for fuels of different sizes: smaller particles lose heat more quickly from convection than larger particles. (Recall the 'leaky bucket analogy' shown in **Figure 3.2**.) The dependence of ignition time on fuel thickness can therefore be present even if the fuel is truly thermally thick (i.e. a true semi-infinite solid).

The effectiveness of convective heat loss for small fuel particles can be so dramatic that it completely shifts the balance between heat gained and heat lost, preventing ignition from occurring at all. **Figure 5.13** shows an experiment in which two square fuel rods are suspended in calm, ambient air and exposed to a radiant heat flux of 41 kW m^{-2} on one side. The 12 mm rod ignites ~30 s after heating begins. In comparison, the 1 mm fuel rod loses heat so fast (even in calm, windless laboratory conditions) that it never gets above 200 °C and thus never ignites. The reason for this difference is that small particles can cool more efficiently by convection than large ones. In the experiment, as the radiant heat flux warms the surface of the fuel, the warm surface heats the air in contact with it. This heated air expands and rises because of buoyancy. Cool ambient air flows in to replace it, cooling the surface of the fuel. However, the cool air heats up readily as it flows along the warm particle surface. The larger the fuel particle, the longer the air is in contact with it and the more the air warms, reducing the ability of the air to cool the particle. For small particles, a larger portion of the surface is exposed to the cooler ambient air, before the air has a chance to warm up. The surface of a large particle cools less quickly than that of a smaller particle because a

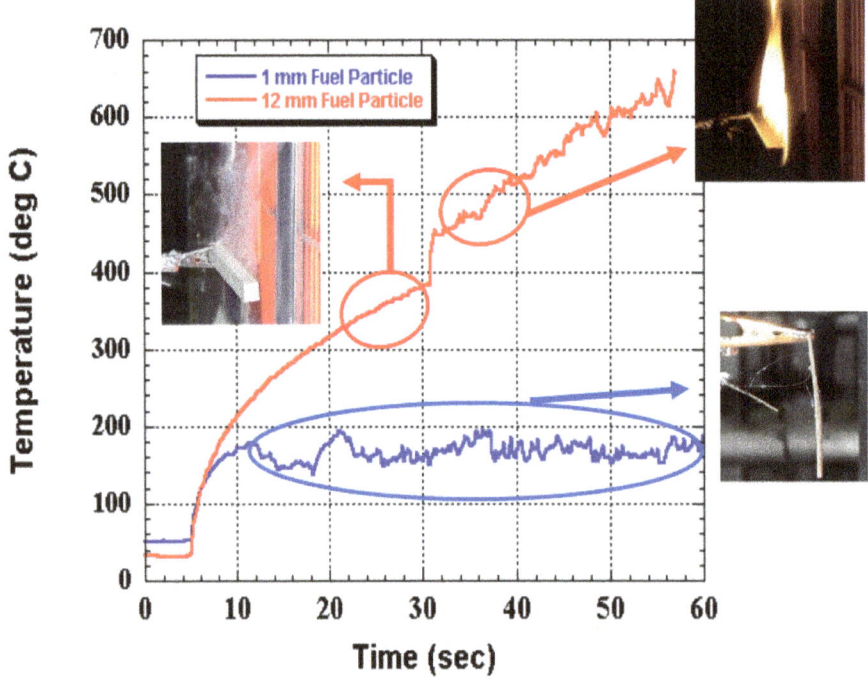

Figure 5.13: The balance between heat gained and heat lost can depend on fuel dimensions, in some cases preventing ignition from occurring. These surface temperature traces demonstrate that a suspended 12 mm square fuel rod heats to ignition in ~30 s with a 41 kW m^{-2} radiant heat flux, whereas a suspended 1 mm square fuel rod never gets above 200 °C and thus never ignites. The radiant heat flux is applied to the vertical right face in all photographs (from Finney *et al.* 2013).

larger proportion of the surface is exposed to warmer air. The converse of this holds true as well. When a flame bathes a relatively cool fuel particle, the flame gases lose heat to the cold surface as they flow over it. The larger the fuel particle, the more those gases cool, reducing the amount of heating a larger particle receives over the majority of its surface. For small particles, the gases stay hot over much more of the surface and the heating rate stays high. Note that in this example, we are considering only natural convection in a windless environment. The differences between the responses of large and small particles would be even greater with a cool wind blowing. Because the efficiency of convective heat transfer can vary so dramatically with fuel particle size, the form of heat transfer that controls ignition and fire spread in wildland fires (i.e. radiation versus convection) changes with the size of the fuel and the environmental conditions. This fact has been unappreciated by many for a long time within the wildland fire community and its relevance to fire spread is a subject of ongoing debate by researchers. As a result, it has not been applied in many fire behaviour models, leading to erroneous assumptions and inaccuracies in the underlying physics.

Live fuels

One unique aspect of wildland fires is the variety of possible fuels involved (refer also to **Chapter 6, Wildland fuel**). Living vegetation is a particularly challenging fuel to address in fire behaviour predictions. Living vegetation, which we refer to as live fuels, can have a complex and constantly changing chemical composition. While these materials contain the same structural carbohydrates as dead fuels (cellulose, hemicellulose, and lignin), they also contain proteins, fats, and other sugars and starches produced from photosynthesis. These compounds can make up more than half of the fuel's dry weight. The chemical composition and proportions of these compounds can change seasonally and may even change over the course of a single day. Some living tissues may also contain liquid resins used for defence against herbivory or insect attack. Because the chemicals that comprise live fuels are diverse and constantly changing, the composition of the gaseous fuel produced by pyrolysis is diverse and constantly changing as well, and this means that fire behaviour in live fuels is different from that in dead fuels.

A second difference between live and dead fuels lies in the ways in which water is stored. The cells of live fuels are full of liquid water–carbohydrate solutions, which are confined within the cell walls. In dead foliage, the cells have collapsed, preventing this mode of water storage. By storing water within cells, live fuels can reach moisture contents of over 100% (expressed as per cent of dry weight), while dead fuel moisture can only top out somewhere between 30% and 40% (the *fibre saturation point*). A third difference lies in the outside surfaces of the leaves of live fuels. For many species, particularly in the hot and dry climates where wildland fires are most frequent, the leaf is coated with a hard, waxy exterior that enables it to resist dehydration. Consider, for example, the textural difference between the soft grass of a green, well-watered urban lawn and the hard, waxy exterior of a chaparral leaf or conifer needle.

The ways in which sugary solutions and water are stored in live fuels probably help to determine whether they will ignite and burn – and which species are more likely to ignite and burn in particular conditions. When living vegetation of some species is heated rapidly in a spreading flame front, it does not evaporate water and release pyrolysis products in the same gentle, continuous way that dead vegetation does. Instead, the process can be quite violent, with jets of moisture and possibly flammable carbohydrates issuing out from ruptures in the surface at high velocities. These jets probably result from high pressures that build up inside the relatively tough cell walls as the liquid solutions are heated, or from high pressures that develop within the hard exteriors of leaves and needles. **Figure 5.14** shows this phenomenon. Side-by-side images taken simultaneously by visual and infrared cameras show dramatic jetting of hot gases from live fuel as it is heated.

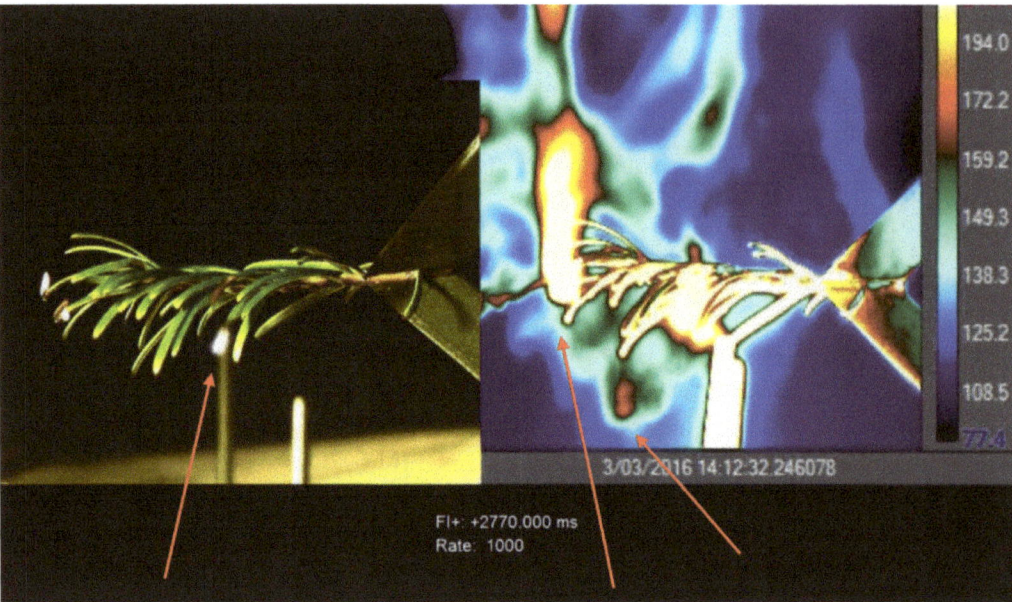

Figure 5.14: Side-by-side images of Douglas-fir needles taken with high-speed visual and infrared cameras. The branches are instantaneously heated by 800 °C air, similar to the heating experienced during a fire. The photographs show dramatic jetting of hot gases (red arrows), some of which are flammable.

In many models, live fuels have been treated simply as thermally thin, wet wood. But making the thermally thin assumption implies that all water has evaporated out of the fuel before ignition. However, experiments have demonstrated that water is indeed still issuing from live fuels at the moment of ignition (see **Figures 5.15** and **5.16**) (McAllister *et al.* 2012; McAllister and Finney 2014; Pickett *et al.* 2010). Additional experiments have demonstrated that the ignition time for several species of live fuels can be predicted equally well with either the thermally thin or thick assumption (McAllister *et al.* 2012), indicating that assuming live fuels are thermally thin is an oversimplification. If living fuels followed the same patterns of drying and heating shown by dead fuels, the ignition time would always increase with moisture content, but this pattern was not evident when ignition times were measured for several species of live fuels whose moisture content varied naturally during the growing season (**Figure 5.17**) (McAllister and Weise 2017).

The chemical and structural differences between live and dead fuels, and among different species of vegetation, help explain why the ignition and burning of live fuels remain poorly understood. Because of the changeable nature of live fuels, it has even been difficult to know their density, thermal conductivity and specific heat at any given time. Research by Jolly *et al.* (2016) provides one of the few studies to track the change in live fuel density over a year. They showed that foliar density ranged from 250 to more than 500 kg m^{-3} during a single growing season. The lowest density was observed in the spring in the newly grown needles. The highest density occurred earlier in the spring in old needles, before the flush of new needles occurred (during the spring dip in moisture content). The densities of the new and old needles gradually approached a density of ~400 kg m^{-3} by the end of the fall, as the old needles transported stored starches to grow the new needles and the new needles matured and hardened.

Another reason for the difficulty of understanding relationships between live fuel moisture and fire behaviour is due to the way moisture content is defined and measured. It is usually reported as the ratio of the mass of water to the dry mass of the fuel

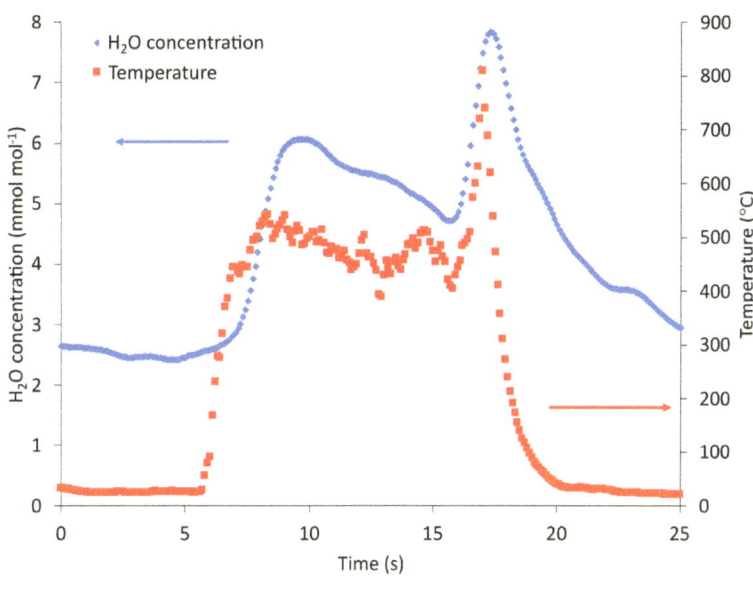

Figure 5.15: Temperature (red, right axis) and water vapour concentration (blue, left axis) measured just above the surface of live lodgepole pine (*Pinus contorta*) needles (moisture content 133%). When the needles were inserted into a 600 °C air stream at ~6 s, both the temperature and the concentration of water vapour concentration increased sharply. At this point, the water vapour was coming from evaporation, and it continued to be released right up until ignition, which occurred at ~16 s. At ignition, both the temperature and water vapour concentration spiked, but at this point the water vapour was a product of combustion (recall **expressions [4.1]** and **[4.2]** in **Chapter 4**). From McAllister and Finney (2014).

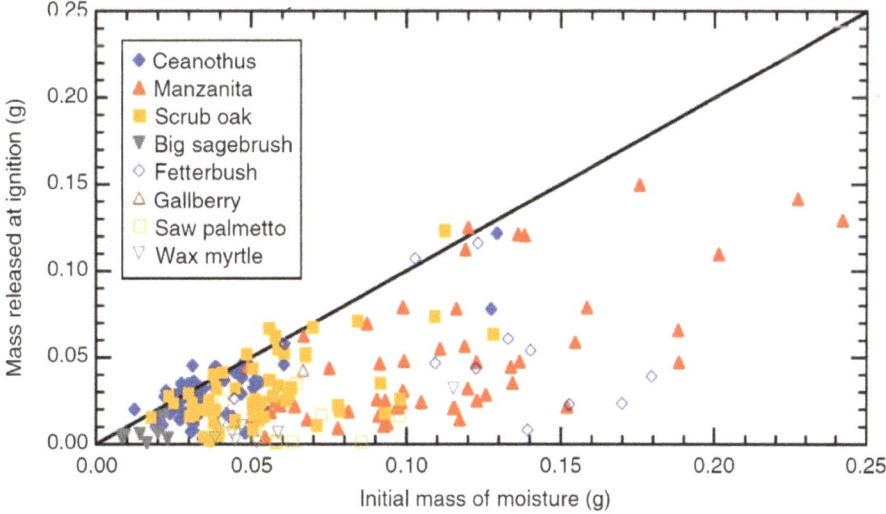

Figure 5.16: Comparison of the initial mass of moisture to the total mass released up to ignition for several species of live fuel burned using a flat flame burner (*Ceanothus crassifolius* Torr., *Arctostaphylos glandulosa* Eastw., *Quercus berberdifolia* Liebm., *Artemisia tridentate* Nutt., *Lyonia lucida* (Lam.), *Ilex glabra* (L.) Gray, *Serenoa repens* (Bartr.) Small, and *Morella cerifera* (L.) Small). Points lying away from the diagonal line indicate that a significant amount of the water initially present in many species of live fuel remains at the moment of ignition. Used with permission from Pickett *et al.* (2010).

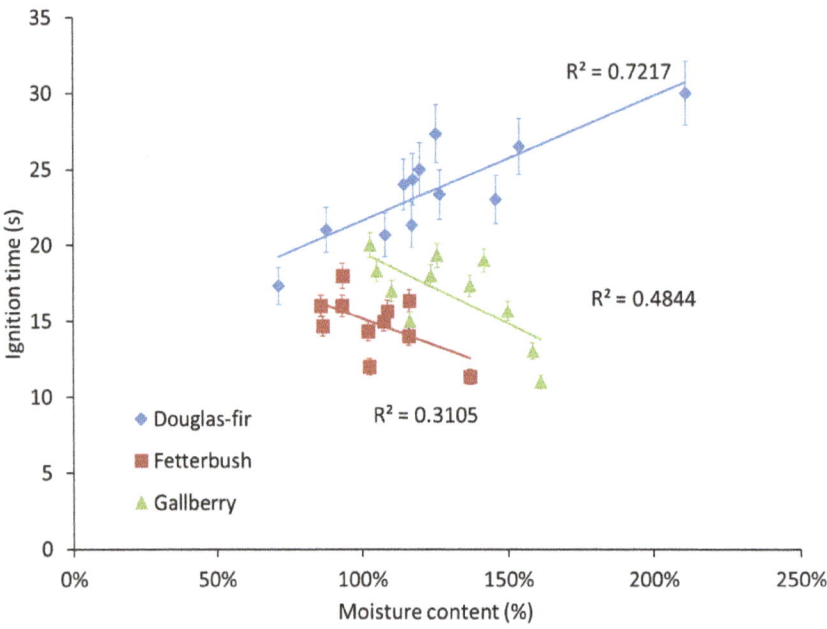

Figure 5.17: Ignition time of three fuel species as a function of moisture content. Moisture content varied naturally during the growing season. Fetterbush and gallberry showed the surprising result that they ignited faster with higher moisture content (reproduced with permission from McAllister and Weise 2017).

(MC = m_{water}/m_{dry}) (see **Chapter 6, Eqn [6.1]** and **Fuel moisture**). In the case of live fuels, however, the dry mass is constantly changing as the plant produces and transports sugars and starches, so the measurement of moisture content can change without a change in the actual amount of water in the fuel.

The ignition and burning of live fuels remain a topic of ongoing research and will require significant creative thinking and the generation of new terms and concepts in the future (e.g. see Jolly and Johnson 2018).

Summary

In this chapter, we waded deep into the topic of ignition and learned precisely how it occurs. We have emphasised that the balance between the heat generated by combustion and the heat lost to the surroundings is the fundamental control for ignition. Respect for the details of the ignition process helps bring perspective to some common assumptions. For example, we have learned that there really is no single 'ignition temperature', but the assumption of one can be very useful for predictions and discussions and does have some basis in the physical processes. In contrast, the assumptions that radiation alone drives the ignition of fine fuels and that live fuels can be treated as thermally thin wet wood are not accurate and are often inappropriately used in some fire behaviour models. The next chapter begins to put the fundamental concepts learned in **Chapters 3–5** together to understand the coupled wildland fire system.

Endnotes

[i] For discussion of flammability, see **Chapter 4**.
[ii] Recall that a *flux* is a rate over a given surface area, so a mass flux is mass per second per area (g m^{-2} s^{-1})
[iii] Technically, a fuel is thermally thick if it is thicker than the thermal penetration depth at a particular time. In order to provide an analytical solution for the ignition time (or other behaviour of interest), however, we employed the semi-infinite solid approximation.
[iv] Here we are again invoking the semi-infinite solid assumption.

References

Babrauskas V (2002) Ignition of wood: a review of the state of the art. *Journal of Fire Protection Engineering* **12**(3), 163–189. doi: 10.1177/10423910260620482

Finney MA, Cohen JD, McAllister SS, Jolly WM (2013) On the need for a theory of wildland fire spread. *International Journal of Wildland Fire* **22**(1), 25–36. doi:10.1071/WF11117

Janssens M (1991) Piloted ignition of wood: a review. *Fire and Materials* **15**(4), 151–167. doi:10.1002/fam.810150402

Jolly WM, Johnson DM (2018) Pyro-ecophysiology: shifting the paradigm of live wildland fuel research. *Fire (Basel, Switzerland)* **1**(1), 8. doi:10.3390/fire1010008

Jolly WM, Hintz J, Linn RL, Kropp RC, Conrad ET, Parsons RA, Winterkamp J (2016) Seasonal variations in red pine (*Pinus resinosa*) and jack pine (*Pinus banksiana*) foliar physio-chemistry and their potential influence on stand-scale wildland fire behavior. *Forest Ecology and Management* **373**, 167–178. doi:10.1016/j.foreco.2016.04.005

McAllister S (2013) Critical mass flux for flaming ignition of wet wood. *Fire Safety Journal* **61**, 200–206. doi:10.1016/j.firesaf.2013.09.002

McAllister S, Finney M (2014) Convective ignition of live forest fuels. *Fire Safety Science* **11**, 1312–1325. doi:10.3801/IAFSS.FSS.11-1312

McAllister S, Weise DR (2017) Effects of season on ignition of live wildland fuels using the forced ignition and flame spread test apparatus. *Combustion Science and Technology* **189**(2), 231–247. doi:10.1080/00102202.2016.1206086

McAllister S, Grenfell I, Hadlow A, Jolly WM, Finney M, Cohen J (2012) Piloted ignition of live forest fuels. *Fire Safety Journal* **51**, 133–142. doi:10.1016/j.firesaf.2012.04.001

Pickett BM, Isackson C, Wunder R, Fletcher TH, Butler BW, Weise DR (2010) Experimental measurements during combustion of moist individual foliage samples. *International Journal of Wildland Fire* **19**(2), 153–162. doi:10.1071/WF07121

6

The environment in wildfire dynamics

This chapter examines the principal characteristics of the wildland fire environment and their role in the physical processes of fire behaviour and spread. Wildland fires are distinguished from fires in other contexts (e.g. buildings, industry) by the character of the fuels, the variability and pervasive influence of the weather, and the presence of highly varied terrain. These factors have comprised a conceptual model widely used for many years to understand the overall behaviour and predictability of fires – the *fire behaviour triangle* (**Figure 2.7**).

While the fire behaviour triangle is valuable as an operational schema, the omission of fire itself from the triangle prevents us from understanding many important fire phenomena (Countryman 1972). Even the simple case of how fires appear to steadily spread through a fuel bed can only be resolved if we understand that spread is driven by the dynamics of a strongly coupled fire system. The characteristics of the fire heat source – its duration, intensity, and geometry – are themselves functions of the environment (components of the fire behaviour triangle). The environment *and* the fire heat source, in turn, strongly affect the processes of heat transfer and ignition. All of these components interact to help explain steady spread. In other words, all of the processes occur within strongly coupled systems.

In this chapter, we discuss fuels, weather, topography *and fire* through their effects on combustion, heat transfer and ignition. This chapter contains some intentional redundancy with **Chapters 3, 4** and **5** so we can address fire processes from the perspective of the fuel or the fire, rather than the perspective of the physical processes alone. In this chapter, we also examine empirical demonstrations of fire processes that extend beyond the theories. For example, we will look at the results of experiments on convective heating, burning rates and particle ignition. The substance of this chapter is the foundation for developing a fire spread model and examining how fire behaviours are affected by environmental factors in **Chapter 7**. Throughout this chapter, we highlight not only what is known about these processes but also what is not known. If we can identify and acknowledge weaknesses in our understanding, we can regard it with a healthy scepticism and bring an open mind to the further study of wildland fire.

Wildland fuel

Fuels for wildland fires are unique in the broader field of fire science because they originate as biomass produced by an enormously diverse biota and they are constantly changing. By using the term 'fuel' instead of 'biomass', we are explicitly assuming that these characteristics relate to the combustion or fire process – else we would just refer to the material as biomass. Plants in different ecosystems and environments grow, reproduce, die and decay at varying rates and amounts and are also transformed by herbivory (**Figure 6.1**). It is not surprising then that the fuel for wildfires is difficult to describe in terms important to fire behaviour, which include height, density, size and spatial pattern. We often think of wildland fuel as mainly dead plant biomass, but that is not always so: some parts of living vegetation also burn, and their properties are far from constant. The chemistry and tissue structure of foliage, for example, changes throughout the season and varies with growing conditions. Plant tissues undergo annual senescence and tissue death. All plants eventually die and decay. Animals, insects and fungi consume biomass and kill plants. All of these processes can greatly alter important fuel properties. Thus, it is challenging not only to describe fuel properties relevant to wildfire behaviour, but also to classify fuels, measure them, and maintain the currency of fuel data over large areas. Yet this information is essential for understanding the behaviour of wildland fires.

Traditionally, variation in wildland fuel has been described within and among three main strata (**Figure 6.1**). *Ground fuels* include decomposed and partly decomposed dead organic material in or below the forest floor; these fuels may be incorporated into the organic soil layers. *Surface fuels* consist of living and dead vegetation close to and above the ground surface; they include leaf and needle litter, grass, shrubs and downed woody material, including logging slash (residual debris after timber harvesting). *Aerial fuels* (sometimes referred to as *crown* or *canopy fuels*) include mainly the living foliage of large shrubs and trees. These three strata have provided an obvious natural basis of fuel classification, and they have strongly structured both fire research and fire modelling because they give rise to representative or archetypal fire behaviours. Thus, wildfires are commonly described by the predominant

Figure 6.1: Fuel components of the wildland environment. Three typical strata are identified: ground fuels, surface fuels and canopy fuels. Ladder fuels consist of bark, branches, moss and lichens, which increase vertical continuity among strata and facilitate ignition from fire below (illustration by Brian Elling).

fuel stratum in which they spread: *ground fires*, *surface fires* and *crown fires*. Some important fuel components do not fit into this three-layer framework, however. These include tree bark, moss, lichens and low dead branches, which may serve as *ladder fuels* that allow fire to climb upward from one stratum to another. One example of a critical ladder fuel is the bark of some Australian eucalypt species; it is not only a ladder fuel but also a primary source of embers for short- and long-range spotting and therefore fire spread.

Ground fires spread slowly (< 1 m min^{-1}), typically through smouldering with intermittent glowing and flaming. *Surface fires* exhibit a wide range of spread rates and intensities, and firefighters directly engage only the mildest fire behaviours in surface fires (refer to **Figure 2.18**). Surface fires in logging debris or litter may spread only 1–20 m min^{-1}, while grass fires in high winds can spread 200–300 m min^{-1}. *Crown fires* spread at intermediate rates, similar to those of brush fires, ~20–100 m min^{-1} at high intensities. However, they are nearly impossible to control directly. All of these kinds of fires can generate embers, which may be transported far ahead of the fire front by strong winds and can cause spot fires that greatly advance the effective rate of fire spread and overcome limits of fuel continuity or geographic barriers (e.g. streams, roads, ridges). Fire behaviour research has often been conducted separately for each fuel stratum and fire type without accounting for interactions with other strata; this approach has led to multiple, irreconcilable model results and difficulty in application to real fires.

Many different schemes for fuel description have been devised, and some are widely used around the world to support operational wildfire prediction systems (Keane 2015). All fuel descriptions are structured by their intended use. Probably the earliest example is work by Hornby (1936), which classified and mapped fuel types explicitly by the kinds of fire behaviour and spread rates expected under 'average worst' weather conditions and also by 'resistance to control' – that is, the effort required to build fire containment line. Spread rates and resistance to control were both classified as low, medium, high or extreme. In Hornby's system, light grass fuels were classified as offering low resistance to control, while large logs in slash fuels were classified as offering high resistance. Barrows (1951) offered a similar fuel description scheme, which classified fuels by categories of expected fire spread rate (low, medium, high) and attempted to account qualitatively for the effect of patchiness (i.e. fuel continuity) on fire spread. A more recent example is found in the fuel hazard guide used in Victoria, Australia, which describes attributes of fuel strata according to their contribution to fire rate of spread and spotting (Hines *et al.* 2010). Ottmar *et al.* (2007) also developed a general system of fuel characterisation for multiple fuel strata within a vegetation profile to supply data for fire behaviour potential classification. Despite the many efforts at fuel classification, wildland fire science still has no common approach or widely accepted method of fuel description that can be applied to fire-related uses of biomass. Most existing fuel descriptions have been developed for practical purposes – mainly to satisfy the requirements of a specific fire behaviour model used for prediction or decision making. The most common applications have been prediction of fire spread and smoke production. While important for their use in specific models, these fuel descriptions generally sacrifice accuracy and precision to achieve practical goals and meet the limitations and assumptions of specific models.

An important example of the pragmatic yet limited nature of fuel descriptions is found in the fuel models used in the United States for fire behaviour prediction and fire danger rating. The Rothermel fire spread equation (Rothermel 1972) lists several fuel characteristics that, when quantified, are combined into a model of a fuel type – called a *fuel model*. The original fuel models (**Table 6.1**) were revised by Anderson (1982) and expanded by Scott and Burgan (2005), but the fuel descriptors remained unchanged. The primary characteristics of each fuel model are amount (i.e. *loading*) (kg m^{-2} or t ac^{-1}) by standardised particle size class and *depth* of the fuel bed (m). The particle size range for

Table 6.1. Fuel particle and fuel bed characteristics as specified in the Rothermel (1972) equation for surface fire spread (σ is particle surface-area-to-volume ratio).

Fuel model	Total load (kg m^{-2})	Dead fuel						Living fuel		Depth (m)
		Fine		Medium		Large				
		σ (m^{-1})	Load (kg m^{-2})	σ (m^{-1})	Load (kg m^{-2})	σ (m^{-1})	Load (kg m^{-2})	σ (m^{-1})	Load (kg m^{-2})	
Short grass	0.17	11483	0.17	–	–	–	0.00	–	–	0.30
Tall grass	0.67	4921	0.67	–	–	–	0.00	–	–	0.76
Brush	1.35	6562	0.22	358	0.11	–	0.00	4921	0.45	0.61
Chaparral	5.60	6562	1.12	358	0.90	98	0.45	4921	1.12	1.83
Timber (grass and understorey)	0.90	9842	0.45	358	0.22	98	0.11	4921	0.11	0.46
Timber (litter)	3.36	6562	0.34	358	0.22	98	0.56	–	–	0.06
Timber (litter and understorey)	6.73	6562	0.67	358	0.45	98	1.12	4921	0.45	0.30
Hardwood (litter)	3.36	8202	0.65	358	0.09	98	0.03	–	–	0.06
Logging slash (light)	8.97	4921	0.34	358	1.01	98	1.24	–	–	0.30
Logging slash (medium)	26.90	4921	0.90	358	3.14	98	3.71	–	–	0.70
Logging slash (heavy)	44.83	4921	1.57	358	5.17	98	6.29	–	–	0.91

each fuel model is represented by the ratio of surface-area-to-volume (σ, m^{-1} or ft^{-1}), which is based on the diameter d of round or square particles ($\sigma = 4/d$). This ratio is intended to characterise the physical effects of particle heat and moisture exchange, but, as we will discuss, other metrics are needed to more accurately characterise the heat gain–heat loss relationships for convective and radiative heating. In the fuel models, the physical properties of particles, such as density and heat of combustion, are assumed constant, partly because they are essentially unknowable across expansive wildlands, but also because the Rothermel fire spread equation is not particularly sensitive to these fuel properties.

Fuel descriptions and models for forest fuels that can burn in crown fires usually include only bulk properties of fine foliage over large areas and a general estimate of the distance of these fuels from the ground (Van Wagner 1977; Keane 2015). Thus they are grossly generalised when we consider the complexity and variability of crown fuels. Fuel models developed for use in the empirical fire modelling systems of Australia (Cruz et al. 2015) and Canada (Forestry Canada Fire Danger Group 1992) contain coefficients and model forms that directly represent specific vegetation types (e.g. mallee–heath shrublands in Australia and mature jack pine (*Pinus banksiana*) or lodgepole pine (*Pinus contorta*) in North America). Several Australian models include loading of surface fuels and bark components for specific tree species. While important for specific applications, these fuel models do not provide an approach that can be widely generalised.

Most of the fuel descriptions currently in operational use are intended to be broadly applicable to vegetation physiognomy – mixtures of grasses, shrubs and forest understorey conditions – but are not explicitly descriptive of the physical fuel properties pertaining to species composition or site-specific variations because, as explained by Van Wagner (1990),

> ... the problem of how to describe a fuel complex in terms that would permit a physical deduction of how fire would spread through it has so far proved intractable.

Operational systems for predicting fire behaviour have yet to address the spatial variability of fuel properties or the effects of fuel heterogeneity on fire behaviour (Anderson *et al.* 2015; Cruz *et al.* 2015).

It is clear that we cannot look to operational fire modelling systems for ways to describe the characteristics of fuels that drive fire behaviour. Instead, let us ask what the fuel properties critical to fire behaviour are – without the constraints from a particular fire behaviour model. Wildland fuels exhibit two natural levels of organisation: *fuel particles* and their material properties (**Table 6.2**) and *fuel beds*, which are collections and arrangements of particles (**Table 6.3**).

Fuel particles and fuel beds

Fuel particles are the elementary constituents of wildland fuel beds, and they present the most

Table 6.2. Fuel particle properties and their relation to physical processes of fire behaviour.

Material property (symbol, units)	Relevance to fire behaviour
Thermal conductivity (k, W m^{-1} K^{-1})	Controls conductive heat transfer into a particle.
Particle density (ρ_p, g m^{-3})	Higher values mean greater conductivity and heat content, which is related to lower spread rate.
Specific heat (c_p, J g^{-1} K^{-1})	Heat required to raise unit mass by 1 K.
Thermal diffusivity (α, $k/\rho c_p$, m^2 s^{-1})	Determines rate of conductive heat transfer through a particle.
Heat content (H_c, J g^{-1})	Amount of thermal energy per unit dry mass
Ash content (% or fraction)	Non-combustible content of particle. Decreases burning rate and heat content.
Moisture content (MC, % or fraction)	Slows ignition, burning rate, and spread but increases burning time.
Size or thickness, indicated by diameter d or cross-section (m)	Small particles are less sensitive to radiant heating; large particles burn longer because higher heat flux needed to ignite.
Shape, surface-area-to-volume ratio (σ, m^{-1}), complexity of leaf edges	Influence heat transfer, ignition and burning rate.
Live or dead, species, physiology, other unknown properties	Many unknowns. Some species ignite and burn well, others do not.

Table 6.3. Fuel bed properties and their general relevance to fire behaviour.

Fuel bed property (symbol, units)	Relevance to fire behaviour
Particle size distribution	Small v. large particles may exhibit distinct ignition and burning behaviours. A mixture of size classes may facilitate fuel consumption.
Bulk density (ρ_b, kg m^{-3})	High values mean slower burning, longer flame residence times.
Packing ratio (β, ρ_b/ρ_p, dimensionless)	High values mean slower burning, longer flame residence times.
Porosity (λ, ϕ, $1/(\sigma\beta)$, m)	Lower values mean slower burning, longer flame residence times.
Fuel bed depth (δ, m)	Deeper beds can mean faster fire spread.
Vertical stratification of fuel properties	Vertical distributions of fuels and gap properties affect ignition and spread.
Dry fuel loading (m_d'', kg m^{-2})	High loadings increase residence time and total stored heat energy, decrease burning rates
Available fuel loading (m_c'', kg m^{-2})	Dry mass of fuel consumed by a fire (in flaming, glowing, or smouldering). Depends on moisture, packing, wind and other factors
Cover, fuels and gaps (%)	Spread thresholds.
Patch sizes, fuels and gaps over 2D area (m^2) or 1D length (m)	Spread thresholds.
Patterns of fuel and gaps (e.g. random, regular, clumpy)	Variability in fire behaviour, spread thresholds.
Live/dead fuel ratio (dimensionless)	More dead fuel tends to increase ability for fire to spread.

important characteristics that distinguish wildland fires from those in industrial or built environments. They are unique in that they are discrete solids, mostly small in size, separated by gaps of air, and they may be living or dead at the time of burning. These characteristics affect their heating, their combustion and their ignition, especially compared to the materials found in building fires (i.e. large surfaces of walls, furniture, liquids or gases, and various synthetic materials). Properties of fuel particles in wildland vegetation are listed in **Table 6.2** along with their relevance to fire behaviour processes.

If we hope to understand how fuel beds affect the physics of wildfire behaviour, we must describe both the particles themselves *and the air gaps between them*. Existing fuel measurement and classification schemes ignore the gaps. For example, the typical applications of line- and planar-intercept methods for sampling woody fuels (Brown 1971; de Vries 1986; Keane 2015) count the intercepted fuel particles over the length of the transect and summarise the intercepted quantities per unit land area (i.e. kg m^{-2}). These measurements quantify average amounts by particle size class but do not characterise particle arrangement (more on fuel and fuel consumption measurement is found in **Chapter 9, Fuel consumption**). Without addressing the arrangement of fuels, these measurements cannot characterise the ways in which particles and fuel bed properties are affected by ignition and burning rates, nor can they characterise the ways in which the fuels, in turn, are affected by fire. It may seem unnecessary to measure gaps in the fuel bed – that is, to measure something that is not present. However, we have seen from the discussion of physical processes in **Chapters 3, 4** and **5** that what is not present, and where it is not present, are critical factors in fire spread. Let us first review the description of fuel particles and the impacts that particle properties have on heat transfer, ignition and combustion. Then we will return to the challenge of characterising gaps – the spatial arrangement of fuels.

Fuel particle geometry

The sizes and shapes of particles strongly affect how fire spreads because these geometries control a particle's response to the heat being transferred through radiation and convection from an approaching fire. In many previous fire models (e.g. Rothermel 1972), fuel particle size was primarily described by the ratio of surface area to volume (σ, m^{-1}). But as we know from the details on radiation and convection heat transfer in **Chapter 3**, the efficiency of convective heat exchange depends not on σ but instead on an empirical *heat transfer coefficient h*, the rate at which a fluid exchanges heat with a particle surface. This coefficient accounts for the ways in which temperature, velocity, thermal boundary layers and the geometry of the surface affect heat transfer. Many models represent the convection heat transfer coefficient as an average for the entire particle, but many particles have shapes that cause substantial variation in the rate of heat transfer (e.g. **Figure 3.21**). For example, thin or spiny edges on a leaf respond more quickly to heat transfer by convection than the centre of the leaf, so a single convective heat transfer coefficient cannot accurately model the ignition of the entire leaf. (The edges tend to ignite first, the centre later (Prince and Fletcher 2014).) The nonlinear inverse relationship between convection heat transfer coefficient and particle diameter shown in **Chapter 3, Figure 3.20** and **Eqn [3.36]** suggests that particles larger than a few millimetres will all respond similarly to even the largest particles or surfaces (e.g. logs) to convective heat exchange and ignition. Particle heat transfer involves cooling as well as heating, so small particles with high values of h both heat and cool efficiently by convection.

A series of simple radiant heating experiments (**Chapter 5, Figure 5.13**) demonstrated that small round or square particles, like grasses and pine needles, are so efficiently cooled by the flow of ambient air around them (natural convection) that they resist heating to ignition temperature (nominally 350 °C) in wildfires. The finest particle (1 mm square wooden rod) cannot be heated to begin pyrolysis or ignition with 41 kW m^{-2} flux of

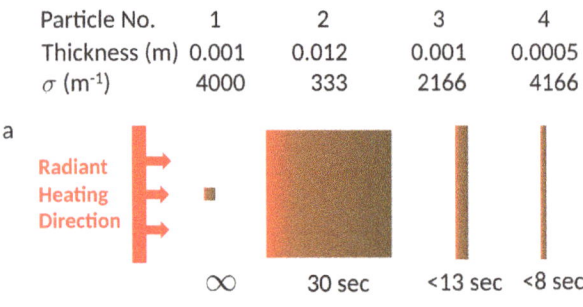

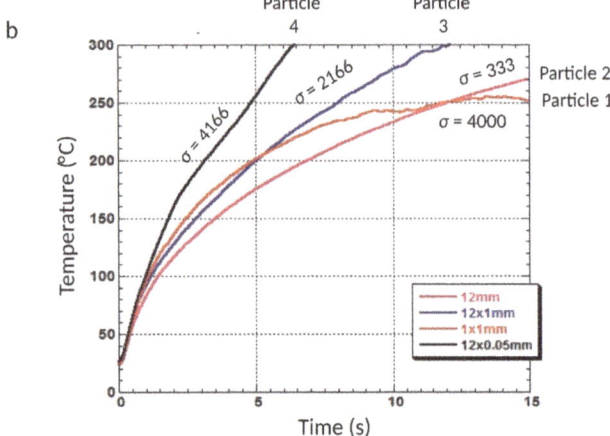

Figure 6.2: Data from laboratory experiments in which fuel particles with varying width and thickness were heated by radiation but cooled by natural convection. Results show (a) that time to reach the pilot ignition temperature for fuel particles receiving a steady flux of thermal radiation (35 kW m^{-2}) depends upon their shape and thickness, not surface-area-to-volume ratio (Cohen 2015). Graph (b) shows the pattern of temperature increase in fuel particles of different dimensions when exposed to radiation and natural convection cooling.

rectangular wooden sheets 12 mm wide and 1 mm (particle 3) and 0.5 mm (particle 4) thick. Both particles 3 and 4 heat to ignition temperature faster than the others, while particle 1 does not reach ignition temperature at all. These results demonstrate some important aspects of the heating process (Cohen 2015). The heating and cooling of the fuel particles depend upon the thickness of both the boundary layer and the particle. As **Figure 3.5** illustrates, the boundary layer thickness depends upon the surface length of the particle in the direction of flow. Bigger particles with larger surfaces have thicker boundary layers, which buffer the surface from contact with cool ambient air flow and thereby make radiant heating more effective. Larger particles thus heat and ignite more readily by radiation than smaller particles. But obviously this is not the whole story, because the particle thickness in the direction away from the heated surface determines the amount of material that can absorb heat by conduction. Compare the ignition time for the 12 mm thick particle (particle 2) and the 1 mm thick particle with the same frontal surface area (particle 3). They both have the same boundary layer thickness, but the 12 mm thick particle has much more internal material per irradiated surface than the 1 mm particle, so it takes longer to reach ignition temperature. Note that σ is irrelevant to the rate at which these fuel particles are heated. Particles 1 and 4 have the greatest σ, yet one is the first to reach ignition temperature and the other does not reach it at all! Neither boundary layer thickness nor particle thickness is described by σ.

Fuel particle properties

Material properties of fuel particles vary considerably according to the biology of plant species and how the environment determines growth rates, sizes, densities, chemistry, and distribution of cells and organs within the plants. These properties affect fine-scale characteristics of particle heating, ignition, combustion, and energy release (**Table 6.2**). Unfortunately, they are thus far unknown and possibly not even measurable across the vast and varied land areas encompassed by wildland fires. Here we

radiation. But, as other experiments have shown (i.e. McAllister and Finney 2017), if heated at the same rate by *convection* with hot gases flowing past the particle, its temperature will rise to ignition in a matter of seconds. A larger particle (a 12 mm square rod) exposed to the same radiation heats to ignition in ~30 s.

Further experiments with radiative heating (**Figure 6.2**) illustrate how the length of the boundary layer and thickness of the particle help account for the rate of heating. The experiments used a 1 × 1 mm wooden rod 12 mm long from end to end (particle 1), a 12 × 12 mm rod (particle 2), and describe some of the research that has characterised physical properties of fuels.

Properties of wood have been intensively researched in relation to commercial uses of wood products, including structure, fire safety and bioenergy. Detailed references include the *Wood Handbook* (Forest Products Laboratory 1999), which reports established values for many physical properties of woods and wood constituents. While this information is helpful for understanding fire behaviour, wood itself is not always the primary fuel for wildland fires, and the physical properties of other fuel materials can differ substantially from those of wood. Grasses, brush, tree bark and living foliage are the main fuels for many wildfires, yet their physical properties have been little investigated. However, some values of thermal properties for oven dry plant materials have been reported (see table 4-5 in Kerr *et al.* 1971; Anderson 1990a). Other information can be found in studies of crop residues or vegetation considered for generating biomass energy. Despite these sources of information, much basic research will be required to determine the thermophysical properties (density, thermal conductivity and specific heat) of plant tissues, both living and dead, before we can understand how or if such variability is important to fire behaviour.

General burning behaviour of fuel has often been termed *flammability*. Flammability is different for solid materials (such as wildland fuel particles) than for gas-phase fuel–air mixtures as defined in **Chapter 4, Types of flames**. Although flammability seems an intuitive concept, it has traditionally been used to encompass four qualitative aspects of combustion of wildland fuels: *ignitability*, *sustainability*, *combustibility* and *consumability* (see review by Varner *et al.* 2015a). Each of these qualities is associated with multiple but correlated physical metrics such as time to ignition and heat release rate, and the qualities are all interdependent with fuel particle and fuel bed properties. Although it is generally accepted that fuel chemistry affects combustion (**Chapter 4, Combustion reactions**), we have seen above that little is known of the chemical variability among species or over time, or how variety in particle properties affects wildfire spread and behaviour. While these properties are difficult to measure precisely under field conditions, they can be measured to some extent under controlled laboratory conditions. One study (**Figure 6.3**) reported considerable differences in burning behaviour of recently fallen foliage of 22 pine and oak species in the south-eastern United States (Varner *et al.* 2015b). The burning behaviours were complex functions of multiple variables, but the dominant trends were explained in terms of flame size, fuel consumption, and duration of flaming and smouldering. Fuel particles of these different species also have different geometries, including different lengths, cross-sectional shapes, and thicknesses, all characteristics that affect their response to heating and to burning, both individually and in small bulk samples, in these kinds of studies.

Many thermophysical properties of fuels have interdependent roles in heat transfer and combustion, so it is difficult to isolate the effect of any single property of a particle on fire behaviour. For example, increased wood density ρ_p generally increases the thermal conductivity and specific heat of the wood. Laboratory experiments that have controlled all fuel properties suggest that particles with higher density burn longer than those with lower density and spread fire more slowly (Fons *et al.* 1963).

Fuel bed properties

Just as fuel particles come in all sizes and shapes, the arrangement of the particles in the three dimensions of space (i.e. the *fuel bed*) is highly variable. Several important properties of the fuel bed are described in **Table 6.3**. A common metric for describing the fuel bed at a specific location is the *bulk density* (ρ_b), the mass per unit volume of particles (kg m^{-3}). To know bulk density requires that we know or measure the vertical depth of the fuel bed (m). In natural fuel beds, bulk density varies considerably vertically and across the ground, giving many fuel beds a heterogeneous

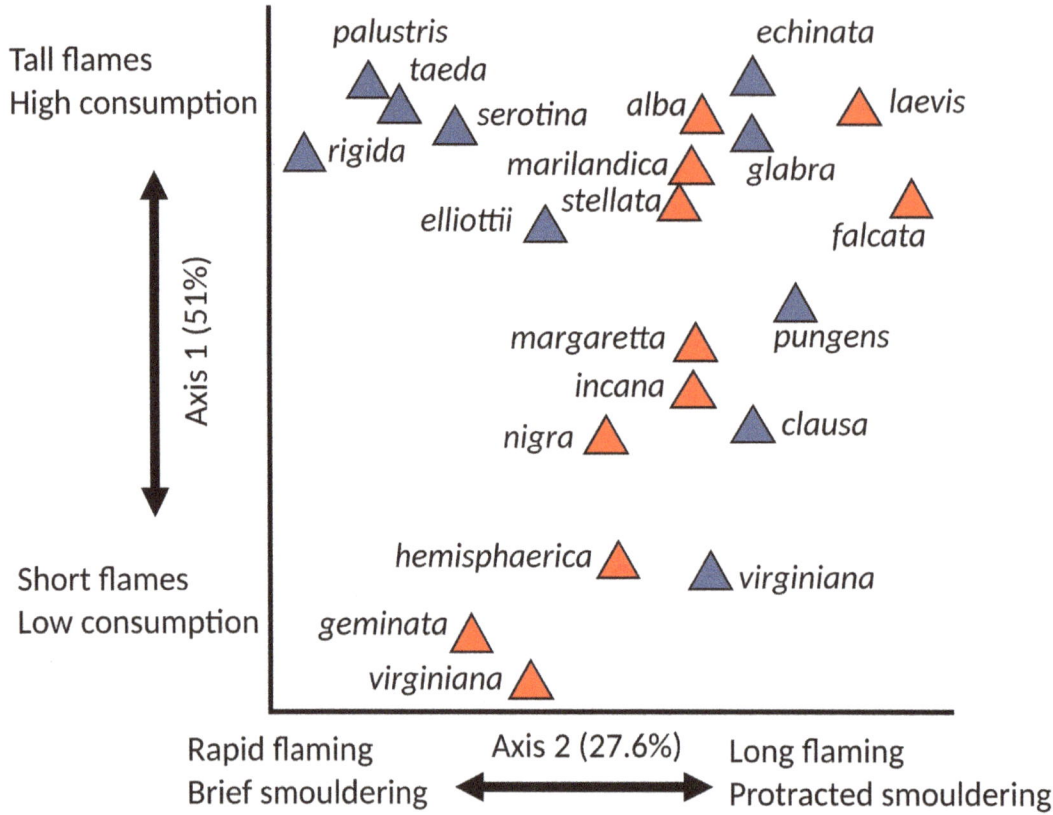

Figure 6.3: Differences in burning characteristics of dead foliage for pine (blue) and oak (red) species (from Varner et al. 2015b) in the south-eastern United States as depicted by principal component analysis. Samples of equivalent dry mass tended to vary in terms of flame length and fuel consumption (y-axis), which accounted for 51.3% of variation, and duration of flaming smouldering (x-axis), which accounted for 27.8% of variation.

appearance. Brown (1981) recognised the vertical and horizontal variation in fuel bed depth, so he used bulk density as a primary means of classifying surface fuel beds and explored the possibility of using mixtures of bulk density categories to represent spatial variation. Bulk density can also be converted to *packing ratio* (β, dimensionless) – which is the fraction of bed volume filled with fuel and can be calculated as the ratio of bulk density ρ_b to particle density ρ_p. Other measures of fuel bed packing can also be calculated but are less commonly referenced in the wildfire literature. One example is the ratio of void volume to fuel surface area, often called porosity λ. While a useful concept, porosity has been defined in different ways: Nelson's (2003) definition, for instance, is different from that used for the burning rate studies described in **Chapter 4, Burning and heat release rate**.

Fuel bed compactness greatly affects heat exchange by radiation and convection, as demonstrated by the simple experiment shown in **Figure 6.4**. Here a mass of ~50 g of dry excelsior (shredded wood) ~1 mm in diameter at two different bulk densities is subjected to a radiant heat flux of ~30 kW m^{-2}. The compact clump (left) experiences autoignition in ~25 s, but the loose clump (right) never ignites. The loose particles do not even discolour or begin to pyrolyse when heated. The reason for this difference in ignition behaviour lies in convection cooling. In both experiments, the fuel particles are exposed to the same radiation, but for the

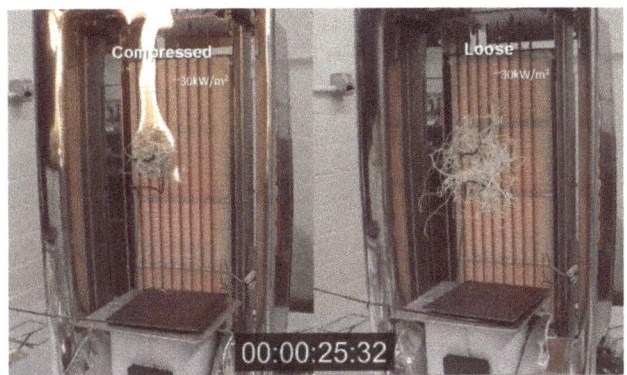

Figure 6.4: Laboratory demonstration showing the effect of fuel compactness on ignition of 1 mm particles exposed to radiant heating at 30 kW m^{-2} in no-wind conditions. The compacted clump (left) ignites in ~20 s because small gaps between particles slow air flow and thus reduces convection cooling. The loosely packed material (right) never pyrolyses or ignites because large gaps between particles allow air flow and effective convection heat exchange.

compact clump the air flow, and hence convection cooling, is inhibited by the small air gaps between particles. As the surface of the fuel particles heats by radiation, the air immediately in contact with their warm surfaces also heats and thus rises, drawing in cool air from below. The compact clump restricts air flow and thus the fuel particles experience less convection cooling and can thus reach higher temperatures sufficient to eventually ignite.

The experiments described above address the rise in temperature of fuel particles in response to a moderate constant flux of radiant heating in quiescent air. Such calm air is rare in wildland fires, however, where there are ambient winds, air currents induced by solar heating, and indrafts caused by the fire itself. As discussed in **Chapter 3, Convection heat transfer**, convective heating and cooling of particles depends upon the temperature of the surrounding fluid, the temperature of the particle surface, and the fluid velocity. Higher velocities mean more efficient cooling or heating, as reflected in the convective heat transfer coefficient (**Figure 3.20**). Higher wind speeds, as illustrated in **Figure 5.7**, increase fuel particle ignition times when exposed to a given radiant flux. This implies that the flow of cool ambient air, which wind forces past fuel particles, will cool individual particles and fuel beds even more efficiently than indicated by these experiments, which used only natural convection. The same principle operates in reverse when fuel particles are exposed to convective heating by flames: flames with higher velocities heat the particles more efficiently and thus reduce ignition time.

The preceding experiments and discussions indicate that convection heating can be the primary means of heating loosely compacted, fine fuel to ignition, irrespective of how much radiation is given off by the fire. Many studies have measured radiation and convection in wildfires and assumed that fire propagation is driven by the *total* amount of heat received (represented by a peak flux or a time-integrated total). Since heat flux from radiation is always present in a burning fuel bed and may often exceed heat flux from convection, this assumption could lead us to assume that radiation is the *dominant* mechanism for heating particles to ignition. There are two problems with this assumption and this perspective. First, heat flux sensors have surfaces too large to accurately measure convection heat transfer to individual small particles in wildland fires, so the convective heat flux is almost always underestimated. (This is discussed in **Chapter 9, Convection**.) Second, even assuming that measurements of convection heat flux are accurate, the relative amounts of radiation and convection do not determine which is the 'dominant' heating mechanism. A more accurate way to determine the dominant heating mechanism would be to determine the process that principally causes the fuel to ignite. If radiant heating is always present, then whether a particle ignites or not may depend on when (or if) convective heating becomes sufficient to cause ignition.

We can understand the important role of rates of heating and cooling (rather than amounts) by using an analogy. Since the process involves the energy balance within the fuel particle, let us revisit the leaky bucket example (**Figure 3.2** in **Chapter 3**). Heating fuel particles to ignition is like trying to fill a bucket that has a hole in the bottom. The flow of water into the bucket represents the flow of heat energy into the fuels, and the flow of water leaking out represents heat loss from the fuels to the environment. To fill the bucket, we must pour water in at a greater rate than the rate at which water is flowing out of the hole in the bottom. In this situation, the flow rates (water pouring in and water leaking out per unit time) are more instructive than the total amount of water added or lost. If water is poured in faster than it leaks out, the bucket will eventually fill up. Let us apply this analogy to the ignition of fuel particles. Heat is simultaneously being added to and lost from the fuel particles, but they will not ignite until the particles have received net heating enough to raise their temperature to ignition (i.e. the bucket is full). Thus the rates of heating and cooling are more important for ignition than the total heat received by the fuels. In wildland fires, rates of heating and cooling can change very quickly as things like turbulent gas fluctuations occur or the flame moves closer to a fuel particle or leans over it.

To take our analogy a little further, the size and shape of the particle are analogous to the size of the hole in the bucket: small particles lose energy more easily than large ones, so they are analogous to a much larger hole in the bucket. This means that more heat must be supplied to reach ignition (just as water must be added at a faster rate to fill the bucket). Many models of fire spread have not considered the effect of cooling rates on ignition, so the process has been described as a dosage-type problem, in which ignition occurred when enough total energy had been transferred to the fuel (i.e. there was no hole in the bucket). With this total-heat approach, ignition was predicted to occur independent of the *rate* of heating and cooling, and thus predictions of ignition and fire spread have remained very challenging.

The relative roles of radiation and convection across a wide range of wildfire behaviours and environmental conditions are not completely understood. Even basic questions remain unanswered. For example, what level of compactness of fine fuel particles interferes sufficiently with convective cooling to cause radiant ignition to be the dominant heating mechanism, given with a particular wind speed and flame zone depth? Compact litter layers, like the tight clump of excelsior in **Figure 6.4**, should respond differently than more open fuels like sparse grass and the loose clump of excelsior in the figure. Yet this concept is not addressed by most fire spread models. Similarly, it is not clear how large woody materials like logs and tree trunks could be cooled sufficiently by wind to prevent radiant ignition. Videos taken inside crown fires (see Taylor *et al.* 2004) show occasions when large, deep flame zones heat and ignite large logs, tree stems and compact litter by radiation, while fine fuel particles elevated above the ground (conifer foliage) require flame contact to ignite (**Figure 6.5**). This suggests that fuel heating and ignition in very intense fires, such as crown fires, are caused by a mixture of heat transfer processes; these depend upon configurations of the particle, fuel bed and the fire itself. Once ignited, the role of fuel particles and fuel beds in continuous flame spread depend on whether their rate of energy release can sustain the flame zone that ignited them.

Fuel moisture

The moisture content of wildland fuel particles has a critical role in all physical processes related to fire behaviour and spread. Moisture content (MC) is traditionally reported on a dry weight basis (i.e. percentage of the oven dry weight):

$$\text{MC\%} = \frac{\text{fresh weight} - \text{dry weight}}{\text{dry weight}} \times 100 \quad [6.1]$$

Moisture content calculated in this way often exceeds 100%, particularly for living material but also for some dead matter. Dead wood that is sound (not rotten) can contain anywhere from 2–~30% water internally, although it may appear to have more if liquid water is on the outer surface (e.g. after rainfall or snowmelt). The fibres of dead wood are usually saturated at around 30% MC. The wood in living trees is far moister, with heartwood having 80–110% water and sapwood having more than 200% because liquid, water-based solutions are held within the cells. Rotten wood, duff and organic soil can have moisture contents as high as 300%, depending on the degree of decomposition. Live foliage can vary from ~50% MC for mature leaves of drought-tolerant species to more than 200% for new foliage. MC for live foliage is typically considered to be around 100% during most of the growing season. The effects of dead fuel moisture on fire spread differ considerably from the effects of live fuels due to the way water is released during heating and ignition (see **Chapter 5, Live fuels**); we discuss these effects and their differences below insofar as the state of knowledge permits.

Dead fuels

The moisture content of dead fuels is constantly changing because the fuel particles are continually exchanging moisture with the atmosphere around them; sometimes they are drying, and at other times they are becoming moister. The rate of change in their moisture content is a strong function of their sizes. These factors make it difficult to know exactly what the moisture of a particular fuel is at any particular time during a fire; hence it is difficult to predict how fires are going to spread and behave.

The responses of dead fuels to changing environmental conditions have been intensively studied. Many detailed synopses describe the processes and models available (Viney 1991; Forest Products Laboratory 1999; Nelson 2001; Matthews 2014), so we provide a survey here.

Without rain or snow causing liquid water to contact the fuel itself, fuel moisture remains primarily in the vapour phase as it is exchanged between the outer surface of the particle and the air. This is primarily assumed to occur through *diffusion*, which is the process by which molecules (water in this case) in high concentration move towards locations with lower concentrations. The locations with high concentration are said to have *high potential*, and the locations with low concentrations have *low potential*. Water potential in the air may be higher or lower than that on the surface of the fuel particle, meaning that water vapour is either taken up by the particle (*adsorption*), thus increasing its moisture content, or it is given off from the particle (*desorption*), decreasing its moisture content. In woody particles, the two processes occur at different rates: particles lose moisture faster than they gain it. These patterns can be seen in *sorption curves*; an example is shown in **Figure 6.6**. The uppermost curve shows 'initial desorption', which occurs in freshly cut wood as it dries out. The other curves show that dead woody particles lose moisture more slowly after they have once been dried, and they gain moisture even more slowly. The different rates of drying and wetting illustrated in **Figure 6.6** may have little relevance to most wildland fire applications because particle moisture varies constantly in wildlands, depending on location, particle condition, microclimate and other variables. Some particles are in contact with the ground and others are suspended above the ground; some are in shade, some in sunlight (which changes throughout the day); some have bark, some do not; particles are in different states of decay; and the structure of the fuel (wood) differs among species. All of these factors influence particle moisture exchange, but they are imprecisely known and hence difficult (essentially impossible) to model with accuracy.

To understand how moisture exchange alters the moisture content of fuels, we need to understand several concepts related to the pressure of water vapour in the air. *Vapour pressure* is the pressure that a gas exerts on a liquid surface. For fuel moisture, vapour pressure is the pressure that water vapour exerts on the liquid water on the surface of the fuel; in practical terms, it tells us how much water vapour is held in the air. In a sealed container half-filled with water, the amount of water that evaporates from the liquid will eventually exert enough pressure on the water surface to prevent further evaporation. This is called the *saturation*

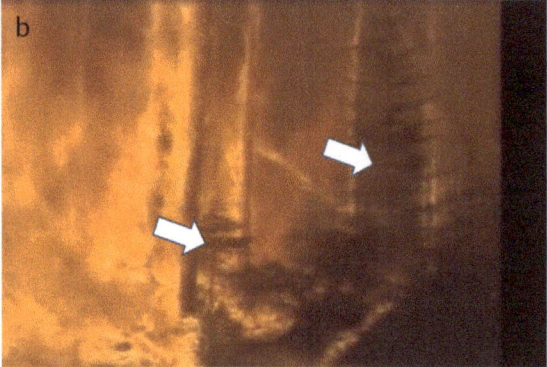

Figure 6.5: Photography showing lack of ignition from radiant heating. Video frames (a) and (b) from in-fire cameras in the International Crown Fire Modelling Experiments (Butler *et al.* 2004) show that fine foliage of black spruce (*Picea mariana*) trees in the understorey (arrows) remain unignited because of convective cooling despite radiant flux above 100 kW m^{-2}; these fuels required flame contact for ignition. Photograph (c) is from a study of ponderosa pine (*Pinus ponderosa*) needles exposed to 35 kW m^{-2} of radiant flux in quiescent air for ~1 min. No pyrolysis or ignition is visible because of cooling by natural convection.

vapour pressure. At saturation vapour pressure, water molecules are evaporating from and condensing on the water surface at exactly the same rate. Saturation vapour pressure is not a fixed number, however; it increases with the temperature of the air, so warm air can hold a lot more water vapour than cold air before it becomes saturated. The temperature at which the air reaches saturation is called the *dewpoint*.

In ambient conditions, the vapour pressure of water in the air is usually less than the saturation pressure. The ratio of the ambient vapour pressure to the saturation vapour pressure is *relative humidity* (RH), expressed as a percentage. An RH of 100% occurs at the saturation vapour pressure; an RH of 50% means that the air currently holds only half as much water vapour as it could at that temperature. As air cools off, the saturation vapour pressure declines, so the air cannot hold as much water vapour as it does when it is warm. When the temperature declines to the dewpoint, excess water vapour in the air condenses and forms dew on objects like wildflowers in the wildlands and grass in the lawn.

We need one more metric of atmospheric moisture to proceed. This is the *vapour pressure deficit*, the difference between the partial pressure of water vapour and the saturation vapour pressure at a given temperature. The drying and wetting rates of fuel particles depend heavily on the vapour pressure deficit because it accounts explicitly for the gradient in vapour pressure near the fuel particle, which drives the rate of water exchange to and from the fuel surface. Warm air at a given relative humidity holds much more moisture than cold air at the same humidity, so its vapour pressure deficit is much greater and it can dry or wet fuels much faster.

Fuel moisture dynamics in dead fuels are best understood in the context of *equilibrium moisture content* (EMC), the moisture content within a fuel particle when it is neither gaining moisture from the air nor losing moisture to it. EMC varies with weather conditions, as shown by models that represent EMC as a function of relative humidity and

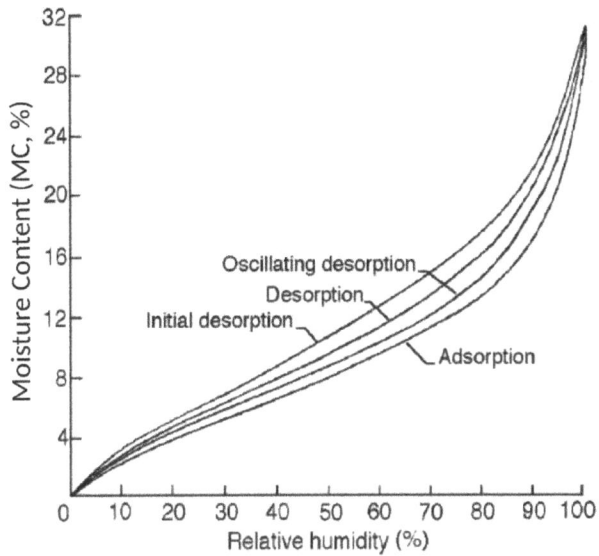

Figure 6.6: Sorption curves for wood in relation to relative humidity. Adsorption of water vapour (i.e. moistening) is slower than desorption (i.e. drying) (Forest Products Laboratory 1999). The 'initial desorption' curve represents drying of fresh, living wood after cutting. Wood in this condition dries more slowly than older cut wood ('desorption' curve).

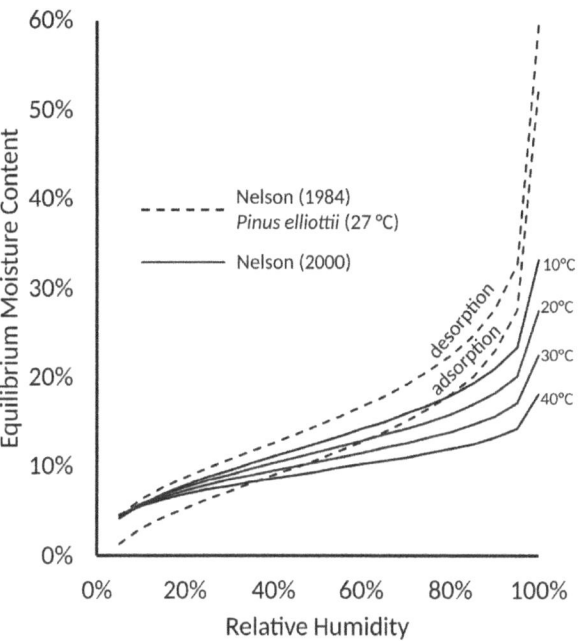

Figure 6.7: Examples of modelled relationships between relative humidity and equilibrium moisture content (EMC) at different temperatures (from Nelson 1984, 2000). At a given relative humidity, warm air has a lower vapour pressure than cool air and thus has a lower EMC.

temperature (**Figure 6.7**). In general, EMC decreases with warmer and drier air.

Fuel particles are at equilibrium moisture content when the vapour pressure of water in the fuel particle is the same as the vapour pressure of the air in contact with the particle surface, so there is no exchange of moisture between the air and the woody fuel. This rarely happens in nature, however, because the moisture in the air is always changing. Air temperature, humidity and solar radiation change constantly throughout the day and night, resulting in constantly changing differences between the actual moisture content of the fuel and the EMC. To understand changes in fuel moisture, then, we must understand how moisture exchange is driven by the external conditions that impinge on the fuel particle surface. These are called the *boundary conditions*. Let us follow these relationships through a simple diurnal cycle: even if the amount of water vapour in the air stays constant, afternoon warming from the sun will increase the air's water holding capacity, which will reduce the RH and increase the vapour pressure deficit, thus drying the woody fuel particles. The reverse will happen through the night: without solar radiation, the air will cool, so its water holding capacity will decline and its RH will increase. The temperature may decline to the dewpoint, at which point the air will be saturated with moisture and excess water vapour will condense into liquid water. The diurnal cycle of solar radiation and temperature can trigger dramatic changes in fire behaviour caused by the constantly changing moisture dynamics of dead fuels.

As with heat transfer to fuel particles (**Chapter 3, Heat transfer**), moisture exchange in fuels is a strong function of the size and shape of the fuel particles: small or thin particles will reach EMC faster than thick particles. This effect has been part of wildland fire management and fire danger rating for many decades. It was introduced in the United States along with the concept of *time lag* (see Fosberg *et al.* 1970). Wood particles demonstrate asymptotic drying, meaning that the rate of drying in response to a change in EMC is initially fast and then tapers off. Thus, as the moisture content approaches the equilibrium value, small changes take a very long time. *Time lag* is defined as the time required for a particle to reach 1 − 1/*e* or 0.6321 (63.21%) of a new equilibrium moisture content (Fosberg et al. 1970).

To satisfy the requirements of fire danger rating – that is, to provide indices of fire activity and potential fire behaviour over a range in time scales from daily to seasonal – fuels were classified by size into four time lags:

- '1 h fuels' were those that would reach 63.21% of a new EMC in less than a day.
- '10 h fuels' were those likely to reach 63.21% in about a day.
- '100 h fuels' were those likely to reach 63.21% in about a week.
- '1000 h fuels' were those likely to reach 63.21% in about a month or over a whole season.

Another time-lag classification of fuels is provided for surface and ground fuels by the Canadian Fire Weather Index System. This system, which is used for rating fire danger and modelling fire behaviour, contains three codes: the Fine Fuel Moisture Code (FFMC) for surface litter less than 2 cm deep, the Duff Moisture Code (DMC) for organic matter on the forest floor 5–10 cm deep, and the Drought Code (DC) for deep organic layers 10–20 cm deep (Van Wagner 1987). The FFMC, DMC and DC have time lags of ~16 h (2/3 day), 288 h (12 days), and 1248 h (52 days), respectively.

Let us return to Fosberg's (1970) time-lag classes to see how well the time-lag concept explains changes in fuel moisture over time. **Figure 6.8a** shows calculated moisture drying patterns for fuel particles with different time lags from a starting condition of 15% to an EMC of 5%. The rapidity of the response of particle moisture varies with EMC, as shown for 1 h time-lag fuel particles in **Figure 6.8b**.

It has often been assumed that all dead fine fuels fit into the 1 h time-lag class, but measurements shown in **Figure 6.9** reveal large differences in drying rates for several kinds of fuels within the 1 h class. In **Figure 6.9a**, dead needles of two conifer species (1 h fuels) respond more slowly than 10 h time-lag pine sticks. In **Figure 6.9b**, foliage of conifer species, grasses, and small twigs show drying patterns that deviate from the pattern of an 'ideal' 1 h particle. The differences were probably

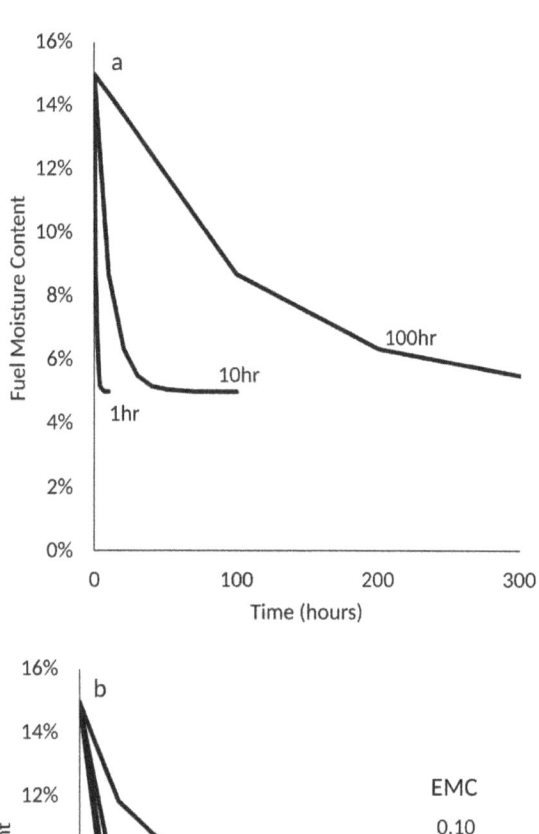

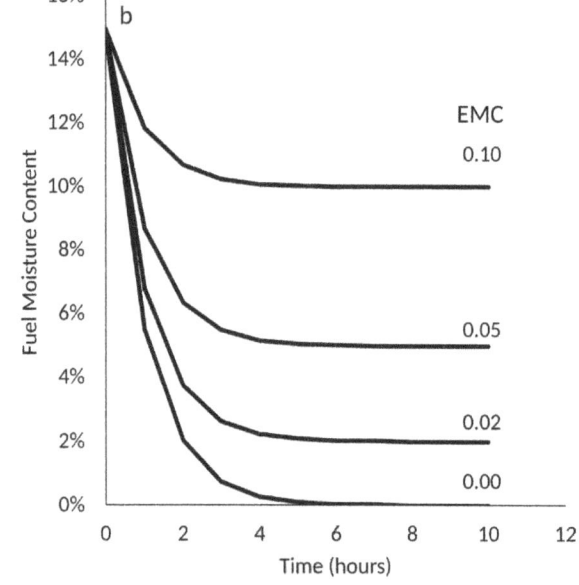

Figure 6.8: Theoretical response of fuel moisture (a) to particle size class (1 h, 10 h, and 100 h fuel particles) from starting moisture content of 15% to an equilibrium moisture content (EMC) of 5% and (b) to EMCs ranging from 0 to 10%, shown only with 1 h fuel particles.

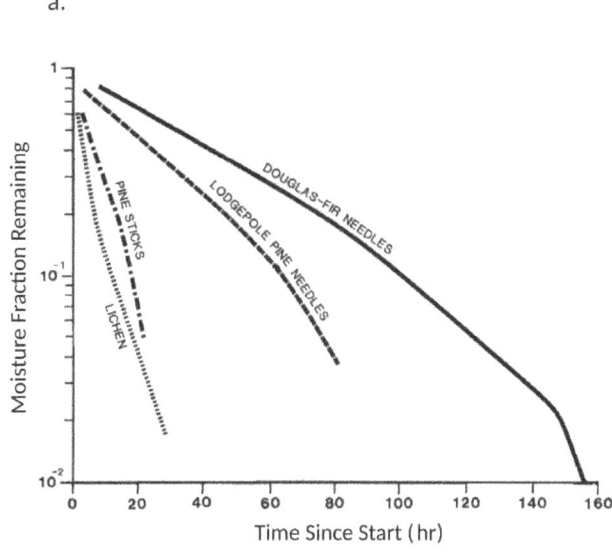

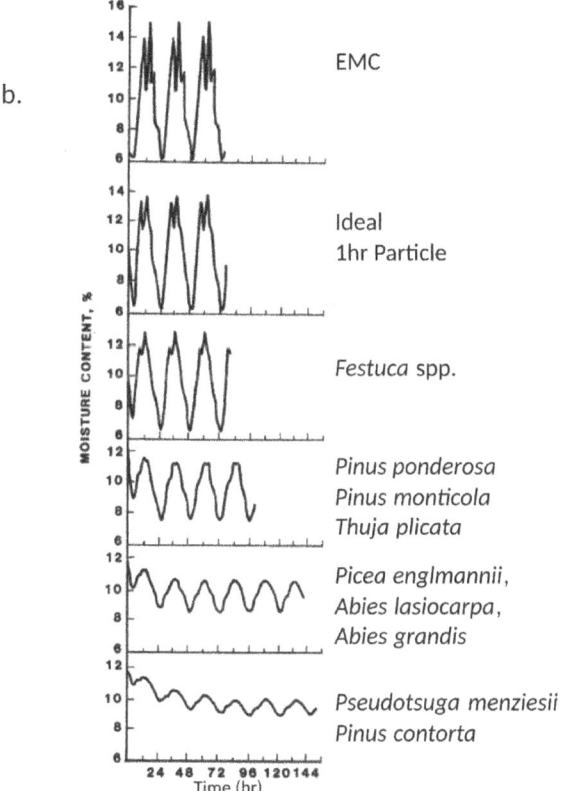

Figure 6.9: Graphs of experimental data from different natural fine fuel particles showed (a) pine needles responded slower than nominal 10 h particles (1.2 cm square pine sticks) but lichen responded faster (Anderson 1990a) and (b) dead needles and grass were less sensitive to changes in EMC than an ideal 1 h fuel particle (Anderson 1985).

caused by variable particle geometry, cellular or material constituents, or morphology of the species and plant parts tested (Anderson 1990a). Thus we cannot assume that the moisture content of fine (1 h) particles of various plant species and organs all behave in the same way to changes in ambient conditions. This means that, in field experiments for studying fire behaviour (**Chapter 9, Fuel moisture**), we must not assume that all dead fine fuels have similar moisture content; to understand fire behaviour, instead, we must measure the moisture contents of all relevant fuels as close as possible to the time of a fire.

Entire fuel particles do not respond completely and instantaneously to changes in exterior conditions. Just as the thickness of fuel particles influences their response to heating from fires (**Figures 5.13** and **6.2**), it also influences their response to changing weather – including the fluctuations that cause drying and wetting. The surface of a fuel particle responds more quickly to these changes than its interior, creating a moisture gradient between the surface and the interior of the fuel. We can characterise the moisture content of the whole particle with a *calculated average*. The time-lag concept is really about this *average* condition. For small fuel particles, most of the solid material in the particle is very close to its outer surface, so the calculated average moisture content changes quickly to changes in boundary conditions. However, for thick particles such as dead logs and deep organic layers of the forest floor, the calculated average is heavily influenced by the internal fuel mass that is located remotely from the exterior, so the overall fuel moisture responds slowly to changes in boundary conditions. In fact, large particles have a relatively small amount of external surface area per unit volume, so each unit of interior mass is responsive to a smaller proportion of the boundary condition.

Because transient environmental conditions create moisture and temperature gradients within a particle, as Nelson (2000) explains, we can think of each particle as consisting of a series of concentric shells that exchange heat and moisture with their neighbouring shells. The long time lag of large particles means that their average moisture content at any given moment is derived from the values at each neighbouring layer, and the large particle contains a physical record of a long series of prior conditions. Thus long time-lag particles can serve as excellent indicators of long-term precipitation trends. Large woody material and deep forest floor layers also differ from small particles if they are resting upon soil or organic ground cover, limiting the underside's exposure to drying weather and increasing the duration of moist boundary conditions at the particle surface.

In the early summer, while the outer layers of large woody particles are subject to drying conditions, the interior layers require weeks or months to dry out substantially. The reverse is generally true after regular precipitation begins in autumn. Monitoring of moisture changes in large logs over a 19-year period (Brackebusch 1975) illustrated the slow seasonal changes in logs 15 cm (6 in), 30 cm (12 in), and 46 cm (18 in) (**Figure 6.10**). During the period of summer drying, moisture gradients in dense duff and organic layers on the ground follow a similar pattern: deeper layers dry more slowly than shallow layers, reach their lowest moisture content later in the summer, and remain dryer than the surface layers longer after precipitation begins in the fall. *Moisture gradients* within dead fuels affect the way they burn, as discussed in the next section.

The dynamics of fuel moisture exchange with the atmosphere can be counterintuitive, similar to the dynamics of particle heating that occur as a fire approaches or it is exposed to a heat source (**Figures 5.13, 6.2, 6.5**). This was demonstrated by experiments conducted in the 1940s by Byram and Jemison (1943). They used a laboratory-scale chamber to expose wooden slats to different levels of humidity, wind 'ventilation' and solar radiation (**Figure 6.11**). Not surprisingly, without solar radiation, wet particles dried faster with wind than without wind, since the dry air flow of the wind increased the rate of water vapour diffusion. However, in the presence of solar heating, wood particles in wind dried more slowly than in still air.

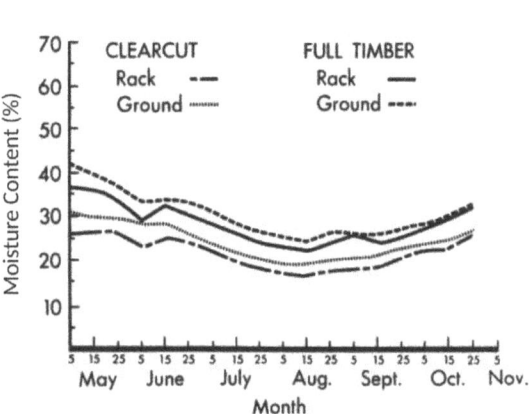

Figure 6.10: Measuring fuel moisture over a 19-year period (a) using western red cedar (*Thuja plicata*) logs with 15 cm, 30 cm and 46 cm diameters. Results (b) show average seasonal moisture, with higher moisture content for logs in contact with the ground and shaded (i.e. under full timber canopy) than logs elevated above ground (resting on a rack) in full sun (Brackebusch 1975).

Convective air flow from the wind cooled the particles, reducing the rate of water vapour diffusion and drying. In other words, when the fuels were exposed to solar radiation, they retained more moisture in windy than in quiet conditions and thus dried out more slowly.

Dead fuel moisture affects the ignition and combustion processes of fire spread, as explained

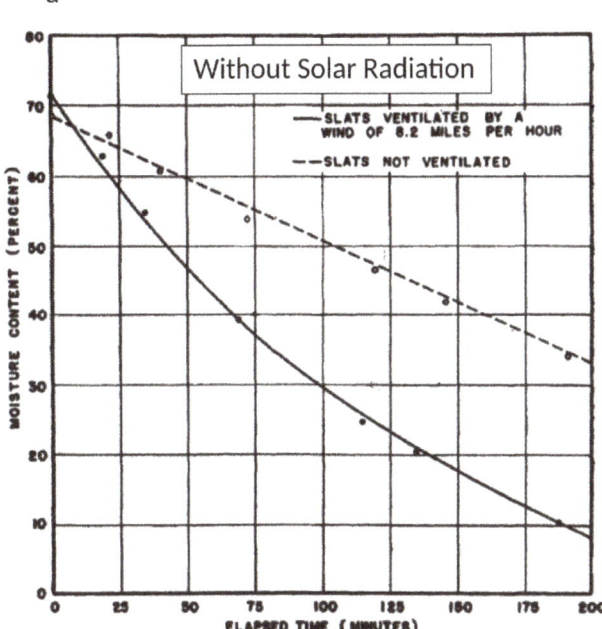

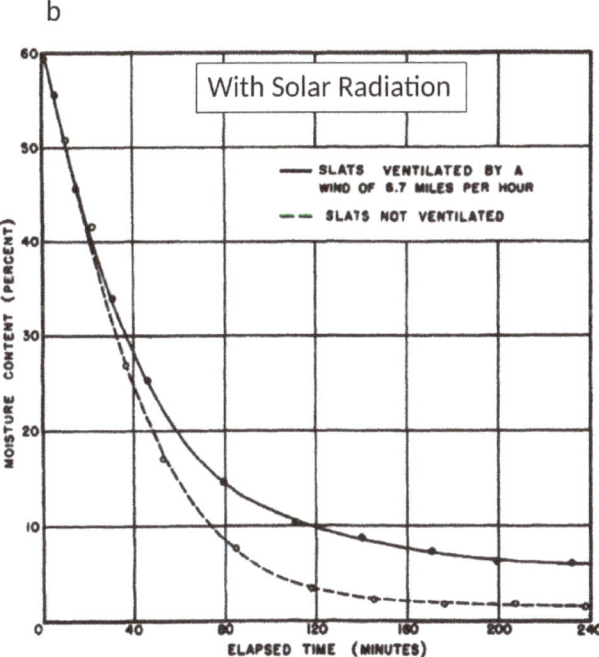

Figure 6.11: Experiments on drying of woody fuels initially immersed in water with varying conditions of wind and solar radiation when exposed to relative humidity of 55% and air temperature of 27 °C. Results are shown for (a) no solar radiation and (b) solar radiation of ~1.2 kW m^{-2}. In (a) wind increases moisture diffusion from the fuel surface and thus increases the drying rate. In (b) wind causes convective cooling and thus reduces particle temperature and drying rate (from Byram and Jemison 1943).

in detail in **Chapters 4** and **5**. It is worth repeating here that the general effect of increased fuel moisture on fire behaviour is to decrease the likelihood that wildfires can start or spread in the first place, and also to reduce the spread rate and intensity of fires that do spread. Another way to think about the effects of moisture on fire spread rate is through the concepts introduced to explain quasi-steady spread by Frandsen (1971), in which the heat produced by the fire is weighed against the heat required for ignition. The heat flux received from the burning fuel is the *source*, and it is weighed against the heat flux required for heating fuel to ignition, the *sink*. Fuel moisture exerts a strong effect on fire spread because it diminishes the heat source (decreased heat release rate and total heat release, and increased flame residence time) and at the same time increases the heat sink (increased requirement for heat in unignited fuels and longer time for ignition), altering the balance between the two and often becoming the determining factor in whether a fire will spread or not.

Fuel burning rates and consumption

From **Chapters 4** and **5**, we know that the properties of fuel particles and fuel beds strongly affect wildfire behaviour, particularly the burning rate and ignition response to heating. Here we summarise relationships between fire behaviour in dead fuels and several other variables: effects of the compactness of the fuel bed on combustion, effects of wind on transitions between smouldering and flaming, effects of moisture on fire spread from one particle to the next, effects of fuel bed compactness and residence time, and relationships between fuel moisture and fuel consumption.

In general, loose packing of fuel beds increases the burning rates of fuel particles and thus decreases the flaming residence time. As beds get more compact, two factors compete in determining the burning rate. Increased radiant heat exchange among closer burning elements elevates the surface temperature of particles, which could increase the rate of combustion. However, smaller air gaps between particles slow the entry of fresh oxygen, which forces fuel gases to rise above the bed before they reach enough oxygen for combustion. Not only is oxygen limited within the fuel bed, but also the absence of flaming inside the bed decreases radiation from soot, decreasing the heat impinging on the solid fuels. Absence of flaming and shortage of oxygen reduce the mean temperature of gases in contact with the fuel surfaces and the reaction rates of glowing combustion and burning char.

The rate and temperature of the surface reaction in combustion depend directly on oxygen contact with the solid fuel surface (**Figure 4.31a, b**). We see this when we blow on coals in a campfire, which increases glowing combustion (brighter colours of burning coals) and more rapid transition from smouldering to flaming combustion. Currently, the effects of wind on the rate of solid fuel combustion and the transition from smouldering to flaming are seldom addressed in wildland fire science and modelling. This may be a serious shortcoming to understanding the behaviour of large fires, which is discussed in **Chapter 8, Mass fires**.

As with fuel burning rates in general, the exact effects of moisture on fire spread are beyond current predictive capabilities for either the flaming or the glowing combustion phase. However, it is clear that higher fuel moisture decreases the burning rate (Anderson 1990b; Nelson 2003) and therefore the heat release rate (**Figure 4.27**). Thus higher fuel moisture reduces the rate of fire spread by diminishing the source of heat, and thus reducing both radiant and convective heat flux to a nearby fuel particle, and increasing the ignition requirements of new fuel particles (which require longer ignition times at a given heat flux to heat and evaporate water) (**Figure 5.11**) (McAllister 2013). As fuel moisture rises beyond some threshold, fires will not spread. This moisture threshold also depends on other factors such as wind and specific fuel characteristics, some of which are explored through modelling in **Chapter 7**.

The flame residence time t_r in a spreading fire (**Chapter 2, Eqn [2.1]** and **Chapter 4, Eqn [4.11]**) determines the time during which flames can transfer heat via convection and radiation. Flame residence time depends on properties of fuel particles and the fuel bed. Higher fuel moisture content increases t_r (Thomas 1971). Large fuel particles have longer t_r than smaller ones. As discussed in **Chapter 4, Burning and heat release rate**, various relationships have been proposed to quantify this pattern; they express residence time as a linear function of particle diameter (Anderson 1969) and as a function of diameter to the 1.5 or 2.0 power (Vaz et al. 1998) or a power in between (Burrows 2001). However, flame residence time for most fuel *beds* is typically much longer than for individual *particles* (t_p). The analysis by Nelson (2003) for fuel beds of depth δ (m) that are dense enough to be classified as ventilation limited suggests the relationship $t_r = t_p \sigma \beta \delta / (1 - \beta)$, where σ is surface-area-to-volume ratio (m^{-1}) and β is packing ratio (dimensionless). This relationship shows that higher packing ratios β increase the flame residence time of the fuel bed for a given particle flaming time t_p. Nelson's equation includes moisture as well. We can see in **Figure 6.12** that his model approximates the trend in the excelsior fuel bed data from Catchpole et al. (1998) but yields consistently longer t_r for equivalent fuel beds than indicated by regression models from data of Wilson (1982) or Finney et al. (2015). All data and models suggest that t_r increases linearly with fuel bed packing ratio and loading but inversely with particle surface-area-to-volume ratio and moisture content (MC). Note that the data for these relations come primarily from single-size fine woody fuel materials in a narrow range of fuel bed properties (loading, depth) and do not represent large woody fuels or fuels from mixtures of particle sizes. The effects of wind and fire spread direction (i.e. heading v. backing) on flame residence time are poorly known, although data reported by Beaufait (1965) suggest both factors affect residence time.

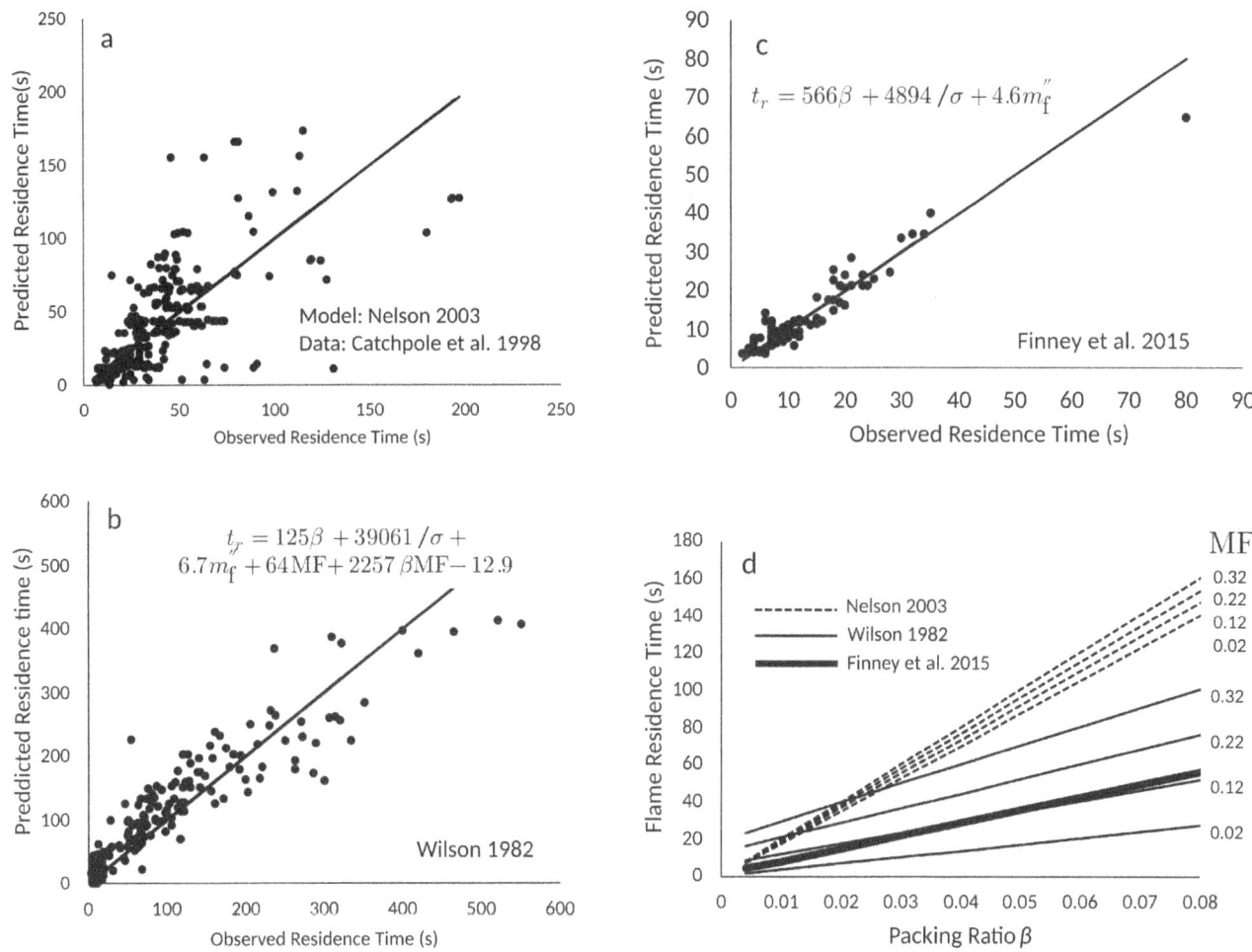

Figure 6.12: Data and models of flame residence time for (a) model results from Nelson (2003) (line) compared to data from Catchpole *et al.* (1998) (dots); (b) a regression model from data by Wilson (1982); (c) a regression model on the cardboard fuel data from Finney *et al.* (2015) with moisture content of ~6%; and (d) comparison of models. β is packing ratio (dimensionless), σ is surface-area-to-volume ratio (m^{-1}), m_f'' is fuel loading (kg m^{-2}) and MF is fuel moisture fraction (mass of water as a fraction of oven dry fuel mass).

Higher moisture content also decreases the total amount of fuel consumed by a fire (Byram *et al.* 1966; Thomas 1971), which is estimated as the difference between pre- and post-fire fuel loading. The amount of large fuel consumed, whether duff or large woody material, decreases with increasing moisture content (**Figure 6.13a**). Total consumption of dead logs lying on the ground, as well as the *rates* of burning and consumption, are strongly tied to the moisture distribution within the fuel material and whether the log is rotten or sound (**Figure 6.13b**) (Reinhardt *et al.* 1991). The distribution of loading among fuels with a variety of fuel particle sizes (e.g. duff and litter) influences the consumption of fuels in any single size class. For example, fires will consume less mass of large logs if they are isolated than if they are embedded among substantial quantities of smaller fuel particles and deep layers of litter and duff (Albini and Reinhardt 1995). Think of this in terms of a campfire where a large solitary log will be difficult to burn except when surrounding by other pieces of burning wood.

We have addressed a few of the relationships between dead fuels and fire behaviour in this section, but many others are poorly understood. Little

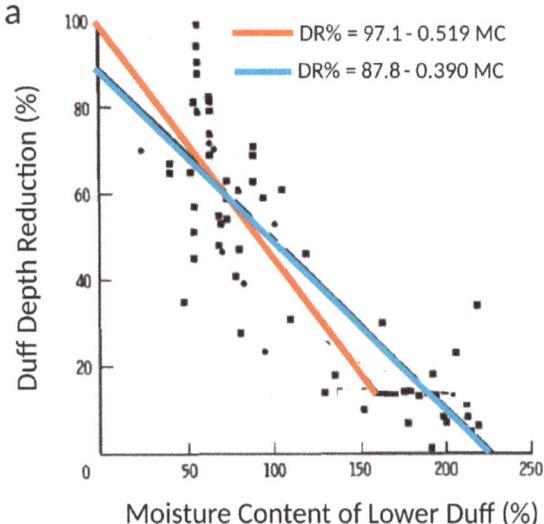

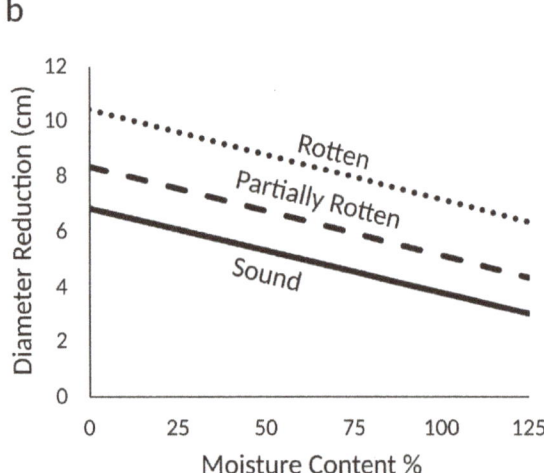

Figure 6.13: Fuel consumption following prescribed burns in logging slash showing dependence on moisture content for (a) duff depth reduction (DR%) (Brown *et al.* 1985) and (b) diameter reduction for sound, rotten and partially rotten logs (Reinhardt *et al.* 1991).

and whether they are suspended or in contact with the ground. Research has also shown that factors associated with the fire itself affect fuel consumption, including whether fires are backing or heading or have thin linear flame zones in a line fire (e.g. a few metres) or constitute very large burning areas (e.g. thousands of hectares). The topic of measuring fuel consumption is covered in more detail in **Chapter 9, Fuel consumption**.

Live fuels

Wildland fires often involve the burning of living vegetation (*live fuels*). However, the state of knowledge regarding ignition and burning of live fuels is abysmally poor. For example, even these three very general questions cannot be answered:

1. How can some live vegetation (e.g. conifer foliage) support sustained flaming spread at 100% moisture content (dry weight basis) when dead grass at 25% moisture cannot?
2. Why does it appear that live foliage from some species of plants (e.g. conifers) can carry fire when live foliage of others cannot (e.g. aspens and other hardwoods)?
3. Why is flaming combustion sustainable on individual dead fuel particles but only in clumps of live fuel particles?

One of the biggest challenges for research into these and related questions is caused by diversity among plant species. Living vegetation has evolved highly variable physiological and morphological characteristics that are critical to their survival, and many of these characteristics cause variation in the species' ignition and burning behaviour. For example, drought-tolerant evergreens in deserts have evolved leaf structures and physiology that limit moisture loss, while deciduous trees in mesic environments have not; and leaves may be needle-shaped or flat, or exhibit complex geometry with edges that are smooth, serrated or spiny – all characteristics that influence their responses to heat. While some of these characteristics can be studied with small fuel samples in controlled laboratory conditions, it is difficult to scale up results of these

literature is available, for example, on the effects of different particle, fuel bed and environmental factors on the amount of fuel consumed in flaming *v.* smouldering combustion. Higher moisture content is often assumed to reduce the amount of fuel consumed in flaming, and it probably decreases consumption during smouldering as well. Other fuel properties affect fire behaviour and consumption of woody fuels, including whether bark is attached to the wood, the state of decay of the fuel particles, studies so we can apply them to the behaviour of wildfires in field contexts (Fernandes and Cruz 2012; Varner *et al.* 2015a), where strong winds and large flaming zones present conditions well beyond those used to test samples in the laboratory. Finally, even as we develop greater understanding of the ignition and combustion of live fuels, we must apply this knowledge to the burning of fuel beds that are characterised by mixtures of live and dead fuel, different species, and varying horizontal and vertical properties (e.g. continuity, bulk density).

For many years, the mysteries of burning in live fuels challenged scientists and practitioners who have tried to explain their ability to carry fire. For example, live fuel has been assumed to be able to burn and carry flaming spread only in the presence of dead fuel (Rothermel 1972). However, evidence from laboratory and field experiments suggests that some species – under some conditions of moisture, wind and slope – can sustain fire spread with no dead fuel (Martin and Sapsis 1987). Another example: crown fires burning in conifer foliage are frequently observed to burn almost entirely in canopies composed of live needles; occasionally they are even observed burning above snow-covered ground, completely deprived of heat released from combustion of dead surface fuels (Agee *et al.* 2002; Mottus and Pengelly 2004). Volatile compounds (e.g. waxes, terpenes) have also been assumed to play a vital role in plant flammability, but there is no unambiguous evidence that these minor plant constituents (by weight or volume) can affect ignition and burning independent of the roles played by leaf physiology, morphology and major plant constituents (Varner *et al.* 2015a).

Although the science is incomplete for many basic questions concerning live fuels, we can see some of their unique features and relationships to wildland fire by comparing their characteristics with those of dead fuels (**Table 6.4**). The moisture content of live fuels is perhaps the most obvious characteristic that differs from that of dead material. Living material is dynamic through the seasonal cycles, so moisture variation occurs almost constantly from the time plants emerge from dormancy until they re-enter it. In high latitudes, this seasonal cycle causes obvious changes in both live moisture content (**Figure 6.14**) and ignition and burning behaviour.

Seasonal and year-to-year moisture dynamics of live foliage reflect several phenomena that influence rates of ignition and fire spread. Many species of shrubs and conifers contain a mixture of needles of varying ages. This causes considerable within-plant variability, particularly in the spring, after new needles have emerged. Young foliage initially has a much higher moisture content than old foliage, with new leaves sometimes exceeding 300% (Chrosciewicz 1986; Agee *et al.* 2002). Moisture

Table 6.4. Comparison of live v. dead fuel properties and their effects on fire behaviour in spreading fires.

Fuel characteristic	Dead	Live
Range of moisture contents	2–30%	50% to more than 300%
Method of moisture loss	Diffusion of water vapour	Explosive release
Ignition time with moisture	Increases	Generally increases, but may decrease in some species under some conditions
Sustained burning	Occurs on individual particles	Occurs in clumps or groups of particles
Primary constituents	Structural carbohydrates (cellulose, hemicellulose, lignin)	Includes structural and non-structural carbohydrates (sugars, starches, pectin), waxes, resins, terpenes
Seasonal dynamics	Respond to short- and long-term ambient moisture. Foliage is added to dead fuels, begins decay.	Dry mass and moisture content change with plant phenology. Foliage moisture varies depending on age.
Particle geometry	Highly varied	Highly varied
Plant species	May include decayed or partially decayed material	Highly varied physiology and chemical constituents

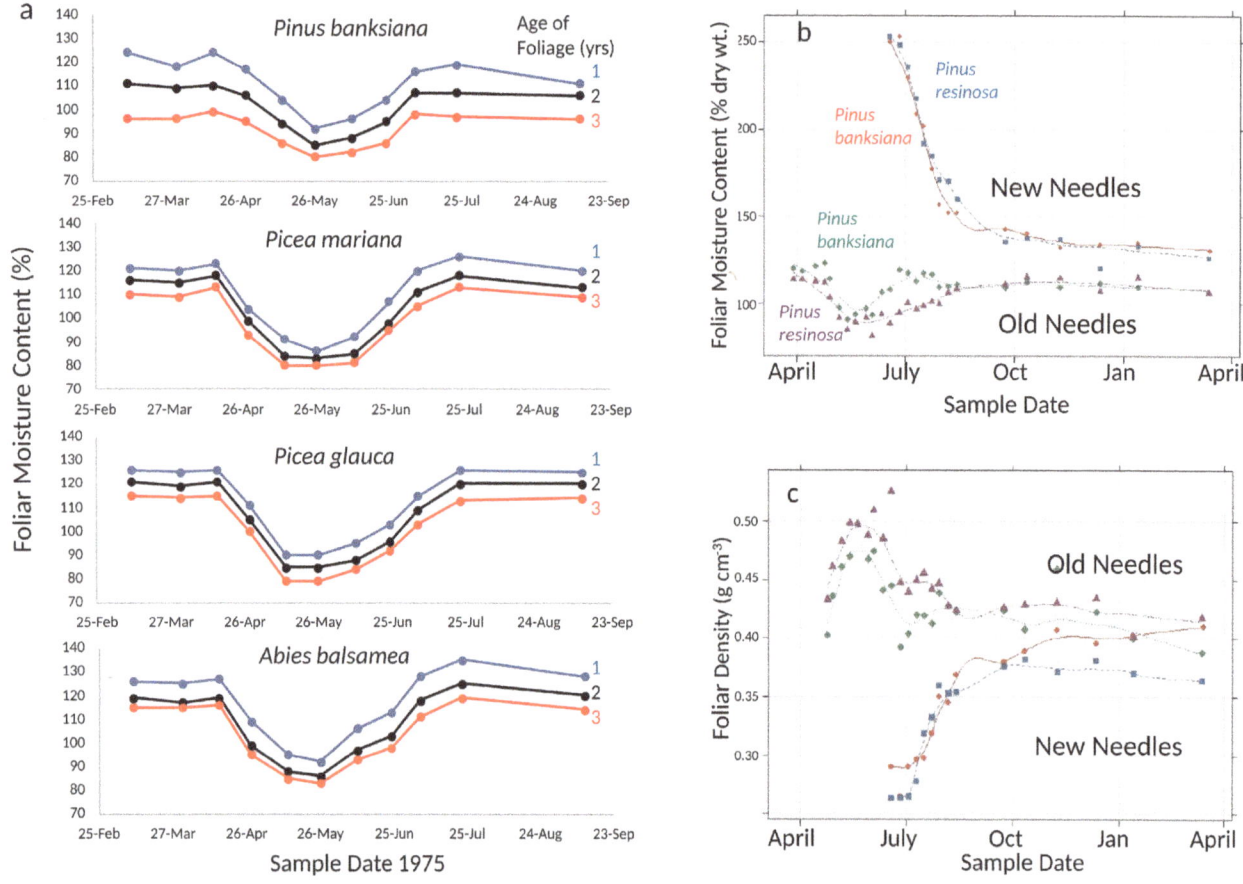

Figure 6.14: Apparent seasonal variation in live needle moisture content calculated on a dry-weight basis (**Eqn [6.1]**) (a) for needles of ages 1, 2 and 3 years for various species in Canada (reproduced with permission from Chrosciewicz 1986), and (b) data showing the moisture variation for new and old foliage is actually caused by (c) increased foliage density (i.e. dry mass of foliage) at release from dormancy (from Jolly et al. 2016).

differences between new and old foliage decline as new leaves mature. Seasonal variation in moisture content occurs in old foliage too, particularly in climates where there is a distinct dormant season. Regardless of their age, leaves of drought-tolerant shrubs such as chamise (*Adenostoma fasciculatum*) may begin the growing season with moisture content at or above 100%, which descends to roughly 50% (Countryman 1974) by late summer. In many conifer species, emergence from winter dormancy in spring typically coincides with minimum moisture of old foliage (**Figure 6.14a**). However, this apparent decrease in foliar moisture is actually an artefact of the way moisture content is calculated: in spring, the dry matter in the leaves (particularly starch) is increasing as the plant prepares to resume springtime physiological activity (**Figure 6.14b, c**) and grow new foliage (Nelson 2001; Jolly et al. 2016). Since moisture content is expressed as a percentage of dry weight (**Eqn [6.1]**), any increase in dry weight (the denominator) without a proportional increase in moist weight (the numerator) will indicate that *per cent* moisture content has declined – even if the *actual* amount of water in the tissues has remained unchanged.

As discussed in **Chapter 5, Predicting ignition times**, ignition time of dead woody fuels typically increases with moisture content for a given heating condition (**Figure 5.11**), but recent experiments on live fuels have found the opposite: some species show decreased ignition time with increased moisture content (**Figure 5.17**) (McAllister and Weise

2017). Such varied ignition responses in live fuel are further evidence that species-specific physiological factors may play a greater role in ignition and fire behaviour than moisture content alone. Similarly, spread rates measured in both wildfires and laboratory experiments have shown little sensitivity to live fuel moisture content within the typical range of 80–120% (Alexander and Cruz 2013; Rossa and Fernandes 2018), and that moisture content is of minimal importance to spread rate in live fuels when compared to environmental variables such as wind and fuel loading (Anderson *et al.* 2015; Weise *et al.* 2016).

When live vegetation ignites and burns in a spreading wildfire, it almost always involves rapid heating by sudden flame impingement rather than long-duration heating at low heat fluxes. This frequently occurs because convective cooling of fine fuel particles, including live foliage, by ambient air offsets radiant heating sufficiently to prevent or delay pyrolysis and ignition until flames impinge directly on the fuel particles (**Figures 3.22, 5.7** and **5.13**). Thus fine fuels, whether live or dead, often have to 'wait' to ignite until they are contacted directly by flames (~1000 °C). We have seen that convective heat transfer to and from dead fuel particles is strongly influenced by their geometry, and the same is true for live foliage. Thus needle-shaped leaves heat up and cool down much faster than the broad faces of flat leaves. When heated by flame or hot air, ignition begins at the outer edges, tips, or serrations of the leaf (Prince and Fletcher 2014). Because the heating of fine live fuels (foliage) is so rapid, however, and because the rapid heating often causes an explosive release of cell contents, combustion of these particles can seem instantaneous. Explosive jets of hot gases and flame propelled from a burning leaf can extend in many directions, expanding the neighbourhood of convective heating to leaves on an entire branch and to other branches above (**Figure 5.14**). Not all moisture is lost from leaves at the time of ignition, however (**Chapter 5, Live fuels**). Studies have shown that live foliage still retains substantial water mass after ignition (Pickett *et al.* 2010), and continues to be released long into burning (McAllister and Finney 2014) (**Figures 5.15** and **5.16**).

As with dead fuels, higher moisture of living foliage is assumed to decrease the ignitability and rate of heat release during burning. However, generalisations about ignitability in live fuels are difficult to justify given the diversity in species, structure and phenology of live vegetation (Varner *et al.* 2015b). Recall the discussion associated with **Figure 5.17** that ignition time shows different responses to live fuel moisture in different species. Full-scale testing of one tree species (small Douglas-fir trees, Babrauskas 2016) suggests an exponential decrease of peak heat release rate with moisture: heat release was highly sensitive to moisture content below ~80–90% but far less sensitive to greater moisture content (**Figure 4.27**). Clearly, much remains to be discovered about live fuels and their heating, ignition and burning behaviours. Advances in wildfire predictions for living vegetation canopies (i.e. crown fires and brush fires) will depend upon these discoveries.

Implications for fuel characterisation and classification

How can we characterise the critical properties of live and dead fuel particles and fuel beds that vary at multiple scales for practical use – that is, for predicting fire behaviour? The main challenge is to accurately and precisely represent those fuel properties that are *essential* to the physical fire processes. Another major challenge is to *classify* fuels – that is, to identify categories of fuel conditions in the simplified form necessary for mapping and describing them across the large landscapes at which wildfires spread. As we have seen from the descriptions of physical processes in prior chapters, such a system must recognise the variability in particle sizes and location, the gaps between particles, and spatial correlations among these (Keane 2015).

In the past decade or so, remote sensing technologies have become common for obtaining detailed measurements of the three-dimensional (3D) locations of fuel particles (Kremens *et al.* 2010). LiDAR (Light Detection and Ranging) is perhaps the most widely used of these techniques (**Figure 6.15**). LiDAR determines the position of reflective objects by the orientation and timing of returned laser pulses. LiDAR data in raw form become 3D 'point clouds' of intercepted solid material. Airborne and ground-based LiDAR vary in their sampling densities, but at the finest resolutions ground-based LiDAR can represent the leaf and branch structures within individual plants. Coarser resolution obtained by airborne sensors is adequate for defining canopy properties, including cover, height and base of the living crowns. Airborne sensors do not penetrate dense canopies well and have limited potential to characterise surface vegetation and fuel structure on the ground compared to ground-based LiDAR. Post-processing is used to derive from LiDAR data the values of parameters that describe fuel properties. For example, relative densities of LiDAR returns can be correlated with bulk properties of forest canopies, and vertical gradients in point density can be used to estimate tree height and canopy base height. To be useful for the broad range of fuel conditions in wildlands, such techniques and post-processing must be able to quantify particle size distributions (fine leaves to large logs), and whether fuels are living or dead.

There are other techniques for remote sensing of fuels. One of these is the use of visible images in 'structure from motion' (SFM) photogrammetry to reconstruct vegetation. The SFM process uses a large number of overlapping digital images from multiple angles to create a 3D representation of the solid surfaces. Images can be obtained from the ground or from airborne platforms such as unmanned aerial vehicles. SFM shares the same limitations as LiDAR in capturing the location or properties of material that is obscured from view by a dense forest canopy, but both techniques may be invaluable for characterising fuels in grasslands, shrublands and open forests. Similar challenges are presented for SFM as for LiDAR – namely, to quantify particle size distributions in various fuel strata, detect fuel materials beneath canopies, and distinguish live from dead material.

Data from remote sensing methods such as LiDAR and SFM contain much more detail on quantity and spatial arrangement of fuels than data from ground-based transect and point-based sampling. Remotely sensed 3D data have been used successfully as inputs to small-scale,

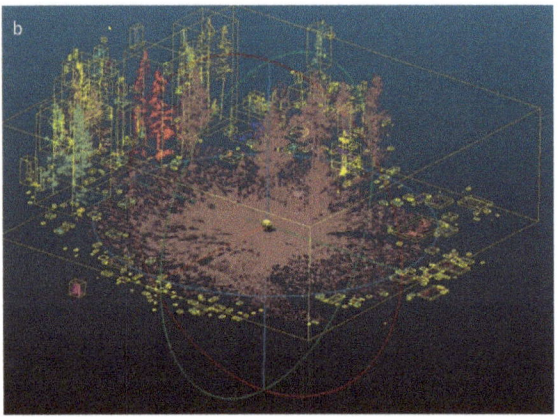

Figure 6.15: 3D point cloud data from a single terrestrial LiDAR scan (TLS) for (a) an individual ponderosa pine tree and (b) a plot of ponderosa pine savannah in south-central Oregon, October 2019. One 3 min scan with the TLS (Leica BLK360) mapped all objects within about a 30 m radius. Spatial resolution of the point cloud is on the order of 1–2 mm. Yellow boxes show first segmentation (database separation of discrete entities) of peripheral trees and other fuel elements, while the single colour in the centre indicates connection of central trees and surface fuel points. Data visualisation and analysis done in CloudCompare v2.10.2 (Zephyrus). Image provided by Russ Parsons, USDA Forest Service.

site-specific fire modelling, but the size and complexity of the databases produced by these techniques prohibit widespread application for operational use in landscape-scale wildfire modelling. The techniques can perhaps be more useful for practical applications if statistical distributions of salient properties and their spatial patterns can be defined and extracted for purposes of classification. The resulting quantitative descriptors of fuel particles and fuel beds (**Tables 6.2** and **6.3**) could include spatial correlations and patterns among clusters, which could then be used as the basis for a practical fuel classification. Then fuel 'archetypes' could be mapped across large landscapes using coarse-scale predictors, just as past fuel models have been mapped, and at the same time represent the fine-scale properties of particles and their spatial properties that are so important to fire behaviour.

An example of this methodology was described by Hiers *et al.* (2009) for small, 4 m square 'cells' based on high-density terrestrial LiDAR data. The LiDAR points were statistically analysed to quantify the loading of fuel particle types (grass, shrub, litter, species), sizes and their vertical position within the fuel profile. A statistical analysis was then used to identify a subset of distinct fuel types (i.e. classes). This approach had an important benefit in that it could contain the inherent spatial resolution of fuel variations. The approach successfully associated the derived fuel types with fine-grained fire behaviour measurements taken from infrared imagery (Loudermilk *et al.* 2012).

If we hope someday to use detailed physical wildfire models to predict fire behaviour and plan fire management activities, we will need new methods to characterise fuels, classify them, and then apply the classifications at the broad scales needed for mapping fuels and modelling wildfire behaviour.

Weather

Unlike fires in homes or industrial settings, wildfires are profoundly captive to the external influences of weather, and sometimes also to weather effects produced by the fire itself. The broad spectrum of time and space scales in which weather interacts with fire include trends of climate and interannual precipitation, seasonal and synoptic weather patterns, lightning storm tracks, and fine scales of wind flow and air turbulence. Temperature, humidity and precipitation patterns have primary influence on fires through fuel moisture as covered earlier. Wildfires are also known colloquially for 'creating their own weather', which can be demonstrated even at laboratory scales; this phenomenon illustrates well the strong coupling between fires and their environment. The disciplines of meteorology and atmospheric science as they apply to fire are far more complex than we can summarise here, so we restrict our focus to the immediate impacts of wind, one of weather's most powerful influences on actively burning fires. Standard texts (Schroeder and Buck 1970) and recent reviews (see Potter 2012a, 2012b) offer more detailed treatments of fire weather.

Winds

Wind refers to movements of ambient air that occur at multiple time scales in three space dimensions. It is seriously difficult to characterise the winds that affect wildland fires. Most fire models consider wind as a single horizontal vector at some height near the ground. Thus *wind speed* (m s^{-1}), as typically used in fire modelling, smooths the gusts and lulls and collapses vertical and horizontal vectors to a single velocity parallel with the ground surface. The inherent variation in wind over space and time makes it difficult to characterise even in the absence of fire but fires themselves strongly affect local characteristics of air flow. It remains especially challenging to characterise wind as it affects a flame zone, partly because the fire, the burned ground and the vegetation structure alter the wind in their vicinity and partly because of the inherent variability of wind at multiple scales. Describing wind effects on the flaming zone is an area of active research with many unknowns (see Potter 2012a for review).

Most fire spread models require wind speed at a height relevant to the flames as an input, since this is the wind that causes flames to tilt and modify the transfer of convective and radiant energy to adjacent fuel particles. Most models have used either the wind speed at midflame height (half the flame height) or the average wind speed over the flames. To produce a wind speed useful in the models, one must be able to measure or forecast the wind at one height and use that to predict the wind at a height relevant to the fire. A common approach in wildland fire is to use models of the vertical distribution of wind speeds (a wind *profile*) to calculate a *wind adjustment factor* (WAF; reviewed by Andrews 2012). The WAF is simply the fractional change in wind speed from one height to the speed at another height. Factors affecting the wind adjustment account for the wind speed at a measured reference height, the canopy structure including cover and vertical foliage distribution, as well as the position on the topography. These factors are often combined into the concept of *sheltering* (Andrews 2012). Greater sheltering implies greater wind reduction near the surface fuels than in unsheltered conditions. The surface fuels and the fires spreading in them are said to be sheltered if there is high canopy cover or terrain that blocks wind flow such as lee sides of hills. The WAFs for sheltered conditions are commonly ~0.1–0.2. Partial sheltering may occur at midslope locations or under moderate canopy cover with WAF of ~0.3. Surface fuels are unsheltered on ridges and with low canopy cover or with trees that have lost leaves in autumn with WAF of 0.4–0.5. The concepts and models that deal with wind profiles, wind interactions with fire and treatment of wind in models are discussed briefly in this section.

Wind profiles without fire

No matter how fast the wind, the air immediately touching the ground surface or along a hard boundary is absolutely calm. In fluid mechanics, we saw that this is referred to as the *no-slip condition*. When there is wind, air movement increases with height above the ground. This is called the *velocity boundary layer* because it encompasses the depth over which the air flow adjusts to the velocity of the free stream (i.e. air unaffected by the surface) (**Chapter 3, Boundary layers**). Many meteorological factors affect wind patterns, and most are beyond the scope of this discussion. We focus here on the nature of the wind speed profiles used in fire prediction and modelling.

The increase in mean wind speed with height within the thermal boundary layer (i.e. 100 m or so of the ground) has been described by various functions that depend upon the condition of the atmosphere and the ground surface. The most commonly used wind profiles reflect conditions of *neutral stability*. Stability describes the rate at which air temperature changes with height above the ground and hence the potential for *vertical* movement of air. *Unstable* and *stable* atmospheres have rates of temperature change with height above the surface that, respectively, encourage and discourage the vertical movement of air. Neutral conditions neither enhance nor inhibit vertical air flow. In each of these stability conditions, the average *horizontal* wind speed increases with height above the ground. Data and models that show the relationship of wind speed to height above the ground are called *wind profiles*, and they are commonly described using a logarithmic profile that depends upon the roughness of the ground surface (**Figure 6.16**).

Smooth surfaces (like water or mowed grass) introduce less drag on air flow than rough surfaces (like forests or cities), so smooth surfaces permit higher winds at lower heights than rough surfaces do. Surface roughness is described by *roughness length* (z_0), which varies from ~0.0002 m for calm water to more than 1.0 m for forests (**Table 6.5**). Roughness length is not the physical height of the roughness elements but rather a scale used in modelling. It is typically around 10% of the physical height of the roughness elements on the ground, but this varies depending on factors such as the number, density, size, shape and orientation of the roughness elements. The average

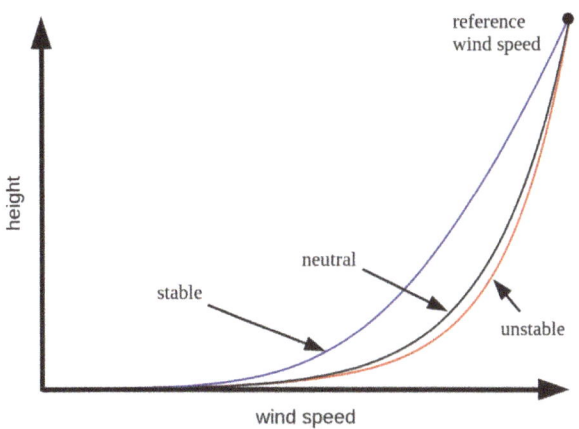

Figure 6.16: Characteristic logarithmic wind profiles for open ground under neutral, stable, and unstable boundary-layer conditions show that mean wind speed near the ground decreases from a given wind velocity at reference height and is higher for unstable conditions and lower for stable conditions than the typical neutral situation.

value of z_0 varies not only with surfaces on the ground but also with the variability of the surrounding terrain, since as features of the terrain itself can act as large roughness elements. Thus areas with hilly and mountainous topography have higher z_0 values than smooth terrain. If vegetation is oriented in rows or strips, the roughness length may vary with wind direction, depending on the alignment of the wind with the landscape elements.

The average wind speed U_2 at an arbitrary height h_2 above ground under neutral stability can be computed from a measured average wind speed U_1 at a known height h_1:

$$U_2 = U_1 \frac{\ln(h_2/z_0)}{\ln(h_1/z_0)} \quad [6.2]$$

The standard height above vegetation (h_1) for measuring U_1 is 20 ft in the United States and 10 m elsewhere. We can use **Eqn [6.2]** and measurements of wind speed at standard height to estimate near-surface winds for fire modelling. Notice from **Figure 6.16** that, for a given reference wind velocity U_1, near-surface winds in stable conditions are slower than they are in neutral conditions (which are assumed in **Eqn [6.2]**), and near-surface winds in unstable conditions are faster than in neutral conditions. Despite this theoretical relationship,

Table 6.5. Examples of roughness length values z_0 for typical vegetation and land cover types (after Hansen 1993).

Roughness length z_0 (m)	Ground surface	Terrain
0.0001	Snow, calm sea	Flat
0.0003–0.0005	Desert	Flat
0.001–0.0014	Mown-short grass	Flat
0.007	Short grass, sparse cover, short prairie	Flat
0.01	Grassy plains, level	Flat
0.026	Low shrubs, grass 18 cm tall	Flat
0.03–0.035	Grass and sparse trees	Flat
0.05–0.15	Grass 50–70 cm tall	Flat
0.16–0.25	Brush (open to dense)	Flat
0.40	Tall grass, forest, clearings and harvested sites	Flat
0.51–0.61	Savannah, 8 m tall trees	Flat
0.75	Grass	Hills
0.70–1.2	Forest	Flat
1.0–1.30	Pine forest, 20 m trees	Flat
2.0–2.5	Forest, sparse buildings	Rolling
3.5	Forested ridges	Hills (150–200 m)
6.0–11.0	Irregular forest	Hills (100–200 m)
1.0–4.0	Urban (various densities)	Flat

fire models and calculations generally ignore stability effects on near-surface wind profiles because measurements of wind speed and direction are already imprecise and highly variable. However, stability effects may need to be incorporated to improve models, especially for winds approaching a large heading fire, because the air flowing over the warm ground or over the smouldering area produces a wind profile more typical of the unstable condition (Beer 1991). We discuss this problem in more detail below, in **Wind in the presence of fire**.

Logarithmic wind profiles like those shown in **Figure 6.16** are appropriate only in open areas. In wildlands, they are most appropriate for the space above the vegetation. They do not help us understand wind speeds within and beneath canopies, which we need to know if a fire is burning in understorey fuels or within a canopy. Beyond the practical concepts of the WAF and sheltering described previously, in forested conditions and dense shrublands, the wind speed profile beneath canopies has been difficult to characterise because of variation in canopy structure. Canopy cover (i.e. the percentage of ground covered by the aerial parts of plants) can vary from a few per cent of the total area for an open savannah to more than 60% for closed forests. Canopies also vary by the mass of foliage and branch material they contain (see Brown 1978; Ter-Mikaelian and Korzukhin 1997) and the horizontal and vertical distributions of these materials (see Scott and Reinhardt 2005; Keane 2015). In addition, canopies themselves are somewhat plastic when exposed to wind. Strong winds can bend trunks and limbs and reorient foliage, creating a speed-dependent drag on air flow. We often see this effect in the swaying of tree branches in high winds and the bending waves of grass in prairies and grain fields.

A practical method for estimating wind profiles throughout the canopy was recently developed by Massman *et al.* (2017). This one-dimensional analytical model uses an actual or idealised vertical foliage density distribution to estimate wind profiles over the full range of canopy cover (**Figure 6.17**). The model is practical for any canopy, from agricultural crops to forests. The wind at any height above the ground with specified roughness characteristics can be used directly or to provide a WAF that modifies the wind speed to a value applicable to fire modelling (assuming we know the height for which a wind value is needed!). The idea of the WAF has practical value similar to that of **Eqn [6.2]**, using a measured value at a known height to obtain wind speed at another height.

Despite what is commonly assumed, wind speed profiles near the ground are often not actually logarithmic. This is particularly true on sloping topography when solar radiation is heating a hillslope (thus inducing stronger uphill air flows) (**Figure 6.18**) and in the evenings, when cold air drains downhill (Albini *et al.* 1982). Variation in wind profiles and their interactions with fire are not well understood. Many different wind profiles have been documented for a deeper section of the atmosphere, up to several thousand meters above ground (Byram 1954). These were classified as to their association with different behaviours of large wildfires. Byram suggested that some wind features, such as when the winds are highest near the ground (sometimes called a jet point) and generally decrease with altitude, were related to the likelihood of 'blow-up' fires. A blow-up fire is defined by rapid increases in energy release and convective interactions with the atmosphere. The high winds near the ground helped increase fire spread rates and calmer winds aloft then aided vertical development of the plume. Effects of different wind profiles on wildfire behaviour have not been proven, however, largely because of the confounding influences of other meteorological factors and variables related to fire size and energy release (Potter 2012a). The role of wind in large fires is further addressed in **Chapter 8, Plumes and pyroconvective atmospheric storms**.

Wind in the presence of fire

All fires affect nearby air movements. Wind and air movements created by the fire itself are called *induced* flows or simply *fire winds* (Smith *et al.* 1975).

Fire winds are caused by buoyancy; the rising hot gases generated by the fire must be replaced by air from the surroundings. Buoyancy causes an area of relatively low pressure to form within the fire. This low pressure occurs for two reasons: (1) the weight of air above the fire is less than the weight of air above the surrounding area (the hydrostatic pressure we learned about in **Chapter 3**), and (2) dynamic pressure that forms due to the vertical movement in the plume. Both cause indrafts to replace the rising air or a vacuum would form. The low pressure 'pulls in' surrounding air. In this area of induced flow, wind speeds are stronger near the ground than above, and they become strongest close to and within the fire. Large fires (**Chapter 8, Plumes and pyroconvective atmospheric storms** and **Vorticity**) may induce very strong winds – exceeding 50 m s^{-1}; these are

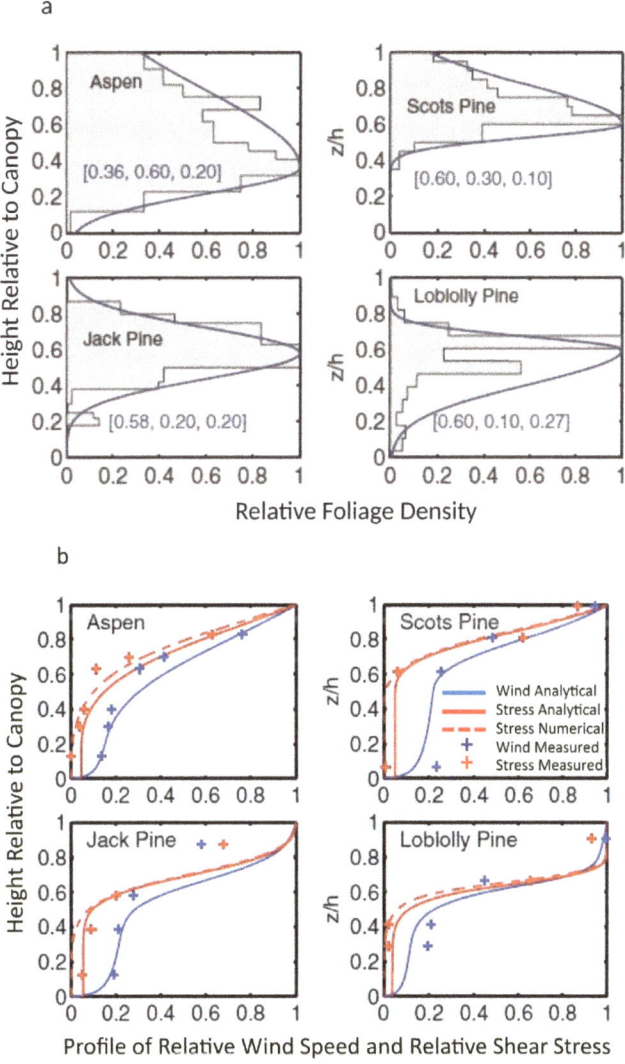

Figure 6.17: Wind profiles within and beneath vegetation canopies are affected by (a) the density and distribution of foliage of different species including *Populus tremuloides* (aspen), *Pinus sylvestris* (Scots pine), *Pinus banksiana* (jack pine), and *Pinus taeda* (loblolly pine). Shown are the actual measured distributions of foliage and a modelled curve fit. Profiles (b) of relative wind speed and relative shear stress can be approximated by practical modelling that calculates the wind speed near the surface for fire modelling to be obtained from wind measured at a reference height above a canopy (reproduced with permission from Massman *et al.* 2017).

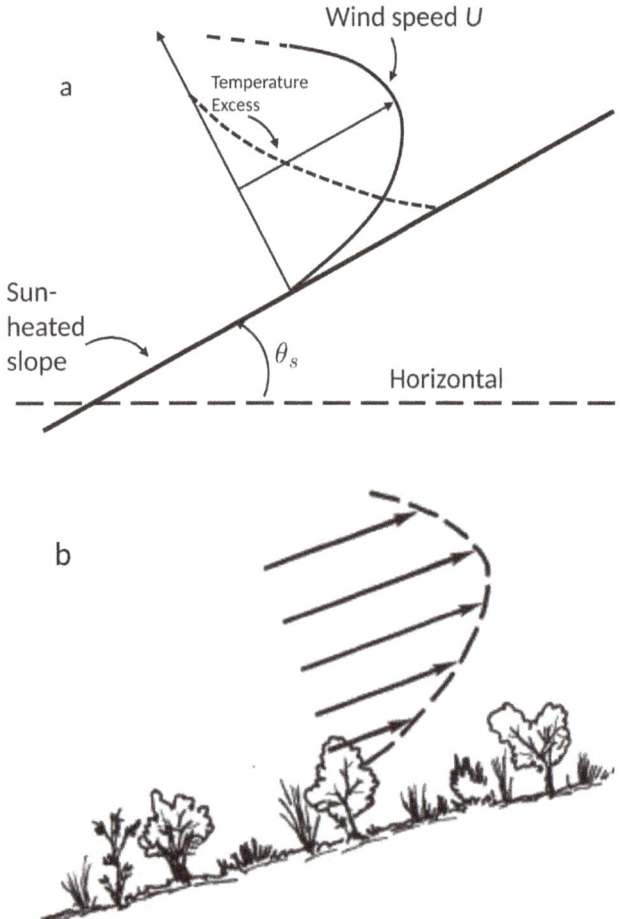

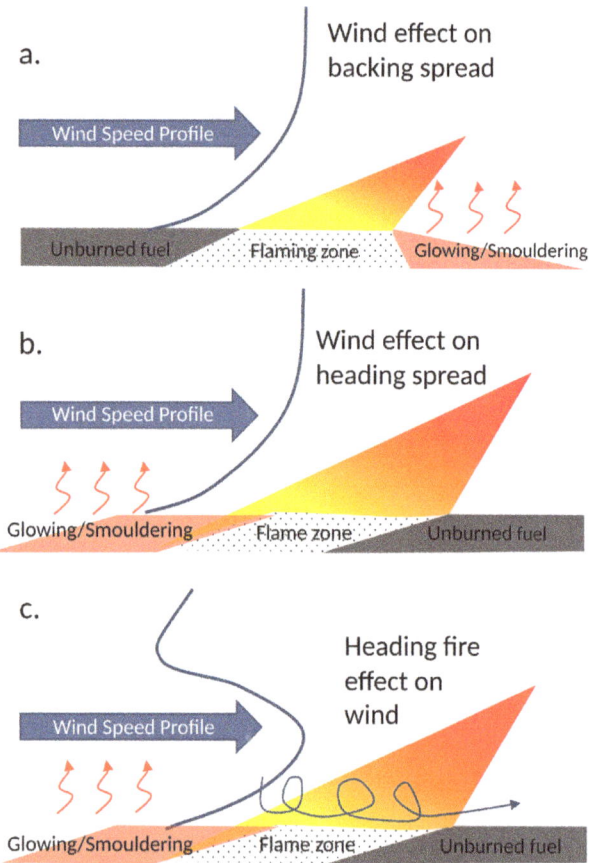

Figure 6.18: Illustrations of characteristic near surface wind profiles with uphill afternoon flows caused by surface heating from solar radiation (a) from Albini *et al.* (1982) and (b) from Rothermel (1983).

Figure 6.19: Wind speed profiles approaching flame zones of spreading fires. (a) Wind flow is little changed by heated ground as it approaches a backing fire. (b) As wind impinges on heading fires, it crosses heated ground and fuels burning behind the flame front, but its profile is assumed unchanged. (c) As wind impinges on heading fires, it crosses heated ground and fuels burning behind the flame front, and its profile is changed to increase wind speeds slightly above the ground.

associated with fire storms ('mass fires') (see reviews by Pitts 1991; Finney and McAllister 2011).

In many fires, the ambient wind is stronger than the fire-induced wind, and it is difficult to separate the two components of wind speed. Nevertheless, interactions involving induced flow and ambient flow strongly affect flame behaviour and heat transfer to fuel particles ahead of a spreading fire. A backing fire (spreading into the wind), for example, is subject to winds flowing over fuel and vegetation that has not yet burned (**Figure 6.19a**). In backing fires, the leading edge of the flame zone receives ambient air that is little modified by the burning area. In contrast, in a heading fire, the air flow impinging on the flame zone must travel over ground where the fuel structure has been altered by burning and has been warmed by the previous passage of the fire. Closer to the rear edge of a heading fire, fuels may still be burning in glowing or smouldering stages, modifying the velocity boundary layer as the ambient wind approaches the fire (**Figure 6.19b**). Beer (1991) examined the theoretical effects of fire on the flow of the boundary layer. He concluded that surface heating for an extended distance behind an advancing fire would warrant the use of a wind profile typical of unstable conditions (see **Figure 6.16**) for performing fire behaviour calculations and this would substantially increase the fire's spread rate.

Passage of a fire also directly changes the fuel structure, which then affects air flow. This is most apparent in canopy fuels (brush and forests), where burning of the canopy foliage reduces vegetation drag and exposes the ground surface. This greatly increases wind speeds entering the back of the fire (Taylor *et al.* 2004). There have been few field-scale studies of the influence of wildland fires on air flow, but there are many reports of downdrafts and increased turbulence behind the passing flame front (see Clements *et al.* 2007; Heilman *et al.* 2017). These field-based point measurements undoubtedly indicate more complex 3D circulations than those suggested by **Figures 6.16** and **6.18**. Some of these have been studied in the laboratory.

Fires directly impact the velocities and temperatures in the boundary layer. Based on small-scale laboratory studies (Nicholl 1970; Hirano and Kanno 1973), we know that the vertical velocities increase near the heat source and the horizontal velocity profile changes from a logarithmic form to one with the highest air speed slightly above the surface, with lower speeds above that gradually increase to the ambient wind speed as they reach above this internal boundary layer (**Figure 6.19c**). Although the surface temperatures of the ground heated by the sun are far lower than those after a passing fire, similar wind profiles are created by solar heating (**Figure 6.18**) and radiative cooling of sloping ground (Albini *et al.* 1982; Clements 2011). Findings like these support the need to study airflow patterns near spreading fires more closely so we may one day understand the interactions between fire and air flow and identify the wind characteristics that most influence fire spread. Current wildfire science is probably missing or incorrectly explaining some very important effects by neglecting the micrometeorology associated with the presence and passage of fire.

Fires and associated surface heating produce complex 3D circulations within the boundary layer, and in flames specifically, that are very important for understanding wildland fire spread. In laboratory studies of heated plates, with both forced air flow and natural convection along inclined surfaces, the velocity boundary layer may develop *longitudinal rolls* (**Figure 6.20a**). *Longitudinal* means that the rolls are oriented parallel with the direction of mean air flow. These rolls alternate in their direction of rotation. When viewed perpendicular to the flow direction (i.e. looking 'downstream' through the fire), the alternating rotation of the rolls produces convergence zones between them that force air to flow either up or down. The repeating pattern of these alternating air flows in the vertical direction is responsible for the peak-and-trough structure of flame fronts (Beer 1991; Finney *et al.* 2015) and the patterns of bright and dark streaks within the flame zone (Miller *et al.* 2017) (**Figure 6.20b, c**). The lateral spacing of the flame peaks is directly proportional to flame length (**Figure 6.20d**), which affects convective heating ahead of the flame zone (discussed below).

The exact origin of longitudinal rolls has not been determined. Possible sources of the phenomenon include the pre-existing structure of the air flow entering the fire and the buoyancy within the combustion zone itself. Regarding the incoming air, small-scale wind tunnel experiments (Miller *et al.* 2017) show that thicker boundary layers, which develop over longer surface distances upwind of the flame zone, enhance the formation of longitudinal rolls within the flame zone (**Figure 6.20c**). Thin boundary layers, which develop from shorter approach distances, result in flame structures that are more symmetrical in both horizontal dimensions rather than being oriented parallel with the wind direction. Buoyancy of the combustion zone itself can also create instabilities that result in fluid circulations similar to instabilities in the boundary layer. An *instability* in gases or liquids is a condition where changes in their behaviour are sensitive to disturbances or variations in the fluid properties caused for example by changes in wind patterns or surface heating.

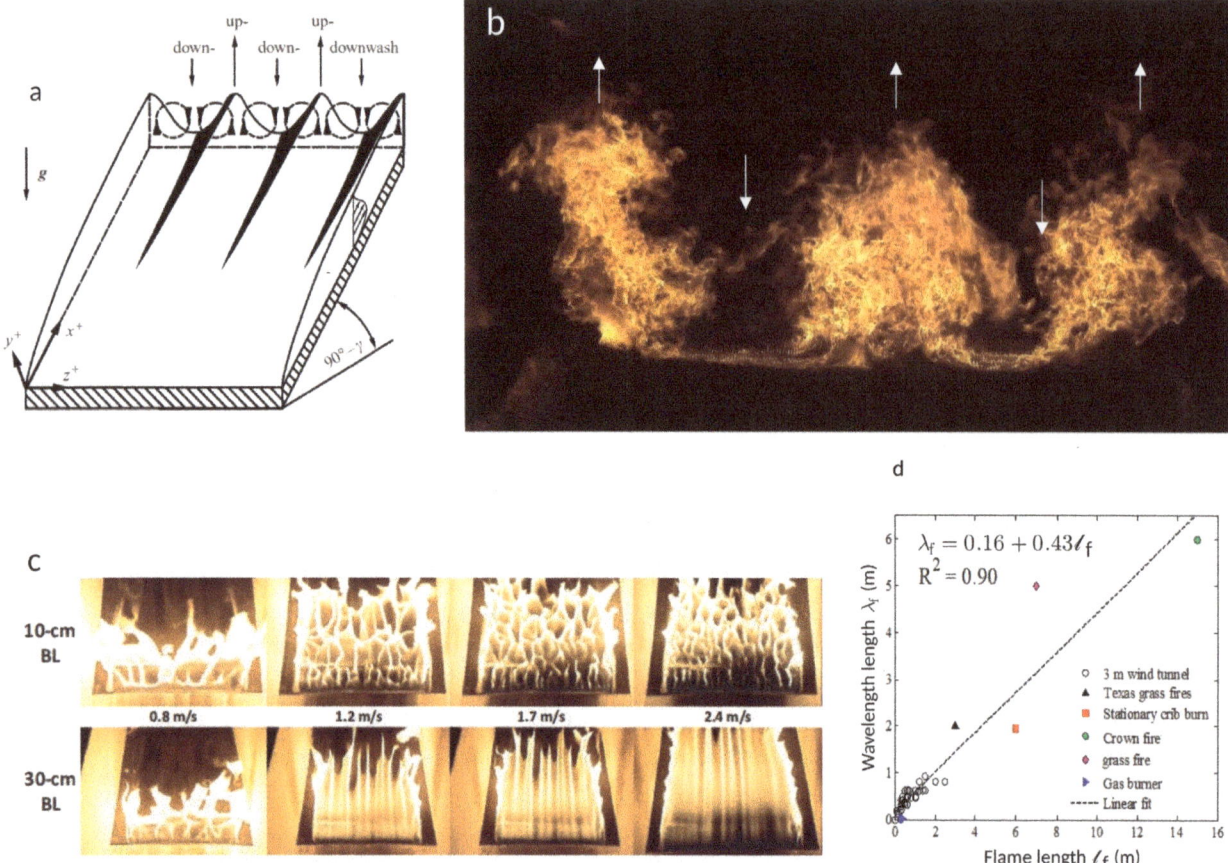

Figure 6.20: Air-flow patterns in the flame zone show (a) longitudinal rolls with alternating rotational directions, which produce a repeating pattern of upward and downward air movement across the flame front (reproduced with permission from Jeschke and Beer 2001). (b) Flame peaks represent ascending flows (up arrows) and descending flows ('troughs') (down arrows) in the flame zone produced by convergence of longitudinal rolls. (c) Flame patterns depend on the wind speed and the length (which controls thickness) of the approaching boundary layer (reproduced with permission from Miller et al. 2017). (d) Laboratory and field data suggest that the wavelength of peak-and-trough flame structures (i.e. the lateral distance between them) depends on flame length for a wide variety of fuel types (reproduced with permission from Finney et al. 2015).

One instability that occurs in wildfires is caused by hot, low-density flames and gases originating under denser ambient air. The layering of high density over low density fluid in the presence of gravity is not stable, meaning it will tend to change to a different, more stable state. Gravity causes the low- and high-density strata to overturn, allowing the low-density gases to escape upward. If the overturning process occurs over a large area, the gases break up into a regular pattern of localised circulations called *Benard cells* (**Figure 6.21a**). Benard cells are often hexagonal, much like the cells in a honeycomb. In liquids, Benard cells commonly circulate with upward flow in the centre and downward flow around the outside edges. In most gases, such as flames, the rotation is reversed, producing downward flow in the centre (Koschmieder 1993) and upward flow around the edges. In photographs of flames under controlled conditions, Benard circulations produced by downward flows can be seen as concave depressions in the flame zones behind the flame front (**Figure 6.21b**). Incoming wind or air forced across these circulations may stretch the cellular

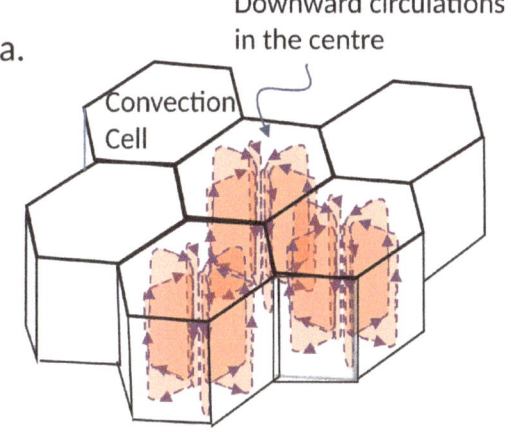
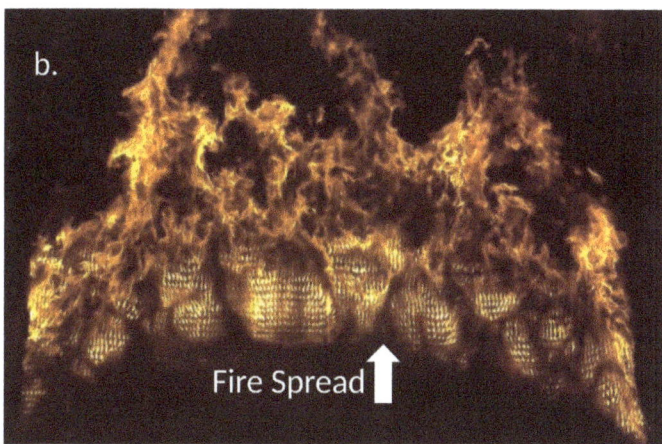

Figure 6.21: Illustration of (a) convection cells for gases with downward circulation direction that appear similar to images of flame zones in spreading fires viewed from above showing concave parcels behind the flame front in (b) laboratory cardboard fuel beds (Finney *et al.* 2015) with flames ~1 m tall; (c) 150 m wide head fire in crop stubble with flames ~5 m tall and flame zone depth of 10 m (photograph courtesy of University of Canterbury, NZ); and (d) 200 m wide head fire in gorse brush fuels with flame length and flame zone depth ~15 m (note vehicles and people in lower left corner for scale; photograph courtesy of Marwan Katurji, University of Canterbury, NZ). These structures maintain coherency as they travel towards the front of the fire.

symmetry in the mean flow direction to form roll-type vortices (i.e. longitudinal rolls). Both Benard cells and roll-type vortices are observed in the flame zones of fires at laboratory (**Figure 6.21b**) and field scales (**Figure 6.21c, d**).

Our discussion of localised air flow, its instabilities and circulations, is vital to understanding fire spread because the fluid motions in flames affect convective heat transfer ahead of the fire. We know from **Chapter 3** that convective heat transfer occurs when a fluid flows past a solid object. In wildland fires, the fluid is the flame, and it is transferring heat to cool, solid fuel particles ahead of the flame zone, eventually heating and igniting them.

High-speed videos of laboratory and field burns reveal that downdrafts from the longitudinal rolls and localised concave cells within the flame zone can force flames down and forward into the fuel bed ahead of the flame zone. These downward movements are critical to convective heating because they maintain contact between the hot flames and the fuels, both inside and ahead of the flame zone (**Figure 6.22**).

Downdrafts associated with the flame zone have been measured in field experiments by Clements *et al.* (2007) and can be readily seen in overhead imagery (**Figure 6.21b, c, d**). The flame circulations (possibly Benard cells) appear to be

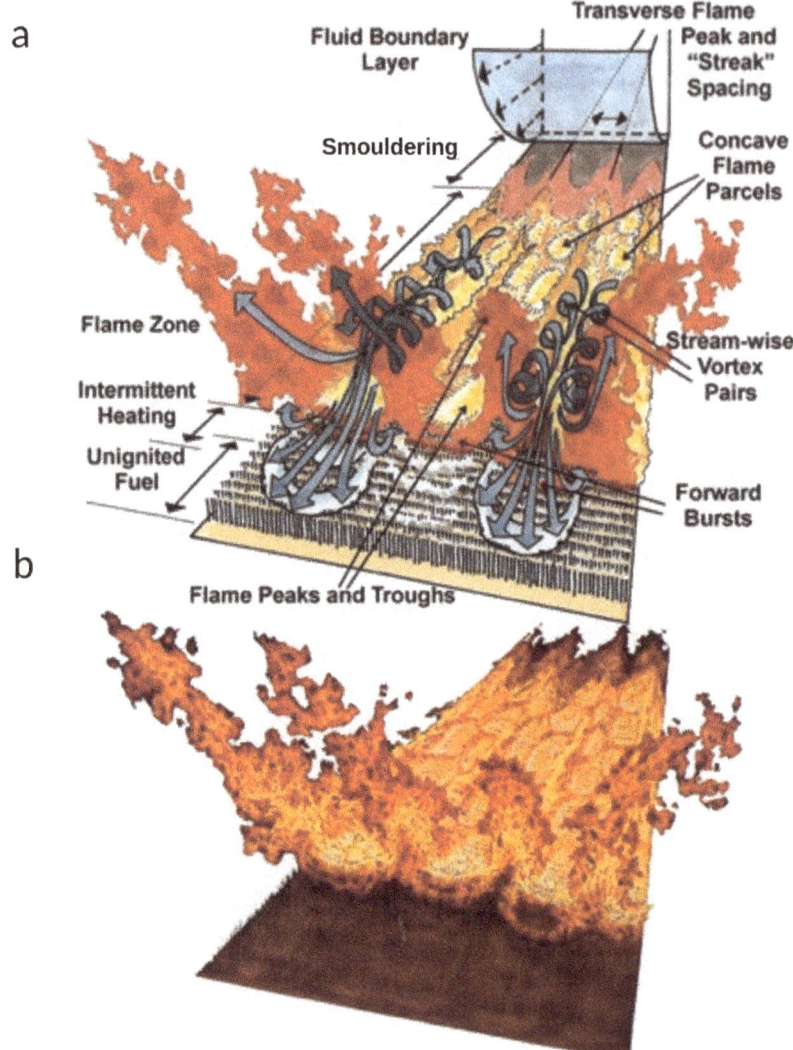

Figure 6.22: Illustration of flame zone dynamics in a section of a spreading line fire showing (a) the main features of flame behaviours, including the streamwise vortex pairs, forward bursting that causes intermittent heating of fuels, and the concave flame parcels near the rear of the flame zone. The counter-rotating circulations produce the peak-and-trough appearance of the flame front and force flames downward into contact with fuel particles ahead of the leading edge of the fire; and (b) a rendering of the appearance of the flame zone. Reprinted with permission from Finney et al. (2015).

coherent (i.e. they remain recognisable over time), and most of them move from the back of the flame zone towards the front. This forward transport of the flame fluid is called *advection*. In fires, advection delivers the flame gases forward so they contact unignited fuels.

The downward flows that create troughs in the flame zone (**Figure 6.22**) can also channel ambient wind through the flames, also forcing the flames ahead. When these troughs occur across wide flame zones, air is being forced *through* the flame zone rather than *around* the edges – an important factor in fire spread that will be discussed in the context of convection heat transfer and spread rate affected by the width of a flame zone. Just as the flames are non-steady and highly variable, so too is the convective heating of fuel particles from these flames.

This phenomenon is called *intermittency*. Intermittency is caused by turbulence and by the instability-caused motions of the flames, one of which results in *forward bursting* of flames into fresh fuels (**Figure 6.22**).

Flame intermittency has an important role in heating and igniting fine fuels (Beer 1991). We can see its influence in both laboratory fires and field burns (**Figure 6.23**). In these experiments, fine thermocouples (25–50 μ) were used to measure air or gas temperatures at high sample rates (50–500 Hz). The data show temperatures oscillating – spiking dramatically – over time. The spikes are produced by intermittent flame contact (see Dupuy *et al.* 2011; Frankman *et al.* 2013; Finney *et al.* 2015) (**Figure 6.23a**). A common way to analyse relationships among multiple physical factors is with dimensionless numbers that allow the primary variables involved in a phenomenon to be combined into a single term. In this case, the flame frequency (f, Hz, or s^{-1}) is represented by the *Strouhal number*, which expresses the ratio of upward acceleration of buoyant flow to crossflow (wind) and is calculated as $f\,\ell_f/U$ where we use flame length ℓ_f and wind velocity U. A second dimensionless number, the *Froude number* commonly correlates well with the Strouhal number for oscillating phenomena. The Froude number represents the balance of inertial force (in this case wind) with the buoyant force of the fire, which involves the vertical acceleration of gravity (g) and a length scale of the flame zone for which flame depth D (m) is used. **Figure 6.23c** shows an application of this *scaling relation* to the data in **Figure 6.23a** and **b**. The graph suggests that the frequency of flame intermittency before ignition consistently decreased with longer flames for both field and laboratory fires. In other words, longer and deeper flames fluctuate at lower frequencies. The application of scaling relations to fire spread may help explain the heating and ignition of fine fuels in fires across a range of fire sizes and energy release rates.

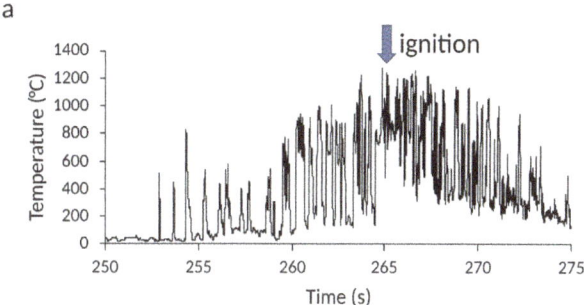

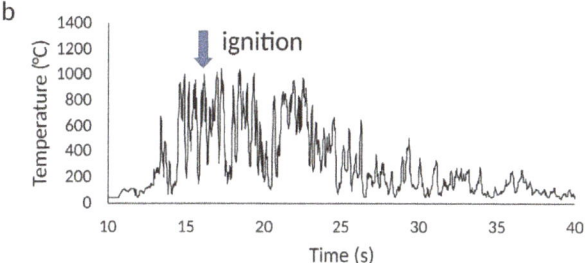

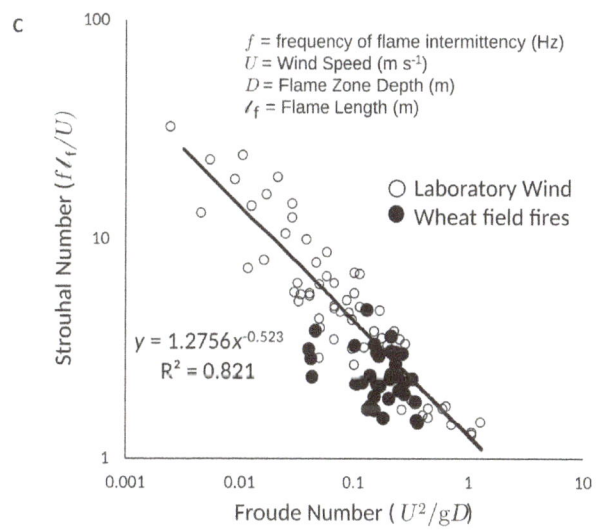

Figure 6.23: Graphs of temperature time series from fine-wire thermocouples for two very different spreading fires show intermittency of flame fluctuations ahead of ignition (arrows) in (a) laboratory fire (wind speed $U = 0.22$ m s^{-1}, flame length $\ell_f = 0.3$ m, flame zone depth $D = 0.10$ m, rate of spread = 0.74 m min^{-1}) and (b) field-scale prescribed fire in wheat straw (wind speed $U = 4.4$ m s^{-1}, flame length $\ell_f = 4.6$ m, flame zone depth $D = 10.4$ m, rate of spread = 88 m min^{-1}). Despite the wide variation in fire behaviour, the frequencies of these intermittent flame contacts show consistent scaling (c) with flame length, flame zone depth and wind speed. The Strouhal number–Froude number scaling used in the graph suggests that the frequency of flame fluctuations decreases with flame length and depth, but increases with wind speed consistently across a wide range of fire sizes and behaviours.

In laboratory experiments using rectangular cardboard fuel particles as the fuel bed, we can see how intermittency of flame contact in a spreading fire affected ignition of a single 1 mm fuel particle placed inside the fuel bed. The surface temperature of the 1 mm particle responded to the intermittent flame contacts by heating in stair-step fashion before ignition (**Figure 6.24a**). The particle heated only when contacted intermittently during flame bursts, and it cooled in the periods between flame bursts, even though radiant heat flux was increasing continually (Finney *et al.* 2015; Cohen 2015). Furthermore, a thermal camera revealed that the corners and edges of rectangular cardboard fuel particles were heating faster than the middle surface (**Figure 6.24b**). This should be expected with convective heating, because the corners and edges have thinner local boundary layers and thus higher convective heat transfer than the middles of the particles (see **Figure 3.21**), just as we saw in regard to the edges of leaves (see **Live fuels** above). **Figure 6.24** illustrates that fine particles can respond to both radiation and convection in spreading fires. As discussed in **Fuel bed properties** above, the dominant heating mechanism – the one that most influences fire spread – may be convection even though radiant heating is intense and continual.

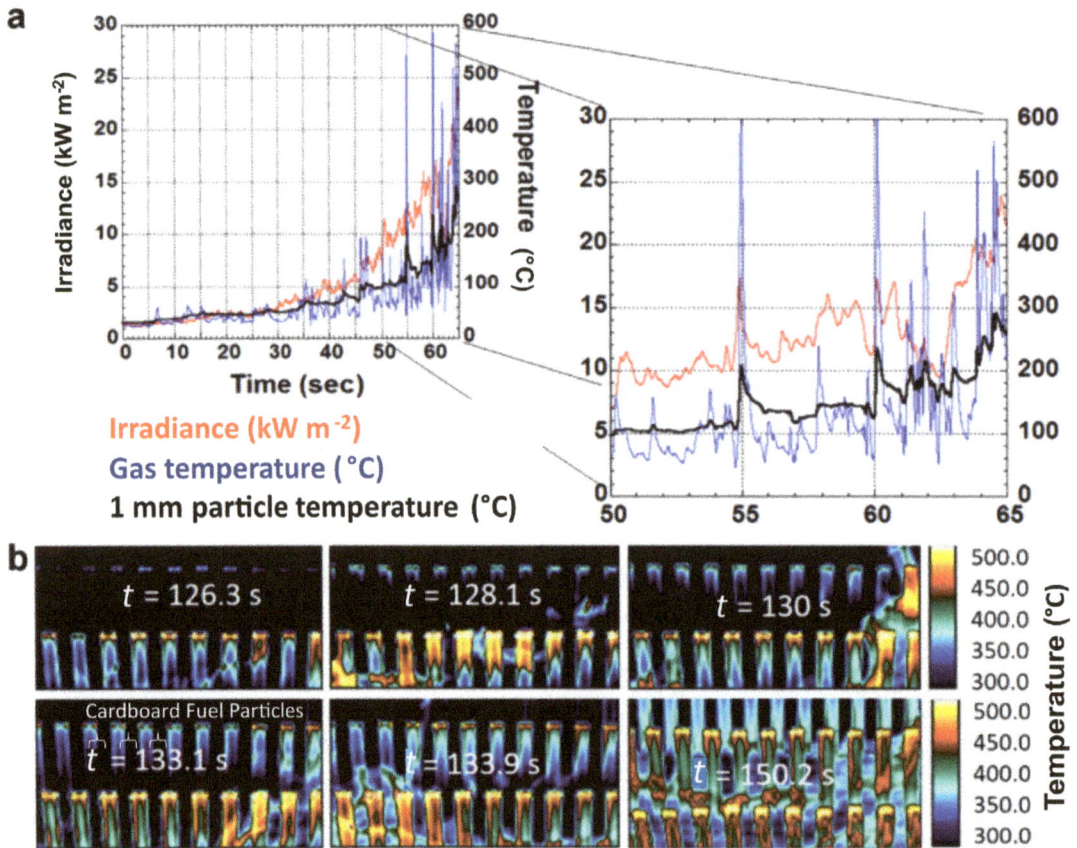

Figure 6.24: (a) Time series of intermittent flame contacts with fuel particles (blue) in a spreading laboratory fire that produces stair-step heating (black) with convective cooling between flame contacts even while radiation is increasing (red). (b) Temperature distribution on rectangular cardboard fuel particles as they heat to ignition in spreading fire. The highest temperatures occur along the thin edges and corners of particles rather than in the centres, indicating convective heating (reproduced with permission from Finney *et al.* 2015).

Wind effect on flame tilt

Many models have relied upon the assumption that wind increases the spread rates of heading fires because radiant heating is enhanced as flames tilt forward and closer to fuel particles (see review by Pitts 1991). Although thermal radiation does increase with proximity to fuels, fine fuels probably are little affected by the increased radiation because of convective cooling. Flame tilt does, however, alter the trajectory of the flame and plume, thus altering the rate at which flames, hot smoke and hot gases from the fire convectively heat and ignite foliage in a tree canopy (triggering torching or crowning). Many laboratory studies have shown flame deflection by wind (*crossflow*), both in spreading fires and with stationary flame sources.

The effect of wind on flame angle is often examined theoretically using a balance between vertical momentum of flame gases and the momentum of horizontal air flow. This concept was introduced above as the Froude number, which is dimensionless. We can easily observe by blowing gently on a candle that the flame tilt depends upon the horizontal wind speed (U, m s^{-1}). It also depends on energy release rate of the fire, correlated with the size of the flames as length ℓ_f (m), or as fireline intensity I_B (kW m^{-1}) and the acceleration of gravity g (9.807 m s^{-2}). These variables determine the vertical velocity of the flame gases achieved by acceleration from buoyancy **(Chapter 4, Non-premixed or diffusion flames)**. The geometry of the flame source (point, line or area) also plays an important role in determining the effect of wind on flame angle. For line-source fires spreading with the wind (heading direction), which is the most common assumption used in the wildland context and in modelling, many relationships have been developed for predicting the flame angle θ (**Figure 6.25a**), and all of them show that, for a given wind speed, longer flames or higher intensity fires affect flame tilt less than shorter flames or lower intensity fires. Two of the more common relationships in wildland fire are:

Putnam (1965): $\tan(\theta) = 1.4\,(U^2/g\ell_f)^{0.5}$ [6.3]

Nelson and Adkins (1986): [6.4]

$\tan(\theta) = 2.58\,(U^3/I_B)^{0.29}$

Similar correlations have been reported by Albini (1981) and Weise and Biging (1996). Fires spreading in the absence of wind show flames standing vertical or tilting slightly away from the

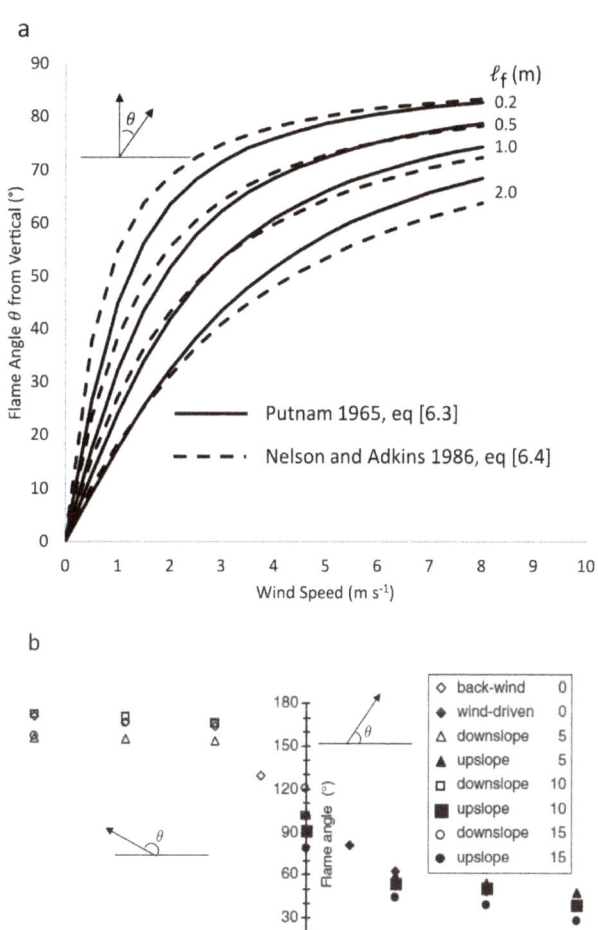

Figure 6.25: Changes in flame tilt angle for line fires (a) as modelled by correlations of wind speed with flame length by Putnam (1965) and Nelson and Adkins (1986) and (b) measured from the horizontal heading direction for spreading line fires in pine needle beds. The graph shows data for backing fires (negative wind speeds) and heading fires (positive wind speeds) on slopes of various angles (reproduced with permission from Mendes-Lopes *et al.* 2003). Note that the flame angle is measured differently in each panel.

unburned fuel. The latter effect occurs when heat released from the interior burned area causes a weak convective flow opposed to the direction of spread into unburned fuel (Viegas 2004). Flames in small incipient fires also display this trajectory (**Figure 2.9a**). As depicted in **Figures 2.9b** and **2.11**, flames from fires backing into the wind and downslope are also oriented towards the burned area (see **Figure 6.25b**), but it is unknown how well existing flame angle models describe flame tilt in backing fires. Flame geometry for fires backing downslope, with or without ambient wind, have been little studied. It does seem that flames in both heading and backing fires attach to steep slopes (see **Topography** below).

The same physical variables – U, ℓ_f or I_B – occur in equations for predicting flame angles from fire sources that have a square or round shape. These equations are usually applied to industrial fires that involve fuel spills or chemical tanks (pools of burning liquid fuel or *pool fires*) rather than wildland fires. Pool fires typically exhibit flaming combustion across the entire surface, and because they are axisymmetric in contrast to rectangular, the wind is able to easily flow around the fire. Equations for these relationships differ from the line-fire equations (**Eqns [6.3]** and **[6.4]** above) because they include the diameter of the flame source in the denominator rather than just the flame dimension or intensity. This means that flame tilt angle from the vertical for a given wind speed decreases as fire area increases in diameter. In wildland fires, such correlations apply primarily to spot fires and point ignition patterns during the short time when energy release occurs uniformly throughout the burning area (see **Figure 2.9a**). As these small fires spread outward, the solid fuel in the interior is eventually consumed and flaming combustion exists only along the outer, expanding edge. At some point, the fire grows large enough that the wind no longer simply 'goes around' the fire and enters behind the flaming front. At this point the flaming edge of the fire acquires characteristics of a line source. Then the wind begins to interact with the flame zone differently and more strongly affects the flame tilt angle. A thorough review of the numerous empirical correlations for pool fires is available in Lam and Weckman (2015a, 2015b).

Wind and flame attachment

Wind influences the attachment of flames to the adjacent ground surface. Flames deflected by wind can attach to the ground downwind both continuously and intermittently. In fire protection engineering, this is referred to as *flame drag* and is important for determining heat transfer from burning liquid fuel spills and ruptured fuel tanks. Flame drag is relevant to wildland fires because flames lapping forward from a wind-driven fire impinge on and heat the fuels. It is not surprising that the length of continuous flame attachment is related to the same variables as those for flame tilt, since the same forces are at work.

The distance of attachment is determined by the balance of vertical momentum of the fire caused by buoyancy and the horizontal momentum of the wind. Attachment is predicted by the diameter of the fire, fuel burning rate (kg m^{-2} s^{-1}) and crossflow velocity. Many empirical relationships have been developed for predicting this relationship (reviewed by Lam and Weckman 2015b; Hu 2017; Tang *et al.* 2017). While most studies of round or square pool fires have limited relevance to the geometry or energy release patterns of spreading wildland fires, studies that investigate the geometries of rectangular fires can help us understand wildfires because their flame zones are much wider than they are deep and combustion is taking place throughout the entire area of the fire. Wide flame zones are likely to increase the flame attachment distance in wind (Tang *et al.* 2019) because wind is forced through the troughs in the flame zone (**Figure 6.22**) and pushes the flames down onto the ground. A broad fire front in a line fire forces more of the wind to penetrate the flame zone because it cannot go around the edges. Where it does travel along the edges, it wraps around them and lifts the flames off the ground. The relationship of flame drag to

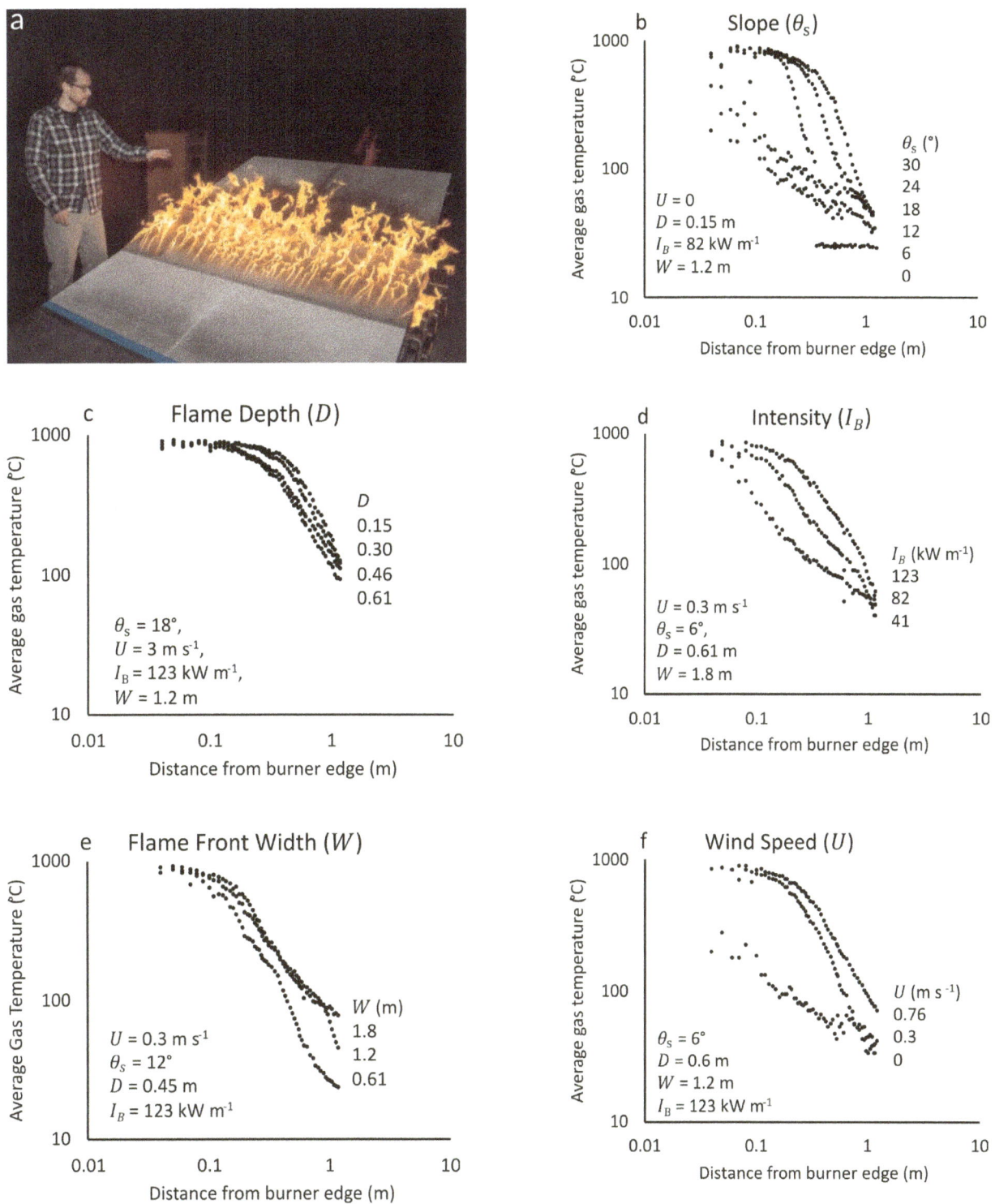

Figure 6.26: Experiments from a stationary sand burner (a) used to examine the gas temperature decay with distance from the burner edge. A rectangular gas burner with variable tilt angle was placed in a wind tunnel to study head fire flame structure and record mean gas temperature distributions at distances along the surface away from the burner edge (see Finney *et al.* 2020). Data reveal that adjacent temperatures decay as a power-law function of surface distance but vary by (b) slope, (c) flame depth, (d) intensity, (e) the width of the flame zone and (f) the wind speed.

intermittency deserves more study to determine its applicability to wildland fire as do patterns in gas temperature around rectangular fires because they both relate to the geometry of fires and convective heating.

In **Chapter 4**, we saw that there is an approximately constant mean temperature of flames above a stationary fire that declines as a power-law function of distance above the flames as they become intermittent (**Figure 4.22**). *Power-law functions* plot as straight lines on log-log axes because the *y*-values are a function of the *x*-values raised to a power (in this case a negative power). Although gas temperatures are known to decrease rapidly from the flame edge (Anderson *et al.* 2010), recent laboratory experiments in a wind tunnel have discovered that the temperature profiles along the ground surface downwind of rectangular flame zones (**Figure 6.26a**) have the same power-law appearance as those above a vertical flame (see Finney *et al.* 2020). However, temperature profiles in the horizontal direction depend upon many additional factors in the wildland environment. Specifically, the profile of the power-law region varies with slope, flame depth, fireline intensity, flame front width and wind speed (**Figure 6.26b–f**). Temperature profiles probably also vary with other factors, including the orientation of the front relative to wind and slope. It is clear that gas temperatures, flame attachment and convective heating adjacent to a fire depend on complex relationships among numerous fire and environment variables. If we can determine the relationships among these variables, we will be better able to represent convective heating from the edges of a wildland fire.

Solar radiation

Solar radiation affects an entire hemisphere of the earth at once, which means that large-scale weather patterns are directly affected by the sun. Solar radiation affects the behaviour of individual wildland fires primarily through infrared heating of the fuel and ground surfaces. This section lists some of the important influences of solar radiation on individual fires, but there are many more detailed sources of scientific information on this topic, for example in meteorology and the solar-power industry, that are directly relevant to wildland fire.

The principles of radiant heating covered in **Chapter 3, Radiation heat transfer** apply to solar radiation, but it is important to note that the flux from solar radiation to the ground surface is much less than that from radiation produced by flames and fires. At the ground surface, the maximum solar flux received anywhere on earth is ~1 kW m^{-2}, much less than fluxes as high as 200 kW m^{-2} in close proximity to large wildland fires. Solar flux at the ground varies considerably by season and synoptic weather, especially cloud cover. Haze, smoke from wildfires and water vapour in the atmosphere all reduce the transmittance of radiant energy to the ground. Radiant heating and cooling of the ground, in turn, affect the stratification and air circulation in the lower atmosphere. One important effect of this pattern is to induce night-time temperature inversions and their afternoon break-up, both of which affect wildfire behaviour substantially.

The orientation of topography modifies the incident sun angle, so north-facing slopes in the Northern Hemisphere receive far less solar radiation than south-facing slopes. The opposite effect occurs in the Southern Hemisphere, where south-facing slopes are less exposed to the sun. Slopes more exposed to the sun tend to have drier fuels during the majority of the fire season and experience stronger upslope convective winds, which are induced by afternoon heating of the ground surface. Slope orientation and its relationship with solar flux also influence vegetation species composition and structure, thus determining many fuel characteristics. In complex topography, these influences create tremendous variation in both fuels and burning conditions.

Vegetation itself affects solar radiation and fuel moisture content, especially in forests, where the vegetation (canopy cover) controls the fine-scale patchiness of the solar flux that reaches the ground.

Solar radiation at the ground surface affects the micrometeorology of temperature and humidity, which in turn affects surface air flow around fuel particles. In the discussion above on moisture dynamics in *dead fuels* (**Figure 6.11**), we saw that the drying of fuel particles is affected by interactions between fuel surface temperature and air flow which are altered by solar radiation.

A curious effect of solar heating has often been observed by fire personnel: fire spread and intensity sometimes change very rapidly – within a minute or so – when the sun emerges from cloud cover. Similarly, fire behaviour becomes less vigorous as soon as the sun is obscured. Because solar radiation produces a low heat flux at the ground (<~1 kW m^{-2}), it seems unlikely that these rapid fire behaviour responses are caused directly by changes in fuel moisture or fuel temperature. Could solar heating of the ground be enhancing the surface air circulation that affects heat transfer or combustion at the flame edge? At the moment, the one experiment that has investigated this question (Butler 2005) supports this explanation, but it also opens many additional questions that should be addressed if we are to develop better explanations and predictions of wildfire behaviour.

Topography

Terrain influences wildland fires both directly and indirectly. At the flame scale, near the edge of a spreading fire, a sloping surface modifies the fluid dynamics and heat transfer to fuel particles and thereby directly changes fire spread and behaviour. Fires then grow asymmetrically by spreading faster upslope than down. More complicated surface topologies of canyons and ridges can affect fires at this flame scale and at the larger scale of the fire itself. The same topographic configurations affect wildfires indirectly as well, by altering fine-scale and general-scale air flows, changing the incident angle of solar heating at the ground, creating elevational gradients in air temperature and humidity, and crucially by altering vegetation growth and ecology, which determine fuel characteristics. Thorough reviews of terrain effects on wildfire behaviours and weather have been published by Whiteman (2000), Sharples (2009), Clements (2011), Potter (2012a) and Sullivan *et al.* (2014). We focus here primarily on the direct effects of terrain on the local processes that influence fire spread and behaviour.

At the flame scale, slope affects fire spread by altering the heat transfer processes. This is most easily observed in the behaviour of the flame and the plume (**Figure 6.27a**). Shadowgraph images of a rectangular burner by Grumstrup *et al.* (2017) illustrate that the flame and the plume tilt uphill more on steep than flat surfaces, eventually attaching for

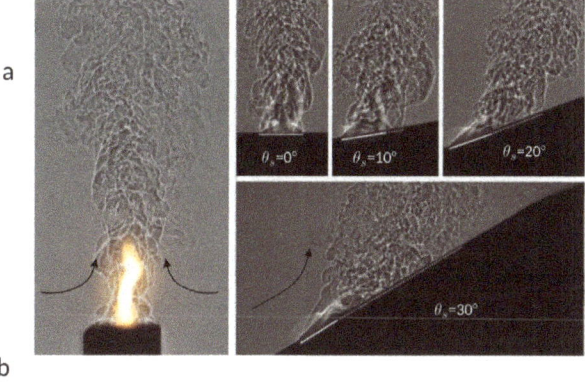

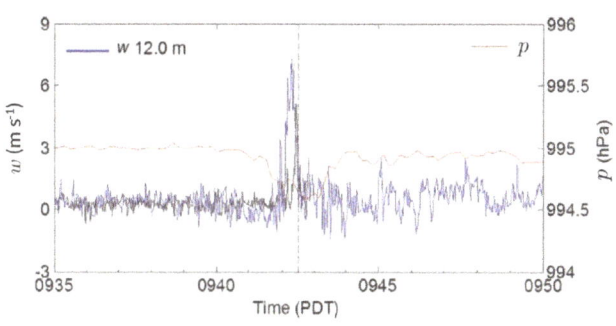

Figure 6.27: Plume attachment (a) visualised from shadowgraph images (produced by the refraction of light when transmitted through a fluid and projected onto a screen) of plume tilt on flat and inclined surfaces from a stationary rectangular flame source (from Grumstrup *et al.* 2017), where arrows show the flow direction of entrained air and dotted lines show the distance of attachment, and (b) as indicated by vertical wind velocity at 12 m above ground (w, blue) and atmospheric pressure changes (p, red) with passage of flame front in a prescribed burn on steep slope (reproduced with permission from Clements and Seto 2015).

some distance. Flow attachment to inclined surfaces has some similarities to the deflection of flame angle and attachment described above for wind (see **Wind and flame attachment**). Thus as flow attachment increases, the fire spread rate increases because flames are forced down and forward, increasing convective heating of fuel particles by direct flame contact (Drysdale and MacMillan 1992). This effect was responsible for the deadly King's Cross Fire in the London Underground train station in 1987, when flames spread rapidly up a wooden escalator trench at a 30° incline (Drysdale *et al.* 1992). Wildfire fatalities have also been linked to the rapid acceleration of upslope fire spread caused by uphill flow attachment (Viegas and Simeoni 2011).

Flow attachment is simple to demonstrate but difficult to predict because it arises through dynamic coupling of the fire with the air flow near the surface. We can understand this by first considering a line fire on flat terrain with equal amounts of cool air entraining into the flame and plume from all sides of the fire (**Figure 6.27a**). The column of hot gases above the fire creates a region of low hydrostatic pressure centred over the fire (**Figure 6.27b**). As the slope angle increases, the entrainment from the uphill side of the fire is reduced compared to that from downhill because the ground itself blocks the flow, so air entrained from the downhill side begins to tilt the flame and plume towards the uphill side. Laboratory experiments demonstrate that flame tilt increases with fire width as well as slope angle, but is largely insensitive to fireline intensity and energy release rate (Smith 1992; Woodburn and Drysdale 1998). Increases in slope and fire width also shift the low-pressure centre uphill, which has been documented in both field conditions by Clements and Seto (2015) and simulations (Linn *et al.* 2007, 2010). Steeper slopes and greater plume tilt eventually shift the low-pressure region in the uphill direction beyond the edge of the flame zone, attaching the flames to the ground by suction. Uphill flame attachment in heading fires is similar to the effect of wind-induced flame drag described above, except here it is caused by an internal coupling rather than an external force. It is not clear what role the heated uphill area has on flame attachment in backing fires (moving downslope), but it could produce different effects than for heading fires and is an issue that remains to be investigated.

Flame and flow attachment are often referred to as the *trench effect* because of the London disaster, although they also occur readily on inclined surfaces without a trench or sidewalls. Laboratory experiments show that flames begin to attach at slopes around 20° (Woodburn and Drysdale 1998; Wu *et al.* 2000; Dupuy *et al.* 2011; Xie *et al.* 2017). Since attachment appears to be independent of intensity, the increased uphill spread rates are probably governed by fluid-mechanical interactions of the fire plume with the surface. Thus the effect of slope on uphill fire spread occurs from a fundamentally different coupling of physical processes than the effects of wind. Both factors tilt flames forward, increasing radiation and convective heating, but they arise from different causes. Slope-induced flame attachment arises from the interaction of the plume with asymmetric entrainment (uphill *v.* downhill). Slope-induced acceleration of fire spread can be limited when lateral flow around the sides of a finite-length flame front narrows the fire's shape (**Figure 6.29**), but research has not identified other negative feedbacks in flow attachment that limit uphill acceleration of fire spread. In contrast, wind is an external force on flame structure and heat transfer that increases spread rate until it is limited by negative feedbacks from increased buoyancy of the flame zone (see the discussion at the end of **Chapter 7**). It is not clear how the effects of wind and slope interact, especially where their forces are not aligned (e.g. with cross-slope winds), but the physical differences mean their components may not be additive as vector sums, assumed in various models (e.g. Rothermel 1972) (Weise and Biging 1997; Nelson 2002; Viegas 2004; Sullivan *et al.* 2014).

The behaviour of trench fires resembles that of wildland fires in steep canyons. Just as the sloping ground surface alters air entrainment to trench fires in one-dimension, sloping canyons alter air entrainment in two dimensions – that is, canyons restrict air from entering the fire from above and from the sides. The restricted air flow can greatly increase up-canyon fire spread rates and growth rates (i.e. perimeter and area). This effect is most obvious for flame zones that stretch across the *water line* of a steeply inclined narrow canyon (i.e. the sloping channel where water flows, **Figure 6.28**). Laboratory and field experiments show that, with fire on both sides of the canyon, air flow that has been induced by the fire is restricted from entraining laterally by the canyon walls and from the up-canyon direction by the slope of the canyon (Viegas and Pita 2004). Thus steeper and narrower canyons channel air to the fire mainly from the down-canyon direction, and air flow from the sides has little effect on narrowing the fire. The interaction of flame-plume tilt with the pressure field described above (**Figure 6.27b**) forces flame attachment at the up-canyon surfaces and increases the fire's spread rate rapidly and – in the case of wildland fires – very dangerously.

Uphill air flow along the flanks of a fire strongly influences convective heating towards the head of

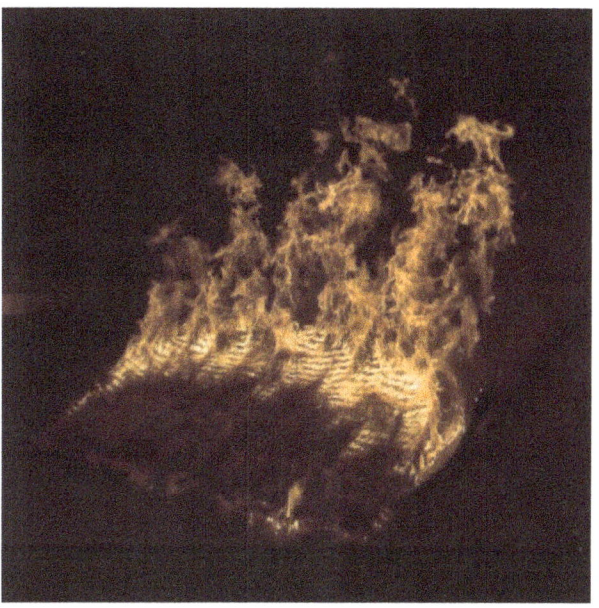

a

b $\theta_s = 20°$ $\theta_s = 30°$

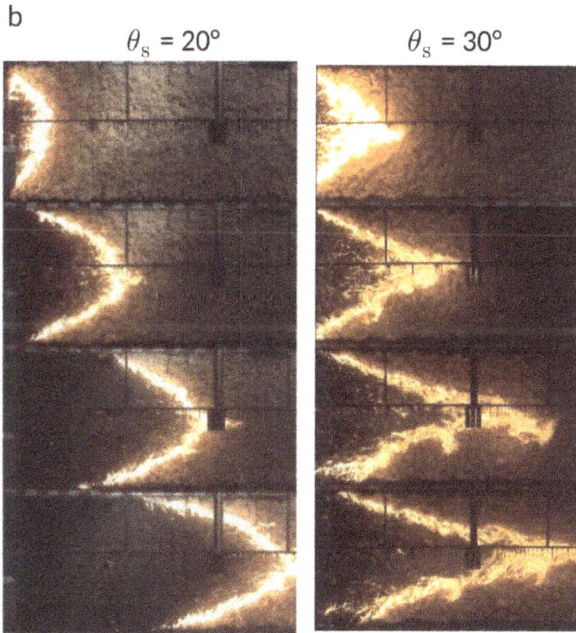

Figure 6.29: Photographs of uphill fire spread in a laboratory experiments shows (a) fluid instabilities and peak-and-trough structures of the flame zone similar to that in wind driven fires and (b) flame shape developing from small line sources at different slope angles. Photographs show oval shapes on moderate slopes (20°) and very pointed shapes on steep slopes (30°) because of convergence of induced air flows and vorticity on flanks (reproduced with permission from Silvani *et al.* 2012). Uphill direction is left to right.

Figure 6.28: Effects of folded topography and canyons (i.e. two-dimensional flow restrictions) on flame spread in a laboratory fuel bed of straw (reproduced with permission from Viegas and Pita 2004).

the fire. These effects were demonstrated in field and laboratory experiments using line ignitions oriented at various angles relative to wind or slope (Viegas 2002). Lines of fire were ignited from the lateral edges of the field plot and a laboratory platform with different slopes and orientations. After ignition, the line fires advanced laterally but spread faster in the flanking direction at uphill portions of the ignition line, resulting in a rotated orientation of the fire edge. This occurred because flame structures are advected uphill parallel to the flame front, progressively increasing convective heat transfer and widening the fire front in the uphill direction. Uphill advection of vortices and coherent flame structures along the fire edge has been observed in other experiments involving fire spread on slopes and in the evolution of curved flame fronts that will be discussed further below.

Linear fire fronts on slopes display the same flow instabilities as wind-driven fire fronts – forward flame bursting, non-steady convective heating, and evidence of concave cellular circulations and stream-wise vorticity, which produce peak-and-trough shaped flame fronts (**Figure 6.29a**). Buoyancy-driven pulsating convection has been documented in upslope fire spread with linear fronts (Atkinson *et al.* 1995), demonstrating that uphill fire spread is dependent upon highly non-steady flame behaviours. Similar to wind-driven fires, fires spreading upslope from point and line ignitions eventually develop bowed fronts on shallow slopes and pointed fronts on steep slopes because of edge effects (**Figure 6.29b**) (Dupuy *et al.* 2011; Silvani *et al.* 2012). The flaming zone on a steep slope shows that air flow and flame trajectory are strongly oriented uphill, with the flattened flame angle oriented nearly parallel to the slope. These conditions enhance non-steady convective heating through intermittent flame contact, thus increasing fire spread rates. Wider line ignitions in both cases must spread farther than narrow ones before they acquire a curved front because edge effects originate farther from the fire centreline.

Steep slopes may also induce fire whirls, particularly along a fire's flanks. Both laboratory fires (**Figure 6.29**) and field-scale fires on slopes often form fire whirls at periodic intervals along their steeply angled flanks (**Figure 6.30a**). These whirls move uphill, spinning flames forward into the fuel bed as they reach the head of the fire (Dupuy *et al.* 2011; Silvani *et al.* 2012). Seen from above as the fire spreads uphill to the right, the whirls on the left flank of a fire rotate anticlockwise, and whirls on the right flank rotate clockwise, suggesting that the spin originates from the shear of rapid uphill air flow through the centre of the fire compared to slower ambient air flow along the edges. Flow shear, as discussed in **Boundary layers** in **Chapter 3**, is produced at the interface of two

Figure 6.30: (a) Lateral vorticity along flank of a wildfire on a steep slope; periodic fire whirls originate downslope and move towards the head of the fire. Photograph from the Mustang Fire, Idaho, 2012. (b) Residual effects of periodic vortex (i.e. fire whirl) formation along the flanks of a fire on a steep slope in Spain. Effects are visible as 'tree streets' within the burned area. The frequency of fire whirl formation and rate of lateral fire spread are unknown for this fire. Photograph from Marc Castellnou.

fluids moving past each other at different rates or directions, and it often causes *vorticity*. Periodic formation and movement of lateral fire whirls sometimes create alternating striped patterns of burned and unburned vegetation parallel to the slope (**Figure 6.30b**). The striping seems to occur when a flank fire, which is advancing laterally at a relatively slow rate across the slope with insufficient intensity to ignite the tree crowns, is interrupted by fire whirls, which appear at somewhat regular intervals and advance rapidly uphill at high intensity, burning a relatively wide swath of canopy. In other words, the flank fires are *underburning* the forest as they spread outward, except when periodic fire whirls cause rapid *crowning* in the uphill direction. This alternating process resembles that which forms tree streets in wind-driven fires (see **Chapter 8, Burn streets**).

Fire configurations

As described previously, wildland fire dynamics heavily depend upon the interactions between fire and the environment. Research into the influences of fuels, weather and topography have typically focused on head fires and their spread and intensity. But heading fire spread does not occur in isolation from other kinds of fire spread. The shape of a fire front is often varied and complex, and its orientation relative to wind and slope have important bearing on the behaviours of the fire and the ways in which fire behaviour responds to moisture. The spatial arrangement of multiple fire fronts and their proximity to each other – the configuration of fire front – is difficult to discuss in generalities. In this section, we summarise some of the known dynamics associated with this topic.

Flame front width and shape

Fires with wide flame zones spread and accelerate faster than fires with narrow ones (see **Chapter 2, Fire acceleration**) and achieve faster spread rates. Fire growth from a line ignition eventually produces a fire front with a bowed, elliptical shape (**Figures 2.10d** and **6.31**). Field experiments in grass fuels covering a range of wind speeds (Cheney and Gould 1995) (**Figure 6.31**) showed that linear ignitions of all widths acquired a convex, elliptically shaped front, but wider linear ignition fronts achieved less curvature within the fixed time of the tests. (We do not know what shapes the fire fronts would have taken if the tests had continued.) Wider fires have been shown in laboratories to spread faster on slopes and in wind (Wolff *et al.* 1991; Dupuy *et al.* 2011). From a linear ignition, evolution of frontal curvature begins along the sides, where incoming wind (combined ambient wind and induced flows) wraps around the edges of the flame zone and deflects the flames towards the middle, standing them more upright. This effect reduces forward convective heating, radiative heating and fire spread rate at the edges of the fire. As the fire progresses farther from the ignition line, the margins of slower spread encroach farther towards the centre of the fire front. Broad ignition lines spread faster than narrow ones because they have a longer linear distance to the edges and thus less edge effect to diminish convective and radiative heating.

As described above, the convective heating from a spreading fire results from the flow dynamics associated with peak-and-trough flame structures (Beer 1991; Finney *et al.* 2015). Wider fire fronts produce greater convective heating because the wind is forced to channel through the flame zone, pushing the flames down into the fuels in the flame troughs (**Figure 6.22**). Although air flow and convective heating cannot easily be observed in the vicinity of these flame zones, simulations can help visualise what happens to winds when flame fronts acquire a bowed shape. Simulations reported by Canfield *et al.* (2014) suggested the wind flow direction entering the back of a fire diverges towards the lateral flanks in the vicinity of the flame zone (**Figure 6.32a**). Greater angles of deflection for local air flows through narrow fires is likely to be responsible for slowing their forward spread rate compared to wider fires and for the slower spread rates along the flanks. The opposite effect is seen with concave-shaped

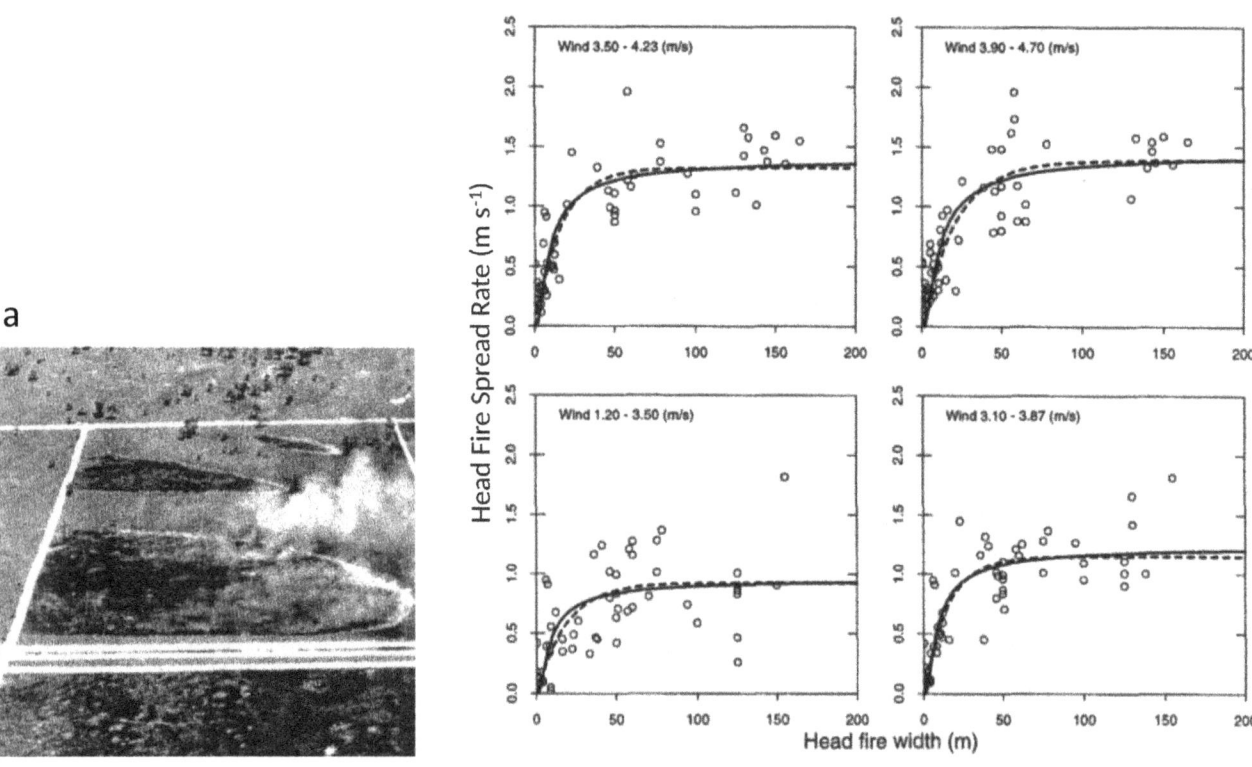

Figure 6.31: Field studies show (a) the effect of flame front width on fire spread rate for fires simultaneously ignited with different line widths and (b) relationships between ignition line width (i.e. 'Head fire width' in the graph) and fire spread rate. Reproduced with permission from Cheney and Gould (1995).

flame zones as seen in **Figure 6.23b**. Concave flame zones rapidly converge as the air flow pulled towards each flank becomes restricted by proximity to adjacent flames (**Figure 6.32b**). This has been demonstrated in small scale laboratory fires on steep slopes by Raposo *et al.* (2018) and in wind driven fires by Hilton *et al.* (2016) and Sullivan *et al.* (2019). The subject of flame zone interaction is also discussed below in the context of multiple flame fronts.

The role of flame zone width in convective heat transfer probably also involves the size of the flames. We saw in **Figure 6.20d** that longer flames have wider spaces between flame peaks and troughs, meaning that short flames would have dozens of peak-and-trough structures in a flame zone with a given width, while longer flames would have fewer of such structures in the same width. Thus a particular flame length would be needed for a line fire to reach a maximum convective heat transfer rate: long flames would require wider flame zones than short ones. A certain number of peak-and-trough structures would be needed to force sufficient wind through the flame zone to cause convective heating. In other words, there may be a minimum ratio of frontal width to flame length, perhaps on the order of 10, before substantial convective heating will occur at the fire front. This concept was discussed by Van Wagner (1971) as vital to designing the scale of laboratory experiments for application to larger fires. It has important bearing on how we interpret past studies of fire spread, particularly laboratory studies with narrow fuel beds and tall flames, and how we design future fire spread experiments.

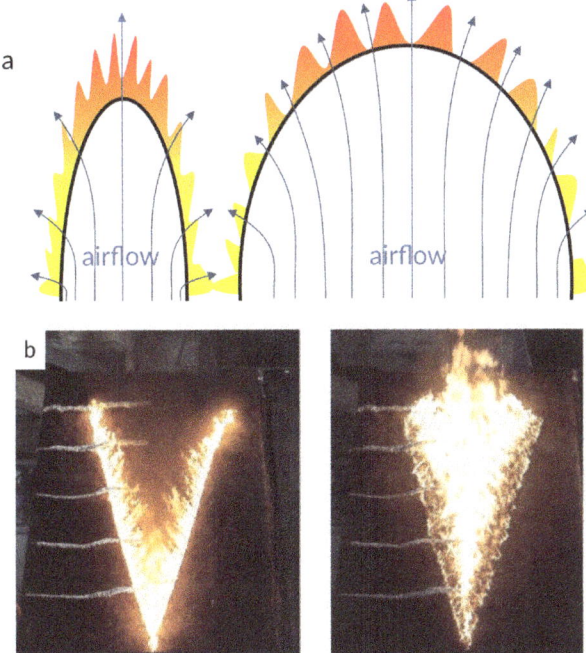

Figure 6.32: Flame zone curvature affects fire spread rates as shown for (a) convex fronts of both narrow and wide fires and the paths of air flows through the peak-and-trough structure of the flame zones that diverge towards the flanks (adapted with permission from Canfield *et al.* 2014) and (b) convex flame fronts, such as this V-shape, which are rapidly occluded by converging spread (reprinted with permission from Raposo *et al.* 2018).

As discussed in **Chapter 3, Radiation heat transfer**, the view factor (ratio of thermal radiation incident on a target to total thermal radiation produced by the emitter) determines the radiant flux on particular objects and depends on the vertical and lateral dimensions of the radiating source (which, in wildland fire, is flame). Thus an increase in the width of the flame front should increase the radiant heat flux to objects some distance away and could therefore increase the fire's spread rate. However, experiments on surface fires in pine litter show that radiant heat flux changes little in fires with small flames, as long as they have fronts wider than 2 m (Wotton *et al.* 1999). Furthermore, the fine-fuel response to convection *v.* radiation (see **Fuel bed properties** above) suggests that radiation may have a minimal role in explaining the effect of frontal width on spread rate. For tall and deep flame zones from crown fires, however, a larger radiation view factor and higher flame zone emissivity is likely to contribute to increased radiant heating of thick woody materials (logs, tree stems) and compact litter, and hence increased fire spread and acceleration.

Backing fires and flanking

Previously in this chapter, we have focused on the forward heat transfer associated with heading fires – that is, those spreading in the same direction as wind or slope (or their combination). Here we address what is known about the configuration of backing fires, which burn downslope and/or upwind. These have been little studied because they generally spread more slowly than heading fires and burn a relatively small area. Because backing fires burn with relatively small flames, they are important for suppression by direct attack (i.e. actions to directly extinguish the fire), for achieving low intensities in prescribed burning (see **Chapter 10, Backing fires**) and for determining the ignition point in two-dimensional fire growth (see **Chapter 8, Fire shapes and growth patterns**).

Backing fires follow a few general patterns: studies conducted and summarised by several authors (Weise and Biging 1997; Fernandes *et al.* 2009; Mendes-Lopes *et al.* 2003; Rossa *et al.* 2015) show that wind speed has little effect on spread rate for line fires backing into the wind. In contrast, increasing moisture content decreases backing spread rate (Fernandes *et al.* 2009). Slope has a more complex effect on backing fires than wind. Spread rates in fires that are backing downslope through beds of loose pine needles or straw decreases with steeper slopes (Van Wagner 1988; Mendes-Lopes *et al.* 2003) until the slope reached about −20 to −25°; as slopes became even steeper, spread rates increased (**Figure 6.33a, b**). Various reasons have been proposed for the increase. A logical explanation is that small, burning fuel elements slough or roll downhill into unburned fuels, particularly as they degrade during burning. Another explanation

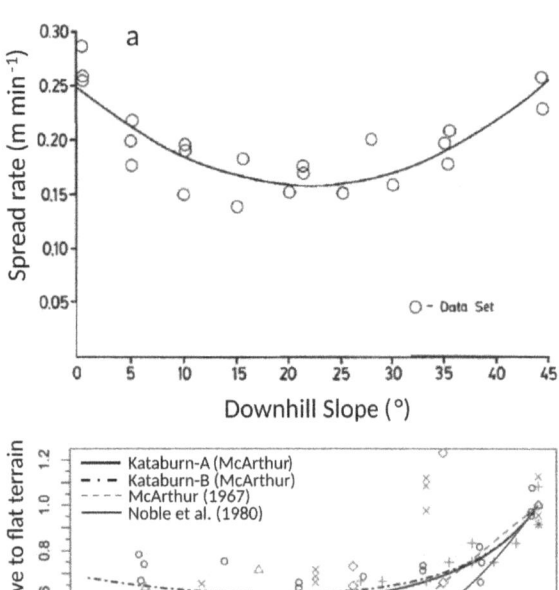

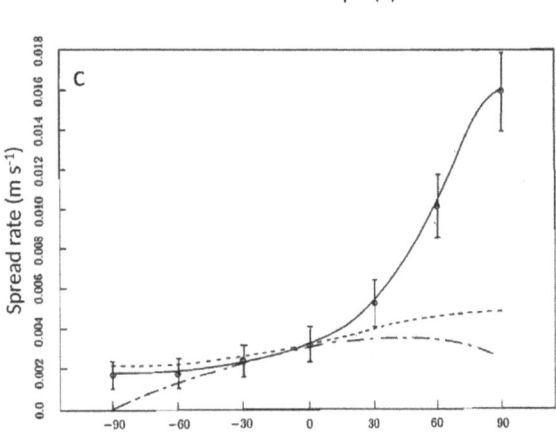

Figure 6.33: Experimental data that describe spread rate of fires (a) backing downslope without wind in fuel beds of pine needles, showing that spread rate declines slightly with increasing slope angle to ~20–30° but increases at steeper slopes, probably from sloughing of fuels as they burn (reprinted with permission from Van Wagner 1988); (b) from four experiments and comparison with predictions of backing spread rate models (lines) (reprinted with permission from Sullivan et al. 2014); and (c) in both uphill and downhill directions on a single pine needle as a function of its angle (reproduced with permission from Weber and de Mestre 1990).

is that there is greater radiation from the fuel bed on steeper slopes, perhaps enhanced by ventilation of glowing combustion by induced air flow (see **Chapter 4, Smouldering and glowing**). Downslope transport of burning particles would help explain why backing fires on the steepest slopes have increased spread rates, while no such increase is observed with fires on flat ground that are backing into the wind. Likewise, experiments in which individual pine needles held at different angles were ignited at the top end showed no increase in spread rate as the 'downhill' tilt angle of the pine needle increased (**Figure 6.33c**) (Weber and de Mestre 1990). This suggests that the physical processes of fire spread must be determined, at least in part, at the scale of the fuel bed rather than solely by orientation of individual particles.

Studies of heat flux in spreading fires have often concluded that convection becomes the dominant heat transfer mechanism in heading fires as stronger winds and steeper slopes force flames forward to impinge on and heat fuel particles. However, radiation is routinely assumed to be the dominant heat transfer mechanism for backing fires and fires spreading without any wind or slope. This assumption is based on the idea that convection caused by buoyancy is pulling cool air from the environment towards the flame zone of a backing (or no-slope, no-wind) fire (see Rothermel 1972), thus cooling the fuels and the flame zone (see **Figure 6.19a**); if convection is a cooling force, then radiation must be the main heat transfer process in these fires. This argument has been presented even for forward spread of crown fires (Butler et al. 2004). However, physical evidence has never confirmed whether the mechanisms of heat transfer vary with spread direction, at different wind speeds or at different slope angles. We see from some focused experiments that the same heat transfer mechanism – convection – governs both heading and backing fires but simply operates at much-reduced length scales in backing fires.

Several fine-scale experiments have addressed the roles of convection v. radiation in fire spread by studying flame spread along continuous solid fuel materials under conditions of *concurrent flow* and *opposed flow*. Concurrent flow describes fires in which the flame is moving along the fuel in the direction of air movement (downwind) or the buoyant force (uphill). Thus concurrent flow describes heading fires. Opposed flow describes fires in which the flame is moving against the direction of air movement (upwind or downhill), so it describes backing fires. Hirano *et al.* (1974) investigated the heating and spread of fire on paper strips under conditions with both concurrent flow and opposed flow. They determined that convection at very small scales accounted for heat transfer to and ignition of materials immediately ahead of the flaming edge, regardless of the direction of air flow.

The process of convective heating at very small scales is often called gas conduction, meaning that the flame gas directly contacts the fuel surface at very low flow speeds. Hirano *et al.* (1974) showed that small flames of fires burning along both sides of paper strips at multiple orientations had low radiant emissivity, insufficient to raise the temperature of the surface to ignition (**Figure 6.34a**). Instead, the flame gas volume expanded at the flame base to make overlapping direct contact with fresh paper on the bottom of the sheet only a few millimetres beyond the pyrolysis edge (**Figure 6.34b**). This is consistent with the finding by Weber and deMestre (1990), who showed that convective heating along a horizontal pine needle came from the underside of the attached flame (**Figure 6.34c**). They demonstrated that flame spread would continue, even when a physical obstacle (e.g. a knife blade) blocked all radiation and flame contact above the particle. This experiment provides an excellent way to visually demonstrate that backing fires and those spreading without wind may still depend primarily on convection for heat transfer and spread at very short length scales (< 1 mm). Short-range heat transfer at the sub-particle scale may help explain sustained spread of backing and flanking fires in compact fuels, such as leaf or needle litter, where the fuel complex involves

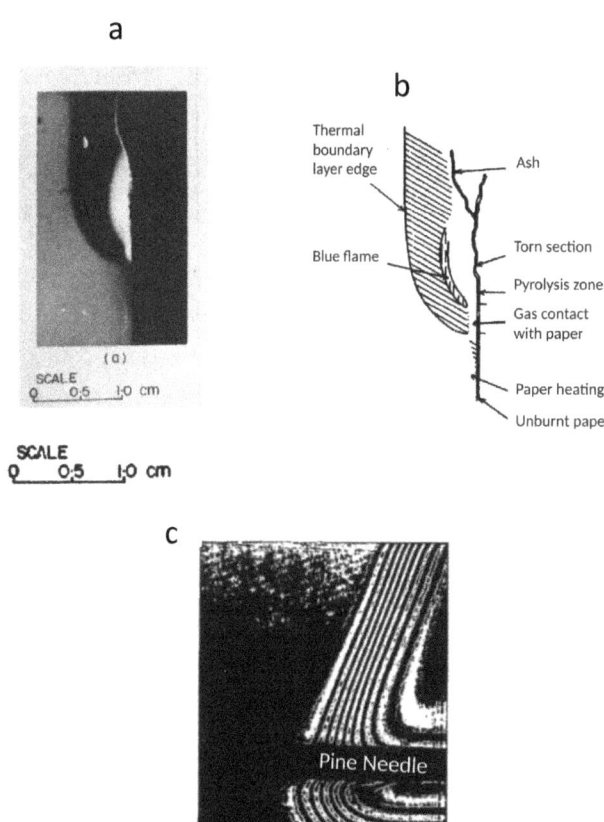

Figure 6.34: Flame spread on thin fuel materials investigated (a) using schlieren imaging (reproduced with permission from Hirano *et al.* (1974)) showed that fuel heating in downward spread was caused by contact with hot gases ahead of the flame as illustrated in (b), and in (c) using interferometry (reproduced with permission from Weber and deMestre (1990)) shows a flame spreading horizontally across a single pine needle. In both studies, flame expansion along the underside of a thin fuel element provides short-range (< 1 mm) convective heating sufficient for fire spread.

both intra-particle flame propagation and sequential ignitions of new particles where they intersect with burning material.

Backing and flanking fires are especially difficult to sustain if fuels are not continuous. The very tiny scale of convection heat transfer responsible for within-particle flame propagation will be insufficient to support fire spread across a fuel bed unless the fuel particles physically intersect. In many discontinuous fuel types, fire will not spread without considerable wind or on steep slopes, and then it spreads mainly in the heading direction; backing spread cannot be sustained. Convective heat transfer in backing and flanking fires may, however, be sufficient for fire spread where wind oscillations and local turbulence widen the trajectory of flame contacts. This may be another example of fire spread that can only be explained by non-steadiness in the environment rather than the mean characteristics of wind or flame.

To understand the roles of radiation and convection in fuel with significant discontinuity, we must consider fire spread in the systems context. Many studies report radiant fluxes incident on a fuel bed of less than 11 kW m^{-2}, but this is a threshold at which large slabs of solid fuels may require infinitely long exposure to heat in order to ignite (**Chapter 5, Figure 5.7**), and infinite time is not available in wildland fuels. If ignition and fuel pyrolysis are caused by radiation, it is most likely coming from the flaming zone, but flaming combustion in wildlands endures for only the length of the residence time. Radiation could be coming from another source, such as glowing particles within the fuel bed, as indicated in several laboratory and field experiments (McCarter and Broido 1965; Thomas 1971). But at such low fluxes, the glowing residence time must meet or exceed the long ignition times to be relevant to ignition. Most backing and flanking fires have flame residence times of 5–20 s and would require exposure to constant radiant heat fluxes of 35–50 kW m^{-2} to ignite new fuel particles, even if they were large solid slabs in quiescent ambient air. But in wildfires, as we have seen, most fuel particles are discrete fine materials such as pine needles and grass, which are sufficiently cooled by natural and forced convection to offset heating by much higher radiant heat fluxes. Furthermore, the peak radiant fluxes of 35–50 kW m^{-2} measured in many surface fires only last for a few seconds as the fire approaches very near the fresh fuel particles. It seems that radiation alone is unlikely to sustain fire spread in discontinuous fuels, but we do not yet know exactly what flow velocity of cool air is able to offset radiant heating at a particular flux for the complex fuel structures present in wildland fuels, thus stopping combustion. Future research will be needed to answer this question.

Multiple flame zones and air-flow interactions

Just as the ambient wind field is modified by the strong buoyancy of a single flame zone, more complex air-flow patterns are created by the competition of air flows among multiple nearby fires which, in turn, affect the behaviour of those fires. Interactions of heat transfer and air flow among separate fires are difficult to model or to quantitatively predict, especially for very large fires (e.g. kilometres across), but it is important to recognise them conceptually because they routinely affect fire behaviours. Interactions are commonly observed and demonstrated at a wide range of scales. For example, the merging of flames between two adjacent candles (**Figure 2.5**) occurs at a scale of centimetres. Interactions at much larger scales are taught to students as important techniques for prescribed burning (**Figure 6.35a**) and for use in suppressing large fires. Backfires, for example, can be ignited at locations and times that take advantage of the indraft ahead of a wildfire; the indraft may be strong enough to counter the ambient wind and draw the spread of secondary ignitions towards the main fire front (**Figure 6.35b**). The air flows among multiple fires have been examined by modelling (**Figure 6.35c, d**), which suggests that restrictions to indrafts would cause spread from nearby flame zones to rapidly converge and increase local intensities at the junction zones. Fine-scale interactions among multiple fires are seen routinely under wildfire conditions when embers lofted ahead of the main fire create numerous spot ignitions that all grow, develop flame fronts, and eventually merge. The processes of spotting in fire behaviour and the aggregated burning characteristics of large-area ignitions are discussed in **Chapter 8, Spotting and spot fires** and **Mass fires**.

Fire interactions occur when flames from separate fires get sufficiently close to tilt towards each other. As flames come closer and closer together,

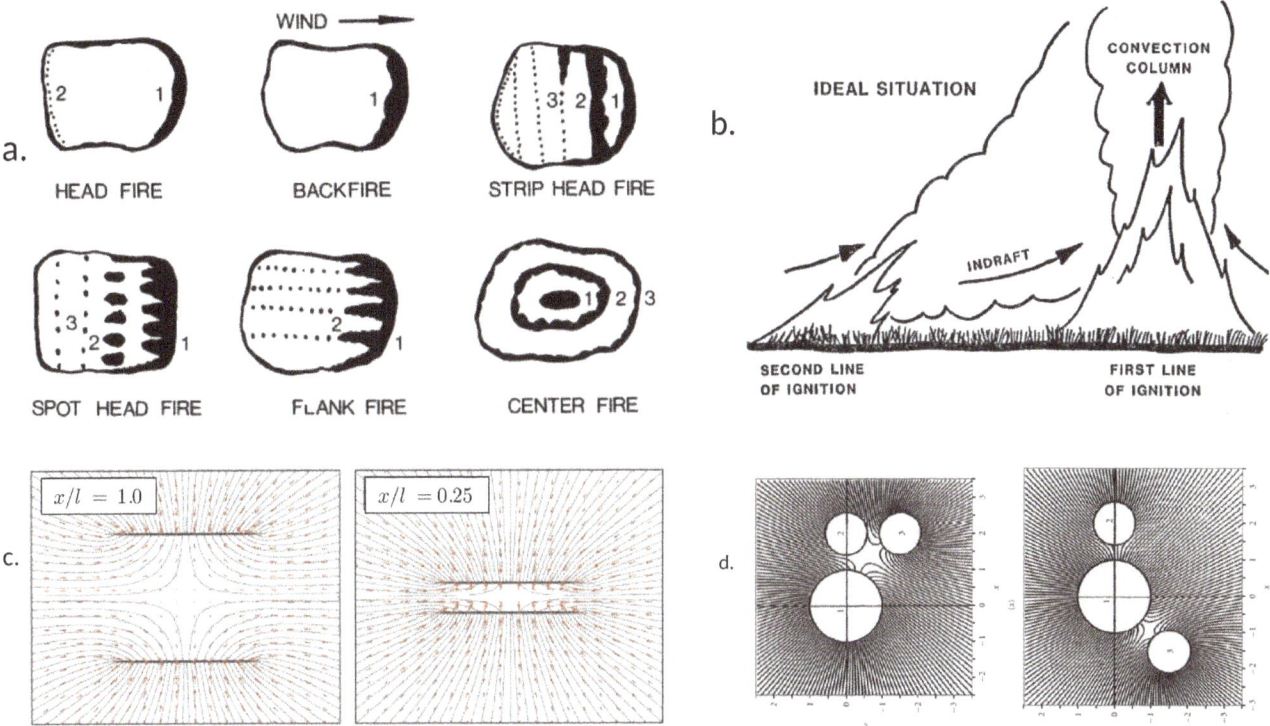

Figure 6.35: Interactions among multiple flame zones are shown for various practical and theoretical configurations, including (a) the standard ignition patterns used in prescribed burning, which rely upon flame zone interaction to control effects and behaviours (from Martin and Dell 1978); (b) interaction of induced air flow to the flame zone of a wildfire used in 'backfire' technique for fire suppression (from Rothermel 1984); (c) calculations of induced air-flow paths (red arrows and streamlines) by nearby linear flame sources (indicated by the dark parallel lines) at different spacings indicated by the ratio of x/l (reprinted with permission from Maynard et al. 2016); and (d) simulated interaction of air flows induced by separate large circular fires (from Weihs and Small 1986).

they eventually begin to touch and then they merge. It is difficult to know what distance is 'sufficiently close' for flames to merge, but laboratory experiments provide many quantitative insights into the changes in fire behaviour that occur as flames merge (Finney and McAllister 2011). For example, **Figure 6.36** shows the change in burning rate that occurs as an array of pools of burning liquid fuel approach each other (Huffman et al. 1969). When the ratio between space width and pool diameter is around 8, fuel is consumed at a low rate. As the pools come closer together, the flames begin to interact and the burning rate increases. When the pools are close enough for the flames to merge (at around $x/d = 2$), the burning rate peaks – an increase of more than 400% (**Figure 6.36**). However, as the pools become even

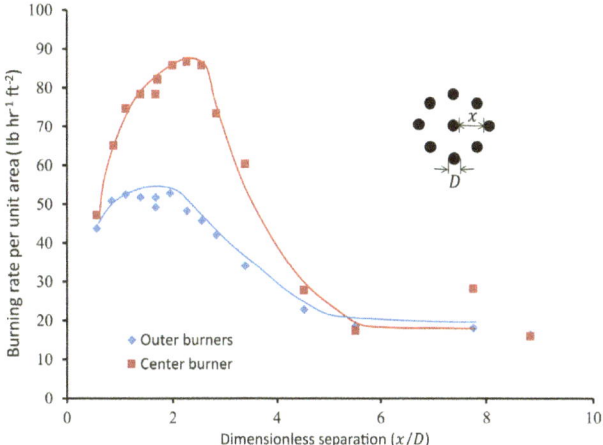

Figure 6.36: Burning rate as a function of separation distance for arrangements of cyclohexane burners each 10.1 cm in diameter, suggesting optimal combinations of oxygen availability and radiant heating (from Huffman et al. 1969).

closer, the burning rate begins to decline, even though it remains higher than when the pools were burning individually. The same trends occur when flames from solid fuels interact and merge.

Because heat release rate and intensity are closely related to the burning rate, they follow the same trends when flames from separate fires interact (see **Chapter 4, Burning and heat release rate**). This trend demonstrates two competing physical processes that occur as flames interact and merge. On the one hand, as the flames tilt towards each other, more radiant and convective heat is transferred to the burning fuel. As heat transfer to the fuel increases, the rate of pyrolysis increases, thus increasing the burning rate. On the other hand, as the flames get closer and closer to each other, there is less and less room for entraining fresh air into regions between the flames. The flames then become fuel-rich, and there is not enough air to fully combust the fuel. This reduces the heat feedback to the fuel, leading to an overall decrease in the burning rate. Before the flames merge, however, the reduced air entrainment causes the temperatures inside the flames to decay more slowly with height than they would in a *single* fire plume, so the decay in flame temperature and vertical velocity described in **Chapter 4** within a single plume do not apply to merging flames. As the number of interacting fires increases, both enhancement of heat transfer and restriction of air entrainment become more pronounced, amplifying their effects on fire behaviours.

Prescribed burning is perhaps the most common environment in which patterns and densities of ignitions occur, since they are often used to control fire spread rates and intensities (see **Chapter 10**). Point or line ignitions are chosen, depending on the desired fireline intensity, with the knowledge that line ignitions accelerate spread and intensity much faster than points (**Figure 6.35a**). When a forward-spreading head fire encounters the backing edge of an adjacent ignition, its intensity is reduced. Thus, the distance between ignitions and the timing of ignitions can be used to prevent the intensity of a prescribed fire from building up to an undesired level. As the flame zones of two or more fires approach each other, the air flowing towards each front becomes restricted by the proximity of the adjacent flames in the area between them, which is called the *junction zone* (see **Figure 6.32b**). As the flames tilt towards each other across the junction zone and then join, fireline intensities increase because the flames represent the conjoined flow of combustion products from multiple flame zones. As described for pools of liquid fuel above, the increased flame zone enhances the heat feedback to the burning material, increasing the pyrolysis rate, and so providing more fuel and increasing the fireline intensity. Because the physical interactions are complex and poorly predicted from models, the desired interactions are mostly guided by experienced personnel, who adjust the ignition patterns as prescribed burning proceeds, based on fuel conditions, ambient wind, slope steepness and other considerations.

When a solid object obstructs air entrainment into a flame zone, it can change fire behaviour dramatically. A wall or board placed at different distances from a burning pool of liquid fuel demonstrates how interruptions of air flow pulled towards the flame tends to increase the flame size and attach the flames to the wall surface. In structure fires, these effects are seen where fires occur at the base of a flat wall or in a corner of a room (**Figure 6.37**). The flame angle and flame length on wide rectangular flame zones (**Figure 6.37a, b**) oriented parallel to a wall are more sensitive to the obstructed inflow than square fires (**Figure 6.37c**) because induced air flowing to the fire can circumvent the square burner more easily. A corner, however, obstructs air reaching a square burner from two sides (**Figure 6.37d**). Flame size increases because the oxygen supply is restricted by the walls, so the hot flame gases must rise farther before enough oxygen is available to mix with the gaseous fuel. In wildland fires, these effects are commonly seen at a small scale, where tree stems restrict air flow into an adjacent flame zone, and at larger scales discussed earlier, when steep slopes or canyons restrict air from entering and cause rapid acceleration of the flame front.

Figure 6.37: Photographs of flames from trays of burning isopropyl alcohol show the effects of reduced entrainment on flame angle and increasing flame height with different burner shapes and proximity to vertical walls for (a) a rectangular fire 0.2 m × 1.2 m; (b) a rectangular fire 0.2 m × 0.6 m; (c) a square fire 0.4 m × 0.4 m; and (d) a square fire 0.4 m × 0.4 m in a corner. The flames from rectangular fires oriented parallel to the wall are more responsive to reduced entrainment (photographs by Ian Grob, USDA Forest Service).

Summary

The variability and diversity of environmental conditions distinguish wildland fires from fires in buildings and industry. Wildland fuels are remarkable for their complexity and resistance to physically succinct descriptions because they display innumerable permutations of fine, coarse, living and dead particles and their spatial arrangements. These particles, discrete and separated by air, provide the context for heat transfer and combustion. Fuel moisture influences all of the physical processes that control fire spread. Wind and topography interact with heat release to alter the entire physical system in ways that are just beginning to be understood. And, of course, the many configurations of the fire itself determine how these environmental factors are coupled with dynamical behaviours of the fire and flame zone. In the next chapter, we explore how these environmental factors interact with physical processes to produce one-dimensional fire spread and associated fire behaviours.

References

Agee JK, Wright CS, Williamson N, Huff MH (2002) Foliar moisture content of Pacific Northwest vegetation and its relation to wildland fire behavior. *Forest Ecology and Management* **167**, 57–66. doi:10.1016/S0378-1127(01)00690-9

Albini FA (1981) A model for the wind-blown flame from a line fire. *Combustion and Flame* **43**, 155–174. doi:10.1016/0010-2180(81)90014-6

Albini FA, Reinhardt ED (1995) Modeling ignition and burning rate of large woody natural fuels. *International Journal of Wildland Fire* **5**(2), 81–91. doi:10.1071/WF9950081

Albini FA, Latham DJ, Baughman RG (1982) 'Estimating upslope convective windspeeds for predicting wildland fire behavior'. General Technical Report INT-257. USDA Forest Service, Intermountain Forest and Range Experiment Station, Ogden, UT.

Alexander ME, Cruz MG (2013) Corrigendum to: Assessing the effect of foliar moisture on the spread rate of crown fires. *International Journal of Wildland Fire* **22**(6), 869–870. doi:10.1071/WF12008_CO

Anderson HE (1969) 'Heat transfer and fire spread'. Research Paper INT-69. USDA Forest Service, Intermountain Research Station, , Ogden, UT.

Anderson HE (1982) 'Aids to determining fuel models for estimating fire behavior'. General Technical Report INT-122, p. 22. USDA Forest Service, Intermountain Forest and Range Experiment Station, Ogden, UT.

Anderson HE (1985) Moisture and fine forest fuel response. In *Proceedings of the 8th Conference on Fire and Forest Meteorology*. 29 April – 2 May 1985, Detroit, MI. pp. 192–199. Society of American Foresters, Bethesda, MT.

Anderson HE (1990a) Moisture diffusivity and response time in fine forest fuels. *Canadian Journal of Forest Research* **20**(3), 315–325. doi:10.1139/x90-046

Anderson HE (1990b) 'Relationship of fuel size and spacing to combustion characteristics of laboratory fuel cribs'. Research paper INT–424. USDA Forest Service, Intermountain Research Station, Ogden, UT.

Anderson WR, Catchpole EA, Butler BW (2010) Convective heat transfer in fire spread through fine fuel beds. *International Journal of Wildland Fire* **19**, 284–298. doi:10.1071/WF09021

Anderson WR, Cruz MG, Fernandes PM, McCaw L, Vega JA, Bradstock RA, Fogarty L, Gould J, McCarthy G, Marsden-Smedley JB, Matthews S (2015) A generic, empirical-based model for predicting rate of fire spread in shrublands. *International Journal of Wildland Fire* **24**(4), 443–460. doi:10.1071/WF14130

Andrews PL (2012) 'Modeling wind adjustment factor and midflame wind speed for Rothermel's surface fire spread model'. General Technical Report RMRS-GTR-266. USDA Forest Service, Rocky Mountain Research Station, Fort Collins, CO.

Atkinson GT, Drysdale DD, Wu Y (1995) Fire driven flow in an inclined trench. *Fire Safety Journal* **25**, 141–158. doi:10.1016/0379-7112(95)00039-9

Babrauskas V (2016) Heat release rates. In *SFPE Handbook of Fire Protection Engineering*. (Eds MJ Hurley et al.) pp. 799–904. Springer, New York, NY.

Barrows JS (1951) 'Fire behavior in northern Rocky Mountain forests'. Station Paper No. 29. USDA Forest Service, Northern Rocky Mountain Forest and Range Experiment Station, Missoula, MT.

Beaufait WR (1965) 'Characteristics of backfires and headfires in a pine needle fuel bed'. Research Note INT-39. USDA Forest Service, Intermountain Forest and Range Experiment Station, Ogden, UT.

Beer T (1991) The interaction of wind and fire. *Boundary-Layer Meteorology* **54**(3), 287–308. doi:10.1007/BF00183958

Brackebusch AP (1975) Gain and loss of moisture in large forest fuels. Research Paper INT-173. US Department of Agriculture, Forest Service, Intermountain Forest and Range Experiment Station, Ogden, UT.

Brown JK (1971) A planar intersect method for sampling fuel volume and surface area. *Forest Science* **17**(1), 96–102.

Brown JK (1978) 'Weight and density of crowns of Rocky Mountain conifers'. Research Paper INT-197. USDA Forest Service, Intermountain Forest and Range Experiment Station, Ogden, UT.

Brown JK (1981) Bulk densities of nonuniform surface fuels and their application to fire modeling. *Forest Science* **27**(4), 667–683.

Brown JK, Marsden MA, Ryan KC, Reinhardt ED (1985) 'Predicting duff and woody fuel consumed by prescribed fire in the Northern Rocky Mountains'. Research Paper INT-337. USDA Forest Service, Intermountain Forest and Range Experiment Station, Ogden, UT.

Burrows ND (2001) Flame residence times and rates of weight loss of eucalypt forest fuel particles. *International Journal of Wildland Fire* **10**(2), 137–143. doi:10.1071/WF01005

Butler BW (2005) The effect of solar insolation on the burning rate of shallow fuel beds. In *Sixth Symposium on Fire and Forest Meteorology*. 25–27 October 2005, Canmore, Alberta.

Butler BW, Finney MA, Andrews PL, Albini FA (2004) A radiation-driven model for crown fire spread. *Canadian Journal of Forest Research* **34**(8), 1588–1599. doi:10.1139/x04-074.

Byram GM (1954) 'Atmospheric conditions related to blowup fires'. Station Paper SE-SP-35. USDA Forest Service Southeastern Forest Experiment Station, Asheville, NC.

Byram GM, Jemison GM (1943) Solar radiation and forest fuel moisture. *Journal of Agricultural Research* **67**(4), 149–176.

Byram GM, Clements HB, Bishop ME, Nelson RM, Jr (1966) *Project Fire Model: An Experimental Study of Model Fires*. Southern Forest Fire Lab, Macon, GA.

Canfield JM, Linn RR, Sauer JA, Finney MA, Forthofer J (2014) A numerical investigation of the interplay between fireline length, geometry, and rate of spread. *Agricultural and Forest Meteorology* **189**, 48–59. doi:10.1016/j.agrformet.2014.01.007

Catchpole WR, Catchpole EA, Butler BW, Rothermel RC, Morris GA, Latham DJ (1998) Rate of spread of free-burning fires in woody fuels in a wind tunnel. *Combustion Science and Technology* **131**(1–6), 1–37. doi:10.1080/00102209808935753

Cheney NP, Gould JS (1995) Fire growth in grassland fuels. *International Journal of Wildland Fire* **5**(4), 237–247. doi:10.1071/WF9950237

Chrosciewicz Z (1986) Foliar moisture content variations in four coniferous tree species of central Alberta. *Canadian Journal of Forest Research* **16**(1), 157–162. doi:10.1139/x86-029

Clements CB (2011) Effects of complex terrain on extreme fire behavior. General Technical Report PNW-GTR-854. In *Synthesis of Knowledge of Extreme Fire Behavior: Volume I for Fire Managers*. (Eds PA Werth, BE Potter, CB Clements, MA Finney, JA Forthofer, SS McAllister, SL Goodrick, ME Alexander, MG Cruz) pp. 5–23. USDA Forest Service, Pacific Northwest Research Station, Portland, OR.

Clements CB, Seto D (2015) Observations of fire–atmosphere interactions and near-surface heat transport on a slope. *Boundary-Layer Meteorology* **154**, 409–426. doi:10.1007/s10546-014-9982-7.

Clements CB, Zhong S, Goodrick S, Li J, Potter BE, Bian X, Heilman WE, Charney JJ, Perna R, Jang M, Lee D, Patel M, Street S, Aumann G (2007) Observing the dynamics of wildland grass fires: FireFlux – a field validation experiment. *Bulletin of the American Meteorological Society* **88**(9), 1369–1382. doi:10.1175/BAMS-88-9-1369

Cohen JD (2015) Fuel particle heat exchange during wildland fire spread. PhD thesis. University of Idaho, Moscow, ID.

Countryman CM (1972) *The Fire Environment Concept*. Pacific Southwest Forest and Range Experiment Station, Berkeley, CA.

Countryman CM (1974) Moisture in living fuels affects fire behavior. *Fire Management* **35**, 10–14.

Cruz MG, Gould JS, Alexander ME, Sullivan AL, McCaw WL, Matthews S (2015) Empirical-based models for predicting head-fire rate of spread in Australian fuel types. *Australian Forestry* **78**(3), 118–158. doi:10.1080/00049158.2015.1055063

de Vries PG (1986) Line intersect sampling. In *Sampling Theory for Forest Inventory*. pp. 242–279. Springer, Berlin, Heidelberg.

Drysdale DD, MacMillan AJR (1992) Flame spread on inclined surfaces. *Fire Safety Journal* **18**, 245–254. doi:10.1016/0379-7112(92)90018-8

Drysdale DD, Macmillan AJR, Shilitto D (1992) The King's Cross fire: experimental verification of the 'trench effect'. *Fire Safety Journal* **18**(1), 75–82. doi:10.1016/0379-7112(92)90048-H

Dupuy JL, Maréchal J, Portier D, Valette JC (2011) The effects of slope and fuel bed width on laboratory fire behaviour. *International Journal of Wildland Fire* **20**(2), 272–288. doi:10.1071/WF09075

Fernandes PM, Cruz MG (2012) Plant flammability experiments offer limited insight into vegetation–fire dynamics interactions. *New Phytologist* **194**(3), 606–609. doi:10.1111/j.1469-8137.2012.04065.x

Fernandes PM, Botelho HS, Rego FC, Loureiro C (2009) Empirical modelling of surface fire behaviour in maritime pine stands. *International Journal of Wildland Fire* **18**, 698–710. doi:10.1071/WF08023

Finney MA, McAllister SS (2011) A review of fire interactions and mass fires. *Journal of Combustion* **2011**, 548328. doi:10.1155/2011/548328

Finney MA, Cohen JD, Forthofer JM, McAllister SS, Gollner MJ, Gorham DJ, Saito K, Akafuah NK, Adam BA, English JD (2015) Role of buoyant flame dynamics in wildfire spread. *Proceedings of the National Academy of Sciences of the United States of America* **112**(32), 9833–9838. www.pnas.org/cgi/doi/10.1073/pnas.1504498112. doi:10.1073/pnas.1504498112

Finney MA, Grumstrup TP, Grenfell IC (2020) Flame characteristics adjacent to a stationary line fire. *Combustion Science and Technology* doi:10.1080/00102202.2020.1863952

Fons WL, Clements HB, George PM (1963) Scale effects on propagation rate of laboratory crib fires. *Symposium (International) on Combustion* **9**, 860–866.

Forest Products Laboratory (1999) 'Wood handbook – wood as an engineering material'. General Technical Report FPL-GTR-113. USDA Forest Service, Forest Products Laboratory, Madison, WI.

Forestry Canada Fire Danger Group (1992) 'Development and structure of the Canadian forest fire behavior prediction system'. Information Report, ST-X-3. Forestry Canada, Ottawa.

Fosberg MA, Lancaster JW, Schroeder MJ (1970) Fuel moisture response – drying relationships under standard and field conditions. *Forest Science* **16**(1), 121–128.

Frandsen WH (1971) Fire spread through porous fuels from the conservation of energy. *Combustion and Flame* **16**(1), 9–16. doi:10.1016/S0010-2180(71)80005-6

Frankman D, Webb BW, Butler BW, Jimenez D, Forthofer JM, Sopko P, Shannon KS, Hiers JK, Ottmar RD (2013) Measurements of convective and radiative heating in wildland fires. *International Journal of Wildland Fire* **22**, 157–167. doi:10.1071/WF11097.

Grumstrup TP, McAllister SS, Finney MA (2017) Qualitative flow visualization of flame attachment on slopes. Presented at the 10th US National Combustion Meeting Organized by the Eastern States Section of the Combustion Institute, 23–26 April 2017, College Park, MD. The Combustion Institute, Pittsburgh, PA:

Hansen FV (1993) 'Surface roughness lengths'. No. ARL-TR-61. Army Research Lab, White Sands Missile Range, NM.

Heilman WE, Bian X, Clark KL, Skowronski NS, Hom JL, Gallagher MR (2017) Atmospheric turbulence observations in the vicinity of surface fires in forested environments. *Journal of Applied Meteorology and Climatology* **56**(12), 3133–3150. doi:10.1175/JAMC-D-17-0146.1

Hiers JK, O'Brien JJ, Mitchell RJ, Grego JM, Loudermilk EL (2009) The wildland fuel cell concept: an approach to characterize fine-scale variation in fuels and fire in frequently burned longleaf pine forests. *International Journal of Wildland Fire* **18**(3), 315–325. doi:10.1071/WF08084

Hilton JE, Miller C, Sharples JJ, Sullivan AL (2016) Curvature effects in the dynamic propagation of wildfires. *International Journal of Wildland Fire* **25**(12), 1238–1251. doi:10.1071/WF16070

Hines F, Tolhurst KG, Wilson AAG, McCarthy GJ (2010) 'Overall fuel hazard assessment guide'. 4th edn. Fire and Adaptive Management Report No. 82. Department of Sustainability and Environment, Melbourne.

Hirano T, Kanno Y (1973) Aerodynamic and thermal structures of the laminar boundary layer over a flat plate with a diffusion flame. *Symposium (International) on Combustion* **14**, 391–398.

Hirano T, Noreikis SE, Waterman TE (1974) Postulations of flame spread mechanism. *Combustion and Flame* **22**, 353–363. doi:10.1016/0010-2180(74)90050-9

Hornby LG (1936) Fire control planning in the Northern Rocky Mountain region. Progress Report No. 1.: USDA Forest Service, Northern Rocky Mountain Forest and Range Experiment Station, Missoula, MT.

Hu L (2017) A review of physics and correlation of pool fire behavior in wind and future challenges. *Fire Safety Journal* **91**, 41–55. doi:10.1016/j.firesaf.2017.05.008

Huffman KG, Welker JR, Sliepcevich CM (1969) Interaction effects of multiple pool fires. *Fire Technology* **5**(3), 225–232. doi:10.1007/BF02591520

Jeschke P, Beer H (2001) Longitudinal vortices in a laminar natural convection boundary layer flow on an inclined flat late and their influence on heat transfer. *Journal of Fluid Mechanics* **432**, 313–339. doi:10.1017/S0022112000003190

Jolly WM, Hintz J, Linn RL, Kropp RC, Conrad ET, Parsons RA, Winterkamp J (2016) Seasonal variations in red pine (*Pinus resinosa*) and jack pine (*Pinus banksiana*) foliar physio-chemistry and their potential influence on stand-scale wildland fire behavior. *Forest Ecology and Management* **373**, 167–178. doi:10.1016/j.foreco.2016.04.005

Keane RE (2015) *Wildland Fuel Fundamentals and Application*. Springer International Publishing, Cham.

Kerr JW, Buck CC, Cline WE, Martin S, Nelson WD (1971) *Nuclear Weapons Effects in a Forest Environment. Thermal and Fire* (No. DASIAC-SR-112). General Electric Company, Santa Barbara CA.

Koschmieder EEL (1993) *Benard Cells and Taylor Vortices*. Cambridge Monographs on Mechanics. Cambridge University Press, Cambridge, UK.

Kremens RL, Smith AM, Dickinson MB (2010) Fire metrology: current and future directions in physics-based measurements. *Fire Ecology* **6**(1), 13–35. doi:10.4996/fireecology.0601013

Lam CS, Weckman EJ (2015a) Wind-blown pool fire, Part I: experimental characterization of the thermal field. *Fire Safety Journal* **75**, 1–13. doi:10.1016/j.firesaf.2015.04.009

Lam CS, Weckman EJ (2015b) Wind-blown pool fire, Part II: comparison of measured flame geometry with semi-empirical correlations. *Fire Safety Journal* **78**, 130–141. doi:10.1016/j.firesaf.2015.08.004

Linn RR, Winterkamp J, Edminster C, Colman JJ, Smith WS (2007) Coupled influences of topography and wind on wildland fire behavior. *International Journal of Wildland Fire* **16**, 183–195. doi:10.1071/WF06078

Linn RR, Winterkamp JL, Weise DR, Edminster C (2010) A numerical study of slope and fuel structure effects on coupled wildfire behavior. *International Journal of Wildland Fire* **19**, 179–201. doi:10.1071/WF07120

Loudermilk EL, O'Brien JJ, Mitchell RJ, Cropper WP, Hiers JK, Grunwald S, Grego J, Fernandez-Diaz JC (2012) Linking complex forest fuel structure and fire behaviour at fine scales. *International Journal of Wildland Fire* **21**(7), 882–893. doi:10.1071/WF10116

Martin RE, Dell JD (1978) General Technical Report PNW-GTR-076. USDA Forest Service, Pacific Northwest Experiment Station, Portland, OR.

Martin RE, Sapsis DB (1987) A method for measuring flame sustainability of live fuels. In *Proceedings of the 9th Conference on Fire and Forest Meteorology*. 21–24 April 1987, San Diego, CA. pp. 71–74. American Meteorological Society, Boston, MA.

Massman WJ, Forthofer JM, Finney MA (2017) An improved canopy wind model for predicting wind adjustment factors and wildland fire behavior. *Canadian Journal of Forest Research* **47**(5), 594–603. doi:10.1139/cjfr-2016-0354

Matthews S (2014) Dead fuel moisture research: 1991–2012. *International Journal of Wildland Fire* **23**(1), 78–92. doi:10.1071/WF13005

Maynard T, Princevac M, Weise DR (2016) A study of the flow field surrounding interacting line fires. *Journal of Combustion* **2016**, 6927482. doi:10.1155/2016/6927482

McAllister S (2013) Critical mass flux for flaming ignition of wet wood. *Fire Safety Journal* **61**, 200–206. doi:10.1016/j.firesaf.2013.09.002

McAllister S, Finney M (2014) Convection ignition of live forest fuels. *Fire Safety Science* **11**, 1312–1325. doi:10.3801/IAFSS.FSS.11-1312

McAllister S, Finney M (2017) Autoignition of wood under combined convective and radiative heating. *Proceedings of the Combustion Institute* **36**(2), 3073–3080. doi:10.1016/j.proci.2016.06.110

McAllister S, Weise DR (2017) Effects of season on ignition of live wildland fuels using the forced ignition and flame spread test apparatus. *Combustion Science and Technology* **189**(2), 231–247. doi:10.1080/00102202.2016.1206086

McCarter RJ, Broido A (1965) Radiative and convective energy from wood crib fires. *Pyrodynamics* **2**(1), 65–85.

Mendes-Lopes JM, Ventura JM, Amaral JM (2003) Flame characteristics, temperature–time curves, and rate of spread in fires propagating in a bed of *Pinus pinaster* needles. *International Journal of Wildland Fire* **12**(1), 67–84. doi:10.1071/WF02063

Miller CH, Tang W, Finney MA, McAllister SS, Forthofer JM, Gollner MJ (2017) An investigation of coherent structures in laminar boundary layer flames. *Combustion and Flame* **181**, 123–135. doi:10.1016/j.combustflame.2017.03.007

Mottus B, Pengelly I (2004) An analysis of the meteorological conditions associated with the 1999 Panther River Fire in Banff National Park [abstract]. In *Proceedings of the 22nd Tall Timbers Fire Ecology Conference: Fire in Temperate, Boreal, and Montane Ecosystems*. 15–18 October 2001, Kanakaskis, Alberta. (Eds RT Engstrom, KEM Galley, WJ de Groot) p. 239. Tall Timbers Research Station, Tallahassee, FL.

Nelson RM, Jr (1984) A method for describing equilibrium moisture content of forest fuels. *Canadian Journal of Forest Research* **14**(4), 597–600. doi:10.1139/x84-108

Nelson RM (2000) Prediction of diurnal change in 10-h fuel stick moisture content. *Canadian Journal of Forest Research* **30**, 1071–1087. doi:10.1139/x00-032

Nelson RM, Jr (2001) Water relations of forest fuels. In *Forest Fires: Behaviour and Ecological Effects*. (Eds E Johnson, K Miyanishi K) pp. 79–149. Academic Press, San Diego, CA.

Nelson RM, Jr (2002) An effective wind speed for models of fire spread. *International Journal of Wildland Fire* **11**(2), 153–161. doi:10.1071/WF02031

Nelson RM, Jr (2003) Reaction times and burning rates for wind tunnel headfires. *International Journal of Wildland Fire* **12**(2), 195–211. doi:10.1071/WF02041

Nelson RM, Adkins CW (1986) Flame characteristics from wind-driven surface fires. *Canadian Journal of Forest Research* **16**, 1293–1300. doi:10.1139/x86-229

Nicholl CIH (1970) Some dynamical effects of heat on a turbulent boundary layer. *Journal of Fluid Mechanics* **40**, 361–384. doi:10.1017/S0022112070000216

Ottmar RD, Sandberg DV, Riccardi CL, Prichard SJ (2007) An overview of the fuel characteristic classification system—quantifying, classifying, and creating fuelbeds for resource planning. *Canadian Journal of Forest Research* **37**(12), 2383–2393. doi:10.1139/X07-077

Pickett BM, Isackson C, Wunder R, Fletcher TH, Butler BW, Weise DR (2010) Experimental measurements during combustion of moist individual foliage samples. *International Journal of Wildland Fire* **19**(2), 153–162. doi:10.1071/WF07121

Pitts WM (1991) Wind effects on fires. *Progress in Energy and Combustion Science* **17**(2), 83–134. doi:10.1016/0360-1285(91)90017-H

Potter BE (2012a) Atmospheric interactions with wildland fire behaviour – I. Basic surface interactions, vertical profiles and synoptic structures. *International Journal of Wildland Fire* **21**, 779–801. doi:10.1071/WF11128

Potter BE (2012b) Atmospheric interactions with wildland fire behaviour – II. Plume and vortex dynamics. *International Journal of Wildland Fire* **21**, 802–817. doi:10.1071/WF11129

Prince DR, Fletcher TH (2014) Differences in burning behavior of live and dead leaves, part 1: measurements. *Combustion Science and Technology* **186**(12), 1844–1857. doi:10.1080/00102202.2014.923412

Putnam AA (1965) A model study of wind-blown free-burning fires. *Symposium (International) on Combustion* **10**(1), 1039–1046. doi:10.1016/S0082-0784(65)80245-4

Raposo JR, Viegas DX, Xie X, Almeida M, Figueiredo AR, Porto L, Sharples J (2018) Analysis of the physical processes associated with junction fires at laboratory and field scales. *International Journal of Wildland Fire* **27**(1), 52–68. doi:10.1071/WF16173

Reinhardt ED, Brown JK, Fischer WC, Graham RT (1991) 'Woody fuel and duff consumption by prescribed fire in northern Idaho mixed conifer logging slash'. Research Paper INT-443. USDA Forest Service, Intermountain Research Station Fire Sciences Laboratory, Missoula, MT.

Rossa CG, Fernandes PM (2018) Live fuel moisture content: The 'pea under the mattress' of fire spread rate modeling? *Fire (Basel, Switzerland)* **1**(3), 43. doi:10.3390/fire1030043

Rossa CG, Davim DA, Viegas DX (2015) Behaviour of slope and wind backing fires. *International Journal of Wildland Fire* **24**(8), 1085–1097. doi:10.1071/WF14215

Rothermel RC (1972) 'A mathematical model for predicting fire spread in wildland fuels'. Research Paper INT-115. USDA Forest Service, Intermountain Research Station, Ogden, UT.

Rothermel RC (1983) How to predict the spread and intensity of forest and range fires. General Technical Report INT-143. USDA Forest Service, Intermountain Forest and Range Experiment Station, Ogden, UT.

Rothermel RC (1984) Fire behavior considerations of aerial ignition. In *Proceedings of a Workshop on Prescribed Fire by Aerial Ignition*. 30 October – 1 November 1984, Missoula, MT. (Ed. RW Mutch) pp. 143–158. Intermountain Fire Council, Missoula, MO.

Schroeder MJ, Buck CC (1970) *Fire Weather: A Guide for Application of Meteorological Information to Forest Fire Control Operations (No. 360)*. USDA Forest Service, Washington DC.

Scott JH, Burgan RE (2005) 'Standard fire behavior fuel models: a comprehensive set for use with Rothermel's surface fire spread model'. General Technical Report RMRS-GTR-153. USDA Forest Service, Rocky Mountain Research Station, Fort Collins, CO.

Scott JH, Reinhardt ED (2005) 'Stereo photo guide for estimating canopy fuel characteristics in conifer stands'. General Technical Report RMRS-GTR-145. USDA Forest Service, Rocky Mountain Research Station, Ogden, UT.

Sharples JJ (2009) An overview of mountain meteorological effects relevant to fire behaviour and bushfire risk. *International Journal of Wildland Fire* **18**(7), 737–754 doi:10.1071/WF08041.

Silvani X, Morandini F, Dupuy JL (2012) Effects of slope on fire spread observed through video images and multiple-point thermal measurements. *Experimental Thermal and Fluid Science* **41**, 99–111. doi:10.1016/j.expthermflusci.2012.03.021

Smith DA (1992) Measurements of flame length and flame angle in an inclined trench. *Fire Safety Journal* **18**(3), 231–244. doi:10.1016/0379-7112(92)90017-7

Smith RK, Morton BR, Leslie LM (1975) The role of dynamic pressure in generating fire wind. *Journal of Fluid Mechanics* **68**(1), 1–19. doi:10.1017/S0022112075000651

Sullivan AL, Sharples JJ, Matthews S, Plucinski MP (2014) A downslope fire spread correction factor based on landscape-scale fire behaviour. *Environmental Modelling & Software* **62**, 153–163. doi:10.1016/j.envsoft.2014.08.024

Sullivan AL, Swedosh W, Hurley RJ, Sharples JJ, Hilton JE (2019) Investigation of the effects of interactions of intersecting oblique fire lines with and without wind in a combustion wind tunnel. *International Journal of Wildland Fire* **28**(9), 704–719. doi:10.1071/WF18217

Tang W, Gorham DJ, Finney MA, Mcallister S, Cohen J, Forthofer J, Gollner MJ (2017) An experimental study on the intermittent extension of flames in wind-driven fires. *Fire Safety Journal* **91**, 742–748 doi:10.1016/j.firesaf.2017.03.030.

Tang F, He Q, Wen J (2019) Effects of crosswind and burner aspect ratio on flame characteristics and flame base drag length of diffusion flames. *Combustion and Flame* **200**, 265–275. doi:10.1016/j.combustflame.2018.11.011

Taylor SW, Wotton BM, Alexander ME, Dalrymple GN (2004) Variation in wind and crown fire behaviour in a northern jack pine black spruce forest. *Canadian Journal of Forest Research* **34**(8), 1561–1576. doi:10.1139/x04-116

Ter-Mikaelian MT, Korzukhin MD (1997) Biomass equations for sixty-five North American tree species. *Forest Ecology and Management* **97**(1), 1–24. doi:10.1016/S0378-1127(97)00019-4

Thomas PH (1971) Rates of spread of some wind-driven fires. *Forestry: An International Journal of Forest Research* **44**(2), 155–175. doi:10.1093/forestry/44.2.155

Van Wagner CE (1971) 'Two solitudes in forest fire research'. Information Report PS-X-29. Canadian Forestry Service, Petawawa Forest Experiment Station, Chalk River, Ontario.

Van Wagner CE (1977) Conditions for the start and spread of crown fire. *Canadian Journal of Forest Research* **7**(1), 23–34. doi:10.1139/x77-004

Van Wagner CE (1987) 'Development and structure of the Canadian Forest Fire Weather Index System'. Forestry Technical Report 35. Canadian Forestry Service, Ottawa, ON.

Van Wagner CE (1988) Effect of slope on fires spreading downhill. *Canadian Journal of Forest Research* **18**(6), 820–822. doi:10.1139/x88-125

Van Wagner CE (1990) Six decades of forest fire science in Canada. *Forestry Chronicle* **66**(2), 133–137. doi:10.5558/tfc66133-2

Varner JM, Kane JM, Kreye JK, Engber E (2015a) The flammability of forest and woodland litter: a synthesis. *Current Forestry Reports* **1**(2), 91–99. doi:10.1007/s40725-015-0012-x

Varner JM, Kane JM, Banwell JM, Kreye JK (2015b) 'Flammability of litter from southeastern trees: a preliminary assessment'. Gen. Tech. Rep. SRS–203. In *Proceedings of the 17th Biennial Southern Silvicultural Research Conference*. 5–7 March 2013, Shreveport, Louisiana, LA. (Eds AG Holley et al.) pp. 183–187. USDA Forest Service, Southern Research Station, Asheville, NC.

Vaz GC, André JCS, Viegas DC (1998) Predicting the residence time of the fire front of surface forest fires. In *Proceedings of III International Conference on Forest Fire Research/14th Conference on Fire and Forest Meteorology*. 16–20 November 1998, Luso, Coimbra (Ed. DX Viegas) pp. 547–564. ADAI, Coimbra.

Viegas DX (2002) Fire line rotation as a mechanism for fire spread on a uniform slope. *International Journal of Wildland Fire* **11**(1), 11–23. doi:10.1071/WF01049

Viegas DX (2004) Slope and wind effects on fire propagation. *International Journal of Wildland Fire* **13**(2), 143–156. doi:10.1071/WF03046

Viegas DX, Pita LP (2004) Fire spread in canyons. *International Journal of Wildland Fire* **13**(3), 253–274. doi:10.1071/WF03050

Viegas DX, Simeoni A (2011) Eruptive behaviour of forest fires. *Fire Technology* **47**(2), 303–320. doi:10.1007/s10694-010-0193-6

Viney NR (1991) A review of fine fuel moisture modelling. *International Journal of Wildland Fire* **1**(4), 215–234. doi:10.1071/WF9910215

Weber RO, de Mestre NJ (1990) Flame spread measurements on single ponderosa pine needles: effect of sample orientation and concurrent external flow. *Combustion Science and Technology* **70**, 17–32. doi:10.1080/00102209008951609

Weihs D, Small RD (1986) 'Interactions and spreading of adjacent large area fires'. Technical Report DNA-TR-86–214. Pacific-Sierra Research Corporation, Los Angeles, CA.

Weise DR, Biging GS (1996) Effects of wind velocity and slope on flame properties. *Canadian Journal of Forest Research* **26**, 1849–1858. doi:10.1139/x26-210

Weise DR, Biging GS (1997) A qualitative comparison of fire spread models incorporating wind and slope effects. *Forest Science* **43**(2), 170–180.

Weise DR, Koo E, Zhou X, Mahalingam S, Morandini F, Balbi JH (2016) Fire spread in chaparral – a comparison of laboratory data and model predictions in burning live fuels. *International Journal of Wildland Fire* **25**(9), 980–994. doi:10.1071/WF15177

Whiteman CD (2000) *Mountain Meteorology Fundamentals and Applications*. Oxford University Press, New York, NY.

Wilson RA (1982) 'A reexamination of fire spread in free-burning porous fuel beds'. Research Paper INT-289. USDA Forest Service, Intermountain Forest and Range Experiment Station, Ogden, UT.

Wolff MF, Carrier GF, Fendell FE (1991) Wind-aided fire spread across arrays of discrete fuel elements. II. Experiment. *Combustion Science and Technology* **77**(4–6), 261–289. doi:10.1080/00102209108951731

Woodburn PJ, Drysdale DD (1998) The dependence of the critical angle on trench and burner geometry. *Fire Safety Journal* **31**, 143–164. doi:10.1016/S0379-7112(98)00004-6

Wotton BM, McAlpine RS, Hobbs MW (1999) The effect of fire front width on surface fire behaviour. *International Journal of Wildland Fire* **9**(4), 247–253. doi:10.1071/WF00021

Wu Y, Xing HJ, Atkinson G (2000) Interaction of fire plume with inclined surface. *Fire Safety Journal* **35**, 391–403. doi:10.1016/S0379-7112(00)00032-1

Xie X, Liu N, Lei J, Shan Y, Zhang L, Chen H, Yuan X, Li H (2017) Upslope fire spread over a pine needle fuel bed in a trench associated with eruptive fire. *Proceedings of the Combustion Institute* **36**(2), 3037–3044. doi:10.1016/j.proci.2016.07.091

7
Wildfire spread

In this chapter, we examine how the physical processes of combustion, heat transfer and ignition are organised to produce a spreading wildland fire. We will formulate a simplified model to study the interactive behaviours in this complex system. Although relatively simple, the model will provide an understanding of the most important interactions in the system and their effects on the behaviours of line fires. We have already covered the key processes involved in fire spread, so we understand the kinds of fuels and environmental conditions that drive wildfires. We have also learned that there are coupled behaviours and feedbacks among some processes, which we will examine more fully in this chapter. Our task here is to first assemble these concepts into a simplified model and then use that model to demonstrate how wildfires spread as a strongly coupled, nonlinear, dynamical system.

System behaviour

What is a 'strongly coupled, nonlinear, dynamical system'? Let us dissect this phrase one word at a time, from back to front:

A *system* is a group of individual processes (in this case ignition, heat transfer and combustion) that work in concert to produce a behaviour (in this case, fire spread).

Dynamical means that the system can change over time. For example, fire behaviour from a small-point ignition can change to produce an area fire with accelerating spread, and the spread rate of a fire changes constantly in the presence of gusty winds. Just as in real fires, results from our system will change over time in response to variations in environmental inputs, even if these inputs are steady, because the system has features that take time to act. These features, such as the time it takes to heat up a fuel particle or the time it takes hot gases to move from one position to another, are called *time lags*. Natural lags in a system can cause changes to not take effect until sometime in the future.

Nonlinear means that the effect one variable has on another is not proportional across the entire range of values. You may remember that the equation for a line is $y = mx$, where m is a constant and x and y are variables. The relationship between y and x is called 'linear' because x has no exponent which really means that 'x is raised to the first power' (e.g. x^1). In a linear system, changes in x cause proportional changes in y. Thus if we double the value of x, the value of y doubles. If the relationship is anything *other* than x to the first power, we call it a *nonlinear* relationship, which means that changes to x do not cause proportional changes in y. Radiation provides a good example of a nonlinear relationship. The radiant heat flux emitted from a fire is related to the fire's temperature *to the fourth power* (**Eqn [3.16]** in **Chapter 3**), so small changes in temperature can cause very large changes in the radiant heat flux. Doubling of the temperature, for instance, would cause a 16-fold increase in radiant heat flux!

Strongly coupled means that some of the components of the system are interdependent – that is, a change in the value of Component A causes a change in the value of Component B, but changes to B also cause changes to A. A classic example of a coupled system is the population dynamics of the predator/prey relationship, in which the number of predators affects the population of prey animals, *and* the number of prey animals affects the number of predators. Coupled systems sometimes generate dramatic and unexpected behaviour in a system even when the behaviour of a single component seems fairly simple, because the behaviour of that component can generate nonlinear changes, oscillations and *positive and negative feedback loops*.

A *positive feedback loop* is a chain-reaction event, such as a cattle stampede: one cow panics, maybe because it saw a snake. That cow's behaviour alarms its nearby neighbours, and they alarm their neighbours in turn and so on, until the whole herd is stampeding because of a single threat to just one cow. The conditions that trigger a positive feedback loop are sometimes called *runaway conditions* – certainly an appropriate term for a cattle stampede. Another example of a positive feedback loop is a financial crash, in which panic sets in among buyers and sellers and then propagates through the system to cause a major change. Systems like these are sometimes called *unstable*. In contrast to positive feedback loops, *negative feedback loops* work to stabilise a system towards equilibrium. The temperature regulation of the human body works this way. If the body temperature is too hot, sweating occurs to cool it. If the body is too cold, capillary constriction occurs to reduce blood flow in the extremities and thus reduce heat loss; this brings the body temperature back towards equilibrium. Most complex systems have *interacting* positive and negative feedbacks, which produce a variety of stable and unstable behaviours, depending on conditions. Many positive feedback loops and runaway conditions are eventually limited by stabilising negative feedback loops. We will see in this chapter that fire spread systems contain some interesting feedbacks and exhibit many nonlinear behaviours. One way to think of fire spread is that it is a system that uses many interacting processes to seek a state of balance – that is, a steady rate of spread.

In this chapter, we will follow a long tradition in fire science by idealising fire spread from a linear flame zone (reviewed by Sullivan 2009). We will construct a physical model of flaming fire spread in the direction perpendicular to the flame front. The model will incorporate the physical processes and scaling relations that we have examined in the preceding chapters into a simple formulation of the spread of a heading fire – that is, a one-dimensional (1D) model of a fire that is spreading straight forward, in line with wind and/or slope. This model is not intended for producing practical predictions, but it will help us understand how the processes we have examined in previous chapters fit together, and it will supply the inductive reasoning needed to generalise from the specifics to understand the entire system. This is what models excel at; they enable scientists to understand how multiple complex processes interact, identify missing pieces of a system, and examine the system's emergent behaviours.

Model framework

Our model framework is a 1D linear domain of arbitrary length that is oriented in the direction of fire spread. This simple geometry is almost identical to that proposed long ago by Fons (1946). We seek to simulate the spread of a heading fire along this line over time by incorporating simplified relationships that describe combustion and burning rate, heat transfer by radiation and convection, and heating and ignition of individual fuel particles (**Figure 7.1**). We will limit the model to describing a flame zone that is spreading through a single-layer fuel bed of discrete particles. The modelled fire may or may not spread, based on the processes we

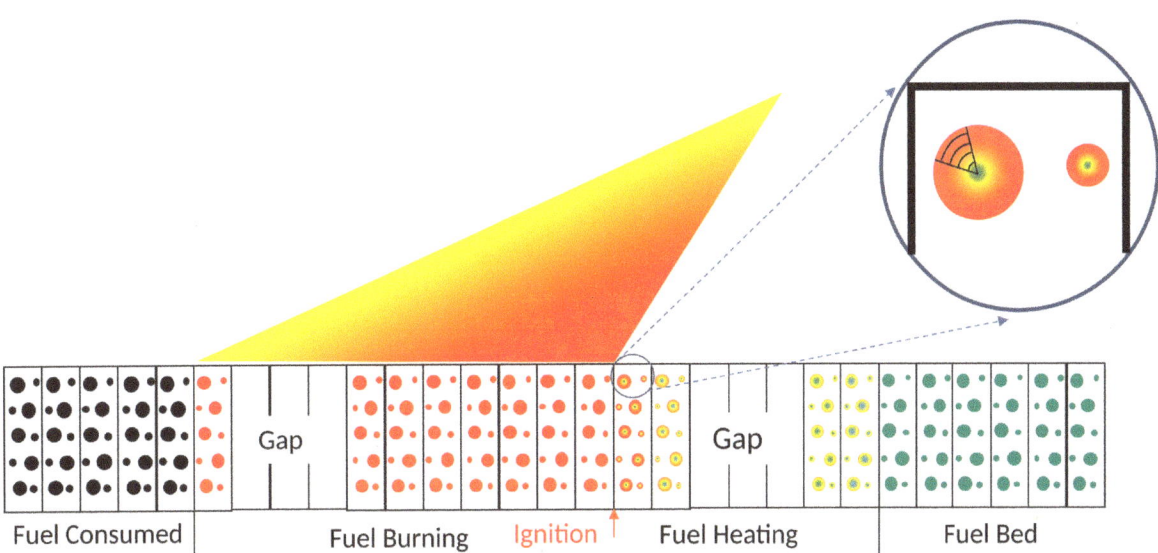

Figure 7.1: Schematic side view of the 1D wildland fire spread model showing a fuel bed composed of cells with discrete particles of specified size and properties or voids (which form fuel gaps). Particle heat and moisture dynamics are calculated ahead of the flame zone as a 1D conduction–diffusion model. The inset shows two particles heating up (warmer colours represent higher temperatures). The left particle also shows how the particle conduction model splits the solid into many pieces (cells) in the radial direction to compute heat flow. Once ignited, particles burn (solid red colour) for a specific time (the flame residence time) and are then designated consumed (black colour).

include and their coupling. The model will contain nonlinear dynamical feedbacks, so the fire may accelerate or decelerate as it spreads. These features make our model diverge in important ways from many past fire models, which have assumed a steady spread rate and thus cannot predict important behaviours like fire extinction or fire acceleration, nor can they respond to changing conditions like fluctuating winds or a sequence of variable fuel properties.

Our 1D model is a straight line broken into 'cells'. Each cell can contain fuel particles, and the model will track the amount and kind of fuel particle in each cell. The size of these cells is somewhat subjective, but it must be small enough to represent the variation in fuel structure along the fuel bed and to resolve heat transfer gradients that occur ahead of the burning fuel. The model must represent fuel variability and explicitly address variation and gaps. Therefore, the cells can have any properties of fuel, including voids, and they will describe individual particles of known number, size and condition. A cell can contain several different types of fuel particles with differences in, for example, size and moisture. To reduce computational effort, the model will explicitly track changes to only one of each kind of particle in a given cell and assume that all similar particles in the cell have the same state. For the tracked fuel particles, the model will compute changes in temperature and moisture over time. The model will step forward in small increments of time (that is, *time steps*, usually around 0.01 s), starting at fire ignition. A critical piece of the model is how we represent the fuel particles.

Fuel particles

The heating of a fuel particle is determined by the heat flux delivered to the particle's surface by the fire and by the particle's thermal response. Much of the energy received at the particle surface is conducted inward, raising the temperature of the solid material. Some of the energy also goes into heating and evaporating moisture from the fuel. As described in **Chapters 3** and **6**, these processes are governed by particle properties such as size, shape, moisture content, density, conductivity and specific heat. As the particle temperature increases, gaseous pyrolysates are released. The pyrolysates eventually become concentrated enough and hot enough to ignite and sustain flaming. For our present model, we assume that particle ignition occurs at a surface temperature of 502 °C (775 K), which is slightly higher than the typical ignition temperature range obtained from laboratory experiments (**Chapter 5**). The rationale for this compromise is to combine into a single value the model functions for (1) indicating the ignition of fuel particles with a nominal temperature criterion irrespective of heating rate, (2) identifying the position of the flame leading edge, and (3) marking the onset of gas phase combustion. The flame leading edge is assumed to be the forward location where flames first become visible (from ~500 to 600 °C), which warrants a higher temperature than particle surface at ignition. Second, the use of a simple constant burning rate omits the incipient phase where little mass is pyrolysed and little energy contributed to the overall flame zone as described below. The final section of the chapter will discuss possible improvements to the model.

We must choose the level of detail to use in representing the heating of the solid fuel to the ignition temperature. There are several possible approaches, each with its own strengths and weaknesses. The simplest approach is to assume that temperature and moisture are constant from the centre of the particle to its outer surface (i.e. the particles are *thermally thin*, as explained in **Chapter 3, Conduction heat transfer** and **Chapter 5, Predicting ignition times**). This assumption is valid when the temperature gradients inside the particle are small. It is the fastest approach for computing the heating of the fuel, and it would be adequate if heating were slow and fuel particles were very small (< 1 mm). However, it would not adequately represent particles – especially large particles – that are experiencing rapid changes in heating and cooling, exactly like what occurs in an approaching fire. Thus it is clear that we must use a more sophisticated approach to represent the heating of solid material. We could use a two- or even a three-dimensional (2D or 3D) model to describe the temperature and moisture gradients inside the particle as conditions change on any of the outer surfaces. Such a model would account for the effects of different fuel shapes and non-uniform surface heating, but it would be computationally demanding and require knowledge of surface conditions that vary around the outside surface of the particle; this knowledge would be very difficult to obtain for complex fuel beds and across large land areas.

A different approach seems most feasible for our model – a 'compromise' approach. It represents the temperature and moisture gradients within a fuel particle as occurring along a single radius inside a cylinder. The cylindrical shape was chosen because it closely approximates the shapes of many real fuel particles, such as needles and branch wood. The compromise approach is computationally efficient, but it has drawbacks. For example, we can simulate the flow of heat and moisture in the radial direction but not in the lengthwise direction, so we will not be able to simulate the difference in heating rates between the tip of a particle, such as a pine needle, and other points along its length (described in **Chapter 3**, **Figure 3.21** and **Chapter 6, Fuel particle geometry, Live fuels**). We also cannot account for different surface conditions around the fuel particle, so we cannot precisely simulate the effects of radiant heat flux that only affects half of the particle circumference; the model will assume that all surfaces are affected.

The model uses a *numerical approach* to compute the conduction of heat from the surface to the centre of the particle. Details of this approach, called the finite volume method (Patankar 1980), are beyond the scope of this book. However, we include a brief description here: the model sets up a line of cells (similar to the way in which we represent the fuel bed) extending from the outside of the fuel particle to its centre. These cells are like boxes stacked with their faces touching along the radius of the particle. Using what is known about the thermal and material properties of the fuels, the model computes the temperature and moisture content of each cell and the flow of heat through that cell's face to the adjacent face of the next cell. The flow through a cell face is a function of the temperature difference between the two adjacent cells because heat is conducted from high to low temperature. The computed heat flow through the face into a cell is then used to recompute its temperature. The process repeats iteratively as the model steps forward in time. Material properties like density and conductivity influence the speed of heat flow through the faces and temperature change of the cell. The whole process is driven by the boundary condition at the outer surface of the fuel particle, through which radiant and convective energy from the fire, the fuel particle and the ambient surroundings are transferred. As the cell heats up, some energy also goes into heating and evaporating moisture. The particle heating/evaporation process progresses until the particle's surface temperature reaches the ignition temperature, at which point the particle ignites and begins to burn.

The model assigns one of three states to each particle at each time step:

1. the *heating* state, in which we compute the heating and drying of the particle up to the point of ignition
2. the *burning* state, in which the particle releases energy that contributes to flaming combustion
3. the *burned* state, in which the particle is considered to have finished releasing gaseous pyrolysates.

The model does not address glowing and smouldering combustion, so it assumes that burned particles are no longer participating in the fire spread process.

Burning rate

The burning rate determines the rate of heat release from the burning fuel, some of which transfers forward to heat the unburned fuel ahead of the fire. As we saw in **Chapter 4, Burning and heat release rate** and **Chapter 6, Burning rates and fuel consumption**, the burning rate of wildland fuels varies with properties of both the fuel bed and the individual particles and it varies over time. But in the absence of reliable general burning rate models, we need to rely on empirical relationships that have been developed from laboratory experiments and we will assume that the burning rate simply remains constant over the entire time that particle is flaming. Thus, we need to determine the flame residence time t_r (s) of the fuel bed and the fuel mass burned in the flaming front. We estimate flame residence time using a relationship developed from laboratory fires spreading in cardboard fuel beds with very small, uniform particles (Finney *et al.* 2015) with and without wind. The relationship fits well the data of Wilson (1982) and is very similar in model form to the findings of Nelson (2003) for other fuel beds of fine woody material (see **Figure 6.12c**). Over the range of residence times observed in the experiments (3–80 s), t_r increased with packing ratio β (dimensionless) and fuel loading m_f'' (kg m^{-2}) and decreased with surface-area-to-volume ratio σ (m^{-1}):

$$t_r = 566\beta + 4894/\sigma + 4.6 m_f'' \qquad [7.1]$$

This relationship will not yield negative values for any combination of fuel properties and is thus robust for use in our fire spread model. However, it ignores many particle properties known to affect ignition and burning rate (e.g. particle density, conductivity and moisture) (see **Chapters 4** and **6**).

The model computes the amount of fuel burned within the flaming front as the total dry fuel mass minus the amount burned in glowing combustion (behind the flaming front) and minus the mineral content of the particle. As discussed in **Chapters 4** and **6**, the amount of fuel consumed in the flaming phase *v.* the glowing phase is a function of many fuel bed and particle properties, as well as moisture and wind. For this model, we make some general assumptions about these properties; at some future time, the assumptions could be replaced by separate models for flaming *v.* glowing combustion. In our model, we assume that 20% of the dry mass is consumed after the flaming phase and 5% of the dry mass is mineral content (which does not participate in combustion). We ignore any effect that glowing combustion behind the flame zone may have on fire spread.

The absence of both theory and models for burning rate greatly limits our ability to model heat transfer. We cannot capture the effects of moisture content, even though (as we saw in **Chapters 4** and **6**), moisture content should increase residence time and decrease the heat release rate and total heat release. We cannot address the burning rates of large woody material, duff or litter. We cannot include the effect of ventilation on either burning rate or the transition from smouldering to flaming. Last, we cannot model the ways in which live fuels burn. Thus the empirical model that we use for residence time (**Eqn [7.1]**) is a first approximation of how fine dead fuel particles burn in a uniform bed in the flaming phase, which should be improved upon in future research.

To calculate the rate of heat transfer in our model, we combine information about the burning of individual fuel particles with information about flame geometry, flame zone depth and flame length. Flame zone depth D (m) is defined as the distance between the farthest-forward flaming particle and the farthest-back flaming particle. Fireline intensity I_B or $\dot{q}'$ (the energy released by flaming combustion per unit time per unit fireline length (kW m^{-1})) is computed by adding up the total energy released by all the burning particles for a time step and dividing by the length of the time step. From fireline intensity, we can compute flame length ℓ_f (m) using an empirical formulation from Byram (1959) and reproduced here as:

$$\ell_f = 0.0775 \dot{q}'^{0.46} \quad [7.2]$$

We also need to know the flame tilt angle to calculate burning rate and heat transfer. Flame tilt angle due to wind θ_w, measured in degrees from vertical, is computed using an empirical model (Putnam 1965) with the ambient wind speed U (m s^{-1}) and acceleration of gravity g (9.807 m s^{-2}):

$$\theta_w = \tan^{-1}(1.4U(g\ell_f)^{-\frac{1}{2}}) \quad [7.3]$$

The flame length and tilt angle provide information on flame geometry that we need to estimate the rate of heat transfer forward to unburned fuel particles.

The steps above describe how we obtain values for residence time, burning rate and flame characteristics to be used in our model, but additional information is needed on heat transfer (radiation and convection) to estimate spread rate. To explain heat transfer, we must examine several equations. We need to understand these physical relationships if we are to understand how the model works, but – perhaps more important – this understanding will also help us comprehend the reasoning behind the model, the primary processes included, and the model's limitations.

Flame radiation heat transfer

Radiant heat transfer from the burning zone to the unburned fuel particles originates in the flame and in the solid glowing fuels within the flaming front (**Figure 7.2**). Our approach to modelling both sources is adapted from that of Koo *et al.* (2005). The fuel particles also exchange radiant energy with the ambient surroundings, which we describe below (under **Ambient environment radiation heat transfer**).

The flame above the fuel is assumed to be a 2D sheet radiating energy to the unburned fuel

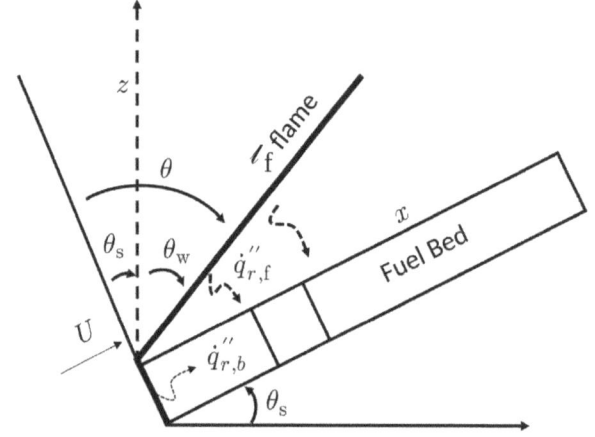

Figure 7.2: Schematic of fuel bed and flame geometry used for radiation heat transfer calculations (adapted from Koo *et al.* 2005).

particles. To simplify calculations, our model assumes that the flame sheet has a constant, uniform temperature and uniform emissivity. The flame sheet emissivity ε_f is computed as a function of the flame length (ℓ_f). The equation uses an empirically derived effective total absorption coefficient of 0.6 m^{-1}:

$$\varepsilon_f = 1 - e^{-0.6\ell_f} \quad [7.4]$$

The uniform radiant emissive power $\dot{E}''_f$, which is the radiant energy released by the flame per unit surface area, is:

$$\dot{E}''_f = \varepsilon_f \sigma_B T_f^4 \quad [7.5]$$

where σ_B is the Stefan-Boltzmann constant (5.67 × 10^{-8} W m^{-2} K^{-4}) and T_f is the radiating temperature of the flame, which we set to 1300 K. This makes T_f^4 a constant and so $\dot{E}''_f$ is only a function of ε_f, and therefore ℓ_f.

Incorporating these relationships and the geometry of the flame and fuel bed, the radiant flux $\dot{q}''_{r,f}$ (W m^{-2}) from the flame that is incident on a fuel particle at distance x in front of the fire is:

$$\dot{q}''_{r,f} = \frac{\zeta_b \dot{E}''_f}{2\delta\sigma}\left(1 - \frac{Z}{(1+Z^2)^{\frac{1}{2}}}\right)\tanh\left(\frac{2}{3}\left(\frac{W}{\ell_f}\right)^{\frac{1}{3}}\right) \quad [7.6]$$

where ζ_b is the bed absorptivity (equal to 1), δ is the bed depth (m), W is the bed width (m, equal to fireline length in a direction perpendicular to the spread direction), and Z is:

$$Z = \frac{\frac{x}{\ell_f} - \sin\theta}{\cos\theta} \quad [7.7]$$

The distance x is measured forward to the particle from the leading edge of the flame zone, which is the position at which a fuel particle just reaches the nominal minimum visible flame temperature of 502 °C (775 K), which the model also uses as the nominal *ignition temperature*. The variable θ is the angle between the flame sheet and the line perpendicular to the upper surface of the fuel bed (see **Figure 7.2**). This is the sum of slope angle θ_s and wind tilt angle θ_w. For now, we must assume that θ_w is independent of θ_s, which – as discussed in **Chapter 6, Topography** – is often not realistic, since flames often tilt upslope due to imbalances in air entrainment. We use **Eqn [7.6]** despite its limitations, since no better model currently exists.

Solid glowing radiation

Radiation from glowing solids within the burning zone transfers heat forward through the fuel bed. Because unburned fuel particles near the leading edge of the burning zone block radiation from those farther away, the incident radiant flux $\dot{q}''_{r,b}$ (kw m^{-2}) on a fuel particle at distance x from the leading edge can be approximated by an exponential function:

$$\dot{q}''_{r,b} = 0.25\dot{E}''_b e^{-0.25Sx} \quad [7.8]$$

where

$$\dot{E}''_b = \varepsilon_b \sigma_B T_b^4 \quad [7.9]$$

and T_b is the bed ember radiating temperature set to 900 K, ε_b is the bed emissivity, which is set to 1.0, and S is the total fuel particle surface area per unit bed volume (m^2 m^{-3}). Notice that **Eqns [7.9]** and **[7.5]** have the same form. While the emissive power of flaming combustion ($\dot{E}''_f$) is dependent on flame length, the emissive power of glowing combustion ($\dot{E}''_b$) is constant because ε_b is fixed, so $\dot{q}''_{r,b}$ depends only on the distance of the particle from the fire (x) and a fuel bed density parameter (S).

Ambient environment radiation heat transfer

The fuel particles in the bed also exchange radiant energy with the cooler surrounding ambient environment. We assume that the view factor between the particle and environment is 1.0, so the net radiant flux exchanged with the environment $\dot{q}''_{r,e}$ (kW m^{-2}) is:

$$\dot{q}''_{r,e} = \varepsilon_e \sigma_B T_e^4 - \varepsilon_p \sigma_B T_s^4 \quad [7.10]$$

where ε_p is the fuel particle emissivity (equal to 1), T_s is the fuel particle surface temperature (K), ε_e is the environment emissivity (equal to 1.0), and T_e is the environment temperature (K). Thus $\dot{q}''_{r,e}$ varies with the temperature difference between the environment and the surface of the fuel particle (temperatures to the fourth power). The greater this difference, the greater the radiant flux exchanged with the environment. Notice that this represents a loss of energy (negative $\dot{q}''_{r,e}$) from the fuel particle, since the environment temperature is normally less than the fuel particle surface temperature.

Convection heat transfer

Chapters 3, 5 and **6** demonstrate that convective heat transfer from flame contact is important to heat and ignite fine particles. However, little experimental or theoretical work is available for characterising the convective environment ahead of a flame zone from a spreading wildland fire – even though this is essential for predicting the rate of heat transfer to unburned fuels. To compute convective heat transfer we need to know the gas temperature and gas velocity in the vicinity of each fuel particle. As presented in **Chapter 3, Convection heat transfer**, the equation for convective heat transfer is:

$$\dot{q}''_c = h(T_g - T_s) \quad [7.11]$$

where $\dot{q}''_c$ is the rate of convective heat transfer (W m^{-2}), T_g (K) is the gas temperature near the fuel particle, T_s is the fuel surface temperature, and h is the convection heat transfer coefficient (W m^{-2} K^{-1}) (discussed further regarding **Eqn [3.28]**, **Chapter 3**). Notice that convective heat transfer can be positive (heat transferred to the fuel) or negative (cooling of the fuel), depending on the sign of $(T_g - T_s)$. The convection heat transfer coefficient can be computed from the Nusselt number Nu, gas conductivity k (W m^{-1} K^{-1}) and fuel particle diameter d (m):

$$h = \frac{Nu\,k}{d} \quad [7.12]$$

The Nusselt number, the dimensionless ratio of convective to conductive heat transfer across the fuel particle surface (see discussion of **Eqn [3.29]**), can be approximated by empirical models for a cylinder. We use two equations for the Nusselt number, one for natural convection (weak to no ambient wind) and one for forced convection (significant wind). Several variables must be calculated to determine which equation to use at a specific time and place: Richardson number, Grashof number, Reynolds number and Prandtl number.

The Richardson number is a dimensionless ratio that compares buoyancy to horizontal ambient wind speed. If the Richardson number is greater than 1, buoyancy is dominating the process of convective heat transfer, so we use the Nusselt number equation for natural convection (**Eqn [7.15]** below). If the Richardson number is equal to or less than 1, ambient wind is dominating the process, so we use the Nusselt number equation for forced convection (**Eqn [7.17]** below). The equation for the Richardson number is:

$$Ri = \frac{Gr_d}{Re_d^2} \quad [7.13]$$

To compute the Richardson number, we need the Grashof number Gr_d, which is the dimensionless ratio of buoyancy to viscous force in a fluid (see discussion of **Eqn [3.34]**):

$$Gr_d = \frac{g\tau(T_s - T_g)d^3}{\nu^2} \quad [7.14]$$

In this equation, g is acceleration due to gravity (m s^{-2}), τ is the coefficient of thermal expansion for an ideal gas ($\tau = 1/T_g$), and ν is the kinematic viscosity of the gas (m^2 s^{-1}).

To solve **Eqn [7.13]**, we also need the Reynolds number for a cylinder in crossflow. The Reynolds number is $Re_d = Ud/\nu$ where U is ambient wind speed (see discussion of **Eqn [3.32]**).

If the Richardson number is greater than 1, so buoyancy dominates the process of convective heat transfer, we compute the Nusselt number using the empirical relation from Churchill and Chu (1975):

$$Nu = \left(0.6 + \frac{0.387 Ra_d^{\frac{1}{6}}}{\left(1 + \left(\frac{0.559}{Pr}\right)^{\frac{9}{16}}\right)^{\frac{8}{27}}}\right)^2 \quad [7.15]$$

This equation requires two more values. The first is the Rayleigh number $\mathrm{Ra_d}$, which describes the relationship between buoyancy, viscosity and diffusivities:

$$\mathrm{Ra_d} = \frac{g\tau(T_s - T_g)d^3}{\nu\alpha} \quad [7.16]$$

where α is thermal diffusivity (m² s⁻¹).

The second value needed for **Eqn [7.15]** is the Prandtl number Pr, which describes the relationship between the diffusion of momentum and the diffusion of heat (see the discussion of **Eqn [3.30]**). The value used in our model is that of air (found in **Appendix B**).

If the Richardson number is less than or equal to 1, so wind dominates the process of convective heat transfer, we use the empirical relation from Churchill and Bernstein (1977) for forced convection:

$$\mathrm{Nu} = 0.3 + \frac{0.62\,\mathrm{Re}_d^{\frac{1}{2}}\mathrm{Pr}^{\frac{1}{3}}}{\left(1 + \left(\frac{0.4}{\mathrm{Pr}}\right)^{\frac{2}{3}}\right)^{\frac{1}{4}}}\left(1 + \left(\frac{\mathrm{Re}_d}{282000}\right)^{\frac{5}{8}}\right)^{\frac{4}{5}} \quad [7.17]$$

All gas properties for convection calculations are evaluated at the film temperature $T_{\mathrm{film}} = (T_s + T_g)/2$ which approximates the mean temperature of the fluid within the boundary layer of the particle.

Now that we have the equations needed to compute convective heat transfer to a fuel particle, we need to know the gas temperature and velocity distribution ahead of the spreading fire. The methods generally used to approximate these are numerical simulations using fluid dynamics models that explicitly represent gas flow using a 2D or 3D gridded domain and information about the flame and fuel structure. This approach is too computationally expensive (for operational use) for us to apply at the resolution needed to estimate convective heat transfer to particles (often mm to cm in diameter) very near the fire where flames can be present (sometimes called the *near field*).

Therefore, we use a method that is relatively simple and is calculated rapidly. In the past, such approaches have been pursued but have required speculation as to how gas temperature decays with distance from the flame front under a variety of wind, slope, energy release and fire geometries. Sometimes this has been assumed to be a Gaussian function or an exponential function (e.g. Hottel *et al.* 1965; Pagni and Peterson 1973), but recent experiments that were specifically designed to characterise this temperature profile found it to be a power-law function of distance.

Our approach relies on an empirical power-law model of mean gas temperature as it declines with distance along the upper surface of a fuel bed extending away from the flame front (Finney *et al.* 2020). We developed the model from temperature measurements taken at regular distances from a rectangular propane burner (**Figure 6.26**). The burner controlled the flame zone aspect ratio (fireline length and flame zone depth) and fireline intensities subject to different slope and wind conditions. Thermocouple measurements near the ground surface revealed that gas temperature was relatively constant for the region of flame attachment (**Figure 6.26**) but then decreased as a power-law function of distance from the flame, like the well-documented distribution above stationary flame sources (**Chapter 4, Figure 4.22**). Therefore our gas temperature model consists of two parts: a constant temperature region near the flame zone, and a decaying temperature region further away from the flames. The model identifies the *flame detachment distance* (x_d) as the point separating these two regions, which is where the average temperature of convective gases drops below 800 °C. It predicts the decaying region as a power-law function of dimensionless variables for distance, wind, slope, aspect ratio and flame zone depth (**Table 7.1**). We used dimensionless terms so the predictions might be scalable to fires larger than the small dimensions of the experimental apparatus.

Normalised distance is computed from the relationship between the distance to the flame leading edge $x + x_d$ and the characteristic plume length

Table 7.1. Terms used in the model to compute the mean gas temperature profile at a normal angle to a flame zone in the heading direction.

Predictor	Dimensionless term
Normalised distance	$\ell^* = ln((x + x_d)/\ell_p)$
Momentum balance between wind U and flame v_f	$U^* = \sin(atan(U\rho_a/w_f\rho_f))$
Sine function of slope	$s^* = \sin^2(2\theta_s)$
Aspect ratio of flame zone	$B^* = D/W$
Normalised flame zone depth	$D^* = D/\ell_p$

scale ℓ_p, which is very similar to the scaling of flame length from fireline intensity (see **Chapter 4, Eqn 4.7**), calculated following Quintiere and Grove (1998) for line plumes:

$$\ell_p = \left[\dot{q}'\left(\rho_a c_p T_a g^{1/2}\right)^{-1}\right]^{2/3} \qquad [7.18]$$

where $\dot{q}'$ is the fireline intensity (kW m^{-1}), ρ_a is the ambient air density (kg m^{-3}), c_p is specific heat of the air (J kg^{-1} K^{-1}), T_a is air temperature (K), and g is gravitational acceleration (9.807 m s^{-2}). Normalised distance grows with distance from the flame zone and increases with air density, specific heat and temperature. It decreases as fireline intensity increases (as flames get bigger).

The effect of wind on gas temperature U^* is represented by the balance between the horizontal flow of ambient air at high density ($U\rho_a$) and the vertical flow of flame gas at low density ($w_f\rho_f$). Stronger winds tilt the flames further forward (**Eqn [7.3]**), increasing the temperature of gases at a given distance. To calculate U^*, we must calculate the vertical gas velocity w_f at the flame tip (Nelson 2002):

$$w_f = \left(\frac{2g\dot{q}'}{1000\rho_a c_p T_a}\right)^{1/3} \qquad [7.19]$$

The vertical gas velocity increases with increasing fireline intensity and decreases with increasing ambient air density, specific heat and temperature.

The effect of slope on gas temperature is estimated from experimental data (**Chapter 6, Figure 6.26**). Increasing slope steepness (θ_s) is associated with increasing flame spread rate and flame attachment, an effect that rapidly changes when slope reaches ~20° (indicating a nonlinear relationship). The function s^* in **Table 7.1** does not apply to slopes greater than 45°, which are not modelled here.

Our model uses two other terms from **Table 7.1**: the flame zone aspect ratio (B^*) and normalised flame zone depth (D^*). The aspect ratio B^* is calculated as the flame zone depth D divided by the flame front width W, and D^* is D divided by ℓ_p. The regression model produced for the decaying region of gas temperature (T_g, °C) (**Figure 7.3**) is:

$$\ln(T_g) = 4.77 - 0.68\ell^* + 2.60s^* + 2.02U^* - 0.72B^* - \\ 0.098D^* - 1.91s^*U^* - 0.401\ell^*s^* - \\ 0.353\ell^*U^* + 0.057\ell^*B^* \qquad [7.20]$$

The environmental and fire-dependent terms in this equation show that the location where the flame detaches from the fuel surface (at 800 °C) moves farther from the flame base for greater slope (s^*) and wind speed (U^*), and closer to the flame base with greater aspect ratio (B^*) and flame zone depth (D^*). The steepness of the temperature profile – that is, the rate at which temperatures increase with distance – also varies with these terms, as shown by their significant interactions with distance (e.g. in the ℓ^*s, ℓ^*U^*, and ℓ^*B^* terms).

To compute the rate of convective heating ahead of a line fire, we also need to know the velocity of the gas impinging on the fuel particle. No models are available for predicting the distribution of velocity adjacent to line fires, and so our model

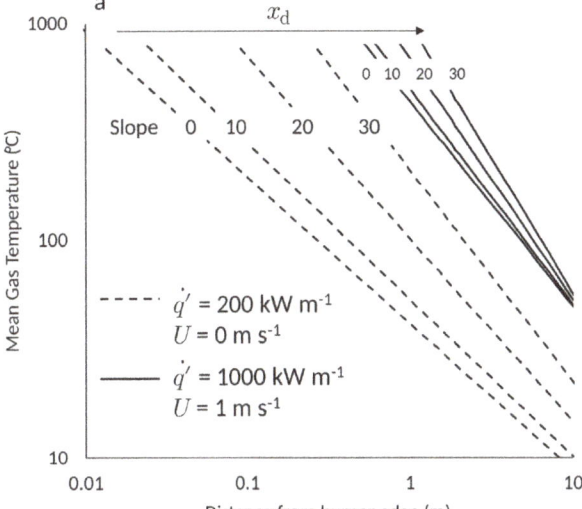

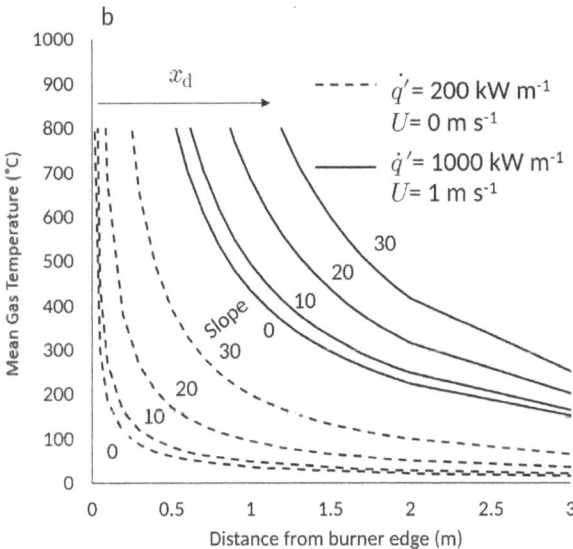

Figure 7.3: Model of gas temperature profile with distance from burner edge for conditions with different fireline intensities, winds and slopes, shown on (a) log-log axes and (b) linear axes. Continuous flaming was assumed above an average gas temperature of 800 °C for the distance ξ_d. The model was developed from laboratory measurements adjacent to a gas burner on a surface where wind and slope were controlled (Finney *et al.* 2020). A stationary flame source was used to control the energy release rate and flame zone geometry. (**Chapter 6, Figure 6.26** describes some of these experiments.)

simply assumes that the velocity within one flame height of the flame's leading edge is the same as the vertical velocity w_f above a line fire (**Eqn [7.19]**). In the region beyond one flame height, the model assumes that the velocity of the gas impinging on the fuel particle is the same as the ambient wind speed. Luckily, a sensitivity analysis of this crude approximation for convective velocity revealed that the fire's spread rate is not as sensitive to velocity as the temperature distribution (not shown) and so this approach will be adequate.

To complete the modelling of convective heating of fuel particles, we make the following simplifications:

- The flame leading edge has a mean gas temperature of 800 °C (1073 K).
- The power-law temperature decay from 800 °C (**Eqn [7.20]**) begins at the flame leading edge (where flames are no longer permanently attached to the adjacent surface), so gas temperature decreases rapidly with distance from the leading edge. This is consistent with experimental data that separate the intermittent and continuous flame regions in both stationary and spreading fires (refer to the laminar/turbulent flame discussion in **Chapter 3, Boundary layers** and **Figures 4.20, 4.22 and 4.23**).
- At a given distance from the flame leading edge, the convective heat flux to a fuel particle surface is calculated using the average gas temperature and the convection heat transfer coefficient (h). This simplification ignores the intermittent heating that occurs in real fires (described in **Chapter 6, Wind in the presence of fire**), but it should be sufficient for our model to explore the coupling in the fire spread system.

Model function

The fire simulation begins with an ignition of one or more cells at the edge of the fuel bed and then steps forward in time. In the initial time steps (set to 0.01 s for all model runs here), the fire is not spreading because it takes time to heat new fuel particles ahead of the fire front from ambient temperature to ignition temperature. Over time, the unburned fuel particles just ahead of the fire dry out and heat up (calculated with the 1D conduction model described under **Fuel particles** above). They ignite when the particle surface temperature reaches 502 °C (775 K). When a particle in a particular cell reaches this temperature, the cell is marked as the position of the leading edge of the flame, since this is roughly the temperature at which flames become visible. Our model simplifies the ignition and spread processes by assuming that ignition and flame front location are coincident. It also assumes that there is no delay in subsequent fire spread once particles are ignited. Although this is not universally true, we have chosen to use 502 °C (775 K) for both ignition and spread. Cells may contain more than one particle type, but particles of varying sizes and properties will heat up differently and each type will only ignite when it reaches the critical surface temperature. This means a cell could contain both ignited and unignited particles.

Meanwhile, at the back of the fire, fuel particles are burning out as their flame residence times expire, based on **Eqn [7.1]** and the time ignited. At each new time step, the flame geometry is calculated from the spatial distribution of currently burning particles. From this flame geometry and its location, heat transfer to the unburned fuel ahead of the fire is computed, and this continues to dry and heat the particles and thus propagate the fire. This cycle continues until the fire reaches the end of the fuel bed or burns out due to insufficient forward heat transfer. The *fire spread rate* is represented here in units of m min^{-1} (rather than SI m s^{-1}) because this is more conventional and easier to envision in both laboratory and field settings. Fire spread rate is *not* a fundamental variable in our model, as it is in many previous models, but rather a natural result of the ignition time sequence of the fuel particles. Therefore, spread rate is recalculated when a new forward particle achieves ignition temperature; it is simply the distance that the leading edge of the fire has moved forward divided by the time elapsed since the last forward movement.

The sections above summarise the logic that underlies our relatively simple model of the forward spread of a line fire. Now we will use the model over a range of fuel, weather and topographic conditions to help us understand the operating characteristics of the 'strongly-coupled, nonlinear, dynamical system' that describes the spread of wildland fires.

Modelled fire spread and behaviour

Before we examine the model results, let us review what can be expected from this or any model. It is tempting to develop a model and immediately evaluate it by comparing its predictions with observations from the real world. However, while a model should indeed represent key characteristics of the real phenomena over important ranges of conditions, a more important use for it may be to help us *understand* the phenomena that we are trying to predict. Given the difficulty of validation discussed in **Chapter 1**, **The challenge of validation**, agreement is not necessarily an indication of a 'good' model, and lack of agreement is not indicative of a 'bad' model. If we are using a model to understand how a system works, then we will evaluate it by examining whether predicted fire characteristics are produced *for the right, logical reasons* or just fortuitously. A badly fitting model with the right formulation may be more useful than a spuriously well-fitting model. The model we have described in this chapter is intended to help us examine and understand dynamical behaviours of wildland fire spread rather than to make predictions, so we may be content with 'looser' fits between observations and predictions than if we were trying to use the model for on-the-ground predictions. The most important consideration in assessing our model is whether the model is functioning in a physically consistent way – that is, with the combustion, heat transfer, and ignition processes operating as we understand them to work in real spreading fires.

We have compared the results from our model with data from well-controlled laboratory burns in three kinds of fuels – laser-cut cardboard (Finney *et al.* 2013, 2015) and pine needles and excelsior (Catchpole *et al.* 1998). Each dataset covered a variety of fuel characteristics (particle sizes, loading, packing, moisture), wind speed and slope conditions. We observed maximum spread rates of up to several m min^{-1} and flame heights of up to ~2 m tall. We included field data from nine fires in crop stubble in New Zealand (Pearce *et al.* 2019), which exhibited much faster spread (up to 100 m min^{-1}) and longer flames (up to ~5 m) than the laboratory data.

Our experimental results showed substantial variation ('scatter'). This can be caused by data error, which occurs even with the most reliable measurements of fire spread because of limited precision of instruments, variability in the environment (e.g. fuels, wind), and dynamic effects of the fire itself (e.g. acceleration). It can also be caused by model error, which arises if the processes specified in the model are biased or improperly represented. It is difficult to distinguish model error from data error, so we must remain cautious when assessing models based only on the apparent fit to observations; we must also assess them based on the physical processes described, to make sure the model describes them appropriately.

Our comparison of observations with predictions showed that modelled predictions of spread rate and flame length agreed with laboratory and field experiments in both trend and magnitude (**Figure 7.4**). Scatter was apparent in all comparisons, but the consistent agreement of observed phenomena with predictions across a wide range of fuel and environmental conditions is reassuring; it indicates that the approximations and simplifications chosen for the model are appropriate for our main purpose – that is, to understand the spread of a line fire at multiple scales. The agreement between observations and predictions suggests

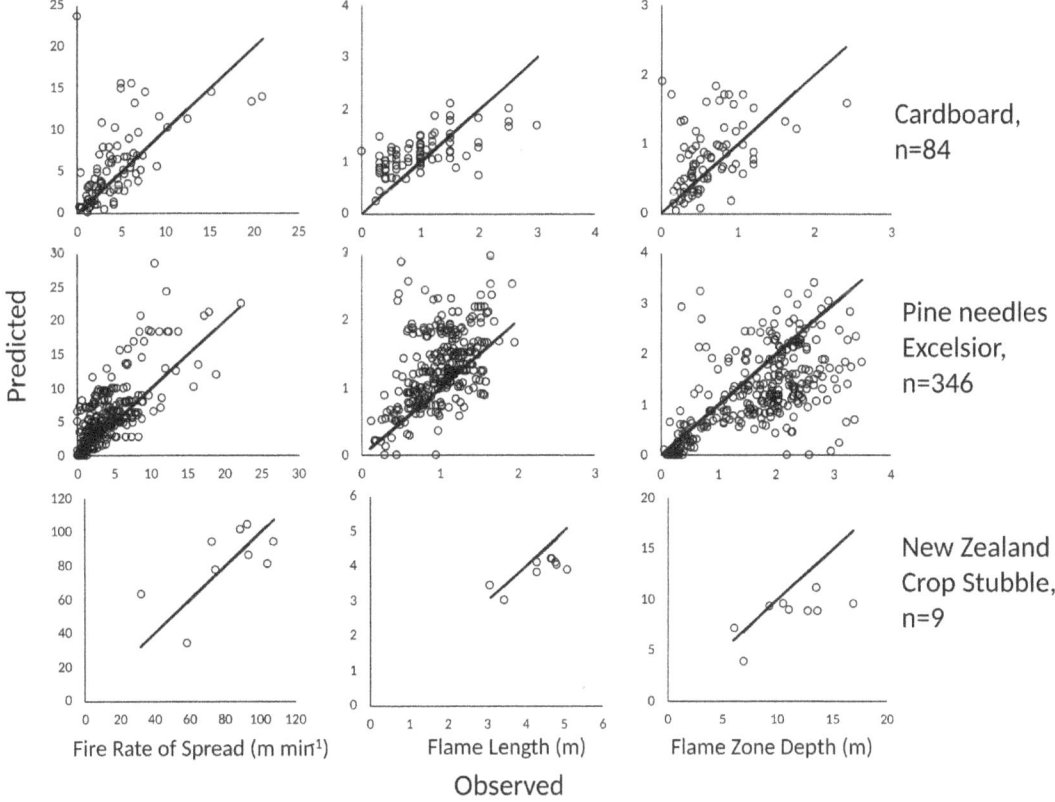

Figure 7.4: Comparison of modelled flame zone characteristics with three datasets from laboratory and field experiments on spreading fires. The number of observations is given by *n*. The agreement of the model for both fine-scale and large-scale fires indicates that a simplified approach to fire spread using a nonlinear 1D model of flame spread can produce insights into the dynamical coupling of physical processes in fire behaviour.

that our simplified 1D model captures the essential physical processes that drive the flaming spread of line fires. If we have represented these processes correctly, we may not need more complicated, computationally demanding methods to develop greater understanding. Instead, we may need further research and testing to verify our findings and determine what aspects of the model need to be added, removed or refined. The remaining part of this chapter delves further into the specific roles of the modelled physical processes by examining specific examples of model results.

Simple fire spread dynamics

The first example examines model results from simulated burning of a simple fuel that we refer to as our *standard fuel bed* for testing different behaviours (**Table 7.2**). For the simple case, wind speed is set to 1 m s^{-1} on flat terrain. **Figure 7.5** shows a series of simulated fire characteristics from the time of ignition. The graphs show the forward progress of the flame front, the instantaneous spread rate, the flame length and heating distance relative to time (top) and distance (bottom).

The trajectory that spread rate follows over time and distance is similar to the trajectories for flame zone depth and flame length (**Figure 7.5**). This reveals the strong coupling that exists within the dynamical fire spread system. All flame zone characteristics are interconnected by the coupling of heat release, heat transfer and ignition. The trends for one variable are indicative of the trends in the others because of the interdependency of intensity with spread rate and energy density of the fuel bed, as we have seen in the equations for calculating fireline intensity from **Chapters 2** and **4** (shown here for review):

$$\ell_f = 0.0775 I_B^{0.46} = 0.0775 \left(H_c m_c'' r \right)^{0.46} \quad [4.6]$$

The most obvious characteristic of the fire simulation shown in **Figure 7.5** is that it is *not steady*. Upon ignition of the first cell (0.01 m) along the upwind edge of the fuel bed, the position of the leading edge of the flame front advances forward over time, not linearly but at an increasing rate. An increasing rate of spread means that the fire is accelerating. But the acceleration phase lasts only for a short time, and the spread rate, flame length and flame zone depth level off to a steady plateau after spreading ~2 m (at ~30 s); spread continues with behaviours that appear to be at equilibrium. Equilibrium behaviour is characteristic of a coupled system that evolves from the interactions of positive and negative feedbacks. In the case of wildland fire, these feedbacks involve numerous interactions of wind, buoyancy, energy release rate, heat transfer and many other conditions. Let us use the model to examine these interactions in more detail.

At the time of ignition, the fireline intensity is very low and flames are small, so little buoyancy is being produced by the flame zone. Therefore, the ambient wind forces the flames to tilt strongly into the fuel bed; the fire ignites nearby fuel particles rapidly because convection is pushing the flames into contact with the fuels. (At this phase, the flames are too small and thin to contribute substantial radiation to the ignition process.) After each particle ignites, it burns at a rate and for a length of time fixed by its size and the bed density (**Eqn [7.1]**). Because the model tracks the ignition and burning of each particle, it has *memory* of the ignition history of all particles in the flame zone, and the particles first ignited will be the first to burn out. As the number of burning particles increases, the flame zone grows in depth and fireline intensity, and radiation begins to influence the forward spread of the fire. Higher intensities

Table 7.2. List of fuel bed characteristics for the 'standard' case.

Modifications to these characteristics are made to illustrate different fire behaviours as described in each section of the text.

Fuel bed characteristics	Value
Fuel bed width (m)	3.0
Fuel particle diameter (mm)	1.0
Moisture content (%)	5.0
Fuel bed loading (kg m^{-2})	0.5
Fuel bed depth (m)	0.1
Fuel bed length (m)	10.0

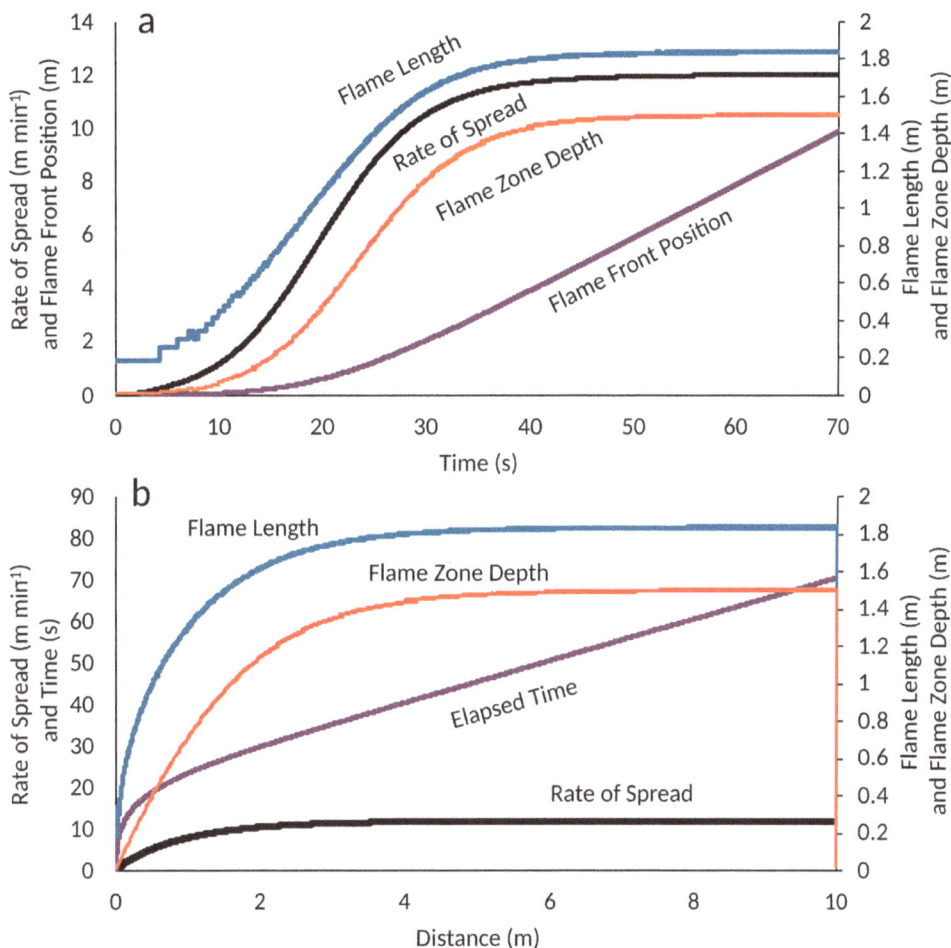

Figure 7.5: Dynamical model behaviour displayed over (a) time and (b) spread distance. The simple case has a 3 m wide fuel bed, a 0.5 kg m^{-2} loading of a single size class of 1 mm fuel particles uniformly distributed at 0.1 m depth over a linear domain of 10 m, a constant wind speed of 1 m s^{-1}, and no slope inclination. Fire spread rate, flame zone depth and flame length all exhibit similar trends because they are closely related.

increase the forward heat transfer by radiation and convection to new fuel particles, further accelerating fire spread. This is an important positive feedback process in the system. However, important negative feedbacks are operating at the same time. As intensity and spread rate increase, greater buoyancy makes flames stand up more. This reduces the forward heat transfer and slows the rate of spread. The interacting feedbacks change fire behaviour gradually rather than instantaneously because the particles in the flame zone take some time to burn out. In other words, the flame zone memory causes a lag in the effects of flame zone changes on new ignitions.

Eventually, as the fire spreads through the fuel bed, the balance of feedback processes in the system produces steady behaviours in spread rate, flame length and flame zone depth. Notice that, even though the behaviours are in a steady state, it is *produced* by a dynamical coupled system that includes strong feedbacks. The steady state can only occur when new particles at the front of the flame zone ignite at exactly the time when particles at the back of the flame zone burn out, so the back of the flame zone is maintaining a constant distance from the leading edge of the fire. Put another way, the steady condition means that the energy released from the fire is just the right amount for net heat transfer to ignite new fuel particles at the same rate as they are consumed. These processes and dynamical interactions explain why it's difficult to solve for the steady spread rate with information only on the fuels, topography and weather (i.e. the wildfire behaviour triangle described in **Chapter 2**); it is because a steady rate of spread depends upon many aspects of the *fire behaviour itself*.

Fuel particle heating and ignition

The spread of a wildland fire requires that discrete fuel particles be heated and ignited. To see how these processes work, we compare results from two simulations with widely varying spread rates (5 m min^{-1} and 26 m min^{-1}). The simulations used the standard fuel bed (**Table 7.2**) but 20 m in length; one simulation used a wind speed of 0.5 m s^{-1} and the other used a wind of 5 m s^{-1}. The model tracked the history of temperature and heat fluxes for a single fuel particle at a position of 16 m, which allowed the fires to reach a steady-state before reaching the particle. Results are compared in terms of the time to ignition (**Figure 7.6a** and **c**) and distance from the leading edge of the fire front to the cell at the time of ignition (**Figure 7.6b** and **d**). In both cases, the particle starts at ambient surface temperature (27 °C or 300 K). As the flame

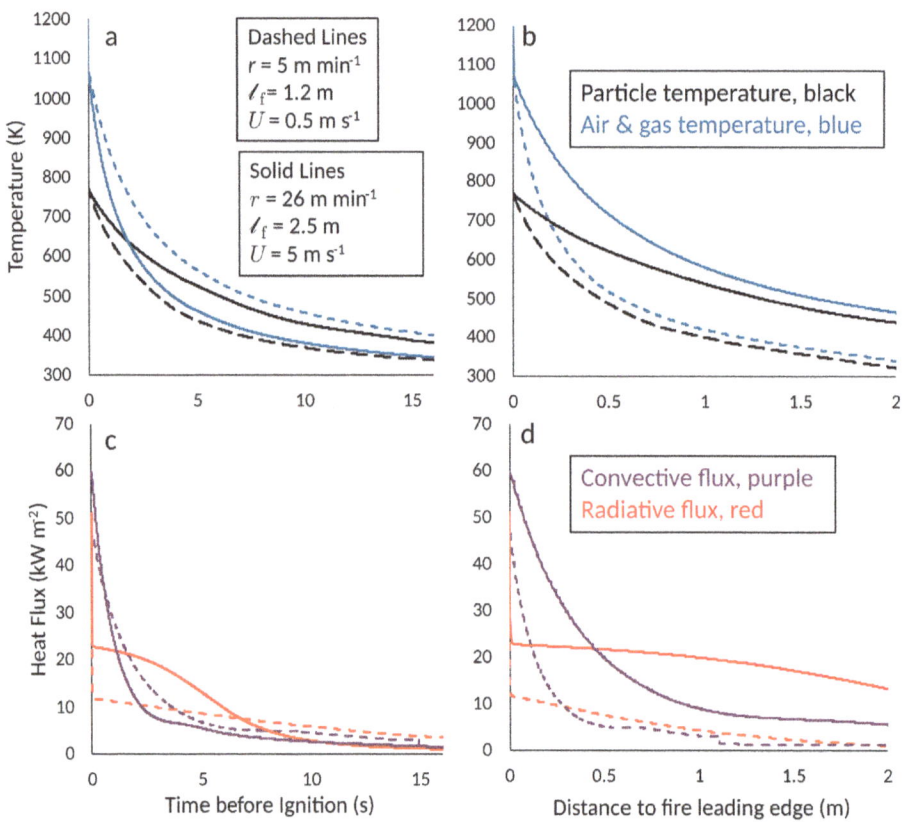

Figure 7.6: Dynamics of fuel particle heating for two fires with different spread rates due to different wind speeds. Dashed lines show the slower spreading fire (slower wind) and solid lines show the faster spreading fire (higher wind). Read the graphs from right (time or distance *before ignition* of the particle) to left (time and distance *at ignition* (zero)). (a) Surface and air temperature for 1 mm particle as a function of time to ignition; (b) surface and air temperature as a function of distance from leading edge of flame; (c) heat fluxes at surface of particle as a function of time to ignition; (d) heat fluxes at surface of particle as a function of distance from leading edge of flame. Remarkably similar heating patterns occur in both slow-spreading and fast-spreading fires, with ignition only occurring in the final seconds (a and c) and centimetres (b and d). Radiant flux is substantially larger in the faster fire because of taller and thicker flames.

zone approaches, the particle begins experiencing a low flux of radiation from flames above the fuel bed. This raises the particle's surface temperature slowly until the final seconds and centimetres. Only then, as the gas temperature reaches and then exceeds the threshold temperature for visible flame (roughly 502 °C or 775 K), does the particle's surface temperature reach the threshold for ignition. These particle heating curves are very similar to the experimental data reported by Fang and Steward (1969), Rothermel (1972), de Mestre *et al.* (1989) and Finney *et al.* (2013), which show that rapid rise in air temperature and particle surface temperature occurs primarily in the final moments and centimetres before ignition (**Figure 6.24**).

Some differences between the slow and fast fires are notable. We see in **Figure 7.6a** and **b** that the particle in the fast-moving fire heats up earlier in time and farther away in distance than in the slow-moving fire, but each reaches the ignition temperature only when the flame edge is very near. Although the average temperature of the air and gases rises rapidly close to the flame for both fires, the temperature in the slow-moving fire is *higher at a given time* but *lower for a given distance* than in the fast-moving fire. Even though the fast-moving fire has higher gas temperatures from longer flames at a given distance from the flame edge, the slower-moving fire has a longer period during which the fuel particle experiences high gas temperatures. This seems counterintuitive, but we must remember that the heating of particles in wildland fires involves a moving frame of reference as the fire spreads towards them, and fuel particles are responding in both time and space to the heat transfer processes that determine ignition. Put another way, the fire's closing speed will affect particle heating.

A similar picture emerges from the behaviour of the convective and radiant heat fluxes received by the 1 mm fuel particle (**Figure 7.6c, d**). Despite higher intensities and flame lengths in the fast-moving fire, the convective heat flux is lower at a given time than in the slow-moving fire but greater at a given distance. Radiant flux remains much higher in the fast-moving than the slow-moving fire (**Figure 7.6c, d**) because the flames are twice as tall (1.2 m *v.* 2.5 m). But the radiant flux only rises above 40 kW m^{-2} in the final second, when radiation from glowing combustion in the fuel bed becomes visible to the target fuel particle and is thus added to radiation from the flames. From **Chapter 5** we saw that time to ignition decreases in response to increasing radiant heat flux, while – because of convective cooling – time to ignition increases with increasing wind speed (**Figure 5.7**). Furthermore, even in the absence of wind, convective cooling can completely prevent ignition of fine particles heated by radiation (**Figure 5.13**). This is another way in which negative feedbacks control the processes that trigger ignition. The radiation produced by our simulated fire would not likely ignite the 1 mm fuel particles because the required ignition time, even at peak radiant flux, is much longer than the flame residence time in the fuel bed.

Fuel loading

The pattern of simple acceleration to a steady-state seen in the case above (**Figure 7.5**) is one of many possible patterns of non-steady spread. Variation in fuel loading can cause different patterns. The model was run with loadings of 1.0, 2.0 and 3.0 kg m^{-2} at constant bed depth in a 10 m wide fuel bed 40 m long to illustrate this. Constant fuel bed depth means that the packing ratio increases with loading and thus the flame residence time as per **Eqn [7.1]**. Results are shown in **Figure 7.7**. Even though a steady 2 m s^{-1} wind was used for all simulations, the non-steady spread phase demonstrates a *damped oscillation* for fuel beds with higher loading. Damped oscillation means that the fire spread rate (and intensity and flame length) exhibits a sequence of 'overshoot' and 'undershoot' cycles before it reaches a final equilibrium rate, with the amplitude of the oscillations reduced in each cycle.

Damped oscillation probably occurs at higher fuel loadings because longer flame residence times (**Eqn [7.1]**) increase the memory and thus the time lag in the system. For all fires, the burning time of

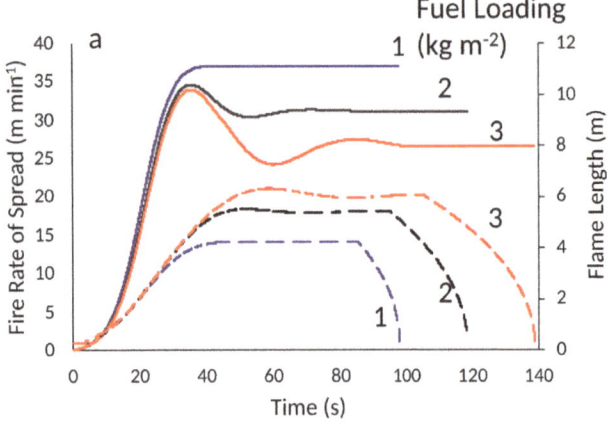

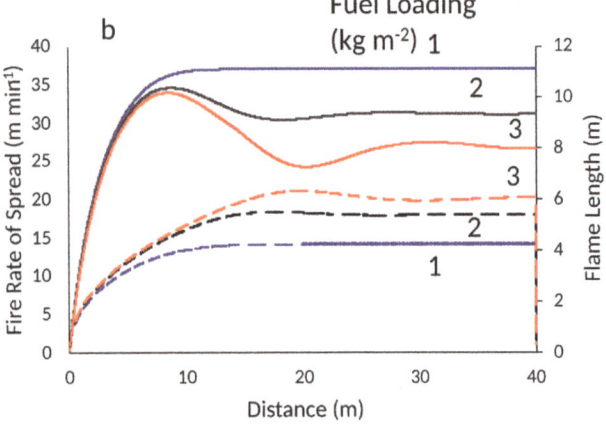

Figure 7.7: Effects of fuel loading on non-steady spread before the spread rate reaches equilibrium displayed over (a) time and (b) fire spread distance. Greater fuel loadings at a constant depth decrease the rate of spread and increase the flame residence time, thus increasing the length of ignition history (i.e. memory) in the burning zone. This produces damped oscillation in spread rate (solid lines) as spread rate approaches equilibrium. Similar oscillation can be seen in the flame length pattern for heavy fuels (dashed lines).

each particle within the flame zone is constant and begins at ignition, thus storing the ignition history of all particles within the active flame zone. All fires that start small accelerate rapidly from ignition as they ignite more and more fuel ahead of the flame front. Intensity and buoyancy are low in this initial phase, so the flames are blown towards the unburned fuels and particle ignitions occur rapidly. For sparse fuel beds (1.0 kg m^{-2}), the residence time and total energy density of the fuel bed are relatively small, so the flame zone has less memory than longer burning fuel beds. For fuel beds with heavy loading, the flame zone is releasing energy from areas ignited longer in the past; this means it will take longer to elicit the negative feedback that will eventually reduce the rate of forward heat transfer and ignition. (We discuss positive and negative feedbacks in more detail below.) We do not have data on this damped oscillation behaviour from field or laboratory fires to compare against the model results, but the behaviour is similar to results from a theoretical model by Albini (1983), in which changes in intensity lagged behind the relatively faster response of fire spread rate to fluctuating winds.

Notice also from **Figure 7.7** that flame length (dashed lines) is greater for fires with higher loading, with a slight oscillation response from heavier fuel loadings. The peak in flame length oscillation occurs at the trough of spread rate oscillation, indicating flames are standing up reducing forward heat transfer and therefore spread rate. All three trends of flame length descend rapidly once the fire reaches the end of the modelled fuel bed, because the fire is no longer igniting new fuel. This reduces the intensity until the trailing edge of the flame zone also reaches the end of the bed, and then the fire goes out.

Flame front width

Wide fires should spread faster than narrow ones because of enhanced radiant and convective heat transfer from the flame front. Radiant flux to fuel particles increases with flame front width (**Eqn [7.6]**). Convective heating also increases with flame front width in the power-law gas temperature profile (**Eqn [7.20]**) by reducing the aspect ratio of the flame zone. As we would expect from these equations, empirical observations of wildfires also suggest that wider fires spread faster (**Chapter 6, Flame front width and shape**). However, this effect is one aspect of our fire model that is difficult to directly compare with real wildland fire behaviours, since real fires have essentially an infinite bed width, but the laboratory data and the 1D formulation of our model implicitly assume the fuel bed (and thus flame front) has a constant width. As discussed in **Chapter 2, Fire shapes**, real wildfires are not normally limited by bed width but instead grow and expand in two dimensions, and for some period of time and size, they spread faster because of the wider flame front at the head of the fire. Some experimental data from both laboratory fires (Wolff *et al.* 1991; Dupuy *et al.* 2011) and field tests (Cheney and Gould 1995; Anderson *et al.* 2015) showed that wider fires spread faster. Laboratory data also suggest that fires spreading from a line ignition may slow down after ignition as they achieve more curvature (Viegas 2004), but these effects are not represented by our 1D model.

For purposes of understanding fire behaviour, agreement of observations with model predictions may not be as informative as a discussion of *why* spread rate responds differently to flame zones of different widths. When we run our model for the standard fuel bed (**Table 7.2**) but with a length of 40 m and a range of flame front widths (fuel bed widths), we see that the modelled fire spreads faster in wider fuel beds than in narrow ones (**Figure 7.8a**). Wider fuel beds enable wider flame fronts to form, thus increasing convection and radiation, as discussed above. But this effect on spread rate diminishes as the fuel bed reaches a width of ~50 m. Further changes in spread rate with further increases in flame front width may be caused by a change in the way wind engages with the flaming zone. In **Chapter 6**, we discussed how wind is affected by the presence of fire: The oncoming air in wind can easily circumvent a narrow fire front of a few metres, thus reducing forward convective heat transfer to the middle of the flame front. But the oncoming air in wind cannot circumvent the entire flame zone of a wide fire front, so it is forced through the peak-and-trough structure of the flame zone, forcing the flames out ahead of the fire front, heating the fuels ahead of the fire and increasing the fire spread rate.

The model suggests that fires require a longer distance and time to reach a steady state in wider fuel beds (**Figure 7.8b, c**). In this coupled system, a faster equilibrium spread rate means it takes a

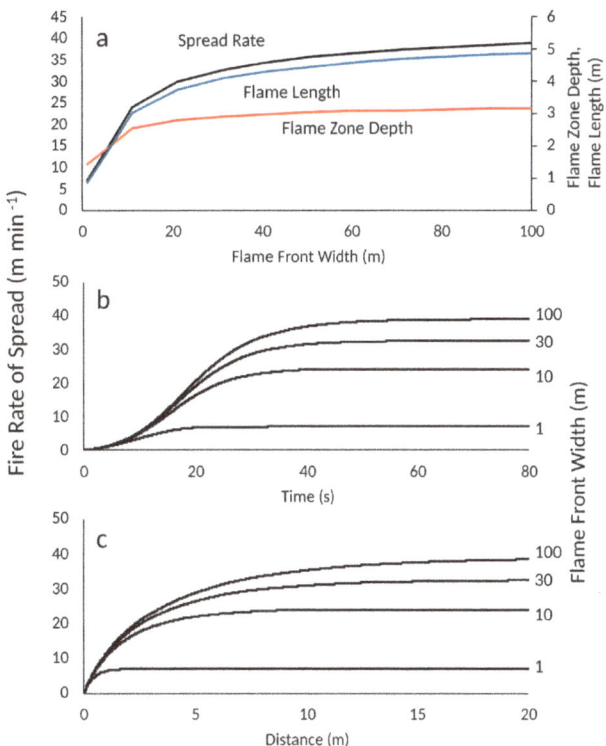

Figure 7.8: Simulations showing the effect of increasing width of the flame front on (a) steady spread rate; (b) acceleration phase in time; and (c) acceleration phase in distance. Wider fire fronts produce faster spread rates mainly because of the greater convective heating that they produce.

longer time and distance for the fire to build the greater flame zone depths and intensities required to spread that fast. This is partly a product of the burning rates of the particles, which operate the same irrespective of fuel bed dimensions, and which introduce memory of ignition and burning history into the flame zone and increase the time needed for the fire to adjust to changes in the system, either created by the fire or by external factors such as wind.

Effects of wind

Next let us look at the effects of wind, but to do this we will also include variations in flame front width (for reasons apparent later). We simulated fire spread with constant wind speeds from 0 to 5 m s^{-1} using the standard fuel bed (**Table 7.2**) but with different widths of the flame front (**Figure 7.9a**). As we saw previously, wider fires spread faster for a given wind speed. For example, with a 5 m s⁻¹ wind, a 1 m wide fire spreads at 11 m min⁻¹, while a 100 m wide front spreads at 150 m min⁻¹. For all widths and wind speeds, the model produces a final steady-state fire spread rate – that is, a spread rate that approaches a maximum value. The pattern of spread rate accelerating and then levelling off with increasing wind speed **Figure 7.9b** is consistent with some observations from laboratory fires (Wolff *et al.* 1991), field-scale fires in grass fuels (Cheney and Gould 1995; Cheney *et al.* 1998), and field-scale fires in understorey litter and shrubs in eucalyptus forests (Cheney *et al.* 2012).

The shapes and causes of relationships between fire spread rate and wind have been the subject of considerable discussion and conflicting results in fire behaviour research. The various functions summarised by Sullivan (2009) (shown in **Figure 2.12**) include forms that increase at an increasing rate (positive curvature), linear functions, and forms that increase spread but at a declining rate (like that exhibited in **Figure 7.9**). These empirical relationships have been obtained from observational data in laboratory and field conditions that include a variety of fuel types, ignition configurations, fire sizes and ranges of wind speed. Some of the fuel types are characterised by large gaps or discontinuities that limit spread below a particular threshold, others by fixed-width fuel beds. (The influence of fuel discontinuities and spread thresholds is discussed in the **Effects of fuel continuity** section below.)

Conflicting evidence and explanations for the relationship between wind and fire spread are not surprising, since it is very difficult to collect data for verifying these relationships. This is partly true in a laboratory setting because of limitations on the sizes of fires, fuel types, and ranges of wind, but it is especially true in field conditions. Although no theoretical functional form for these relationships that applies across all wildfire conditions has been developed, our model suggests it is unlikely that spread rate always increases with wind. Eventually, wind must be fast enough to remove pyrolysates from the fuel bed more rapidly than they can combust, thus reducing the energy release rate and spread rate (see **Brief discussion of chemical kinetics** in **Chapter 4**). However, stronger winds may cause the fire to spread faster by physical processes outside the flaming zone – especially ember transport and spotting.

Other factors further complicate efforts to collect observations for describing the relationship between wind and spread rate. For example, fire spread rate responds to the size of flames and the width of the fire front, so even if wind could be held constant in an experiment, fires would tend to become wider and spread faster over time. Furthermore, the acceleration phase for fire spread may last from several minutes to tens of minutes for small fires, so observations made during the

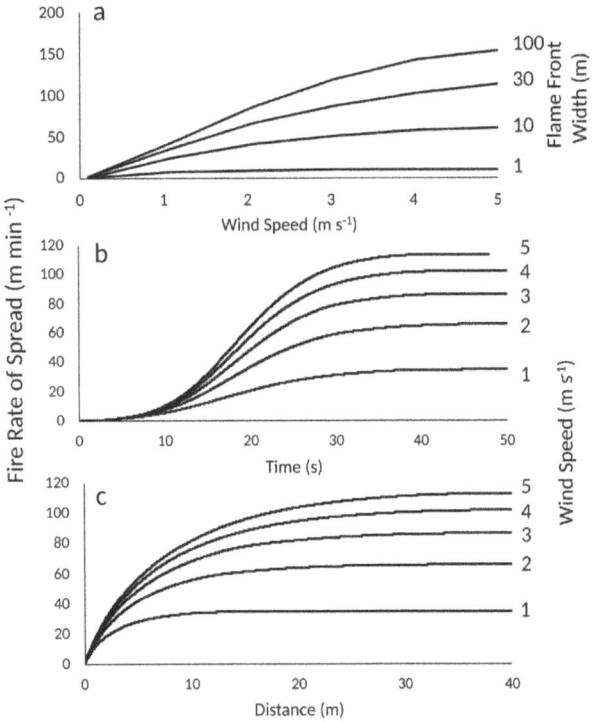

Figure 7.9: Simulations showing the effects of higher wind speeds on (a) final spread rate with different flame front widths; (b) time to steady spread rate; and (c) distance required to achieve steady spread rate. For (b) and (c) a fixed flame front width of 30 m was used. The results show that the increase in fire spread rate with wind diminishes with higher velocity winds.

accelerating phase would not reflect a steady-state condition. We can see the combined effects of wind and flame front width on spread in **Figure 7.9**. At a given flame width or time or distance, fires in stronger winds spread faster than those in slower winds. These graphs suggest that the time required to achieve equilibrium spread is about the same for all wind speeds (**Figure 7.9b**), but the equilibrium rate is achieved much farther downwind with higher winds (**Figure 7.9c**).

Non-steady wind

Most fire spread models rely upon input of a constant wind speed, but winds are never constant in speed or direction in a wildland fire. With our dynamical model, we can examine the effects of time-varying wind speed on fire behaviour. **Figure 7.10** shows fire behaviour in the standard fuel bed (**Table 7.2**) with an oscillating (sinusoidal) wind speed that is cut off at the lower end to avoid negative values (mean of 2.0 m s^{-1}, amplitude of 3.0 s^{-1}, and period of 8 s). The graphs show that wind fluctuations produce fire spread rates that oscillate with similar frequency to that of wind speed but lagging slightly behind. The amplitude of spread rate increases over time and distance, varying from ~1 m min^{-1} to ~20 m min^{-1}. The time lag in the response of spread rate is caused by the delay in the particle's response to heat flux from the fire, which also produces a sawtooth wave form in flame length and flame zone depth (**Figure 7.10a**). The time lag in particle ignition, and thus numbers and mass of burning particles in the fuel bed, produces a counterintuitive pattern in which the longest flames occur during the lowest wind speeds. The time lag is similar to that produced by wind gusts as described by Albini (1983), in which wind gusts cause a rapid increase in spread rate followed by a slightly later increase in intensity. The effect is most noticeable when plotted against fire travel distance (**Figure 7.10b**). Fire progress slows during the wind lulls, compressing fire behaviour values plotted on the *x*-axis during these slow spread periods and accentuating them during the time of rapid spread.

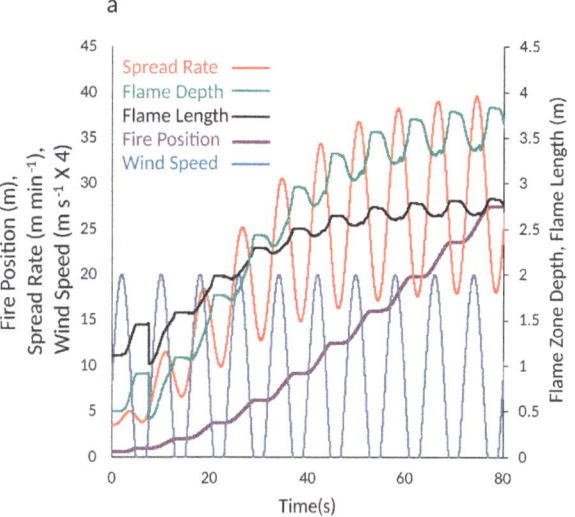

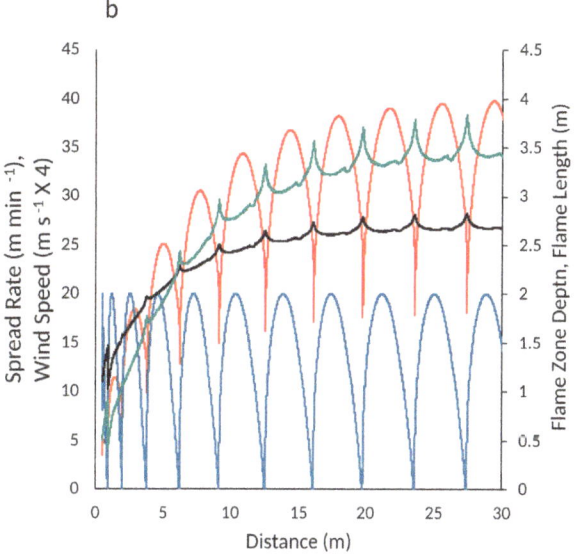

Figure 7.10: Modelled dynamics of fire spread characteristics produced by time-varying wind. (a) Plotted over time, a periodic fluctuation in wind speed (blue) produces a nearly identical but lagged response in fire spread rate (red). (b) Plotted over distance, a similar trend is evident, but the variation is altered by the varying spread rate of the fire. Note that the peak intensities (flame length in black) co-occur with the minimum spread rate (red) and wind speed (blue) because of the lag in fire response to changing conditions.

Effects of slope

Surface inclination affects uphill fire spread in numerous ways, as discussed in **Chapter 6, Topography**. Heat transfer from radiation and convection increases with slope, but convection increases strongly in a nonlinear fashion as flames attach to slopes steeper than ~20°. The trend in spread rate from our model (**Figure 7.11**) reflects the sine function that is used in the equation for calculating mean gas temperature (**Table 7.1**). This function increases most steeply after ~20° slope and plateaus beyond 45°. Like most slope functions for fire spread (Sullivan *et al.* 2014) it is not based on the dynamical processes of plume tilt or flame attachment. Many challenges remain in understanding and modelling how slope affects convective heating. Our model does illustrate that slope interacts with other variables in the way it influences rate of spread – particularly the width of the flame zone. Just as wider flame fronts produce faster spread in wind-driven fires, wider flame fronts also make flame attachment more likely on a given slope because the centre of the front is isolated from lateral air flow that would lift the flames above the slope surface and away from the fuel. Our 1D model cannot reflect the full 2D effects of fire shape on uphill spread rate, but in reality, we would expect that fires with wider flame fronts would spread faster uphill until the 2D effects of fire growth produce strongly pointed fires and thus reduce the length of fire front where flame attachment occurs (see **Chapter 6, Topography**).

Effects of dead fuel moisture

Moisture content affects the physics of fire spread in many ways, but only a few of these are represented in our model. The model includes the effects of moisture on the thermophysical properties of dead fuel particles (conductivity, density, heat capacity, latent heat transfer) and consequently the time required to heat and ignite them. However, our equation for flame residence time (**Eqn [7.1]**) does not address the fact that high fuel moisture will reduce heat release (rate and total) as was described in **Chapter 6** (see **Fuel burning rates and consumption** and **Figure 4.27**). For this reason, we expect the model to underestimate the sensitivity of fire spread to fuel moisture. Nevertheless, we can interpret some of the modelled effects of moisture that result from the increasing requirement for heating dead, moist fuels.

The model results shown in **Figure 7.12a** reflect the standard fuel bed (**Table 7.2**) but with varying winds and fuel moistures. As moisture content increases, the fire spreads more slowly, with less intensity (not shown), and it goes out below a particular wind threshold (represented by the dot terminus of each curve). This behaviour reveals an important concept in threshold behaviours – that wildfire spread thresholds are functions of *all* environmental factors, not just moisture, and have been reported by numerous laboratory and field studies (see Anderson *et al.* 2015; Cheney *et al.* 1998). The linear to weakly negative exponential decrease in fire spread rate with moisture content (**Figure 7.12b**) is consistent with many of the moisture trends observed in both field and laboratory burns in dead fuels (Anderson 1964; Cheney *et al.* 1998; Sullivan 2009), even though there are wide variations in the equations and models reported. All operational fire behaviour modelling systems reflect decreases in spread rate and intensity because of dead fuel moisture (Rothermel 1972; Stocks *et al.* 1989; Cruz *et al.* 2015).

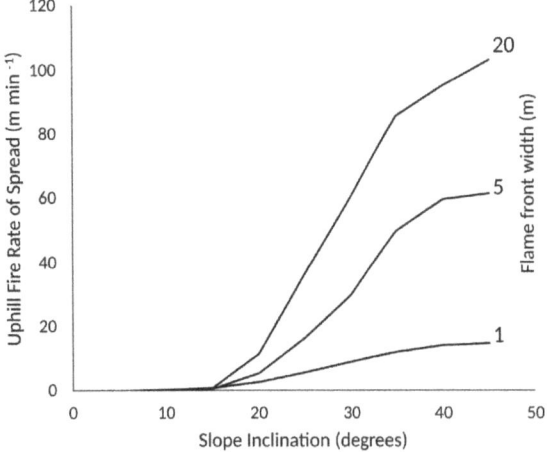

Figure 7.11: Modelled effect of slope on fire spread rates for different flame front widths. This pattern directly reflects the assumption used to represent mean gas temperature in the convective heating model (see **Table 7.1**).

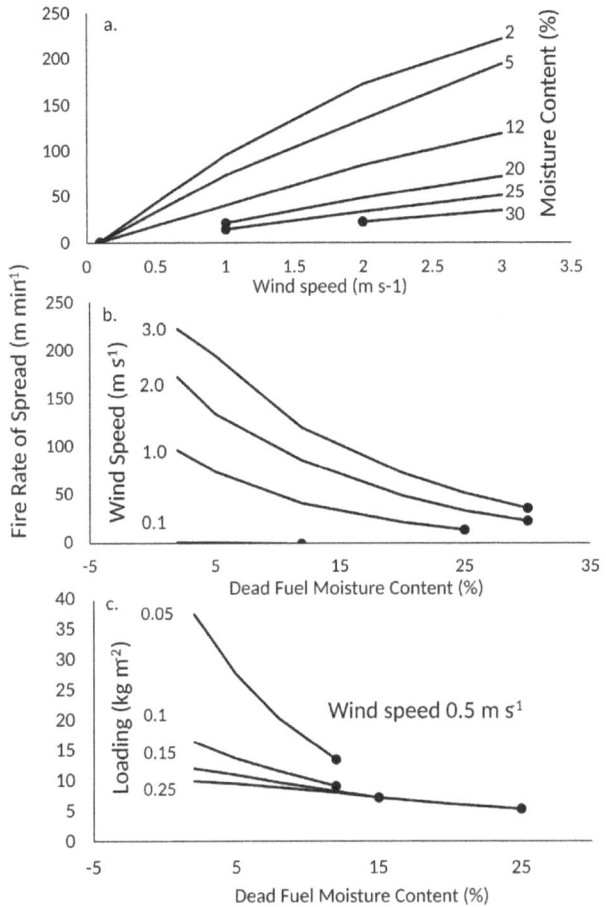

Figure 7.12: Modelled effect of increasing moisture content on fire spread rate shows trends and thresholds in spread (a) as a function of wind speed for moisture content of 2–30%; (b) as a function of moisture content for different wind speeds; and (c) for different loadings, demonstrating that spread rate and spread thresholds are far more sensitive to moisture content in sparse fuels than in heavy fuel loads.

Moisture-related thresholds (also called *extinction points*) depend upon all of the environmental factors involved in wildfire spread, although only some of these interactions are reflected in our model. In other words, there is no 'extinction moisture' that is independent of the complete environmental context of the fire (fuel, topography, weather and the fire itself). The filled circle at the terminus of each curve in **Figure 7.12** identifies the threshold of fuel moisture above which fire cannot spread. As moisture content increases, faster winds are needed to sustain spread. Fires in lightly loaded fuel beds (**Figure 7.12c**) are very sensitive to moisture content, as indicated by the steep trend and low threshold moisture for loadings of 0.05 kg m^{-2}. Fire spreads more slowly with higher loadings of fine fuels, such as 0.25 kg m^{-2}, and exhibits a fire spread threshold at a greater moisture content (around 25%). This is consistent with the ranges of extinction moistures that are predetermined for standard fuel types in the United States (Anderson 1982); grasses have a set extinction moisture of 12–15%, while timber fuels have a set extinction moisture of 25–30%. In other words, fires in timber can continue to spread when fuel moistures are far greater than the moisture at which fires in grassy fuels go out. Laboratory experiments have similarly found that fuel types with larger fuel particles are capable of supporting fire spread at higher moisture contents (Wilson 1985, 1990).

Effects of fuel continuity

The discreteness of fuel particles comprising wildland fuel beds means that fire spread only occurs if heat transfer across the gaps between fuel particles is sufficient for ignition. Although most fire spread models assume that fuels are *continuous,* many fuel beds are obviously patchy at multiple spatial scales. Sometimes this matters for fire spread, at other times it does not. Let us now use our model to examine how fuel gaps affect fire spread thresholds and how those thresholds depend upon the combined effects of environmental factors and the fire itself. To explore these effects, we will look at a series of simulations made on the standard fuel bed (**Table 7.2**) modified to be 20 m long and 10 m wide and with a repeating pattern of fuel patches and gaps of no fuel (**Figure 7.13a**). Wind speed was varied from 0 to 5 m s^{-1}.

The model runs demonstrate that fire spread thresholds depend on fuel patch lengths, gap lengths and wind speeds (**Figure 7.13b**). The results contain wavering trends in fire spread rate as a function of gap distance that are artefacts of the numerical modelling technique. At the limits of fire spread, the model is very sensitive to cell size and gap distance, so the heating dynamics of fuel

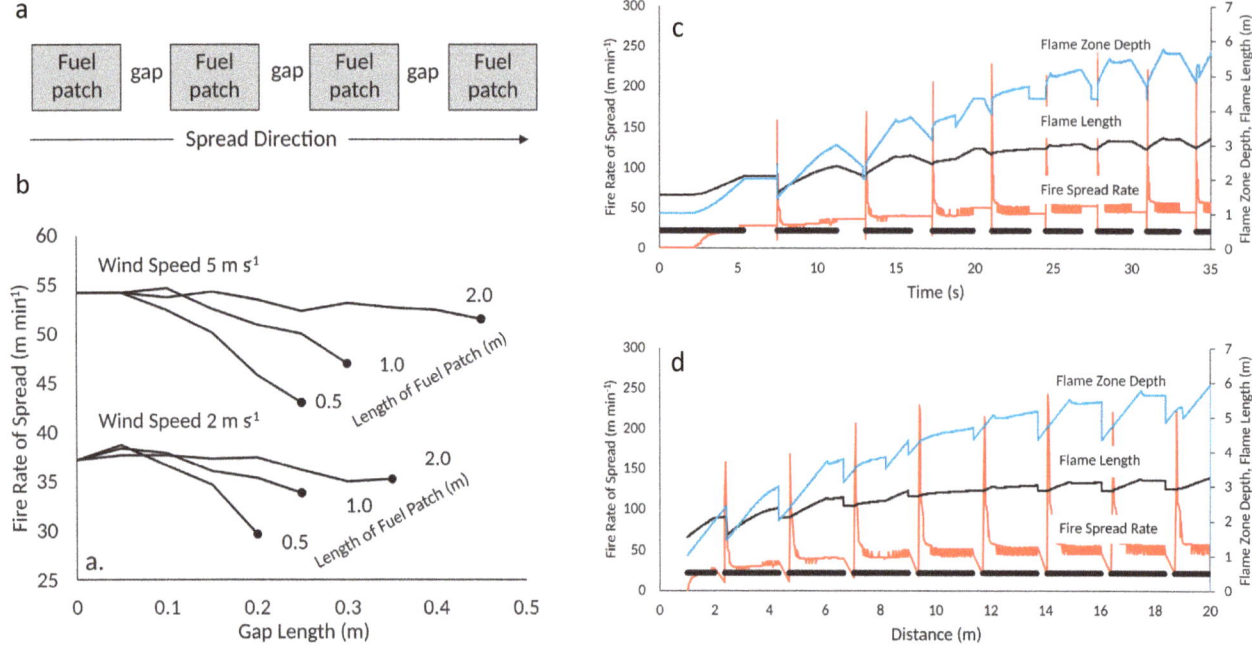

Figure 7.13: Illustrations of (a) fuel patch distance and gap distance used in the model, which alternate in the spread direction, and (b) modelled effects of fuel discontinuities, suggesting that the length of the fuel patch between gaps is of considerable importance to allow acceleration of the fire and build-up of intensity sufficient to ignite fuels across gaps. In (c) and (d) the black horizontal lines represent the locations of 2 m fuel patches with 0.35 m gaps in between. The modelled progression of fire spread through these discontinuous fuels is graphed over (c) time and (d) distance, showing that flame zone characteristics are highly non-steady as the intensity, heat transfer and ignition processes are repeatedly interrupted. Fire spread rate and flame depth and length are 'instantaneous' values calculated at the time and space resolution of the simulation.

particles are affected at the precision of cells in the model and are displayed as 'bumps' in the model results. This can be seen in the threshold behaviour when the model is run at smaller cell sizes. Without the stark thresholds in spread, these numerical effects become minor compared to the physical processes being simulated and thus are not noticeable.

Figure 7.13b shows that higher winds enable the fire to spread across longer gaps, as do longer patches of continuous fuels. Higher winds push the flames towards the unburned fuels, increasing convective heat transfer across the fuel-free gap. Longer continuous patches of fuel allow the fire intensity to increase within the patch and thus transfer more heat across the gap. The graph also indicates that the average spread rate decreases with gap length because of the delay repeatedly introduced as the fire builds up enough intensity to transfer sufficient heat across gaps. These results are qualitatively comparable with research findings in laboratory and field experiments that show spread in discontinuous fuel beds only at higher wind speeds and lower moisture contents (see, e.g., Britton and Clark 1981; Burrows et al. 2009; Weise et al. 2005).

An individual simulation shows the dramatic effect of fuel gaps on the 'instantaneous' calculated spread rate, flame length and flame zone depth (**Figure 7.13c, d**), which depend upon the time and space resolutions of the simulations. The fuel bed used for this simulation had 2 m fuel patches with 0.35 m gaps in a 2 m s^{-1} wind. Thick black lines indicate fuel patches, and the blank areas between indicate fuel gaps. When we examine the trace of spread rate (red lines in **Figure 7.13c, d**), we see a repeating pattern: a downward spike followed by a larger upward spike. The downward spike marks the location of the windward edge of the fuel gap. When the fire reaches the edge of a fuel gap, there is a delay in igniting the fuel on the opposed side. Within that time step, the spread rate drops to near zero because the fire front does not advance while the next particle is heated. At the moment when a particle beyond the gap is ignited, the spread rate immediately becomes very fast (more than 200 m min^{-1}) because the fire crosses the long gap in one time step. Although we can calculate an average spread rate for the fuel bed, as we did for **Figure 7.13b**, the average is actually composed of many fast and slow local, near-instantaneous rates. This illustrates the difficulty with measuring fire spread rate without specifying a time or space scale for averaging, because the fire front could be nearly stationary at one moment, then speeding forward at excessive instantaneous rates the next.

Trends in flame length and flame zone depth (**Figure 7.13c, d**) display an uneven pattern of fluctuations that reach maximum values at the end of each fuel patch. This pattern is caused by the interplay in time and space between particles at the back of the fire burning out and particles at the front of the fire igniting. This causes the flame zone depth to quickly grow and shrink as the front and back of the flame zone cross the gaps. Flame length follows a similar, less dramatic pattern because it is closely linked to flame zone depth.

Fire spread thresholds are the ultimate nonlinear behaviour. When they occur, fire spreads or fails to spread at a discrete point, which comprises a discontinuity in the spread function. Because of this, the spread thresholds caused by fuel gaps are highly sensitive to the nonlinear interactions of many factors, including the fuel moisture, wind speed and fuel load, which we have already addressed. In addition, thresholds are affected by the depth of the simulated ignition line ('ignition configuration') (**Figure 7.14**). The ignition configuration controls the initial amount of fuel burning and thus the intensity of fire at its inception, so a short flame zone depth at ignition requires a long distance for acceleration to produce sufficient

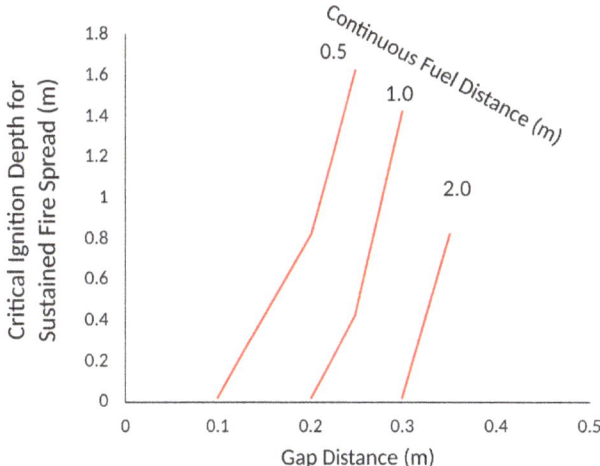

Figure 7.14: Graph showing the dependency of sustained fire spread on the combination of fuel patch length and gap length. Wind speed for the simulation was 2 m s^{-1} and fuel depth was 0.1 m, with a loading of 1 kg m^{-2} in a fuel bed 10 m wide and 30 m long. Larger gaps in general require deeper flame zone ignition lines to produce enough heat for continued flame spread across gaps. Similarly, shorter fuel patch lengths for a given gap size require more energy at ignition to sustain spread across gaps.

intensity and heat transfer to sustain spread across the initial gap. **Figure 7.14** illustrates this effect: longer ignition depths are needed to spread fire across greater gap distances and, for a given flame zone depth at ignition, longer patches of continuous fuels can spread fire over wider gaps. Patch length limits the fuel available for building up intensity, which is needed to transfer heat across the gap. Point and line ignitions are both used in prescribed fires, but the effects of these different ignition strategies on sustained spread have been little studied.

The concept that threshold behaviours reflect the combined effects of moisture, wind and fuel variables is well supported by field experience. As we have seen throughout this chapter, all environmental variables are important to fire behaviour and can affect thresholds. In addition, we have seen that properties of the fire itself – flame front width, ignition configuration, slope and fuel discontinuity – also influence thresholds. However, no models that dynamically address threshold behaviours are currently available for operational use. Some models offer a user-defined setting to identify the maximum moisture of extinction (e.g. Rothermel 1972), and other models predict the probabilities of spread based on moisture, wind, and/or estimates of fuel continuity or per cent of fuel cover. As our understanding of threshold behaviours grows with further experiments and dynamic modelling of coupled physical processes, we may be better able to predict thresholds in practical applications.

Positive and negative feedbacks

As we saw by exercising our fire spread model, positive and negative feedback processes interact to affect behaviours of spreading line fires. The feedbacks inherent in wildland fire spread are difficult to study directly, partly because fire spread processes have not been clearly characterised, but also because their role in the fire spread system cannot be seen without explaining the entire system itself. Another challenge, which is characteristic of dynamical coupled systems, is that many factors play a role in both positive and negative feedbacks, making it difficult to clearly isolate the effects of one factor independent of the others. While we acknowledge these obstacles, we also believe that a discussion of the ways in which feedbacks seem to influence wildland fire is important for understanding the dynamical behaviours that we observe in fire spread and for modelling them.

Positive feedbacks drive the system farther from its current state. By using our understanding of the fire spread process, we can identify three main positive feedback processes that operate through increasing heat transfer (**Table 7.3**).

The first important positive feedback involves fireline intensity, because it increases as the fire spreads faster. Higher intensities increase the size of the flames, which increases the flux of radiation as influenced by the radiation view factor (the amount of flame a fuel particle can 'see') and the emissivity. Both increase the rates of fuel heating and fire spread. Higher intensities and bigger flames also increase convective heat transfer because they create higher average gas temperatures at a given distance ahead of the flame zone. The increases in heat flux from radiation and convection, fuel heating and faster spread further increase the fireline intensity, thus creating a positive feedback.

The second positive feedback involves the width of the fire front. As small fires grow wider there is an increase in convective and radiant heat transfer causing the fire to spread faster. The increased convection occurs because wider fires force the wind through the flame zone (**Figure 7.8**). Wider fires also increase the radiation view factor causing positive feedback. Eventually many of these positive feedbacks reduce as the fire grows larger. This occurs because intensity increases heat transfer primarily through its effect on flame size (see **Eqns [4.6]** and **[4.7]** in **Chapter 4**), and the emissivity of flames reaches a maximum at a thickness of ~3 m (**Chapter 3, Figure 3.14**). Since flame emissivity has an upper limit, its influence on heat transfer and rate of spread as the fire grows wider is also limited (**Figure 7.8a**). Because the positive feedbacks from increasing fire width and increasing convection through the flame zone both decline as the fire grows, their influences are strongest in small fires. Thus the positive feedback of fire width on fire spread, at least for line fires at the flame scale, is limited and the fires cannot grow indefinitely or 'run away.' As fires increase in size, their behaviours are driven by many additional processes at the whole-fire scale; these are discussed in **Chapter 8**.

A third positive feedback process involves flame attachment on steep slopes. This effect is poorly understood, both from a modelling standpoint and in general. As described in **Chapter 6, Topography**, beyond some critical slope angle (~20°), flames attach to the slope, driving convective heating directly forward and increasing fire intensity and flame size. This pattern of fire growth is seen in the pointed shapes of fires on steep slopes (**Figure 6.29b**). If the fire front remains linear or there are sidewalls or a steep canyon, the positive feedback caused by uphill fire spread may build for a long time, possibly until the increasingly curved or even pointed shape of the fire front allows lateral air to wrap around the edges, detaching the flames and lifting the plume off the ground.

Negative feedbacks also help explain why fires do not run away indefinitely. **Chapter 6** suggests four possible negative feedbacks (**Table 7.3**), only two of which are included in our simple model. The first is flame buoyancy, which increases with fireline intensity. As intensity increases, buoyancy reduces the influence of wind on their tilt angle. Higher fire intensity reduces flame tilt as reflected in **Eqns [6.3]**, **[6.4]** and **[7.3]**, thus reducing the rate of forward heat transfer, which in turn reduces fireline intensity. The dual role of intensity in both positive and negative feedbacks exemplifies how complex and strongly coupled the fire spread system is.

The second negative feedback is the depth of the flame zone; if flame zone depth increases, the forward convective heat transfer may be reduced. The physical mechanism is poorly understood in the wildland fire context but is observed in our experiments on flame zones (described in **Chapter 6, Figure 6.26**). Flame zone depth influences the temperature profile in the convection plume through the fire's aspect ratio (ratio of flame zone depth to width, **Table 7.1**). If the aspect ratio increases in a fire with constant fireline intensity and flame front width, the gas temperature at a given distance from the leading edge decreases. There is evidence from circular fires for this effect. For a range of energy release rates from axisymmetric flame sources, flame height is inversely dependent on the source diameter (Zukoski *et al.* 1985); that is, as the fire diameter increases, the flame height declines. This principle applies to line fires because, as the depth of the flame zone increases without changing its width (i.e. aspect ratio increases), a line fire is gradually transformed towards a circular or square shape, which decreases the flame height and energy release rate (Quintiere and Grove 1998). Consequently, with wind or slope, the deeper the flame zone, the lower the temperature at a given distance from the flame front (**Eqn [7.20]**). As a fire spreads and intensity builds, the flame zone increases in depth, which reduces the mean gas temperature for a given intensity at a given distance ahead of the fire, thus reducing convective heat transfer to fuel particles.

Based on other research, two additional negative feedback processes could be operating. The first is that pulsation frequency decreases with flame zone size (**Eqn [4.5]** and **Figure 6.23**). As intensity builds and flame zones increase with faster fires, the frequency of forward bursts of hot gases could decline and thus increase the interval between flame contacts with the fuels. At the same time, however, the size of the flame structures is growing larger, so it could increase the duration of flame contact with fuel particles. Longer flames also allow for increasing velocity due to longer flow lengths for gravitational acceleration and thus more efficient

Table 7.3. List and brief description of positive and negative feedback processes in fire spread.

	Function in fire spread
Positive feedback processes	
Fireline intensity	Higher intensity increases heat transfer by radiation and convection.
Fire front width	Wider fronts can increase convective heat transfer and spread rate.
Slope	Flame attachment to slopes steeper than ~20° increases convective heat transfer, intensity.
Negative feedback processes	
Flame angle/buoyancy	Higher intensity increases buoyant forces, decreasing heat transfer.
Flame zone depth	Deeper flame zones for a given intensity decrease convective heating.
Flame pulsation	Lower frequency pulsation of taller flames may decrease convective heating rate, but may also extend further and last longer – the net effect on ignition and spread is unknown.
Frontal curvature	Curved fronts decrease forward convective heating.

convective heat transfer during flame contact. Thus, the net consequences of flame intermittency on ignition time and spread rate remain uncertain.

A final negative feedback process in some fires could result from the increasing frontal curvature of a line fire as it grows (see **Figures. 2.10 and 6.31a**). A linear flame zone with no curvature eventually becomes bowed, and the increasingly convex front means a greater fraction of the perimeter exists as flanking fires which are oriented away from the heading direction. Simulations suggest that the direction of wind approaching from behind a flame zone becomes oriented normal to the flame front. As the front becomes more bowed, convective heating may be directed at increasing angles away from the direction of forward spread (Canfield *et al.* 2014; see **Figure 6.32a**). For fires with very pointed shapes, radiation in the forward direction can be reduced as well. Both of these effects could explain the deceleration of fire spread from line sources observed by Viegas (2004).

Model improvements

Many of the system behaviours of wildland fires are demonstrated by the simple 1D model that we have described in this chapter. The model shows, for example, that fire spread accelerates from ignition to a steady-state, that wider flame fronts spread faster (to a point), that time-varying wind speed produces complex oscillating and lagged behaviours, and that spread/no-spread thresholds are caused by fuel discontinuities interacting with environmental conditions. These behaviours are seen every day in real fires, but they are not currently addressed by any operational fire modelling system. By exploring the model developed here, we have been able to discuss *how* heat transfer, combustion and ignition combine to produce fire spread. However, as we have noted throughout, we have had to make assumptions and simplifications regarding how the model represents some of the physical processes of fire behaviour, and these compromises limit the range of conditions and behaviours that the model can accurately represent.

While the model could be improved in many ways, we will focus here on just a few changes and additions that could substantially increase the breadth of applications for the model, help us develop even deeper insight into the fire spread process, and move the field of fire science closer to providing an operational model that is based on the strongly coupled, dynamical system behaviours of wildland fires.

Combustion

We could improve the model's representation of fuel combustion by replacing the fixed-flame residence time equation (**Eqn [7.1]**) with a time-dependent burning rate model for particles and fuel beds. We know from **Chapter 4** that burning rate is not constant over the flame zone, so our assumption of uniform burning rate may cause the model to misrepresent the flame zone's leading and trailing edges and forward heat transfer. Another improvement would be to include the effects of fuel moisture content on burning rate and total heat release, both of which should decrease with increasing moisture and thereby produce more dramatic thresholds in spread conditions. Ideally, the model should also address the effects of solid-phase combustion. This would enable the model to reflect changes in burning rate as a function of wind ventilation, predict the transition from smouldering to flaming combustion, estimate the proportion of fuel consumed in flaming *v.* smouldering, and generate heat from all phases of combustion for estimating fire effects on plants and soils.

Ignition

In this model, we have assumed that particles begin burning at the flame front once the surface temperature exceeds a nominal value of 502 °C (775 K). As discussed in **Chapter 5**, use of a set ignition temperature is a simplified way to address a dynamic process of a critical rate of pyrolysis. More physically accurate modelling of ignition would be based on determining the rate of heat release from the mixture of oxygen and the products of pyrolysis that is sufficient to initiate sustained combustion of the solid fuel. Because this approach would be based on the heat release rate, it could improve the model's ability to determine when coherent flame zones are possible in very high winds and in sparse or discontinuous fuel beds. The model could also be improved by separating the particle ignition temperature from the criterion for marking the position of the flame leading edge.

Wind

As discussed in **Chapter 6**, the effects of wind on fire, and the effects of fire on wind, are areas that need much scientific investigation. An improved model could address the effect of the heated ground behind the flame front on the wind field. It could also represent the effects of high wind speeds that blow pyrolysates away from the flaming zone, reducing both heat release and spread rate.

Flame zone orientation

Our 1D model only represents the propagation of a heading fire and cannot represent flame zones that are oriented at various angles to the oncoming wind or uphill direction. Fires grow in 2D, of course, but a 2D model may not be needed to better understand the fire spread characteristics of flanking and backing fires. Instead, the 1D model framework could address these behaviours if we understood how heat transfer varies with the orientation of the fire front. Considerable research will be needed to determine how convective heating works when the flame zone is not aligned with the wind and/or slope, but this improvement to the model would greatly increase its applicability to 2D fire spread.

Heat transfer

Our 1D model runs quickly because it represents convective heating as a mean gas temperature profile from a purely empirical model. A model based on fluid mechanics principles would be more robust, but it would also be computationally intensive to the point where it is currently infeasible. We will have to await better and perhaps different methods for representing convection adjacent to a flaming front before we can address this limitation of the model. To better represent fire-induced air flow, its effects on convective heat exchange with fuel particles and its effects on solid combustion, we also need information on the velocity distribution of hot gases near flame zones that is not currently available. Finally, we could improve the modelling of heat transfer by adjusting convection of fuel particles based on the density of the fuel bed.

Crown fire

The fire behaviour simulated with the 1D model is intended to approximate a single-stratum fire – that is, a surface fire. However, it could be extended to include a second elevated stratum, which would make it comparable to current models that include multiple strata; this improvement could eventually enable us to use our 1D model to address the behaviour of crown fires.

Summary

In this chapter, we have developed a 1D model for spread of a line fire through a bed of discrete particles. Then we used this model to discuss the dynamical and coupled behaviours of wildland fire. Although we made many simplifications as we developed the model, we have been able to use it to better understand the physical causes of many realistic fire behaviours. A more comprehensive model could address a great deal more of the complexity of fire behaviours and physical processes that is covered in **Chapters 1–6**, but that level of complexity may be unnecessary for helping us understand common fire behaviours. The next chapter will examine some large-fire behaviours, many of which are well beyond the assumptions of the spreading line fire.

References

Albini FA (1983) The variability of wind-aided free-burning fires. *Combustion Science and Technology* **31**, 303–311. doi:10.1080/00102208308923648

Anderson HE (1964) 'Mechanisms of fire spread'. Research Paper INT-8. USDA Forest Service, Intermountain Forest and Range Experiment Station, Ogden, UT.

Anderson HE (1982) 'Aids to determining fuel models for estimating fire behavior'. General Technical Report INT-122. USDA Forest Service, Intermountain Forest and Range Experiment Station, Odgen, UT..

Anderson WR, Cruz MG, Fernandes PM, McCaw L, Vega JA, Bradstock RA, Fogarty L, Gould J, McCarthy G, Marsden-Smedley JB, Matthews S (2015) A generic, empirical-based model for predicting rate of fire spread in shrublands. *International Journal of Wildland Fire* **24**(4), 443–460. doi:10.1071/WF14130

Britton CM, Clark FA (1981) Will your sagebrush range burn? *Rangelands* **3**(5), 207–208.

Burrows ND, Ward B, Robinson A (2009) Fuel dynamics and fire spread in spinifex grasslands of the Western Desert. *Proceedings of the Royal Society of Queensland* **115**, 69–76.

Byram GM (1959) Combustion of forest fuels. In *Forest Fire: Control and Use*. (Ed. KP Davis) pp. 61–89. McGraw-Hill, New York, NY.

Canfield JM, Linn RR, Sauer JA, Finney M, Forthofer J (2014) A numerical investigation of the interplay between fireline length, geometry, and rate of spread. *Agricultural and Forest Meteorology* **189**, 48–59. doi:10.1016/j.agrformet.2014.01.007

Catchpole WR, Catchpole EA, Butler BW, Rothermel RC, Morris GA, Latham DJ (1998) Rate of spread of free-burning fires in woody fuels in a wind tunnel. *Combustion Science and Technology* **131**(1–6), 1–37. doi:10.1080/00102209808935753

Cheney NP, Gould JS (1995) Fire growth in grassland fuels. *International Journal of Wildland Fire* **5**(4), 237–247. doi:10.1071/WF9950237

Cheney NP, Gould JS, Catchpole WR (1998) Prediction of fire spread in grasslands. *International Journal of Wildland Fire* **8**(1), 1–13. doi:10.1071/WF9980001

Cheney NP, Gould JS, McCaw WL, Anderson WR (2012) Predicting fire behaviour in dry eucalypt forest in southern Australia. *Forest Ecology and Management* **280**, 120–131. doi:10.1016/j.foreco.2012.06.012

Churchill SW, Bernstein M (1977) A correlating equation for forced convection from gases and liquids to a circular cylinder in crossflow. *Journal of Heat Transfer* **99**, 300–306. doi:10.1115/1.3450685

Churchill SW, Chu HHS (1975) Correlating equations for laminar and turbulent free convection from a horizontal cylinder. *International Journal of Heat and Mass Transfer* **18**(9), 1049–1053. doi:10.1016/0017-9310(75)90222-7

Cruz MG, Gould JS, Alexander ME, Sullivan AL, McCaw WL, Matthews S (2015) Empirical-based models for predicting head-fire rate of spread in Australian fuel types. *Australian Forestry* **78**(3), 118–158. doi:10.1080/00049158.2015.1055063

de Mestre NJ, Catchpole EA, Anderson DH, Rothermel RC (1989) Uniform propagation of a planar fire front without wind. *Combustion Science and Technology* **65**, 231–244. doi:10.1080/00102208908924051

Dupuy JL, Maréchal J, Portier D, Valette JC (2011) The effects of slope and fuel bed width on laboratory fire behaviour. *International Journal of Wildland Fire* **20**(2), 272–288. doi:10.1071/WF09075

Fang JB, Steward FR (1969) Flame spread through randomly packed fuel particles. *Combustion and Flame* **13**(4), 392–398. doi:10.1016/0010-2180(69)90108-4

Finney MA, Forthofer J, Grenfell IC, Adam BA, Akafuah NK, Saito K (2013) A study of flame spread in engineered cardboard fuelbeds: Part I: Correlations and observations. In *Seventh International Symposium on Scale Modeling (ISSM-7)*. 6–9 August 2013, Hirosaki, Japan. International Scale Modeling Committee.

Finney MA, Cohen JD, Forthofer JM, McAllister SS, Gollner MJ, Gorham DJ, Saito K, Akafuah NK, Adam BA, English JD (2015) Role of buoyant flame dynamics in wildfire spread. *Proceedings of the National Academy of Sciences of the United States of America* **112**(32), 9833–9838. doi:10.1073/pnas.1504498112

Finney MA, Grumstrup TP, Grenfell IC (2020) Flame characteristics adjacent to a stationary line fire. *Combustion Science and Technology*. doi: 10.1080/00102202.2020.1863952

Fons WL (1946) Analysis of fire spread in light forest fuels. *Journal of Agricultural Research* **72**(3), 93–122.

Hottel HC, Williams GC, Steward FR (1965) The modeling of fire spread through a fuel bed. *Symposium (International) on Combustion* **10**(1), 997–1007.

Koo E, Pagni PJ, Stephens SL, Huff J, Woycheese J, Weise DR (2005) A simple physical model for forest fire spread rate. *Fire Safety Science* **8**, 851–862. doi:10.3801/IAFSS.FSS.8-851

Nelson RM, Jr (2002) An effective wind speed for models of fire spread. *International Journal of Wildland Fire* **11**, 153–161. doi:10.1071/WF02031

Nelson RM, Jr (2003) Reaction times and burning rates for wind tunnel headfires. *International Journal of Wildland Fire* **12**(2), 195–211. doi:10.1071/WF02041

Pagni PJ, Peterson TG (1973) Flame spread through porous fuels. In *Symposium (International) on Combustion* **14**(1), 1099–1107.

Patankar S (1980) *Numerical Heat Transfer and Fluid Flow*. Hemisphere Publishing Corporation, McGraw Hill Book Company, New York, NY.

Pearce HG, Finney MA, Strand T, Katurji M, Clements C (2019) New Zealand field-scale fire experiments to test convective heat transfer in wildland fires. In *Proceedings for the 6th International Fire Behavior and Fuels Conference*. 29 April – 3 May 2019, Sydney, Australia. pp. 90–95. International Association of Wildland Fire, Missoula, MO.

Putnam AA (1965) A model study of wind-blown free-burning fires. *Proceedings of Tenth International Symposium on Combustion* **10**, 1039–1046.

Quintiere JG, Grove BS (1998) A unified analysis for fire plumes. *Symposium (International) on Combustion* **27**, 2757–2766.

Rothermel RC (1972) 'A mathematical model for predicting fire spread in wildland fuels'. Research Paper INT-115. USDA Forestry Service, Intermountain Forest and Range Experiment Station, Ogden, UT.

Stocks BJ, Lynham TJ, Lawson BD, Alexander ME, Wagner CV, McAlpine RS, Dube DE (1989) Canadian forest fire danger rating system: an overview. *Forestry Chronicle* **65**(4), 258–265. doi:10.5558/tfc65258-4

Sullivan AL (2009) Wildland surface fire spread modelling, 1990–2007. 2: Empirical and quasi-empirical models. *International Journal of Wildland Fire* **18**(4), 369–386. doi:10.1071/WF06142

Sullivan AL, Sharples JJ, Matthews S, Plucinski MP (2014) A downslope fire spread correction factor based on landscape-scale fire behaviour. *Environmental Modelling & Software* **62**, 153–163. doi:10.1016/j.envsoft.2014.08.024

Viegas DX (2004) On the existence of a steady state regime for slope and wind driven fires. *International Journal of Wildland Fire* **13**, 101–117. doi:10.1071/WF03008

Weise DR, Zhou X, Sun L, Mahalingam S (2005) Fire spread in chaparral – 'go or no-go?'. *International Journal of Wildland Fire* **14**, 99–106. doi:10.1071/WF04049

Wilson RA (1982) 'A reexamination of fire spread in free-burning porous fuel beds'. Research Paper INT-289. USDA Forest Service, Intermountain Forest and Range Experiment Station, Odgen, UT.

Wilson RA (1985) Observations of extinction and marginal burning states in free burning porous fuel beds. *Combustion Science and Technology* **44**(3–4), 179–193. doi:10.1080/00102208508960302

Wilson RA (1990) 'Reexamination of Rothermel's fire spread equations in no-wind and no-slope conditions'. Research Paper INT-434. USDA Forest Service, Intermountain Research Station, Odgen, UT.

Wolff MF, Carrier GF, Fendell FE (1991) Wind-aided fire spread across arrays of discrete fuel elements. II. Experiment. *Combustion Science and Technology* **77**(4–6), 261–289. doi:10.1080/00102209108951731

Zukoski EE, Cetegen BM, Kubota T (1985) Visible structure of buoyant diffusion flames. *Symposium (International) on Combustion* **20**(1), 361–366.

8
Behaviours of large fires

In this chapter we examine some behaviours associated with large wildfires that may be manifested over the entire burning area. Some of these phenomena are consequences of the flame-scale physical processes presented earlier and modelled in **Chapter 7**, but others are caused by atmospheric factors at the whole-fire scale, which often depend upon feedbacks between the fire and the environment. Although we refer to these behaviours as 'large-fire' behaviours, many are not a function of the size of the fire. Some can be replicated in a laboratory because, while not considered large, they occur at the scale of the entire fire. These behaviours tend to be associated with large fires because that is when they threaten life and property. When such behaviours occur at flame scales, as in the laboratory, they are typically irrelevant to the overall growth or impact of the fire, but they are useful for demonstration and research. In this chapter, we will compare large-fire behaviours with the line-fire behaviours considered in **Chapter 2, Line fire concept** and the modelling presented in **Chapter 7**, in which one segment of a linear flame zone has been assumed to behave independently from the rest of the fire. We will also compare large-fire behaviour with behaviour of the small 'point' fires that we have considered, such as the candle. For each topic covered in this chapter, we will also discuss the uncertainties in current understanding, and highlight some research questions and approaches that might improve the physical explanations of these behaviours.

Crown fire

A crown fire involves ignition and flaming spread of a fire through a tree crown or other plant canopy that is comprised mostly of living foliage elevated above the ground (i.e. aerial fuel). Crown fires can technically occur in canopies of both forests and shrubs, but the term is most commonly applied to forests. Native eucalyptus forests in Australia and native conifer forests in North America, Europe and Asia can support crown fires. In addition, exotic conifer and eucalypt plantations throughout the world are highly susceptible to crown fire because the individual tree canopies are closely spaced, and the trees are spatially more uniform than natural forests. Arctic boreal forests represent a classic natural crown fire regime because the species' ecological dynamics depend upon periodic destruction and renewal by landscape-scale fire. Some plant and tree species have foliage that does not readily ignite and burn – many broad-leaved deciduous species in North America, for example – but research has not yet provided a complete explanation of the reasons for their relative immunity from crown fire.

Crown fires typically exhibit the highest fireline intensities shown by wildland fire (20–100 MW m^{-1}), although they generally spread at moderate rates (30–60 m min^{-1}, and greater) (Van Wagner 1977; Rothermel 1991; Stocks *et al.* 2004a) when compared to shrub and grass fires. The high intensities of crown fires are caused by rapid consumption of live fine fuel particles, which may constitute 1–2 kg m^{-2} in addition to the fine fuel loading on the ground. Fine foliage is distributed vertically at relatively low packing ratios (i.e. the fraction of volume that is filled with fuel), so it burns rapidly, producing high heat fluxes and very tall flames (~10–50 m). When driven by strong winds, steep slopes, and pyroconvective activity (described below), crown fires can burn vast areas (10^5 ha or more). As they burn large quantities of live vegetation, they produce large quantities of dark or black smoke because combustion is less complete than that of slower burning fuels (see **Chapter 4, Basic structure and characteristics**).

The physical explanations for wildland fire spread and behaviour are remarkably limited for many aspects of fire behaviour, but crown fire is the exemplar of a fire phenomenon much modelled but little studied. One reason for our poor understanding of physics in crown fires is that it is extremely difficult and dangerous to study such high-intensity fires at full scale, when they are spreading over large areas during extreme weather conditions. Laboratory experiments are essential for discovering how live fuels ignite and combust (i.e. **Chapters 4–6**), but laboratories cannot easily replicate the radiant fluxes and fluid mechanics of flames the size of those produced in actual crown fires. Scientists have attempted to instrument and study crown fire behaviour in wildfire conditions, but such fires are logistically uncontrollable, so they present great challenges to those who must anticipate their movement and behaviour in order to obtain measurements. Planned, field-based experiments using prescribed fire are better controlled than wildfire experiments, but they are difficult and expensive to conduct. Most of the field experiments on crown fire behaviour were conducted in Canadian conifer forests during the 1970s and 1980s (Van Wagner 1977, 1993; Stocks 1987); the results were used in development of the Canadian Fire Behaviour Prediction System. More recently, the International Crown Fire Modelling Experiments (Stocks *et al.* 2004b) enabled scientists to use intensive instrumentation to measure heat fluxes, spread rate and flame behaviour in crown fires. However, even these experiments were insufficient to answer basic questions about the physical processes that control crown fire initiation and spread (Butler *et al.* 2004a).

Characterisation of canopy fuels presents another challenge to researchers studying crown fires. It is difficult to measure and map canopy fuels because they are explicitly three-dimensional and highly variable, both at the level of individual trees and at the stand level (Stocks *et al.* 2004a); however, remote sensing is now offering techniques to help address this problem (see **Implications for fuel characterisation and classification** in **Chapter 6** and **Ladder and canopy fuels** in **Chapter 9**). Crown fires of any kind usually have high rates of canopy consumption, meaning that post-fire observations of burned canopy offer little information to distinguish how crown fire spread and behaviour occurred.

Existing concepts of crown fire behaviour are almost entirely derived from Van Wagner's (1977) classification (**Figure 8.1**). Even in recent modelling efforts, his intuitive concepts have been incorporated with only minor refinement (Scott and Reinhardt 2001; Alexander and Cruz 2011). Van Wagner distinguished three types of crown fire:

1. *Passive crown fire* occurs when flames and predominantly convective heat from a surface fire ignite the fuels at the base of the tree canopy. Passive crown fire may also be called *torching* and range from a single tree to groups of trees. Torching occurs when flames and predominantly convective heat from a surface fire ignites the lower extents of elevated or aerial fuel. *Ladder fuels* facilitate initiation of torching and provide fuel continuity between surface and canopy strata because they serve as ladders for fire to 'climb up' and ignite the

canopy. Ladder fuels often include bark, limbs, small trees, shrubs and lichens. Van Wagner's criterion for the transition from surface fire to crown fire requires that a critical air temperature be reached at the height of the canopy base above the surface line fire that provides the heat source for torching.

2. After a passive crown fire is initiated, it may transition to an *active crown fire* (a.k.a. *running crown fire*, *continuous crown fire*). This is also called a *dependent crown fire* because it depends on heat release from an existing surface fire to persist. Van Wagner suggested that an active crown fire can maintain its coherent flame zone only if it acquires its fuel supply at some minimum rate, which he called the *critical mass flow rate*. The critical mass flow rate is related to the relationship between the bulk density of aerial fuels and the spread rate of the fire itself. The rapid burning of fine fuel in a crown fire can only be sustained when it spreads fast enough that new fuel becomes involved in the flame front before the old fuel burns out. In other words, a continuous steady flame zone can exist only when the ignition time for aerial fuels is the same as or less than their flame residence time. This principle is similar to the role of flame residence time in line fire spread (**Chapter 7**). The flame residence time for burning tree crowns has been estimated as 10–40 s (Despain *et al.* 1996; Taylor *et al.* 2004), so crown fires probably must spread quickly or not at all. In this feature, crown fire spread resembles fire spread in the discontinuous fuels of many shrub vegetation types.

3. The final kind of crown fire in Van Wagner's classification is the *independent crown fire* which, once started, does not require any heat from surface burning to be sustained. Crown fires reported to have spread over snow-covered ground would certainly qualify as independent (Agee *et al.* 2002; Mottus and Pengelly 2004). Trees are not the only vegetation that can burn in independent crown fires; desert shrub vegetation can also exhibit this behaviour under high winds because of sparse cover and bare ground beneath and between fuel patches.

The challenge for crown fire research is to find physical explanations for these behaviours that extend beyond the conceptual framework introduced by Van Wagner. There are some key physical

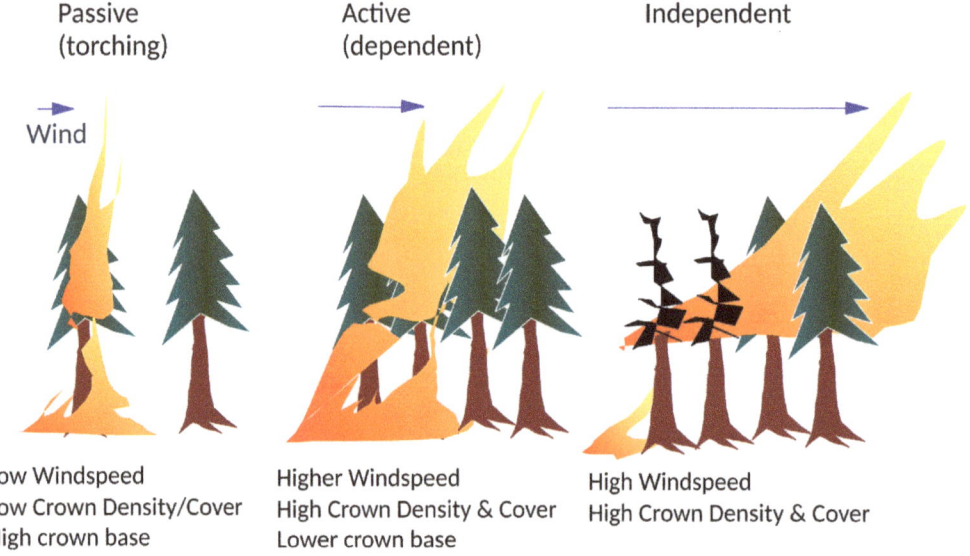

Figure 8.1: Crown fire classification from Van Wagner (1977) showing passive, active and independent crown fires, and the fuel structure and environmental factors associated with each.

processes involved in crown fire spread that will require much more research to fully understand. These processes include ignition, the roles of radiation and convection, the role of tree boles as ladder fuels, and the roles of recently killed crown fuels in fire spread.

Ignition, the essential criterion for initiating and spreading wildland fire is poorly characterised for live aerial fuels. As discussed in the **Live fuels** sections of **Chapters 5** and **6**, ignition and combustion of live fuels are very complex problems, which we understand very little. To understand ignition better in crown fires, we must learn how the characteristics of flame zones affect canopy heating and how fuel particles heat up in low-density aerial fuel beds. Van Wagner's model for initiating a crown fire from a surface fire requires only that aerial fuels reach a critical temperature above a linear flame zone. This is regularly modelled using fireline intensity and flame length of the surface line fire, but the uncertainties associated with these fire behaviour metrics mean imprecision in estimating vertical air temperature profiles (Alexander and Cruz 2012). Also, the geometry of the fire source greatly affects vertical air temperature distributions; fires burning over large areas and with deeper flame zones produce higher plume temperatures above them than fires burning with thin, linear flame zones. The heating of aerial fuel particles depends upon duration as well as heat flux and air/gas temperature. Because of variation in the spatial arrangement of aerial fuels and the geometry of fire spread, wide variability probably occurs in the heating time needed for ignition. One factor that contributes to this variation is the fineness of crown foliage, especially that of conifers. Fine foliage is highly efficient at convective cooling as well as heating (**Chapter 3, Convection heat transfer**), as indicated by high convection heat transfer coefficients (70–170 W m^{-2} °C^{-1}; Michaletz and Johnson 2006; McAllister and Finney 2014). Even though convective heating is critically important to our understanding of ignition in crown fires, few studies have addressed it. One example is modelling by Cruz *et al.* (2006), which attempted to include the time-dependent heating of canopy fuels from a spreading surface fire where the plume and flame zone dimensions were included.

To understand the relative roles of radiation and convection in crown fire spread, we need information on flame temperatures and emissivity in the crown fire flame zones. Some studies have indicated that crown fires can spread through radiant heating alone (Albini and Stocks 1986; Butler *et al.* 2004b), and we know that crown fires can achieve flame zones within the canopy fuel bed itself that are sufficiently thick and tall to transfer radiant heat very efficiently; canopy flame zones can produce radiant emissivity near unity and a large view factor (see **Chapter 3, Radiation heat transfer**). Van Wagner (1977) reasoned that the flame front of crown fires should radiate with a maximum of ~125 kW m^{-2} at flame temperatures of 1000 °C. For radiation alone to heat fuels, it must occur before flame contact. Actual measurements are rare, but a wide-angle radiometer on a crown fire recorded fluxes less than ~150 kW m^{-2} until flame contact which then exceeded 200 kW m^{-2} (Frankman *et al.* 2013) consistent with localised radiometric flame temperatures above 1100 °C (see **Chapter 3, Radiation heat transfer**). Other measurements using a narrow angle radiometer (Butler *et al.* 2004a) also indicated peak radiant fluxes exceeding 200 kW m^{-2} during the International Crown Fire Modelling Experiment; these values detected local regions of flame temperatures higher than thermocouple measurements that fluctuated between 200 and 1000 °C, possibly because of cool air circulations. The uses of both kinds of radiometers is discussed in **Chapter 9, Heat transfer**. At any of these fluxes, radiation can rapidly ignite large woody fuel particles, vertical tree stems and compact forest floor layers, provided the understorey is sufficiently open to allow radiation to penetrate forward (see critical heat flux, **Figure 5.7** in **Chapter 5**). Evidence from cameras inside crown fires shows pyrolysis and ignition of large fuels by radiation in forests with an open understorey (i.e. space between tree stems) (**Figure 8.2a**). Denser understoreys (small trees or brush composed of fine fuel particles) and soot production within the flame zone could attenuate radiation sufficiently to reduce forward radiation heat transfer. However, even if radiant heat flux can ignite large woody particles and the litter surface, these fuels may burn too slowly to generate flame zones in the canopy with the high heat fluxes needed to sustain crown fire spread. If this is so, then convective heating from the burning of fine canopy fuels is essential to produce enough heat for canopy spread.

We have seen that the spread of many wildland fires requires convection to ignite fine fuel particles. In some crown fires, a strong role for convection was suggested from video evidence by Clark *et al.* (1999) and Coen *et al.* (2004). They observed that vorticity of flames in crown fires (**Figure 8.2b, c**) produced intermittent forward bursting of convection ahead of the flame front into fine fuels in both the canopy and the understorey. In fact, the video indicated that live foliage and fine fuels ignite *only* upon flame contact, even with high levels of incident radiation (**Figure 8.2a**). Flame contact may be required because fire-induced air flow pulls cool ambient air in from ahead of the fire, convectively cooling the fine fuels (Frankman *et al.* 2013). In-fire videos also reveal that ember deposition may play a crucial role in fire spread by causing short-range spotting ahead of the flame zone. A 'blizzard' of embers causes high-density spot ignitions in surface fuels, which coalesce into a region of surface fire ahead of the canopy ignition (Van Wagner 1964; Taylor *et al.* 2004); this could indicate that some crown fires are relying on surface spotting to spread. Although these observations indicate that both radiation and convection probably have significant roles in crown fire spread, they are insufficient to help us discern the precise role of each process; indeed, crown fire spread is also likely to

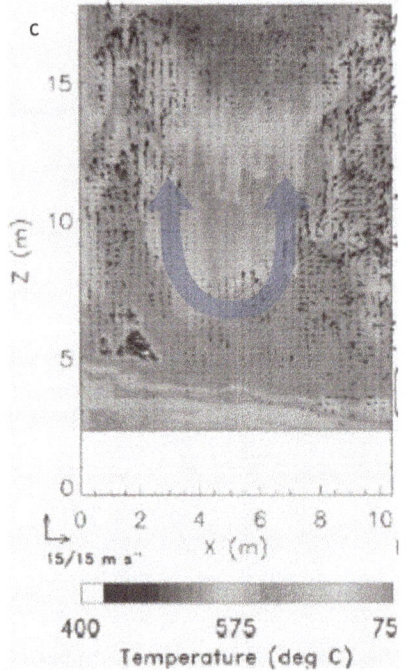

Figure 8.2: Photographs of crown fire in the North-west Territories, Canada, in forests of jack pine (*Pinus banksiana*) and black spruce (*Picea mariana*) in 1999 (a) show pyrolysate released from logs and compact litter by radiant heating but not from foliage on small trees, which are being cooled by ambient convection; (b) peak-and-trough structure of flames at the front of an approaching crown fire, which are created by buoyant dynamics and vorticity (see **Chapter 6, Wind in the presence of fire**); and (c) upward flow vectors (small arrows) from infrared analysis of a crown fire on either side of trough structure in flames (reproduced with permission from Clark *et al.* 1999) that intermittently burst forward within the troughs.

depend upon many additional factors, including the composition and structure of the fuel profile.

Many observers have reported that tree boles influence fire spread, perhaps functioning as ladder fuels, radiant heat sources, and influences on heat flux within the canopy. Most observers of forest fires have seen vortices on the leeward sides of tree stems, aiding the ascent of flames from the ground surface up into the canopy (**Figure 8.3**). Van Wagner (1964) referred to the appearance of trees stems in a small crown fire as 'flaming columns joining two storeys' (i.e. the surface fuel stratum and the crown stratum). Small scale experiments by Gill (1974) with glass rods and a gas burner showed that wind speed and rod diameter both influence the height of flames on the lee side simply through fluid dynamic effects of the rod as an obstruction to flow. This suggests that leeward vorticity can assist fire in reaching and attaching to canopy fuels (**Figure 8.3**). Unlike glass rods, tree stems are likely to have flammable tree bark, lichens and dead fuels suspended on branches that would greatly assist vertical fire spread. So far, there has been no attempt to account for the role that lee-side vorticity induced by tree stems might play in crown fires. Tree stems may also contribute to forward heat transfer when their surfaces are burning in flaming and glowing combustion (**Figure 8.3**). The energy emitted by these solid surfaces could supplement radiant heat from thin flame zones, just as glowing solid fuels augment radiant heating in crib fire experiments (**Figure 4.26** in **Chapter 4**; McCarter and Broido 1965; Thomas 1971). No tests have compared

Figure 8.3: Lee side attachment and ascent of flames along lee or uphill sides of tree stems increases vertical continuity and solid burning surface material for enhancing radiant heat transfer in the understorey of (a) coast redwood forest California (photograph by M.A. Finney); (b) jack pine forest North-west Territories, Canada (photograph from USDA Forest Service); (c) Douglas-fir trees, Montana; and (d) lodgepole pine, Montana, USA (photographs (c) and (d) by M.A. Finney).

radiant heat transfer in dense forests of small-diameter stems with that in open forests of large trees. Similarly, research has not examined the influence of variable tree crown structure on fire spread. Van Wagner's critical mass flow rate depends only on the bulk density of the canopy fuels, even though vertical and horizontal crown structure varies among trees and among species. If bulk properties of the canopy alone can explain crown fire spread, then initiation and spread of crown fires would not differ between canopies composed of large trees with large gaps and canopies composed of small trees with small gaps, as long as their bulk densities were identical. This seems unlikely when we consider the basic principles of heat transfer and ignition.

Let us consider another factor that may strongly influence the behaviour of crown fires – the effects of recently damaged or killed canopy fuels on fire spread, and especially insect-damaged foliage. Much debate and effort in recent decades has focused on how insect-damaged and insect-killed conifer forests affect crown fire behaviour (see review by Hicke et al. 2012). Many forests in North America undergo beetle or moth infestations that either kill various proportions of the trees or temporarily defoliate them. Lodgepole pine (*Pinus contorta*) has been a particular focus of research and discussion, but spruce (*Picea* spp.), ponderosa pine (*Pinus ponderosa*), balsam fir (*Abies balsamea*) and Douglas-fir (*Pseudotsuga menziesii*) have also received considerable attention. Investigations of crown fire behaviour in insect-damaged or -killed tree crowns are especially complicated because the process of insect infestation is itself a complex, dynamical system that couples tree physiology with meteorological, successional and topographic influences.

The complications are illustrated by the variation in fuel conditions according to the kind of insect attack (e.g. bark beetle v. defoliator), the proportion of the trees or foliage that is affected or killed and the period after attack. For *Pinus contorta* and associated forests in North America affected by mountain pine beetle (*Dendroctonous ponderosae*), foliar moisture content on trees suffering a fatal beetle attack will decrease even before needles exhibit a change in colour (green to red) (Jolly et al. 2012). The subsequent 'red stage' dead foliage then remains attached for a few years before falling, leaving the 'grey stage' of barren tree branches. Meteorological conditions of the forest and fuels are changed too, for example by reduced wind sheltering and increased exposure to solar radiation and to precipitation, all of which can affect fire behaviour. For a decade or more following attack, dead trees will weaken by decay, break or topple to the ground, and increase the amount of downed large woody fuel. Surviving trees will grow and new seedlings will establish. Complete removal of foliage from the canopy fuel profile, either during the grey stage or after defoliating insects, should temporarily reduce the fuel continuity and aerial fine fuel loading available for a potential crown fire. However, conclusive fire experiments on these conditions have not been conducted, and most studies have relied on modelling to explore fire behaviour consequences. The complexity of multiple fuel and weather factors varying over time requires much more study to clearly understand the impact of insect damage on fire behaviour.

The fire behaviour consequences of beetle-killed forests in the 'red stage' have perhaps received the most research attention, but studies based on modelling, fuel measurements and observations offer contradictory interpretations of the influence of dead foliage on fire behaviour. Field observations have been reported from fires in forests where foliage is dead and still attached to branches ('red stage' fuels) of some fraction of the canopy (see Perrakis et al. 2014). These reports suggest that crown fires may initiate under milder conditions of wind and dead fuel moisture than fires predicted by Canadian fire spread models for living green forests and may spread about twice as fast as predicted. However, it is remarkable that measured spread rates of crown fire in red-stage dry foliage are within the same 20–60 m min^{-1} range as spread measured in green forests at ~100% moisture content, although the measurements have high variability. We do not have good interpretations of these behaviours, both because they are produced by complex interactions in a coupled system and because we know little about the physics of ignition and burning of live fuels. The role of spotting in crown fire spread is also unclear, but beetle-killed forests affect many of the factors involved in ember production and ignition as described below. Experiments clearly show that dead dry needles ignite faster than green ones, but this does not necessarily mean that crown fires spread faster or have higher intensity in dead than live fuels. It is also difficult to apply results from laboratory tests on insect-killed fuels to the field scale (Fernandes and Cruz 2012; Rossa and Fernandes 2018) when we know so little about the physics of live foliage ignition and burning, especially in natural vegetation configurations (see **Live fuels** in **Chapters 5** and **6**).

A final question about crown fires concerns how to describe and measure it. We often use a single, averaged spread rate to describe forward spread. This is far from a perfect description for any fire, even one in continuous and uniform fuel beds (see **Chapter 1**, **The challenge of validation**), but it seems especially crude for describing crown fires. Van Wagner's model uses thresholds for determining ignition and forward spread, which add some refinement to the concept of a single forward spread rate and help account for variation in forward progress over time and space. However, thresholds are not sufficient to account for the influences of fine-scale variation in surface and canopy fuel structure, time-varying winds, and buoyant dynamics in the flame zone and plume – which all influence the progress of crown fires in both experimental fires (Taylor *et al.* 2004; McRae *et al.* 2005) and wildfires (Simard *et al.* 1983; Wade and Ward 1973; Haines 1982). As mentioned above, the role of spotting in advancing crown fire could be quite important to the overall rate of crown fire spread but has been little examined. Fine-scale spotting definitely adds uncertainty to the problem of defining the leading edge of the flame front. In addition, its role in crown fire spread would depend upon the spread rate of the fire and the delay between ignition and flaming spread of spot fires (Alexander and Cruz 2006), which we do not understand well (see the next section). As a consequence of these many uncertainties, we cannot be sure how to compare spread rates from plot-level experiments with spread rates reported from field-level observations and wildfires, especially when the latter are *averages* of fire progress at different scales of space and time. This concern applies to all fire spread, but the thresholding behaviour of crown fire (vertically and horizontally) may increase spread rate sensitivity over time and space. Thus, when a specific spread rate of, say, 50 m min^{-1} is predicted or reported, we should ask whether this applies to short runs or long distances and for what length of time. If advances in crown fire physics and modelling can eventually characterise spread rates as distributions defined by time and space scales rather than single, averaged values, then practical applications of crown fire propagation rates could be more helpful for planning over large areas (Taylor *et al.* 2004).

Spotting and spot fires

Spotting is the transport of burning solid fuel material beyond the boundaries of an active fire that ignites *spot fires*. Burning materials, called *embers* or *firebrands*, are most often lofted into the air by the flame and plume and transported by ambient winds to locations ahead of the flame zone. Firebrands may disperse across a range of lateral directions either by ejection from large plumes (Thurston *et al.* 2017) or by *saltation* – a process in which particles, like sand on a beach or burning embers, are bounced along the ground surface (see Kok *et al.* 2012). It is difficult to detect and measure firebrand travel or density, but firebrands are easily seen at night or with an infrared camera. Upon landing on suitable fuel, they can start spot fires. Spotting is usually revealed by the presence and locations of spot fires themselves – the final successful stage of the spotting process – rather than by the embers. Spot fires are routinely detected by infrared fire mapping (**Figure 8.4**). Firebrands may be burning in flaming or glowing combustion, but they are most often in a glowing state while they travel and when they land.

Most burning embers travel only short distances – a few tens to hundreds of metres – but they can travel much farther: embers have caused spot fires at distances more than 20 km from the main fire in both Australia (Cruz *et al.* 2012) and the United States (Anderson 1968). Charred (i.e. not burning) materials such as bark, cones, sticks and branches are found at even greater distances. Ember transport is a very effective means of advancing wildfires across barriers and fuel discontinuities such as streams, lakes and barren areas. Spotting across lakes in the Boundary Waters Canoe Area (Minnesota, USA) is so prevalent during wildfires that forest structure and speciation are dictated by fire growth patterns relative to windward and leeward shorelines (Heinselman 1973): windward shores support a fringe of large *Pinus resinosa* and *Pinus strobus* trees protected from high-intensity heading fires and where embers borne across the lakes ignite and burn at low intensity as they accelerate towards the leeward shores. Ember transport into and within urban developments is the predominant way in which wildfires ignite homes (see Cohen 2000; Caton *et al.* 2017). Even short-range spot fires are particularly challenging for fire containment efforts because they breach control lines. When fires are burning in continuous fuels, it is unclear when or if overall fire spread is enhanced by short-range spotting or if the advance of the main fire front envelops spot ignitions before they can accelerate and spread independently. Prolific spot fire occurrence and ignition density may lead to other large-fire behaviours described later in this chapter, including fire whirls (see **Vorticity**) and mass fires (see **Mass fires**).

The dynamics of spotting and spot fire occurrence were described by Tarifa *et al.* (1965) and

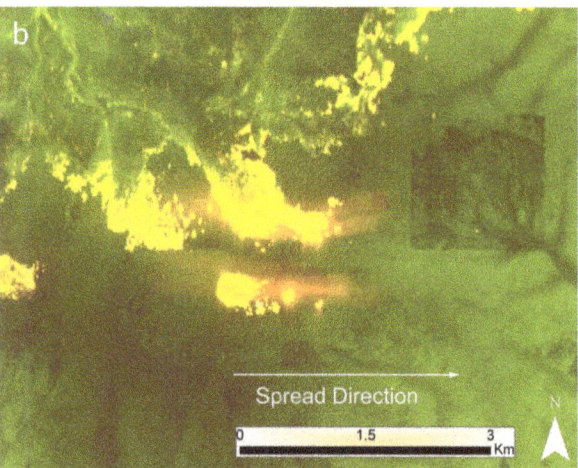

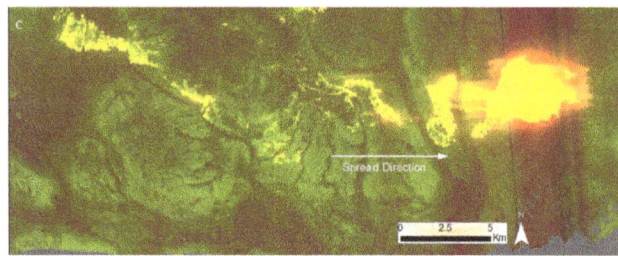

Figure 8.4: Images of spotting from large wildfires: (a) GOES 8 image of the Camp Fire as it encountered the town of Paradise, California, on 8 November 2018, showing spot fires burning inside and beyond the town boundaries; (b) thermal infrared line scan of a small section of the Copper King fire near Thompson Falls, Montana on 23 August 2016, showing spotting from 1–2 km distant from the existing edge of the fire; and (c) thermal line scan from the Rice Ridge fire on 3 September 2017 near Seely Lake, Montana, showing large spot fires on ridges ahead of the main front. (Images (b) and (c) from the National Infrared Operations centre of the USDA Forest Service.)

later by Albini (1979, 1983) as a series of steps (**Figure 8.5**):

1. Firebrands are produced from vegetation materials within the fuel bed.
2. They continue burning for some period of time after separating from the fuel bed.
3. They are lofted and transported (airborne or bounced) into unburned fuels.
4. They ignite the new fuels and transition to flaming spread.

All of these steps involve a great deal of uncertainty and variability; they have been minimally studied and may be impossible to quantify. We will probably never be able to know the numbers, sizes or flight trajectories of embers well enough to predict them in active fires. However, we may not have to quantify them physically in order to anticipate their impact statistically. In other words, the likelihood that firebrands will ignite spot fires at various distances may be approximated based on information about fire behaviour, fuels, wind, humidity and terrain – even though none of the actual processes are physically known. Modelling by Albini (1979), for example, addressed this problem by solving for the maximum distance that spot-fire ignitions could occur when ember burnout equalled transport time in specific terrain and wind conditions.

The raw materials for firebrands are the fuels (i.e. vegetation, living and dead). Effective firebrands must be small enough to be lofted but large enough to continue burning in transport. Those that burn out before contacting fuels at the ground are not viable ignition sources. Fine grasses are not likely to produce firebrands that can start fires at long distances, but they may produce vast quantities of small firebrands. When driven by winds through the air or bounced across fields and roads, they can effectively propagate a sequence of short-range ignitions and spot fires. Brush and conifer trees provide woody materials – small twigs, branches, cones and bark flakes – that can burn much longer in transport. These materials are often only partially burned (to a charred state) in the

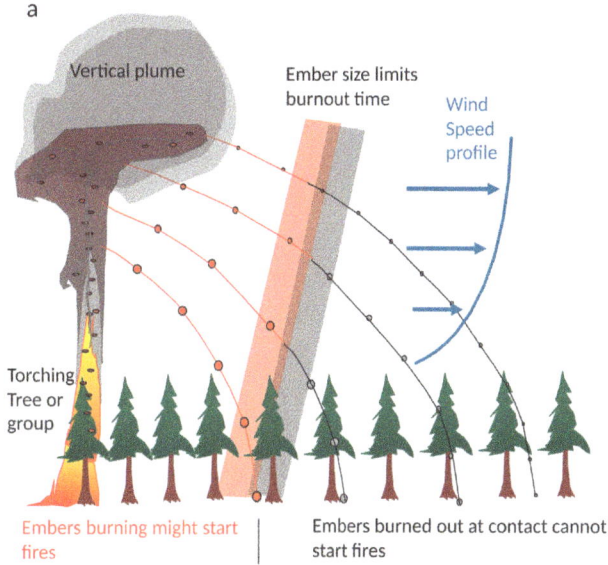

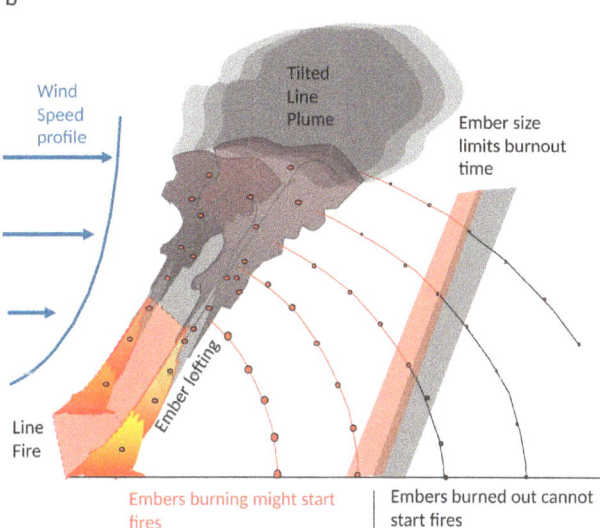

Figure 8.5: Factors involved in spotting from (a) torching trees and (b) from wind driven line fires. Torching trees or groups of trees are assumed to loft embers vertically, but downwind transport depends upon ambient winds. The inclined plume of line fires lofts embers at inclined trajectory. Maximum spotting distance is limited by both ember transport distance and burnout time. Large embers burn for a long time but are not lofted as high or drift as far as small embers, yet small embers burn out before reaching the ground.

flame front. Pyrolysis during flaming may have caused cracking and curling of the char, weakening the materials to the point where they are easily stripped from branches and trunks by the high velocity of flames accelerating upward (Tohidi et al. 2017).

Charred firebrands typically have much lower density than their original source material because most of the mass has volatilised during flaming. Their shapes are often assumed to be small cylinders or disks, a simplification that is helpful in determining aerodynamic properties that affect lofting and transport. However, this simplification is likely to be a poor reflection of reality. Firebrands studied in laboratory wind tunnels display spinning that alters drag and terminal velocity, and they also exhibit spiralling in their descent trajectories, so their paths and even their shapes cannot be assumed static as they ascend and descend through the air. Some kinds of firebrands have unusual properties that are even more difficult to characterise but have a strong effect on flight and burning (Ellis 2011; Hall et al. 2015). These are unique to specific shrubs and trees, particularly eucalyptus. Some eucalyptus species are copious producers of bark strips (from shaggy-bark species) and coiled ribbons (from stringy-bark species). These unusual firebrands can sustain combustion for perhaps tens of minutes of flight time (Hall et al. 2015).

The process of lofting and downstream transport of embers depends on the aerodynamics of firebrands and the air flow in which they are borne. Lofting begins with entrainment in the flame and then continues vertically and often horizontally as firebrands are carried by the rising plume (**Figure 8.5**). As we saw in **Chapter 4, Non-premixed or diffusion flames**, buoyant flames accelerate with vertical length. Surface fires with flames a few metres long have velocities of up to 5 m s^{-1}, which is sufficient force to break partially burned twigs. Tall flames in crown fires can achieve vertical velocities of 20–30 m s^{-1}, so they can lift much larger particles. Albini (1983) differentiated the vertical lofting trajectory from torching trees from the inclined lofting trajectory of embers in wind-driven line fires. Torching trees loft embers vertically, and the embers are assumed to exit the plume at the height where their terminal velocity (at which they are no longer accelerating) equals

their upward velocity. Terminal velocity has been estimated at 2–7 m s^{-1} when embers detach from the fuel bed state (Ellis 2011), but it decreases over time because the ember is consumed during flight (Tarifa et al. 1965; Koo et al. 2010; Ellis 2011). After firebrands exit the plume, they descend within the horizontal wind field as it carries them forward (Tarifa et al. 1965; Albini 1979). Because crown fires have abundant source material for embers and strong buoyancy in their flames, they can transport embers farther than other kinds of fire (**Figure 8.6**) (Albini et al. 2012). Because wind-driven fires transport embers horizontally both as they ascend and descend, they can also increase the travel distance (Albini 1983; Albini et al. 2012; Thurston et al. 2017). In both cases, the velocity profile of the wind field bearing the embers has been assumed to have a logarithmic form above the vegetation (see **Chapter 6, Winds**), but this would obviously vary with terrain or if other wind profiles existed, particularly in the vicinity of a large-fire plume.

The burning behaviour of firebrands is critical to their success in igniting spot fires. Long-burning firebrands are likely to survive long transport

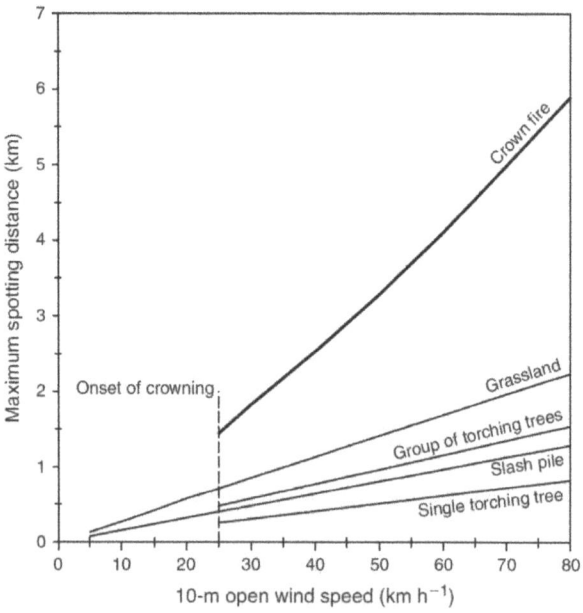

Figure 8.6: Maximum spotting distances modelled for different sources of firebrands (reproduced with permission from Albini et al. 2012).

distances and remain capable of ignition when they land. Large firebrands are likely to contact receptive fuels with greater residual mass and burning time than small firebrands. Both burning duration and size of the firebrand are therefore likely to increase the chances of ignition. Shape may also affect the 'life span' of a firebrand. Spiral bark strips from eucalyptus are credited with long-distance spotting because glowing combustion is sheltered on the interior coil surface (Hall *et al.* 2015). Experiments suggest high variability in ember burning rates and duration due to many possible causes: natural variation in firebrand shape, properties of the source material, the state of ember decomposition at the time of lofting (e.g. char depth) and burning rates during transport. Moisture content of firebrand source material may also affect production of viable embers, although it is difficult to distinguish these effects from those of moisture in receptive fuels on the ultimate likelihood of ignition.

Firebrands can generate a spot fire only if they can ignite receptive fuels away from the main fire front. As we have seen in **Chapter 5**, flaming ignition results from heat transfer at rates sufficient to pyrolyse fuels at some critical rate (see the review of spot ignitions by Fernandez-Pello 2017). However, firebrands generally ignite new fuels with smouldering rather than flaming combustion. To produce smouldering ignition, the hot surface of the firebrand must maintain physical contact with the surface of the new fuel for some period of time. Smouldering ignition can occur over a wide range of time and heating rates, so the criteria for smouldering are considerably more variable than those for flaming ignition (**Chapter 5, Types of ignition**). Moisture content and wind speed also influence ignition time by firebrands (Ellis 2011, 2015). Smaller firebrands have been shown to require higher temperatures for ignition of new fuels than larger ones. In fact, small firebrands may not be able to ignite some fuel materials at all. Observations suggest that spot fires most often originate in rotten wood, dried dung (Bunting and Wright 1974), decomposed forest floor material (duff) and moss. This has not been thoroughly studied, but there are some logical explanations: the fine sizes of particles in decayed materials offer high surface area for contact with small firebrands, and rotten wood is less dense and has lower conductivity than sound wood. This lower conductivity reduces heat lost through the fuel bed, possibly shifting the balance between heat generation and heat loss enough to allow ignition.

For smouldering ignition to produce a spot fire, it must transition to flaming combustion before it can spread. As described by Rein (2009) and Santoso *et al.* (2019) (and introduced in **Chapter 4, Smouldering and glowing**), there are no reliable models for this transition. General observations indicate that the transition is aided by ventilation from wind, exposure to solar heating and reduced daytime humidity. Ventilation affects smouldering in the same way that blowing on coals affects a campfire: the added oxygen increases glowing surface temperature and the fire begins to flame. Laboratory experiments by Ellis (2015) show how wind and fuel bed moisture interact to effect ignition from firebrands (**Figure 8.7**). Very dry fuels exhibited a nearly 100% chance of ignition with or without wind. In no-wind conditions, the probability of ignition declined rapidly when fuel moisture content exceeded ~7%. When 1–2 m s^{-1} air flow was added, the probability of ignition was less sensitive to fuel moisture, remaining above 10% even at fuel moistures around 12–13%. The increased velocity of air flow on smouldering fuels forces more oxygen into contact with the combustion reaction on the surface of the solid fuel; this can dramatically increase the temperature and burning rate. As temperatures rise on the glowing surface, heat flux to adjacent fuels increases through radiation, convection and conduction. Eventually, the increased heat flux is sufficient for autoignition of the pyrolysates, which occurs at ~600 °C (see **Chapter 5, Types of ignition**). Once flaming combustion begins, the sustained spread of the spot fire is subject to the same physical processes that affect any wildland fire.

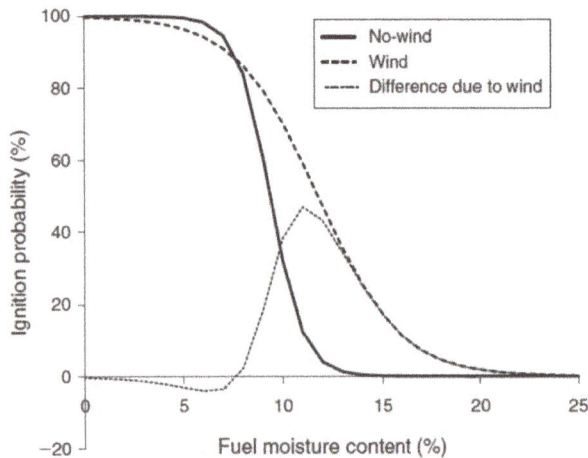

Figure 8.7: Ignition probability of firebrands as a function of moisture and wind speed from laboratory studies on eucalyptus litter (reproduced with permission from Ellis 2015).

Probability of ignition is a term used to describe the likelihood of spot fires in the US National Fire Danger Rating System (NFDRS, Bradshaw *et al.* 1984). As used there, it is not strictly related to the probability of embers igniting and spreading spot fires that is described above. This is because the probability of ignition in the NFDRS is derived from experiments in which flaming matches were forced into contact with pine needles conditioned to different moisture contents (Blackmarr 1972). Thus these experiments determined the probability of successful *flaming* ignitions rather than the probability of smouldering ignitions that transition to flaming. Observations of spotting in actual wildland fire conditions, however, may not be related to either our explanation above or the probability of ignition in NFDRS. A wildland fire can produce so many embers that the probability of spotting may be very high even if the probability of any given ember igniting a spreading spot fire is very low. The probability of ignition as observed in the field is probably more akin to the probability of *detecting* a spreading spot fire than the probability of an ember *igniting* one.

It is currently unknown how fire-scale characteristics would alter ember lofting and transport compared to the flame-scale fire behaviours described above. For example, the size of the fire and the diameter and height of the plume should be important to the spotting processes. Analysis of ember lofting has been mainly based on the flame-scale energy release rates associated with line fires. It seems likely that larger burning areas and associated stronger updrafts would be capable of carrying more embers higher above the ground than indicated by the frontal fireline intensity. This remains an important area of future research.

Fire shapes and growth patterns

The shape that a fire should theoretically achieve under ideal conditions has not been determined, but fires burning in uniform and continuous fuels with constant wind and slope generally tend to produce ellipsoid shapes (**Figure 2.14** in **Chapter 2, Fire shapes**). Various shapes seem to fit this observation (**Figure 8.8**), including the single ellipse (Alexander 1985), double ellipse (Albini 1976; Anderson 1983), and other egg shapes (Peet 1967). Fan shapes and lemnsicates have also been proposed, but the simple ellipse remains the most commonly assumed (Richards 1995). We are uncertain about ideal fire shape partly because it cannot actually be observed, since wildfire and experimental fire conditions are never truly uniform in space or constant in time. This makes it impossible to collect reliable empirical data over extensive areas and time spans. In addition to this problem with observations, we are challenged by our incomplete understanding of the physics of fire spread. We do not yet know how critical processes affect spread in two dimensions, and we do not know if fire shape is a function of locally determined conditions around a fire or a product of the whole fire.

The remarkable fact that wind-driven fires display generally elliptical shapes over a wide range of sizes – from laboratory scales to landscape scales over 100 000 ha – argues that the processes operating at the fine scale must also be important at the large scale – that is, that fire shape is largely

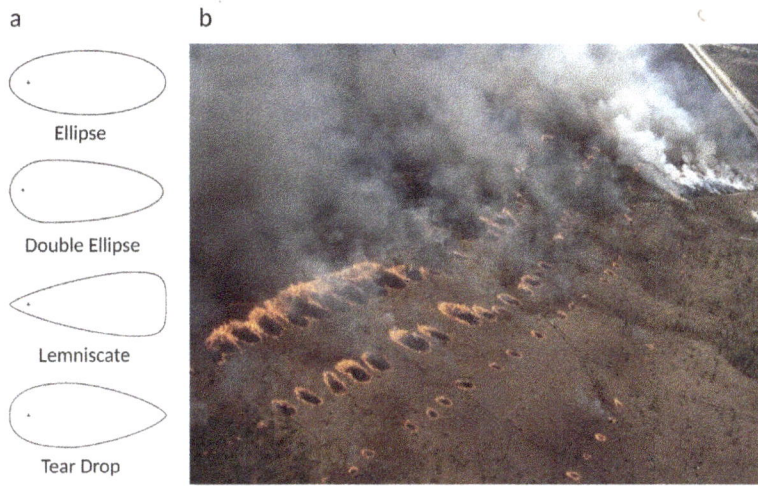

Figure 8.8: Shapes of fires (a) commonly assumed (adapted with permission from Richards 1995) and (b) shown from prescribed burning operation in Florida, USA.

determined at the local scale. Many of the empirical equations for elliptical shapes are essentially the same function of wind speed (**Figure 8.9**), whether derived from laboratory-scale fires (e.g. Albini 1976; Anderson 1983; **Eqn [8.2]**) or field-scale fires (Alexander 1985, e.g. **Eqn [8.3]**). The exception to this generalisation is that the initial growth of a point-source fire may require some time for acceleration before it acquires its ultimate shape.

Other evidence supporting local-scale control of fire shape comes from long experience and success with fire growth modelling based on application of Huygens' principle and Fermat's principle of 'least time'. These principles explicitly require that fire behaviour on individual segments of a fire front be completely independent of fire behaviour on other segments (Anderson *et al.* 1982; Richards 1995). Both principles were originally proposed to explain the travel of light waves. Huygens' principle assumes that any point on the edge of a wave can generate a wave of the same shape. When applied to fire spread, the concept means that small point-source fires ignited along all edges of a fire perimeter will expand as independent ellipses, and together they will define the fire front after a finite time (Anderson *et al.* 1982). For Huygens' principle, then, fire spread distance is calculated as a function of time. Fermat's principle states that a wave must move along the path with the shortest travel time (which is not necessarily the shortest distance). When this principle is applied to modelling of fire growth, fire spread times are calculated throughout a network of points at known locations (e.g. throughout a grid), and the results are then searched for the fastest route, which is then assumed correct. For Fermat's principle, travel time is calculated as a function of distance. Both Huygens' and Fermat's principles produce nearly identical fire growth in computer algorithms, except for numerical artefacts such as spatial resolution (Finney 2002).

The commonness of elliptical shapes for fires of all sizes and fuel types (see **Figures 2.10, 2.13**) means that the ellipse has great practical value for modelling fire growth and behaviour. The shape of the ellipse is typically described by the ratio of length l_y to width l_x (**Figures 2.14, 8.9**), which indicates its *eccentricity*. Larger l_y/l_x ratios indicate greater eccentricity. Empirical data collected in Canada (Simard and Young 1978; Alexander 1985; Forestry Canada Fire Danger Group 1992), Australia (Cheney 1981) and the United States (Anderson 1983) have all shown that l_y/l_x increases with wind speed, but there is no agreement on the exact form of the relationship. The variation in results is illustrated by the following equations, where U_{mf} is the mid-flame wind speed and U_{10} is the open wind speed at 10 m above the vegetation (km h^{-1}) (**Figure 8.9**):

Simard and Young (1978): $l_y/l_x = e^{0.0162 U_{10}^{1.2}}$ [8.1]

Anderson (1983): $l_y/l_x = 0.936 e^{0.071688 U_{mf}} + 0.461 e^{-0.04325 U_{mf}}$ [8.2]

Alexander (1985): $l_y/l_x = 1.0 + 0.00120 U_{10}^{2.154}$ [8.3]

Forestry Canada Fire Danger Group (1992): $l_y/l_x = 1.0 + 8.729(1 - e^{-0.03 U_{10}})^{2.155}$ [8.4]

Cheney (1981): $l_y/l_x = 1.1 U_{10}^{0.464}$ [8.5]

To graph and compare these functions, we assumed that the mid-flame wind speed is half the 10 m wind speed ($U_{mf} = 0.5 U_{10}$). (Anderson's **Eqn [8.2]** yields $l_y/l_x = 1.397$ at $U_{mf} = 0.0$ compared to $l_y/l_x = 1.0$ for all others, because it was developed to describe a double ellipse with different rearward and forward dimensions.)

Eqns [8.1], [8.2] and **[8.3]** increase at an increasing rate with faster winds (**Figure 8.9**). This may reflect an increasing contribution of spotting with higher winds, which would advance ember transport ahead of the main front and cause the heading fire to spread faster than flanking portions of the fire. As discussed in **Chapter 2**, in **Fire area and perimeter**, increasing eccentricity (i.e. increasing l_y/l_x ratio) with wind also occurs if flames are extinguished along the flanks of the fire because of insufficient lateral heat transfer in discontinuous fuels. **Eqns [8.4]** from the Forestry Canada Fire Danger Group (1992) and **[8.5]** from Australian grass fires (Cheney 1981) show a levelling of eccentricity at higher wind speeds. It is not clear, however, whether differences in functional form among these equations represent physical differences in fire growth, ambiguity of the data, or judgement exercised in statistical fitting.

Wind and slope commonly interact in wildland fire spread. These interactions have been combined mathematically (Weise and Biging 1997; Nelson 2002; Viegas 2004; Sharples 2008), but no definitive physical explanation of the physical processes have been provided. **Eqns [8.1]–[8.5]** are intended for application to wind-driven fires, but topographic slope seems to produce similar elliptical fire growth, at least at low inclinations (Viegas 2004; Dupuy *et al.* 2011; Silvani *et al.* 2012). Just as with spread rate, fire shapes on sloping topography must be examined parallel to the slope, even though their effects on fire growth are often displayed on horizontal projections. Laboratory experiments with point ignitions show that, where flames attach on slopes above ~20°, the fires develop strongly pointed shapes (reviewed in **Chapter 6, Topography**). Computer simulations of line ignitions also develop acutely pointed fronts on steep slopes (Linn *et al.* 2007, 2010). This can be explained by air-flow patterns because, when the flame attaches to the uphill surface on a slope greater than 20°, the fire induces strong air flow up through the middle of the burned area. To develop a more complete physical explanation, we will need to better understand relationships of the flame and plume angle with flame zone dimensions, slope angle and energy release as they interact to produce spread on steep slopes.

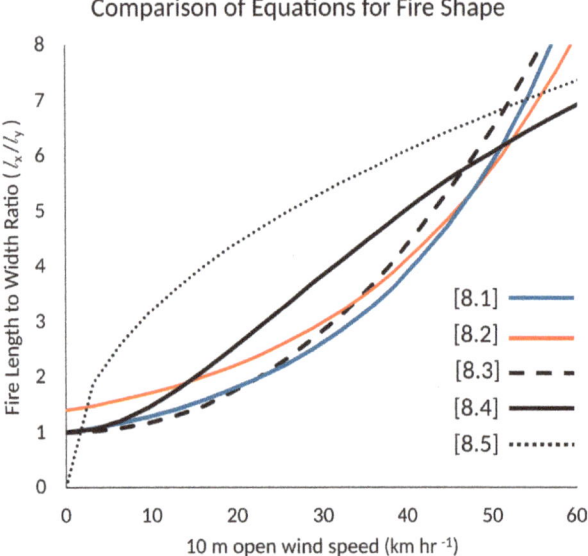

Figure 8.9: Comparison of empirical functions for length/width ratios (l_y/l_x) of elliptical fires as a function of wind speed. See text for equations and sources.

Available evidence suggests that elliptical fire shapes are independent of fuel type and fire behaviour itself (Alexander 1985), although modelling of fire shape requires that wind speed be adjusted to account for the different vertical structures, roughness and shading/sheltering that characterise different wildland fuels (e.g. grass, brush, forests and slash) (see **Chapter 6, Winds**) (Anderson 1983). If fire shape is indeed independent of fuel type and fire behaviour, then the relative fire spread rates and intensity distributions around a two-dimensional fire front must depend only upon wind speed and slope across the whole fire. However, this seems at odds with the known effects of wind on flames at local scales, where flame tilt and drag depend upon local energy release rates and flame zone geometry (**Chapter 6, Wind effect on flame tilt**). If local scales dominate, then each segment of the flame zone is influenced by wind and fire characteristics at that scale, and energy release from the local flame zone dictates the tilt and deflection of flames and heat transfer normal to the fire edge. We do not currently have enough knowledge to distinguish how much local flame and heat release characteristics influence fire growth and overall fire shape.

Another unknown concerning fire growth and shape is the location of the ignition point within an elliptical fire (**Figure 2.14**). Modelling can assume either a constant rate of spread in backing fires (see **Chapter 6, Backing fires and flanking**) (meaning that the ignition point is always a fixed distance from the rear edge of the fire) or a variable rate, with the backing spread at a rate determined by the rear focus of the ellipse as the ignition location (**Chapter 2, Eqn [2.4]**). The second assumption seems consistent with the minor changes in the observed rates with which fires back into the wind or downslope. However, no physical rationale has been proposed to support either assumption. Little area is usually burned by backing fires and rear-facing flanks, so the consequences on fire shape are minor relative to those of spread in more forward directions.

Although the elliptical shape assumes that the environment is constant in time, it is doubtful that this is ever true in real fires. Richards (1994) analysed the effects of varying wind direction on modelled fire growth patterns. He found, not surprisingly, that the l_y/l_x ratio of the ellipse is strongly affected by directional variation in wind. This confirmed observations by Simard and Young (1978) that fires are 'fatter' (i.e. less eccentric) for a given wind speed when the wind direction oscillates randomly over a range of directions. While the fire shapes were affected by directional variation in wind, the area burned was not. In other words, the *size* of the burned area was only affected by the variation in wind *speed*, not its *direction*. Furthermore, Richards found that none of the variation in wind speed or direction acting on an otherwise elliptical fire could produce fire shapes with only one axis of symmetry (e.g. fan-shaped, pointed). Depending on the amplitude of wind variation, simulated fire shapes deviated considerably from the perfect ellipse, but it is an important conclusion from Richards' analysis that if fire is capable of spreading in all directions (i.e. will not extinguish in the backing direction or along rear flanks), then time-varying wind speed and/or direction will not make the fire shape asymmetric.

Elliptically shaped and symmetrical fires will only be produced when fuels are continuous and uniform. Horizontally discontinuous fuels may introduce thresholds for fire spread. In these cases, spread is sustainable only where fireline intensity surpasses a critical level; it will abruptly start or stop at various points along the perimeter, wherever intensity declines or increases radially (i.e. crosses the critical level) as the flaming zone moves away from the heading portion of the ellipse (**Figure 2.17**). Fire shapes on landscapes where the fuels are patchy or clumpy, such as the *Spinifex* vegetation of interior Australia (**Figure 8.10**), exhibit jagged points and angular contours where fuel discontinuity limits fire growth primarily to forward directions (Burrows *et al.* 2009; Bird *et al.* 2012). Fire shape is also likely to be affected when the fire front moves from one fuel type or stratum into another (e.g. grassy fuels to shrubs, surface fuels to canopy fuels), thus dramatically changing the

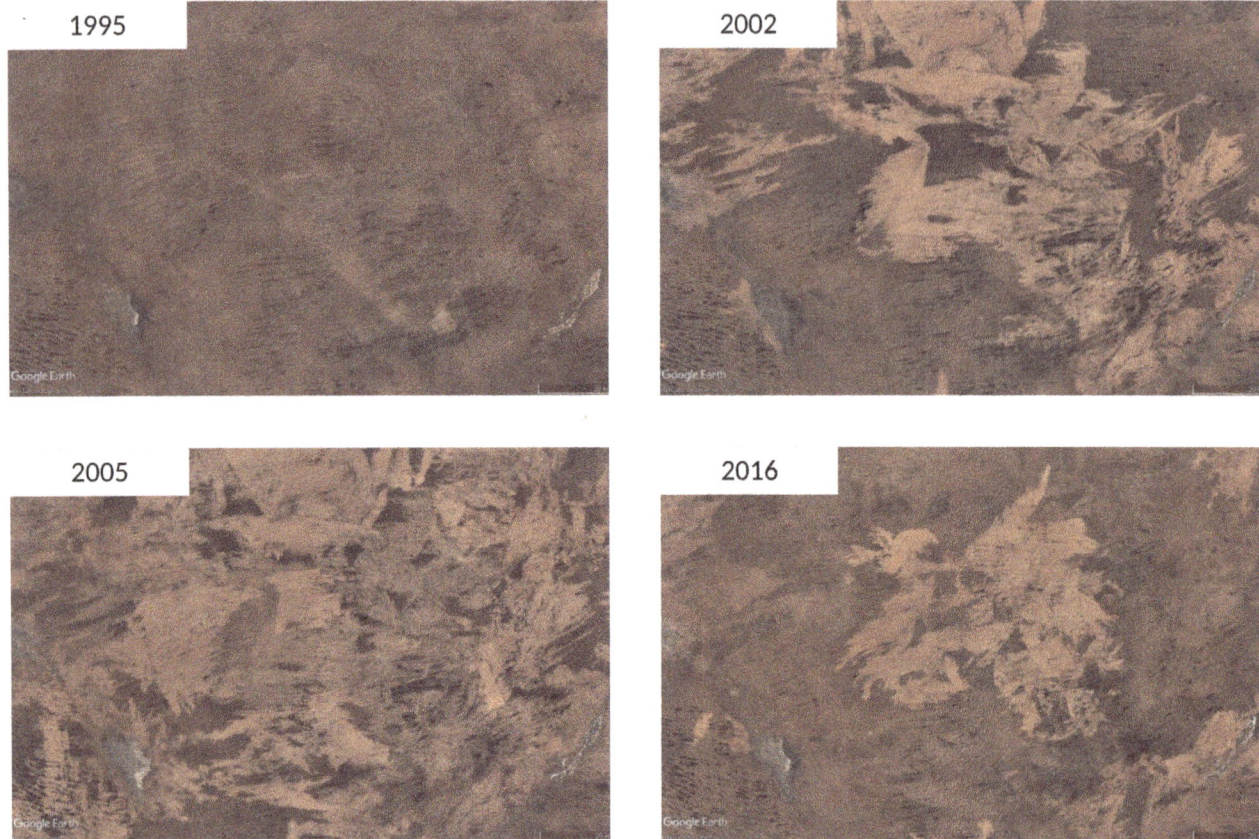

Figure 8.10: Image sequence of Western Australian fires showing repeated patterns of angular fire shapes in discontinuous spinifex shrubland fuel type (images from Google Earth).

spread rate. A fire growth model by Finney (1998) that is based on the assumed elliptical shape can incorporate threshold conditions around the fire edge, including transitions to crown fire, variation in wind and slope direction, and spotting from the heading portion of the fire front (**Figure 8.11**). By defining thresholds in spread or spread rate as functions of fireline intensity, this model produces simulated fires with fan-shaped perimeters and with very pointed shapes, which vary with wind speed, slope and crown properties.

An important characteristic of fire growth patterns is the distribution of fireline intensities within the footprint of the fire. Under uniform and constant conditions, the fire exhibits a radial pattern of intensity and spread rate (**Figure 2.17**). For perfect ellipses, this radial pattern can be calculated as described in **Chapter 2, Eqns [2.14]–[2.18]** (Catchpole *et al*. 1982; Catchpole *et al*. 1992). With fires burning in time-varying winds and angled to the slope, the pattern of intensities varies, depending on the spread direction at the time a given point on the fire front is burned. **Figure 8.12** shows fire growth and intensity patterns simulated with a fire growth model (Finney 1998) that has time-varying winds. We see that the lower and higher intensities shift within the burned area, depending on the wind direction. Complex fire intensity patterns are produced, some of which have stimulated hypothetical explanations, as we will see in the next section.

Burn streets

Sometimes wildfires create burn patterns that contain 'streets' of unburned or unconsumed

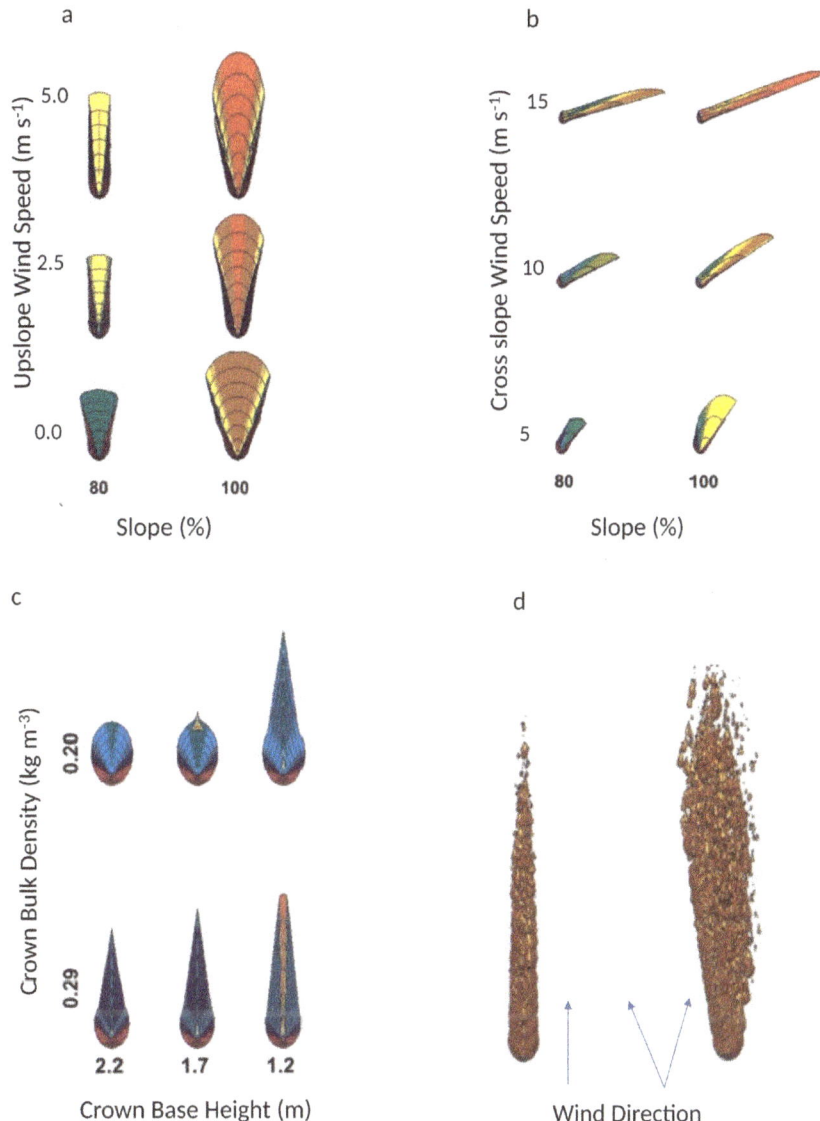

Figure 8.11: Shapes and intensity patterns of simulated fires on topography, where thresholds in spread occur on (a) steep south-facing slopes of varying steepness with upslope wind (0, 2.5 and 5 m s^{-1}); (b) west winds of 5, 10, and 15 m s^{-1} blowing across south-facing slopes of varying steepness; (c) crown fire transition at the head of fires in canopies with different characteristics; and (d) with spotting and constant (left) and varying (right) wind direction (from Finney 1998).

vegetation, which are outlined by abrupt boundaries with areas of near-complete fuel consumption (**Figure 8.13**). In forests, the strips of residual vegetation have been referred to as *tree crown streets* or *tree streets* (Haines 1982). Streets can also occur as alternating strips of black char and residual ash in grass fires (Cheney and Sullivan 2008). The alternating patterns of burned and unburned vegetation in 'streets' may offer some insights into the physical processes that take place on some large fires.

The characteristics of burn streets vary in terms of shape and completeness (Haines 1982). Variations include (1) complete concentric streets of unburned vegetation that outline the entire fire edge (**Figure 8.14a**), (2) curved streets that partially overlap along the forward portion of the fire and alternate right to left (**Figure 8.14b**), or (3) linear streets

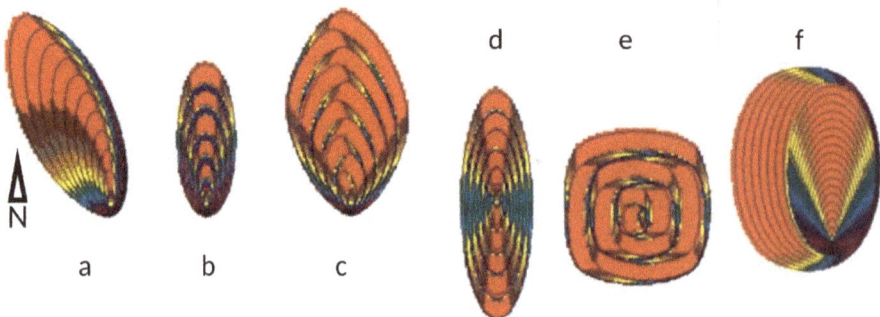

Figure 8.12: Patterns of fireline intensity (from highest (red) to lowest (purple)) within simulated surface fire growth patterns on flat topography produced by wind variations affecting the direction of fire growth: (a) constant wind speed with gradual shift to easterly winds; (b) south wind with alternating speeds; (c) constant speed with alternating directions (40° from south) at a constant interval; (d) alternating directions (180°); (e) anticlockwise rotating wind direction at constant time interval; and (f) instantaneous wind shift from south to east (from Finney 1998).

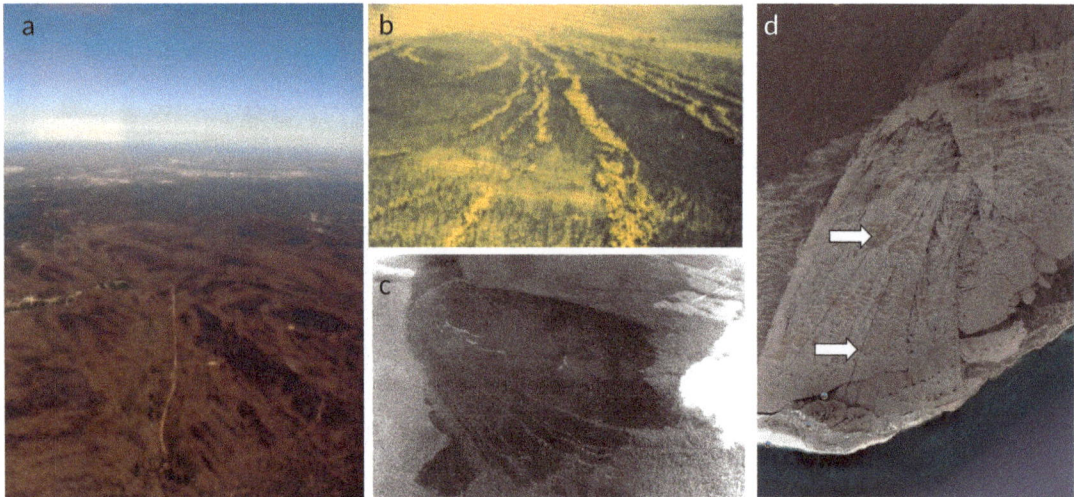

Figure 8.13: Photographs of tree streets that indicate oscillating fire behaviour in (a) Western Australia (from Cheney 2010); (b) New Jersey pine barrens, USA; (c) Exotic Dancer Fire (from Wade and Ward 1973); and (d) long straight streets along flanks of the Cocklebiddy Fire 2018 spreading towards the ocean, Western Australia (image from Google Earth).

that mark the boundary of a flank fire that occurred within a larger fire (not shown). **Figure 8.14a, b** shows that curved streets tend to be wider towards the head of the fire and narrower towards the rear, and that the residual bands of fuel that comprise streets tend to be narrower than the bands in which high fuel consumption occurred. Individual tree stems along the edges of residual forest bands show higher charring on the sides that face the outside of the unburned band (**Figure 8.14c**). Smaller fires often exhibit curved streets, while larger fires seem to contain more straight ones.

Two possible explanations for vegetation streets were offered by Wade and Ward (1973). Their hypothesis of oscillating wind direction could explain the pattern of partially overlapping streets seen in burned forests ((2) above), the shapes of residual bands, and the complex char patterns on tree boles. Consistent with this hypothesis, simulations of the growth of fires with elliptical fire fronts in oscillating wind conditions readily produce alternating bands of intensity. Simulations with oscillating winds also show the tapering observed at the ends of residual bands (Finney

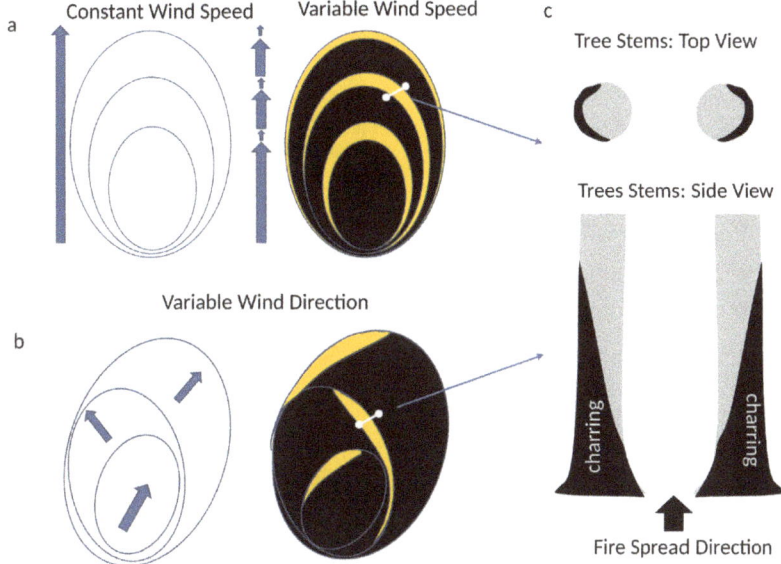

Figure 8.14: Illustrations of tree street formation resulting from variation in (a) wind speed and (b) in wind direction. Wind oscillations alter intensity and relative flame spread directions. Slow spread episodes correspond to reduced consumption of forest canopy (yellow regions). (c) Complex char patterns produced on tree stems along the outsides of the residual bands (adapted with permission from Haines 1982).

1998; Richards and Walberg 1998). In these simulations, a shift in wind speed or direction can reduce the intensity of the head fire to that of a flanking fire, only to resume heading behaviour and higher intensities after the wind shifts back. Fendell (1986) arrived at this same conclusion. Cheney and Sullivan (2008) offered a similar explanation for patterns of fuel consumption and residual ash in Australian grass fires. Oscillations of wind speed and direction could originate not only from ambient winds but also from interactions of the plume with cross winds. As discussed in **Vorticity** below, plumes can obstruct wind flows and cause vortices to 'shed' on the lee side of the plume, but this has not been directly measured in relation to the streets.

We know that heading fires form vortices of flame on the lee sides of tree boles (**Figure 8.3**). Fires often leave evidence of these vortices in the form of high charring on the lee sides of trees (i.e. the sides facing in the direction of heading fire spread). In oscillating wind conditions, however, the 'lee side' of a tree is not constant. Higher lee side charring will occur in the direction the wind is blowing towards, leaving higher charring on the side of trees facing the burned band (**Figure 8.14c**).

The case of complete concentric bands described by Haines (1982) (**Figure 8.14a**) could be simply explained by alternating wind speed surges and lulls in a single direction (**Figure 8.14a**). Straight-line streets are often seen along the flanks of larger fires. These can be caused by temporary wind lulls or shifts in wind direction, as was observed on the Hayman Fire in Colorado (Finney *et al.* 2003) coincident with switching wind direction recorded at a nearby weather station.

Wade and Ward (1973) advanced a second hypothesis to explain burn streets in the forests of the Air Force Bomb Range Fire, which occurred in 1971 in North Carolina, USA. In this case, they suggested that the alternating bands of burned and unburned forest resulted from a repeating cycle of rapid wind-driven spread, spotting and mass ignition, followed by buoyant plume development that countered the ambient wind and slowed forward spread until fuels under the plume burned out. When this occurs, however, it would seem likely to leave ragged edges along bands of residual vegetation rather than the smooth contours seen in aerial photographs (similar to **Figure 8.13**) (and it could not explain the burn streets that occur in fuels with minimal spotting potential, such as grass fires or fires in which sawgrass (*Cladium* spp.) burns over water, where spotting does not occur at all (see description by Haines 1982).

Haines (1982) offered a third explanation for the formation of burn streets, and considerable research has focused on this possibility: that *horizontal roll vortices* paralleling the contours of the outer fire perimeter intermittently impinge on the ground surface and vegetation, forcing flame spread outward from the point of downward air impact and perpendicular to all points along the fire front. Laboratory and field investigations by Haines and others revealed various configurations of plume vorticity but little evidence connecting these circulations to the documented characteristics of residual vegetation bands. While longitudinal vorticity is certainly observed locally in flame zones, as discussed in **Chapter 6** (**Figures 6.20, 6.22**), coherent horizontal vorticity has not been observed at the scale of an entire fire edge. At this point, the simplest and most comprehensive explanation for burn streets seems to be oscillation in wind speed and/or direction (Albini 1984; Fendell 1986), with possible feedback through interaction of the ambient wind with the fire plume.

Plumes and pyroconvective atmospheric storms

Pyroconvective storms are probably the most common behaviour witnessed on the largest wildland fires. They result from vertical development of the convection column into the upper troposphere, 10 km or more above the earth (**Figure 8.15**). These wildfire-induced atmospheric events occur on every vegetated continent. They are associated with the most destructive, deadly fires in recent decades in Australia, Portugal, Canada, Chile and the United States. Pyroconvective storms are as much an atmospheric phenomenon as a wildfire

Figure 8.15: Pyroconvective plumes now classified as a flammagenitus and more commonly called pyrocumulus (PyroCu) or pyrocumulonimbus (PyroCb). Examples (a) from the historic Tillamook Burn (1933) in Oregon, USA, and the (b) Carr Fire in California, USA (2018) (photograph (b) by Jim Mackensen, USDA Forest Service).

process. Several terms are used to describe cloud formation in these storms. The term *flammagenitus* describes a cloud generated by the buoyancy of a surface heat source, such as a wildfire. Small flammageniti are referred to as *pyrocumulus* clouds; these are puffy white clouds that form when the rising air in a fire plume reaches an altitude where the water vapour condenses to liquid. The largest such cloud formations are referred to as *pyrocumulonimbus* or *cumulonimbus flammageniti*. Pyrocumulonimbus clouds may reach 10 000–20 000 m in height as they generate pyroconvective storms. These are essentially fire-created thunderstorms, which exhibit many features familiar in thunderstorms, including an anvil top, strong surface outflow winds, downbursts, rain, hail and lightning – and they can accelerate fire growth in multiple directions simultaneously. Intense pyrocumulonimbus events are important safety issues for firefighters, since they produce hazardous conditions including microbursts, and erratic winds, lightning and possible new fires started by those lightning strikes.

Pyroconvective storms are generated by the buoyancy of hot gases and flames rising from a fire. The rising hot, low-density gas is replaced by air flowing in from nearby. Even the updrafts from small candle flames in **Figure 2.3** and burner flames in **Figure 4.9** induce an inflow of air that feeds fresh oxygen to the combustion reaction and carry the combustion products upward. The plume above a candle flame also affects the air flow through a process called *turbulent entrainment*, the process of mixing cool and slow-moving air into the folds of turbulent eddies in the rising plume. While flames typically have laminar flow at the base, most become increasingly turbulent as they reach lengths beyond ~0.2 m. Entrainment increases with turbulence, so the amount of air entrained by a wildland fire is much greater above the fire than in the laminar regions at the base of the flame. You can see the increased turbulence above a heat source in the shadowgraph image of a candle flame in **Chapter 2** (**Figure 2.4**) and the smoke plume above a smouldering incense stick in **Chapter 3** (**Figure 3.6**).

As fires become larger and release more energy, the plume above them rises higher and becomes involved with a greater volume of the earth's lower atmosphere – the troposphere. This means that the ascending plume interacts with many of the properties and structure of the atmosphere, including the wind and temperature profiles and the moisture already in the air. To understand the behaviour of truly large fires, we must understand how they modify the atmosphere and are in turn modified by the atmosphere.

Large columns and plumes above wildfires look like clouds, and like clouds they are comprised mainly of water (**Figure 8.16**). The moisture comes from several sources: the moisture that was present in the fuel before it burned, the moisture that formed as a product of the combustion reactions, and the moisture already present in the ambient air. As the mixture of warm air and moisture within a fire plume ascends into the lower pressures of higher altitudes, the plume volume expands and consequently cools. Water vapour that cools below the dewpoint changes phase to become liquid droplets. Liquid water at higher altitudes may also freeze, and both liquid and frozen water become visible as white clouds.

The term *latent heat* describes the heat released into or absorbed from the environment when a substance changes phase – from gas to liquid and from liquid to solid. When water undergoes phase changes in the plume from a fire – when it condenses from gas to liquid and from liquid to ice crystals – it releases a great deal of energy. This energy enhances the rise of wildfire plumes, just as it does in standard atmospheric storms (**Figure 8.16**). As moisture ascends in a fire plume,

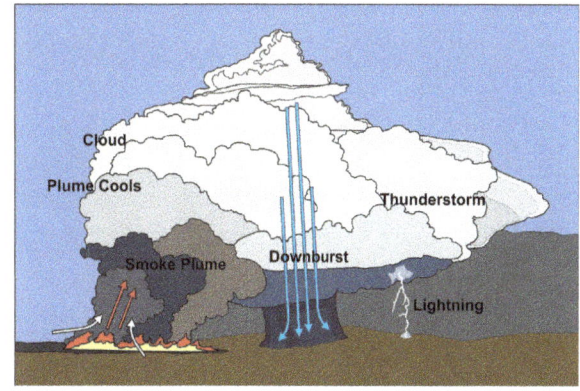

Figure 8.16: Structure of a pyroconvective storm involving pyrocumulonimbus cloud formation, lightning and precipitation (illustration by Brian Elling).

the latent heat from water condensing and freezing adds heat to the atmosphere and thus increases the plume temperature, which enhances its buoyancy. This positive feedback (moisture cooling as it rises and releasing heat with each phase change) causes the air in the plume to continue to rise, which condenses and freezes more water vapour, which releases more heat into the air and further increases the height of the convection column.

Pyroconvective plumes differ in some important ways from 'common' atmospheric storms. The differences are principally caused by the abundance of fine particulates (such as soot) from the fire. These fine particles facilitate condensation of water vapour, so the plumes tend to produce more small droplets and fewer large ones than common thunderstorms produce. Because small droplets weigh less than large ones, drag from the rising air can pull the droplets higher than it could pull large drops, thus limiting precipitation. Since precipitation involves the melting of ice and eventual evaporation of water (phase changes that absorb heat from the atmosphere), less precipitation implies fewer downdrafts of cool air impacting on the ground. Thus, pyrocumulus and pyrocumulonimbus clouds have been suggested to be less likely to produce rain than common thunderstorms (Rosenfeld *et al.* 2007). The abundance of small particles also affects the electric charge of lightning produced from pyrocumulonimbus clouds, making it more likely to be positively charged than lightning from common thunderstorms (Latham 1991; Rosenfeld *et al.* 2007). Although both negatively and positively charged cloud-to-ground lightning can ignite vegetation, a greater proportion of positively charged strikes cause fires (Latham and Williams 2001). Lightning from pyrocumulus and pyrocumulonimbus clouds has been recorded as starting fires 30 km from the main fire.

Not all fires develop tall plumes or pyrocumulus clouds, because atmospheric conditions – the vertical profile of wind speed, moisture, and temperature in the atmosphere – vary among fires. Atmospheric science is well developed on these subjects, so the reader is referred to standard texts and to recent papers by Potter (2012a, b) for details on the subject. The concept that best accounts for differences among fires in this regard is *atmospheric stability*, which is governed by the temperature change with height in the atmosphere for a given amount of humidity. If we move a parcel of air upward in the atmosphere, we expect it to expand because it is gradually escaping the earth's gravity and surrounded by less air pressure at higher altitudes. It is thus becoming less dense, which also means its temperature is declining. Ideally, the parcel's temperature declines at a rate of 9.8 °C per vertical kilometre. As our parcel rises, however, it must interact with the air in the atmosphere, which may or may not have the ideal cooling rate of 9.8 °C per vertical kilometre. If the temperature in the atmosphere is decreasing at the ideal rate, conditions are considered *neutrally stable*; if we move our parcel of air vertically in these conditions, it will experience no buoyant force and will remain in the new position. If the atmospheric temperature is decreasing vertically at a rate *less than* 9.8 °C per kilometre, conditions are is considered *stable*; if we move a parcel of air vertically in these conditions, it will experience a buoyant force pushing it downward, because in its new position it was cooler than the surrounding air. Conversely, if the atmospheric temperature decreases vertically at a rate *greater than* 9.8 °C per kilometre, conditions are considered *unstable*, and if we move a parcel of air vertically, it will cool more slowly than the surrounding atmosphere and therefore continue to move upward. Unstable atmospheric conditions in the environment of a wildland fire will enhance its plume development.

Some of the largest pyrocumulonimbus clouds reach heights that inject smoke into the upper troposphere and even into the lower stratosphere (Fromm *et al.* 2005). Aerosols and particulates injected at these high altitudes can persist in the upper atmosphere for a long time and impact global climate at a hemispheric scale. Peterson *et al.* (2017) have developed a conceptual model that describes the important processes that lead to development of pyrocumulonimbus clouds and pyroconvective storms:

- First, near-surface atmospheric conditions must be dry, hot and somewhat windy, so intense burning – and hence plume buoyancy – is maintained. Fuel loads, fire dynamics and terrain factors may also be important in this step.
- Next, the lower tropospheric layer is usually deep (2000–4000 m), dry and unstable, with minimal wind shear; all of these conditions enhance vertical plume growth to the height where condensation will occur. (Wind shear describes conditions in which air layers are moving in different speeds or directions. Strong wind shear will divert the fire plume sideways and limit vertical development.)
- Finally, in or above the height at which condensation occurs, the atmosphere must have a layer of significant moisture. Through the heat transfer processes of condensation and freezing, this moisture adds considerable buoyancy to the plume and enables it to reach much higher altitudes than without the added moisture.

Another factor in developing pyroconvective storms is the wind, or more properly the vertical profile of wind direction and speed. Wind flow patterns can vary tremendously from layer to layer of the atmosphere. In layers where the winds are calm, plume development is likely to be mostly vertical. When wind speeds increase with altitude – a common pattern – they can tilt the plume or shear its trajectory and thus discourage further vertical development. When wind speeds are constant or decreasing with height, they may enhance vertical development of the plume and increase the likelihood of 'blow-up' behaviour – that is, rapid increases in spread and intensity, which are often associated with strong surface winds (Byram 1954; Potter 2012a). (Whenever the plume takes on a more vertical trajectory, pyroconvective development increases, inducing stronger surface winds because of inflow due to increased combustion rates.) In conditions where wind speeds increase with height, the fire will be 'wind-driven' under a tilted plume, so vertical development may be limited. When the wind profile contains reversals – that is, the wind is moving in different directions or at different speeds in different layers – vertical development of the plume may increase dramatically as soon as it reaches a layer with slower winds, and this may cause rapid, unexpected increases in fire spread and intensity on the ground. While these explanations all seem logical, it is difficult to find clear evidence of when and how they operate, given that many other mesoscale weather variables can cause rapid changes in fire behaviour.

The horizontal forces of wind are constantly interacting with the vertical forces of buoyancy in wildland fires. A relatively simply way to address these interactions in large fires was proposed by Byram (1959) as the ratio of the power of the fire to the power of the wind (P_f/P_w). This ratio allowed comparison of 'the rate of energy flow in the wind field with the rate at which thermal energy is converted into kinetic energy in the convection column over a fire' and was later named the *convection number* C_n (Nelson 1993). The power of the fire (P_f) was derived for an arbitrary segment of a linear flame front using the local fireline intensity (I_B), acceleration due to gravity (g, 9.81 m s^{-2}), the specific heat of air c_p (J kg^{-1} K^{-1}), and the air temperature T_∞ (K):

$$P_f = \frac{gI_B}{c_p T_\infty} \quad [8.6]$$

The power of the wind (P_w) with speed U (m s^{-1}) at a specified height above ground is calculated using the air density ρ (kg m^{-3}). It addresses only the wind acting on the fire, so the fire spread rate r (m s^{-1}) must be subtracted:

$$P_w = \frac{1}{2}\rho(U-r)^3 \quad [8.7]$$

Thus, the convection number (P_f/P_w) is:

$$C_n = \frac{P_f}{P_w} = \frac{2gI_B}{\rho c_p T_\infty (U-r)^3} \quad [8.8]$$

Byram suggested using the convection number as the 'energy criterion' (Byram 1959) to distinguish two regimes of fire behaviour. When $C_n < 1$, the fire is said to be 'wind driven'. When $C_n > 1$, the fire is often described as 'plume driven', meaning that it could experience enhanced effects from a large convection column, including downdrafts and strong induced surface air flows, and it could display 'erratic' fire behaviours. 'Plume driven' is an imprecise description of what happens when the power of the fire is considerably greater than the power of the wind, because it does not help us identify the causes of the condition or its consequences on fire behaviour. One cannot tell just from the characteristics of a convection column whether it would meet Byram's C_n classification for one type of fire or another.

Several simplifying assumptions are made in the calculation of the convection number C_n, but it seems robust to most of them. For example, the number was originally developed for neutrally stable atmospheric conditions, but it was found to be largely insensitive to deviations in atmospheric stability (within 20%) for altitudes below ~1 km (Nelson 2003). Below the atmospheric level at which moisture condenses, C_n can account for shear layers in the wind speed profile through their effect on P_w. However, C_n does not apply in the upper atmosphere, where moisture phase changes and their latent heats affect plume rise. In addition, P_f assumes that turbulent entrainment of air does not affect plume buoyancy. Because the concept of C_n represents fire energy release as fireline intensity (line fire configuration) it does not apply to area-wide fires (see **Mass fires** below) or to large-scale vorticity in the plume (see **Vorticity** below), which would substantially alter plume entrainment and thus plume buoyancy.

The long duration of large fires, which can span multiple days or weeks, means that they are subjected to variations in mesoscale weather systems that move in and out of the geographical area of the fire. The discussion of fire meteorology is far more complex than we can summarise here, so the reader is referred to standard texts on meteorology and fire weather in particular (Schroeder and Buck 1970). We focus instead on a few examples in which fire behaviour changed radically in response to mesoscale weather systems.

One of the most important synoptic weather systems affecting behaviour is the *cold front* because it is associated with a sudden change in both wind direction and speed. A cold front is the leading edge of cool, dense air that arrives with a low-pressure region. In the northern hemisphere, the anticlockwise circulation of air flow associated with low atmospheric pressure means that the wind direction changes from west or south-westerly to north-westerly. This can happen quite rapidly, and substantial turbulence is usually associated with the change. When a cold front in the Northern Hemisphere impinges on a wildfire that has little or no containment on its southern flank, the shift to a west-north-westerly wind can transform the south-facing flank fire into a head fire that increases rapidly in spread rate and intensity (**Figure 8.17a**). In the Southern Hemisphere, the clockwise circulation around low pressure systems means that frontal passage usually causes winds to change from west or north-westerly to south-westerly, with the same rapid and dangerous effects on fire spread (**Figure 8.17b**).

Research is beginning to study relationships between large-scale fire behaviours and mesoscale patterns of wind, moisture, and atmospheric stability. Detailed examinations are needed to investigate the numerous interactions among variables at this scale.

Vorticity

Vorticity is a measure of the spinning motion of a fluid. This rotation occurs at all scales on all fires, even in the plume above a candle. Vorticity can be created through sheared flow near the ground or by density and pressure gradients in the air. For example, as wind moving across the ground encounters drag from the vegetation or topography, it is slowed down, but the higher wind velocity remaining above the ground causes the flow to

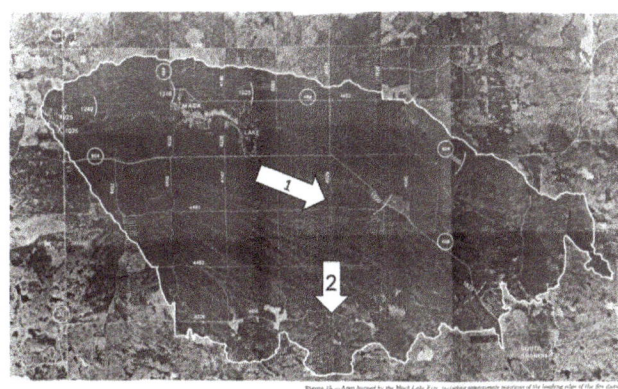

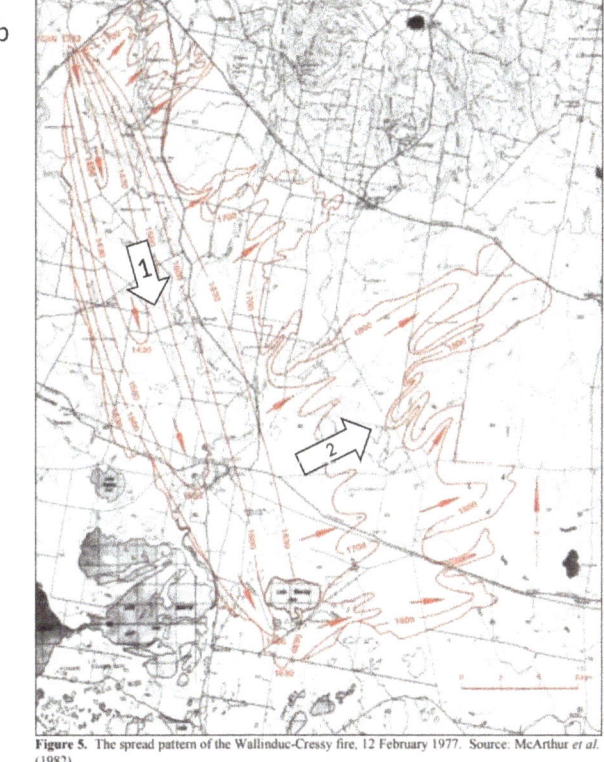

Figure 8.17: Examples of fire growth patterns with wind shifts following passage of cold fronts: (a) Mack Lake Fire, 1980 in Minnesota, USA, showing fire progressing from west to east (arrow 1), then to the south following frontal passage (arrow 2). This was a typical wind shift in the Northern Hemisphere (Simard et al. 1983); and (b) Wallindue-Cressey Fire Australia (1977) showing that fire spread as wind changes from north-north-west (arrow 1) to south-west following a frontal passage (arrow 2) (reproduced with permission from McArthur et al. 1982).

roll downward, thus creating a horizontal circulation – one kind of vorticity. Vorticity can also be concentrated, dissipated, moved around and reoriented. Sometimes vorticity plays a dominant and dangerous role in fire behaviour, such as in a fire whirl. Intense vorticity can cause high winds, strong inflow of air near the ground, and lofting of large firebrands to great heights. In this section, we describe some manifestations of vorticity that concern fire management and safety.

Pulsating or puffing

The subject of puffing fires was introduced in **Chapter 4**, with a discussion of small-scale, laboratory fires. Here we expand the discussion and apply the concept to large fires. We can see evidence of pulsating behaviour in flickering candles, in campfires, and in vertical and horizontal 'stacks' of smoke billows above some large fires (**Figure 8.18a**). This is called *pulsating* or *puffing*. Puffing may have significant effects on wildfire characteristics. Here we discuss how it can periodically strengthen the ground-level wind speed flowing into the fire, generating large forward bursts of flame outward from the edge of the fire (**Figure 8.18b**) and lofting and expelling multitudes of embers.

The subject of pulsating fires has been of interest to those working in combustion science, fire protection engineering and wildland fire for many decades. Byram and Nelson (1970) found that the puffing frequency (f, measured in Hertz (Hz), number of puffs per second) of round or square fires over diameters ranging from 10^{-2} to 100 m is proportional to the negative square root of the fire diameter D (**Figure 8.19a**), which indicate that puffing frequency decreases as fire diameter increases:

$$f \propto d^{-1/2} \qquad [8.9]$$

The principal driver of puffing is buoyancy – that is, the effect of gravity on the density difference between the flame or plume and the surrounding ambient air (see **Chapter 3, Basic**

Figure 8.18: Photographs of (a) periodic pyrocumulus cloud development from a large forest fire in Yellowstone National Park, USA, in 2008 (photograph by Mila Zinkova) and (b) stacked billows in the smoke column from fire in subalpine forests of Idaho (Salt Fire, 2011). In (b) note the outward bursts of flames near the base of the column (from a USDA Forest Service video).

concepts, material properties and terminology). The relationship is largely independent of fuel type because the mean temperature of combustion is essentially the same for all hydrocarbon diffusion flames – around 1000 °C. Since the combustion temperature is the same for all wildland fuels, their flame buoyancy and puffing rate do not vary with fuel type.

Puffing results from a strong coupling between the fire and the surrounding air. It begins when low-density flame gases originate from combustion at ground level. This stratum forms below high-density ambient air and creates an *instability* – often referred to as a Raleigh-Taylor instability because the basic physical processes were described by Lord Raleigh and Sir Geoffrey Taylor early in the 20th century. (See the discussion of stability in the section above.) The stratum of low-density gases must rise, and the high-density air must descend to replace it. As this dynamic begins, the area of hot gases rising from nearly circular or square fires causes replacement air to flow in equally from all sides of the fire. (The ground surface, of course, prevents air from coming in from below.) The symmetry of indrafts around circular or square fires and the upward flow in the middle creates a 'toroidal vortex' – a self-contained internal circulation that is doughnut-shaped, like a smoke ring, with upward flow in the centre and downward flow around the edges (**Figure 8.19b, c**). The gases circulate in toroidal motion because of the shear between updrafts in the middle of the plume and down-flowing or in-flowing ambient air on the edges.

Another factor causing the vorticity is *baroclinic torque*, which occurs when a continuous change in pressure (i.e. a pressure gradient) occurs at an angle to a continuous change in density (i.e. density gradient). In the case of circular and square fires, the density gradient is oriented radially away from the centre of the fire, with low density in the middle increasing to high density on the outside. Perpendicular to the density gradient is the vertical pressure gradient, with higher pressure on the ground that decreases continuously as the plume rises because of gravity. The net effect of these circulations is to draw air rapidly along the ground towards the lower pressure in the centre of the fire. The toroidal vortices form along the outside of the plume near the ground. They are not stationary; instead, influenced by the plume's rising air, they travel upward along the outside of the plume. Once a vortex moves far enough up and away from the ground, a new vortex forms in its place and begins moving upward.

Scientists have tested the idea that buoyancy accounts for puffing frequency in different ways. Cetegen and Ahmed (1993) showed that puffing frequency of both flaming pool fires and helium plumes (no combustion) were related to the

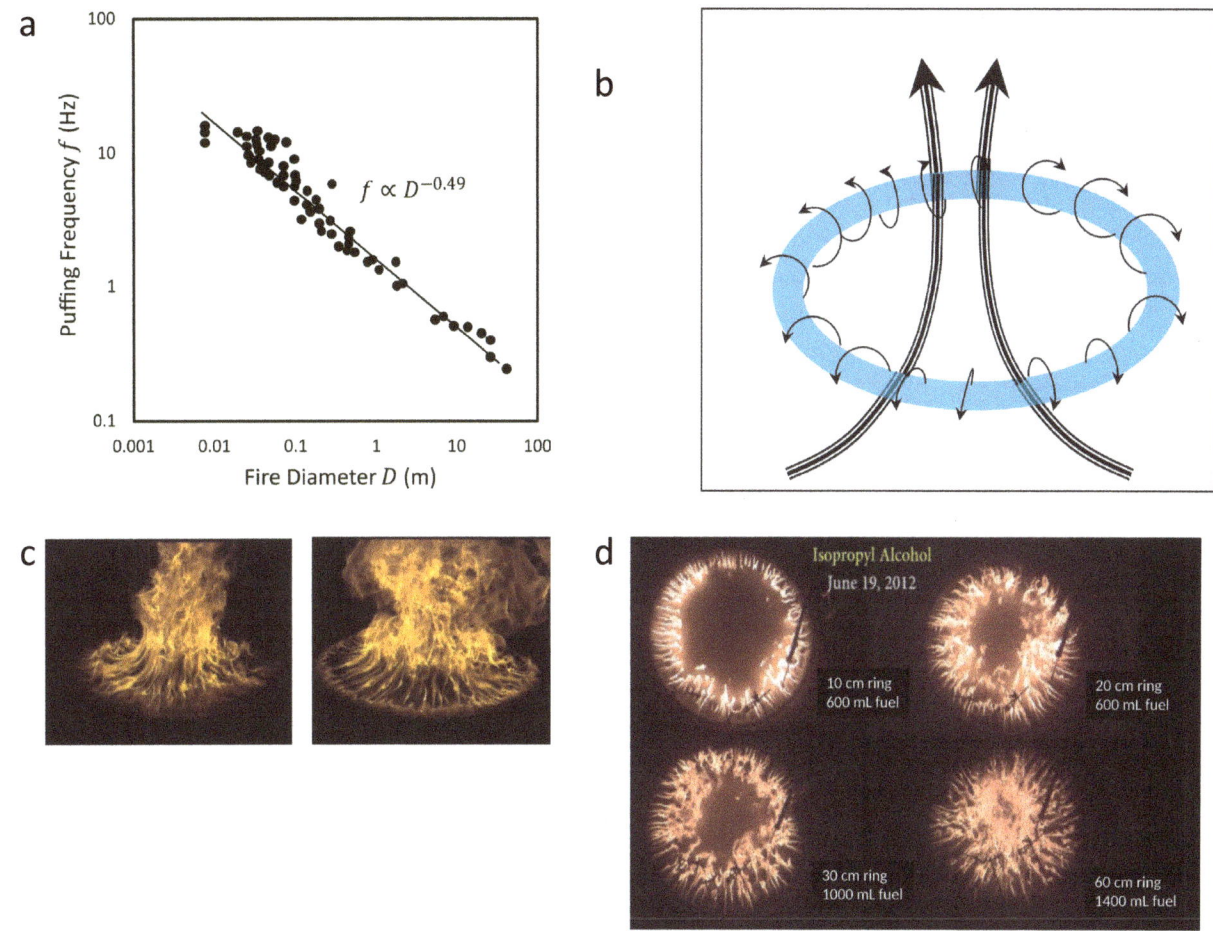

Figure 8.19: (a) Puffing frequency in a pool fire is a function of the diameter of the fire source (reproduced with permission from Hamins *et al.* 1992). (b) Puffing involves the periodic production of toroidal vortices in the fire and plume above (the blue ring with internal circulations). (c) Two stages of plume structure for puffing in a 1 m diameter pool of ethanol (photographs by Ian Grob). (d) Overhead view of experiments (photographs by Ian Grob) with 2 m diameter ring fires showed that puffing behaviour occurred with all thicknesses of outer flaming ring even without combustion in the centre and continued until ring fire became discontinuous (fuel burned out along part of the circumference).

diameter of the buoyant source. Others have found that the puffing frequency could be reduced by diluting the proportion of helium, which increased the density of the air–gas mixture and thus reduced its buoyancy and puffing frequency. Fang *et al.* (2011) compared fires at low atmospheric pressure in the high elevation of Lhasa, Tibet, with fires at sea level. Fires at low atmospheric pressure (high elevations) showed higher puffing frequencies because less soot formed in the flames, which produced higher flame temperatures and thus lower gas densities. These studies indicate that puffing frequency is influenced not only by the size of the heat source but also, at least weakly, by interactions of buoyancy with the density and pressure of ambient air. Regardless of this finding, however, the main factor that controls puffing frequency seems to be fire size.

Puffing fire behaviour can easily be demonstrated with pools of liquid fuel as shown in **Figure 4.17**. Puffing can also be demonstrated with a burning *ring* – that is, a circular-shaped fire with no combustion occurring in the centre – if the ring is continuous (**Figure 8.19d**). A ring shape is relevant to large wildland fires because they rarely release energy at a uniform rate over the entire burning area. As a fire grows larger, the fuels in its centre – especially the fine fuels – burn out relatively quickly, so the centre releases far less energy than the outer flaming edge (see **Figure 2.9**). This creates a ring of fire around a smouldering or burned-out centre. It is not clear how the pulsating behaviour of ring fires is related to the combination of diameter, width of the flaming ring, and energy release rate. However, laboratory experiments indicate that if flaming ceases along any portion of a ring fire, the coordinated puffing of the entire fire ceases. Thus when we observe large, slow pulsation in wildland fires, it probably indicates that a great deal of energy is being released from a large area or at least a ring of unbroken flaming. Such a fire is not behaving like a flame zone of a line fire because the backing, flanking, and heading directions are all releasing a large amount of heat, creating a continuous ring of buoyancy. Therefore, when large fires exhibit puffing behaviour at the scale of the whole fire, fire models that are designed for line-fire situations are unlikely to successfully explain or predict their behaviour.

Fire whirls

Fire whirls are formed when rotating air (*vorticity*) encounters a source of heat at the ground surface which, for our purposes, is the fire itself (Emmons and Ying 1967). This vorticity can originate by any of several possibilities, including ambient wind shear near the ground, wakes behind mountains, weather frontal boundaries, and even fire-induced winds. Formation of a fire whirl requires that the heat source intensify the initial vorticity. This occurs as the natural ascent of buoyant hot gases and flames stretches the circulation around the axis of the vortex upward, producing a rapidly rotating, tube-like structure (**Figure 8.20**). As the buoyant stretching elongates the tube upward, the tube itself narrows in diameter – a positive feedback that forces the whirl to spin faster to conserve its angular momentum (see **Chapter 3, Vortex flows**). Essentially, this process is concentrating a broad area of low vorticity into a much smaller area of strong vorticity.

The high-velocity circulation in a fire whirl generates a *radial force balance* (sometimes called cyclostrophic flow) that reduces energy exchange between the whirl and the surrounding environment. Two forces are involved in the force balance. The first is centrifugal force, which would ordinarily cause the rotating air to be spun outward, just like a person thrown off a merry-go-round. This outward-directed force is balanced by the second force, a pressure gradient that is directed inward. The pressure gradient occurs because (1) the high internal temperatures of the whirl keep the gas densities low, thereby reducing the atmospheric pressure inside the vortex, and (2) the air is moving faster inside the vortex than outside and thus has lower dynamic pressure. The low pressure inside the whirl acts like suction towards its middle, countering the centrifugal force outward from the spinning air. The balance between the pressure gradient pulling inward and centrifugal force pushing outward keeps the flames and air flow in the vortex tube from mixing with the surrounding environment. This limits the loss of both kinetic and thermal energy from the whirl, which increases the flame height in the whirl (by a factor of two or more), since gaseous fuel rises farther up in the whirl before it can mix with enough oxygen to combust completely. Because the thermal plume in a fire whirl is taller and hotter than in a non-rotating fire, the air pressure at the ground is lower, causing stronger entrainment at the surface. Friction with the ground slows the rotation slightly, reducing centrifugal forces near the ground, so the balance between the centrifugal force and the pressure-gradient force is disrupted, causing even greater flow towards the whirl's centre line. Fire whirls create very strong surface winds, which can cause major damage and be very dangerous to people in the vicinity.

Fire whirls provide an illustration of a strongly coupled system caused by fire-induced air flow at the ground surface. Strong winds at the ground

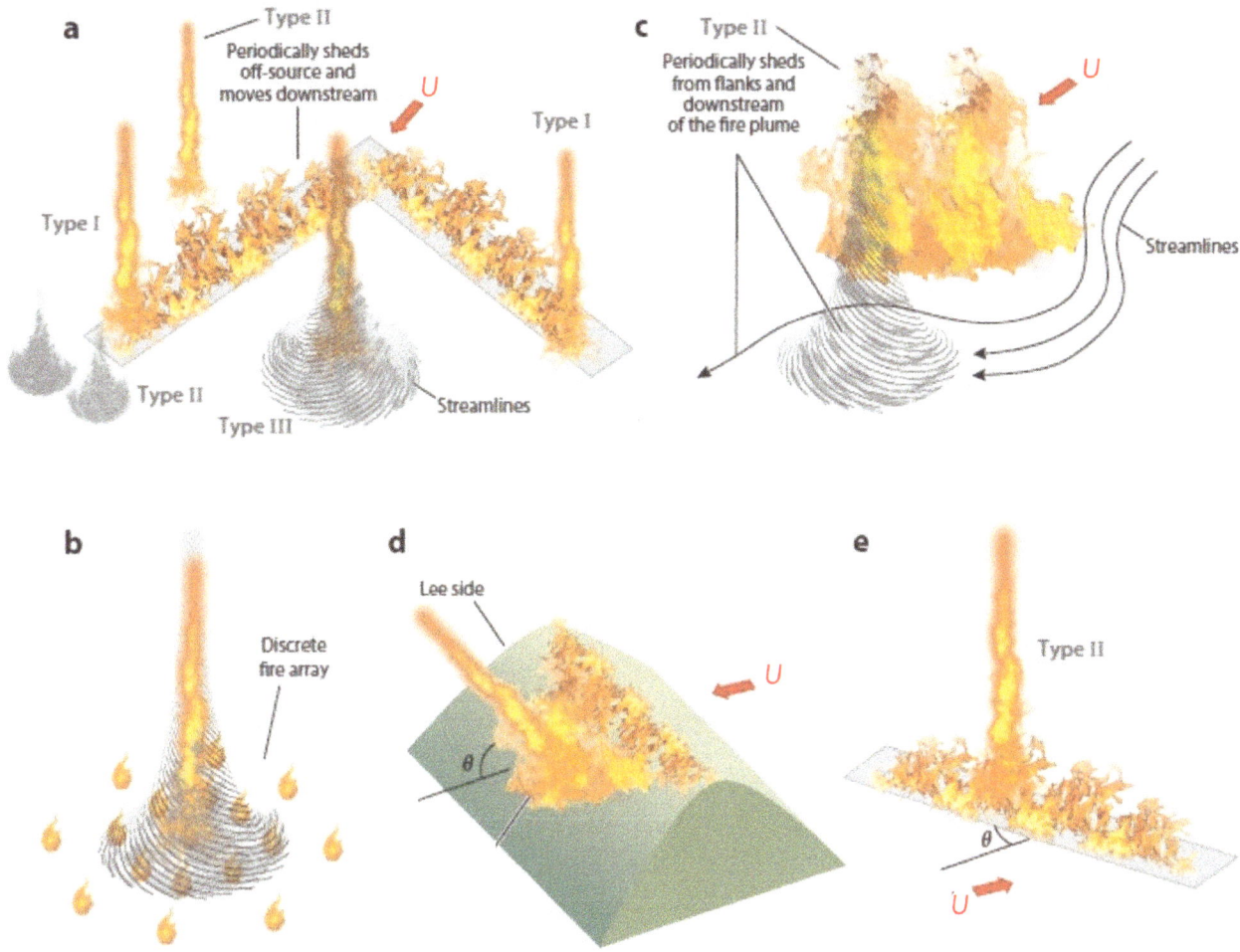

Figure 8.20: Different ways in which fire whirls can form (reproduced with permission from Tohidi *et al.* 2018): (a) an L-shaped concave corner of fire edge; (b) air-flow interactions among multiple point source fires that creates interstitial and fire-scale vorticity; (c) wake vortices shed from lee side of large fire in cross wind; (d) fire on lee side of ridgeline accentuating vorticity that was created when flow separated over crest, leading to fire whirl oriented perpendicular to slope; and (e) periodic fire whirl formation along line fire when oriented diagonal to direction of wind flow. Tohidi *et al.* (2018) distinguish Type I as stationary whirls remaining on the fire source, Type II as whirls periodically developing and moving on and off the fire source, and Type III as whirls developing and staying away from the fire source.

level increase the burning rate (**Chapter 4, Burning and heat release rate**) through three mechanisms:

1. More ventilation increases the surface temperature and thus the rate of solid phase combustion (i.e. burning char).
2. Higher fuel burning temperature and burning rates decrease the density of the gases at heights above the fire and thus lower the atmospheric pressure of the whirl.
3. Strong surface winds can deflect flames laterally, impinging unburned fuels, increasing heat transfer to adjacent fuels and quickly spreading the fire.

Fire whirls can increase the rate of fuel consumption several-fold (Martin *et al.* 1976; Forthofer and Goodrick 2011; Tohidi *et al.* 2018). This demonstrates a positive feedback among the processes that maintain the fire whirl system: air rotation restricts cooling of the plume by entrainment and increases ground-level airspeed, which improves combustion, which increases the heat release rate, which strengthens the low pressure, which increases the rotation, which restricts cooling by entrainment. The processes are interdependent, so interruption of any one of them will weaken or destroy the whirl.

It is difficult to identify the source of rotation that causes a fire whirl. Apparently, many natural configurations can cause sufficient vorticity for fire whirls (Forthofer and Goodrick 2011; Tohidi et al. 2018). One of these is vertical wind shear, which develops when ambient wind flows over the ground surface, producing horizontal vorticity. A tube of horizontal vorticity can be concentrated and reoriented to vertical vorticity by a fire plume, thus forming a fire whirl. Imagine the horizontal vortex tube to be a thick, straight length of rope lying on the ground. If you lift it up in the middle, you elevate a bowed segment of the rope with vertical sections on each side that are still connected to the ground. If a horizontal vortex tube is lifted from the ground by the buoyancy of a fire, its vertical spinning can form vertical fire whirls. When a fire has established a strong fire convection column, ground-generated vorticity often occurs on the lee side of the plume as the air flows around it (see **Figure 8.20c** and **Counter-rotating vortex pairs and wake vortices** below).

Several other situations seem to set up the conditions needed for creating a fire whirl. A wake that forms on the lee side of a hill or mountain can accumulate shear-generated vorticity that develops into fire whirls (**Figure 8.20d**, Countryman 1971). The contours of a fire itself can also cause fire whirls. The 1923 Kanto Fire that followed an earthquake in Japan formed a fire whirl that killed 38 000 people in 15 min. Investigation of this event by Soma and Saito (1988) identified the 'L' (or 'elbow') shape of the fire as a causal factor. As wind impinged on the outside of the elbow, air flows were generated along the edges of the fire and converged on the inside of the elbow, generating a fire whirl (**Figure 8.20a**). Fire-induced winds can cause a fire whirl, even with no ambient wind present, if multiple fires are oriented in a particular way (Zhou and Wu 2007, **Figure 8.20b**). As demonstrated in laboratory tests, the multiple fires compete for air flowing into the flame zones and through the channels between flame zones. This process produces horizontal jet-like structures with substantial shear and vorticity. This situation could develop after extensive spotting has occurred or in the midst of fuel discontinuities. Similarly, single fire perimeters with curved concave or convex shapes could cause competing inflows that result in rotations, which could develop into fire whirls.

In addition to ground shear and local fire-induced winds, large ambient atmospheric processes may cause fire whirls. Atmospheric frontal boundaries, where air masses with different densities collide, often contain significant vorticity, which can feed a fire whirl. Two kinds of frontal boundaries have been suspected of producing fire whirls: cold fronts (Billing and Rawson 1982; Umscheid et al. 2006) and sea-breeze fronts (Seto and Clements 2011). Another atmospheric process that may produce fire whirls is a *hydraulic jump*, which is a breaking wave-like structure that forms when cold, dense air flowing down a slope reaches the bottom. The destructive 2018 Carr Fire whirl may have been caused by an atmospheric hydraulic jump (Forthofer 2019). The resulting vorticity formed a tornado-strength whirl that killed a firefighter by rolling his pickup truck over a distance of 100 m.

Some researchers have speculated that large-fire whirls could be produced by processes similar to those that cause regular tornadoes. There is so little understanding about the formation of fire whirls that there is even significant disagreement as to appropriate use of the terms *fire whirl v. fire tornado*. Some scientists have proposed definitions of fire tornadoes and fire whirls simply based on size and wind speed (Goens 1978) but these have not become standardised. Others define them based on whether they are always attached to the ground (fire whirl) or sometimes leave the ground (fire tornado). McRae *et al.* (2013) claimed that a 2003 Australian event was a true fire tornado because they observed breaks in the swath of wind damage after the fire. While patches and strips of undamaged areas are common in regular tornadoes, it is possible that the undamaged areas in the 2003 event occurred not because the fire whirl lifted off the ground but because it occasionally weakened to a level below the threshold for wind damage, then strengthened again and caused more damage further down its path. Such variable behaviour can be seen in video evidence from the 2008 Indians Fire whirl. The video record shows periodic strengthening and weakening of the fire whirl throughout its ~1 h duration (Moore 2008). Some scientists have speculated that the presence or absence of pyrocumulonimbus clouds (see **Plumes and pyroconvective atmospheric storms** above) might help distinguish a fire tornado from a fire whirl (Lareau *et al.* 2018). As water vapour condenses and freezes high in a pyrocumulonimbus cloud, it increases buoyancy within the plume. This increased buoyancy could enhance the rotation of a fire whirl and stretch it vertically, accentuating the horizontal pressure gradient and wind speeds. At this point, there is no consensus on the difference between a fire whirl and a fire tornado, or even if such a distinction should be made.

Counter-rotating vortex pairs and wake vortices

Flow encountering an obstruction must find a way around it. You have probably watched water in a stream flowing around a rock or other obstacle and generating eddies in its wake (i.e. on the lee side). This can happen in wildland fires as well. The flame and plume are tall columns of fluid that obstruct the flow of wind and generate eddy-like circulations along its sides and in its wake. The science on these vortices in wildland fires is incomplete, but some insight can be gained from knowledge of the somewhat analogous situation known as a *jet in crossflow*. Two general kinds of jets in crossflow are studied in the field of fluid mechanics: momentum-dominated jets and buoyancy-dominated jets. A momentum-dominated jet, like the engine of a jet aircraft, pushes air out at high speed; thus the jet flow has substantial momentum when it is generated at the nozzle. This differs from the kind of jets we consider here, which are buoyancy-dominated jets caused by the movement of low-density fluid (the flame and smoke plume). A buoyancy-dominated jet has little momentum at its origin on the ground. We discuss jets in crossflow here to aid our understanding of counter-rotating vortex pairs and wake vortices, which can significantly affect surface winds and subsequent fire behaviour near large wildland fires.

Round, buoyant and non-buoyant jets of ejecting fluid in crossflow have been well studied for engineering applications (**Figure 8.21**), and observations of some wildfires look remarkably similar to results from experiments and simulations of buoyant jets in crossflow. The experiments in which the jet orifice is flush with the ground or wall produce the behaviours that most closely resemble some of the behaviours observed in wildfires.

Wildfires have several characteristics in common with buoyant jets. For example, their updraft velocity depends upon the fires' energy release rate and its uniformity across the burning zone, and their vorticity is created by baroclinic torque and gravitational effects (discussed in the section above). However, wildfires differ from buoyant jets in other ways: their area is rarely round or symmetric; gas combustion can occurs well above the ground surface, producing flames that add buoyancy with height; and wildland fires are subject to complex and variable wind speed profiles above ground that affect the interactions of flow within the plume with the flow of air around it.

Two vorticity features seen in some large wildfires are remarkably similar to features of jets in crossflow. The first is called a *counter-rotating vortex pair* located within the main plume (**Figure 8.21**). These vortices are represented by time-averaged behaviours (they exist as an average characteristic of the phenomenon) which give the plume a cross-section that resembles a kidney. As the name

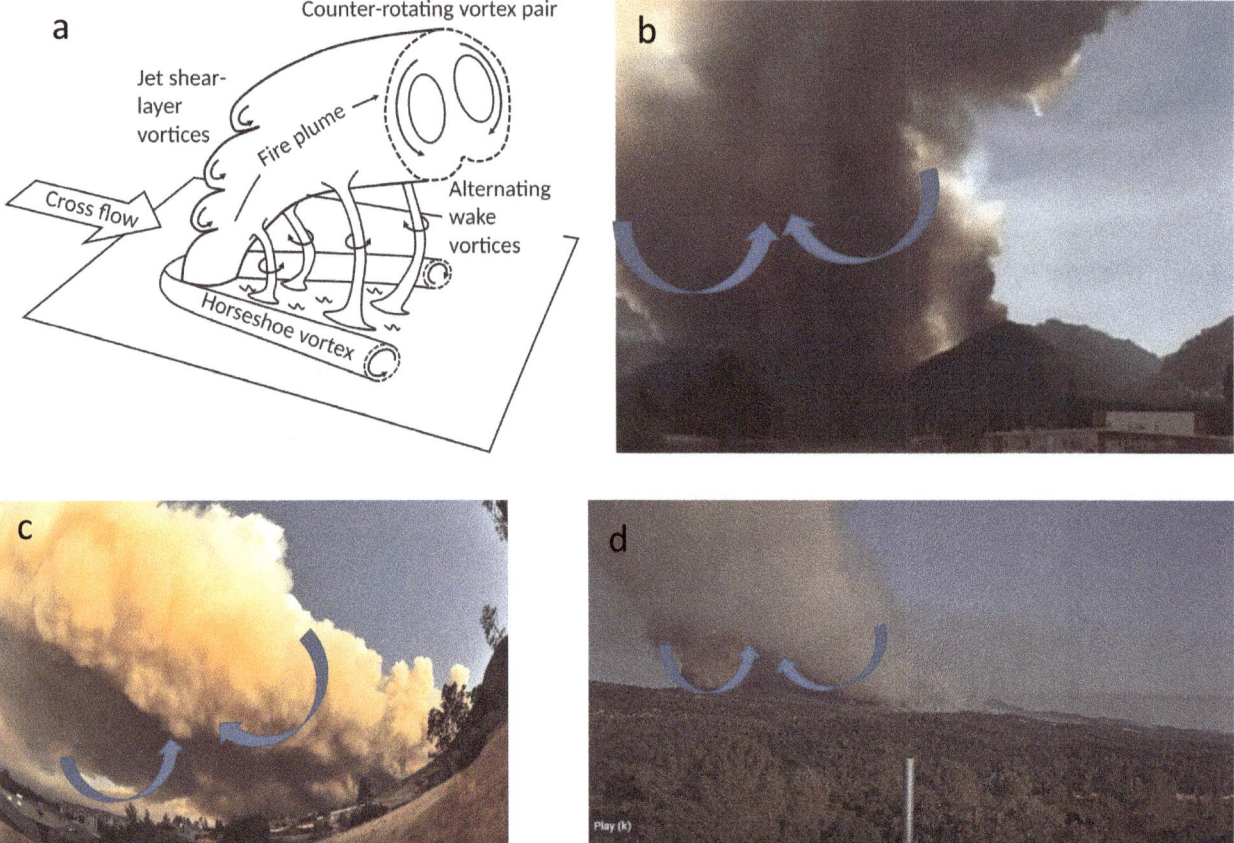

Figure 8.21: Vorticity caused by a fire plume in crossflow (vortices shown by arrows): (a) main features of jets in crossflow include counter-rotating vortex pairs and wake vortices (adapted with permission from Fric and Roshko 1994). Photographs from three fires show counter-rotating vortex pairs: (b) in the Roaring Lion Fire, Hamilton, Montana, USA, 2016 (https://www.youtube.com/watch?v=OxcDxp07okc); (c) in the Los Conches Fire, Los Alamos, New Mexico, USA, 2011; and (d) Wall Fire, Butte County, California, USA, 2017 (https://www.youtube.com/watch?v=mup5-RfADik).

implies, the vortices rotate in opposite directions, with flows converging where they join in the middle of the lee side of the plume. The presence of counter-rotating vortex pairs on fires indicates that most of the wind is blowing around the outside of the fire plume rather than through the middle. Fires with counter-rotating vortex pairs seem to have a deep, coherent zone where the fire is burning at high intensity and a plume with a core that has strong vertical velocity; these two characteristics keep the fire from breaking into a typical line fire at the head. Needless to say, such fires are not similar to the typical line fire assumed by most fire models. Fires that form counter-rotating vortex pairs may develop the necessary deep flaming region by very rapid spread and/or when they are spreading through long-burning fuels (e.g. large-diameter downed woody material and deep duff and litter layers).

The second vortex features of jets in crossflow are called *wake vortices*. They are non-steady vortices, they form repeatedly, and they appear to wander downstream of the main fire. Wake vortices are thus not properties of the time-averaged condition of the plume itself. Wake vortices appear to form at nearly regular time intervals. They may be generated alternately from side to side of the fire, with alternating clockwise and anticlockwise rotations. The frequencies of vortex shedding and their drift speeds seem to depend on properties of the wind and the plume. At any instant, several vortices may be seen moving downwind from a fire (**Figure 8.22**). Laboratory studies reveal that wake vortices originate in the boundary layer along the ground as horizontal vorticity becomes redirected vertically (Fric and Roshko 1994). This is similar to some of the ways in which fire whirls form (see the analogy with a lifted rope described in the previous section). This distinguishes wake vortices in fires from wakes generated behind rocks or other physical obstructions, in which the vorticity originates from the surfaces of the solid bodies rather than from buoyancy along the object's edge or irregularities on the ground.

Similarities of fire plumes and jets in crossflow suggest additional research that could test the scaling relations developed from laboratory studies regarding counter-rotating vortex pairs and wake vortices. These scaling relations and other characteristics of jets in crossflow are summarised by Mahesh (2013). They report that the plume tilt depends upon the relative velocities and densities of the crosswind and the plume jet, just as the same factors determine flame tilt in line and pool fires (**Chapter 6, Wind effect on flame tilt**). Stronger cross-winds relative to the vertical velocity lead to greater plume tilt and more eccentric cross-sections of counter-rotating vortex pairs. Researchers have also tested scaling relationships for characterising the frequency of vortex shedding and the structure of wake vortices (Fric and Roshko 1994). These relationships are based on the relative velocities (wind and plume) and the Reynolds number involving the plume diameter and wind speed (see **Chapter 3, Convection heat transfer, Eqn [3.32]**). Improvement of our understanding of counter-rotating vortex pairs and wake vortices is not just a topic of academic interest; it can help fire managers and citizens recognise the conditions in which these vortices can form on fires because, as with all fire-related vorticity, they can produce strong surface winds and dangerous fire behaviour. They are also well beyond the realm of predictive modelling.

Vorticity-driven lateral spread

Vorticity can affect fire behaviour in a significant, unexpected way when it causes *vorticity-driven lateral spread* on the lee side of a ridge (Simpson *et al.* 2014). This phenomenon occurred in the fire growth patterns of several fires near Canberra, Australia, in 2003. The fires spread laterally along the lee sides of ridges, in directions perpendicular to the wind (Sharples *et al.* 2012). Vorticity-driven lateral spread resembles the kind of fire spread along ridge lines described by Countryman (1971). It is caused by the interactions of wind, terrain and fire itself. Air flowing over the crest of a ridge sometimes separates from the ground surface for some distance. This means that the air does not descend along the ground surface on the lee side; instead, it

Figure 8.22: Wake vortices (rotation indicated by arrows) shed from the lee side of (downwind from) the main fire front in (a) Roaring Lion Fire, Hamilton, Montana, USA, 2016 and (b) bonfire at Burning Man festival, Nevada, USA, 2016 (https://www.youtube.com/watch?v=nSJkydvQTJY). Laboratory studies of jets in cross-flow, which are close analogues of these fires, show that the frequency of shedding depends upon characteristics of plume velocity, wind speed and fire diameter. Wake vortices are created alternately from opposite sides of a fire, with rotations in opposite directions (see Mahesh 2013).

Figure 8.23: Vorticity-driven lateral spread near the crest of a ridge. For some configurations of slope steepness and wind speed, ambient wind flow separates at the ridge to cause lee-side vorticity. When fire is present, this may cause it to spread perpendicular to the wind direction along the lee side of the ridge, loft embers and create spot fires downwind (illustration by Brian Elling based on Sharples *et al.* 2012).

continues to flow straight ahead at the elevation of the ridge top. A horizontal vortex forms on the lee side between the wind and the ground under it, causing the wind to blow uphill along the lee side in opposition to the general wind direction (recirculating flow). If a fire is burning on the lee side at the same time that the wind above is separating from the ridge, the horizontal vortex will drive it laterally along the lee side, parallel to the ridge line (**Figure 8.23**). Flow separation of the wind will interact with uphill flow from the lee-side fire to create and concentrate vorticity and form strong near-surface winds (**Figure 8.23**). The resulting fast lateral fire spread may be unexpected and could endanger firefighters. As the fire spreads laterally along the ridgeline, it is likely to cause significant downwind spotting and possibly conditions in which mass ignition could occur.

Mass fires

The terms *mass fire* and *fire storm* are used to describe behaviours associated with large areas burning all at once or with multiple intense fires burning simultaneously inside the fire. These behaviours include strong fire-induced surface winds (20–50 m s^{-1}) flowing towards the fire, a tall convection column with possible rotation, and increased consumption and enhanced burning rates of fuel (Countryman 1964; Pitts 1991; Finney and McAllister 2011). The notion of mass fire originated with wartime experience and observations of saturation-bombing of urban areas in Germany and Japan during World War II (Bond 1946). Horrific destruction of cities was accomplished by burning them with mixtures of incendiary bombs and high-explosive bombs.

Mass fires are fires in which the whole fire is burning at once, distinguishing them from line fires. This means that heat release in a mass fire is distributed throughout the burning area. While mass fires have high intensity, which implies long flames, the size of their flames relative to the diameter of the fire is small because a very large area is burning. Experiments by Heskestad (1991) suggested that flame height may be shorter in mass fires than in standard axisymmetric fires (e.g. pool fires) because mass fires release heat over such large areas that only localised air circulations are formed compared to large fire–scale circulations characteristic of puffing fires. **Figure 8.24a** shows a laboratory-scale mass fire 7.3 m across. This time series of photographs shows fibreboard soaked in liquid fuel burning at different stages after ignition (Heskestad 1991). The patchy, almost cellular flame structure occurs because air inflow from the outside edges cannot reach the middle of the fire. We discussed similar circulations in flame zones in **Chapter 6, Wind in the presence of fire**. This study suggested that discontinuous flaming over an area fire would occur when the ratio of flame length (ℓ_f, m) relative to diameter of the burning area (*d*, m) was below a certain threshold:

$$\ell_f/d < 0.52 \qquad [8.10]$$

Fires below this ratio would qualify as mass fires. However, the criteria for creating mass fires are not defined precisely but, according to Carrier *et al.* (1981) require the following:

- a burning area of at least several square kilometres
- high loading of long-burning fuels (at least 2–3 h) to respond to fire-induced air flow and ventilation
- relatively little wind that facilitates a vertical plume trajectory
- an unstable atmospheric temperature profile enhancing plume development.

Mass fires can occur naturally in wildlands when a large area becomes ignited in a short time by embers. They could also be created by large prescribed fires, volcanoes, incendiary warfare, and earthquakes that disrupt urban infrastructure over large areas, such as gas transport lines.

Many forests currently contain high loadings of large woody materials and have deep litter and duff layers that can easily meet the fuel requirement for mass fires (Stephens *et al.* 2018). Because these fuel components and their burning behaviour are not included in models of line-fire spread and no physical model has included their contribution to large-fire phenomena, they are easily overlooked as a condition that could trigger mass fires and consequent extreme, violent fire storms.

As is typical of fire phenomena in general, mass fire behaviours depend upon strong coupling between the fire and its environment. The coupling of fire behaviour with wind is critical to the occurrence of mass fires. While mass fires tend to occur when ambient wind speed is relatively low, they are characterised by very strong winds along the ground. Two explanations for these winds have been proposed (Finney and McAllister 2011). Both require a tall, wide convection column standing vertically above a large, relatively round or square burning area. While most large fires have tall plumes, possibly even pyrocumulonimbus clouds (as described above), they do not all have intense positive feedback between the plume and the fire on the ground, as mass fires do. One

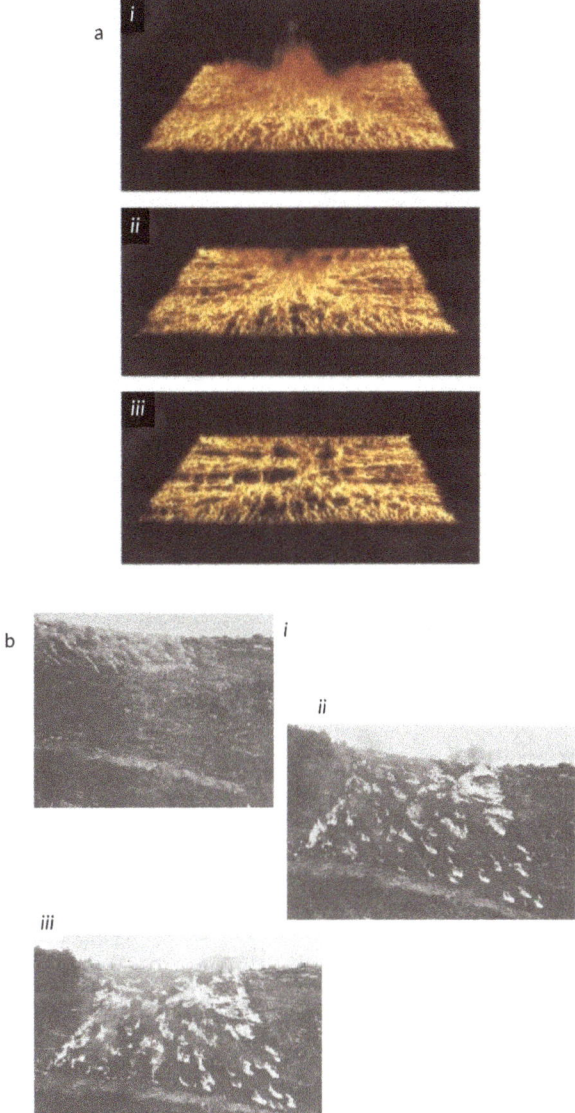

Figure 8.24: Mass fire behaviour from (a) liquid fuel burning on 7.3 m × 7.3 m area showing development of cellular flame structure as intensity diminishes over time (i 18 s after ignition, ii 21 s, iii 26 s) with increasing fuel consumption (reproduced with permission from Heskestad 1991) and (b) field-scale tests involving multiple large slash piles, which show vorticity developing within the array as ignition progressed (from Countryman 1964).

explanation for the strong surface winds in this feedback cycle relates to rotation of the column. As described in **Vorticity** above, rotation creates low hydrostatic pressure in the area over a fire. In the case of mass fires, much of the large area under the plume could be subject to low pressure, which would be maintained and strengthened by the vorticity itself, creating a tremendous pressure difference between the plume and the outside air and thus drawing air into the bottom of the column at high velocity.

The other explanation for strong surface winds lies in the relationship between the width of the burning area and the width of the plume above. When the plume gets very wide relative to the width of the burning area, air is not entrained into the centre as efficiently as when it is smaller. Reduced entrainment of cool air into the fire zone forces gases to rise higher in the plume before they can encounter sufficient oxygen to combust. This increases plume temperatures at high altitudes above the fire. The higher temperatures increase vertical growth of the plume, which further reduces air pressure at the surface, which in turn induces stronger inflow of air at the ground. Byram (1959) described this feedback process as a 'heat engine'.

Whether strong surface winds are produced by plume vorticity or Byram's heat engine or both, the high-velocity surface winds along the ground can enhance the burning rate of solid fuels. Greater air flow increases the oxygen available to glowing combustion (**Chapters 4, Smouldering and glowing**). If more oxygen becomes available to smouldering fuels, they may transition from smouldering to flaming, thus increasing heat release rates substantially. When this occurs, wildland fuel components that are normally considered unavailable for burning in the flaming phase, or even in smouldering, can contribute to increased fire intensity because they are influenced by the powerful winds generated at the scale of the whole fire.

The dependency of solid fuel burning rates on conditions associated with mass fires have been demonstrated in laboratory experiments. McAllister (2021) studied the effects of reduced plume entrainment, which is typical of mass fires, by using a chimney as an analogue for a large-radius fire with minimal entrainment from the sides (**Figure 8.25**). Because the chimney eliminates entrainment of cool air into the plume, the oxygen needed for combustion at the base of the fire (cross-piles of sticks called *cribs*, shown in **Figure 4.26**) could come only from increased air flow at the ground surface. Use of the chimney increased the burning rate up to twofold relative to free-burning conditions, depending on stick size and density of the fuel cribs. In wildland fires, increased ventilation from surface winds can increase consumption of solid fuels that would otherwise smoulder or become extinguished. Comparable results were reported from experiments that used forced ventilation of burning wooden cribs and from measurements that showed high rates of air flow at the bases of fire whirls. Both forced ventilation and fire whirls multiplied the burning rate several-fold relative to free-burning fires (Martin *et al.* 1976; Harmathy 1978). Additional research is needed to understand how coupled fire-wind systems cause mass fires; in particular, studies are needed on the width of fire areas and convection columns needed to produce behaviours like those seen in the laboratory.

Mass fires do not require uniform flaming combustion across an entire area (**Figure 8.24b**); they can also be generated when multiple fires occur simultaneously within a large burn area. As mentioned above (**Eqn [8.10]**), discontinuous flaming will occur if the ratio of the flame length to the fire diameter is below some threshold. However, laboratory and field experiments have shown that, as air is entrained into multiple flame sources at once, the interactions of air flows can produce a unified plume structure and can enhance burning rates. As discussed in **Chapter 6, Multiple flame zones and air-flow interactions**, burning rates of arrays of fires increase relative to their individual burning rates as the spacing between fires decreases to an optimum point, but then burning rates decreased at very tight spacing (**Figure 6.36**) (Finney and McAllister 2011; Huffman *et al.* 1969; Pitts 1991). The increase as fires become closer together is caused by enhanced feedback of radiant heat to the

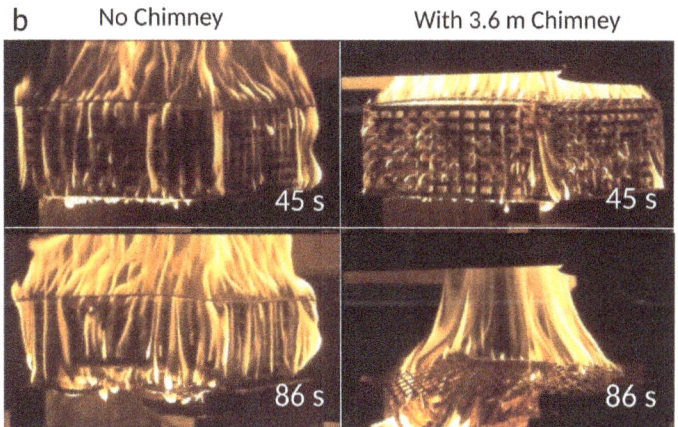
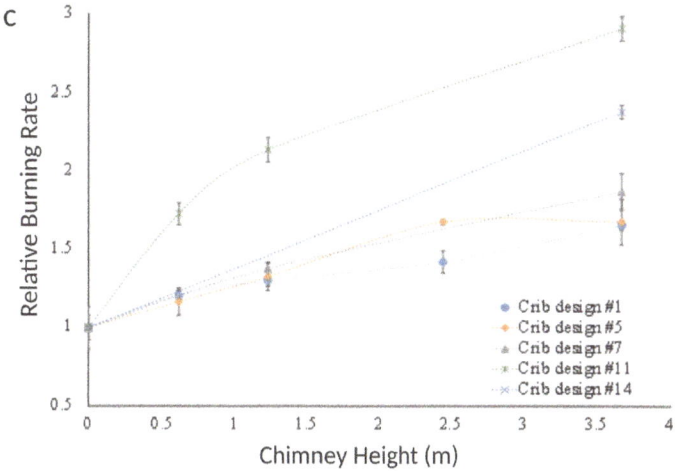

Figure 8.25: Laboratory experiments demonstrated that (a) a chimney placed above a fire reduces entrainment into the plume and generally increases the solid fuel burning rate, as demonstrated by (b) contrast in burning rates for crib design #11 (15 layers of 0.64 cm sticks, 14 sticks per layer, with horizontal dimensions of 25.4 cm × 25.4 cm) with and without a 3.6 m tall chimney (photographs by Ian Grob); and (c) most cribs irrespective of design increased burning rate with the chimney (relative burning rate is compared to free burning). The low hydrostatic pressure in the chimney is similar to that in large-fire plumes and columns, which induce rapid air flow along the ground surface (from McAllister 2021).

burning fuel, and the decrease as fires become 'too close' is caused by reduced air entrainment into the plume.

Summary

It is clear that large wildland fires can exhibit very powerful behaviours that are unique to their size, geometry and energy release rates. Large fires burning with characteristics of crown fires in forest canopies, launching burning material for many kilometres, or flaming across large areas as mass fires and fire storms cannot be well approximated by predictive models that are rooted in assumptions of steady-spreading linear flame zones, which comprise the basis of all fire prediction systems now in use. Recognising behaviours that fall outside of our current prediction ability is an important first step in considering, at least qualitatively, when and how those behaviours occur, how they may impact valuable assets – both human developments and natural resources – and how they might be predicted, avoided or managed. There is strong evidence that management activities can modify fuel and vegetation conditions to make valuable assets defensible and extreme fire behaviours less likely to occur in the first place, but those topics are outside the scope of this book.

References

Agee JK, Wright CS, Williamson N, Huff MH (2002) Foliar moisture content of Pacific Northwest vegetation and its relation to wildland fire behavior. *Forest Ecology and Management* **167**, 57–66. doi:10.1016/S0378-1127(01)00690-9

Albini FA (1976) 'Estimating wildfire behaviour and effects'. General Technical Report INT-30. USDA Forest Service, Intermountain Forest and Range Experiment Station, Ogden, UT.

Albini FA (1979). 'Spot fire distance from burning trees: a predictive model'. General Technical Report INT-56. USDA Forest Service, Intermountain Forest and Range Experiment Station, Ogden, UT.

Albini FA (1983) 'Potential spotting distance from wind-driven surface fires'. Research Paper INT-309. USDA Forest Service, Intermountain Forest and Range Experiment Station, Ogden, UT.

Albini FA (1984) Wildland Fires: Predicting the behavior of wildland fires – among nature's most potent forces – can save lives, money, and natural resources. *American Scientist* **72**(6), 590–597.

Albini FA, Stocks BJ (1986) Predicted and observed rates of spread of crown fires in immature jack pine. *Combustion Science and Technology* **48**(1–2), 65–76. doi:10.1080/00102208608923884

Albini FA, Alexander ME, Cruz MG (2012) A mathematical model for predicting the maximum potential spotting distance from a crown fire. *International Journal of Wildland Fire* **21**(5), 609–627. doi:10.1071/WF11020

Alexander ME (1985) Estimating the length-to-breadth ratio of elliptical forest fire patterns. In *Proceedings of the 8th Conference on Fire and Forest Meteorology*. 29 April – 2 May 1985, Detroit, MI. pp. 287–304. Society of American Foresters, Bethesda, MD.

Alexander ME, Cruz MG (2006) Evaluating a model for predicting active crown fire rate of spread using wildfire observations. *Canadian Journal of Forest Research* **36**(11), 3015–3028. doi:10.1139/x06-174

Alexander ME, Cruz MG (2011) Crown fire dynamics in conifer forests. In *Synthesis of Knowledge of Extreme Fire Behavior: Volume I for Fire Managers*. (Eds PA Werth, BE Potter, CB Clements, MA Finney, JA Forthofer, SS McAllister, SL Goodrick, ME Alexander, MG Cruz) pp. 107–142. USDA Forest Service, Pacific Northwest Research Station, Portland, OR.

Alexander ME, Cruz MG (2012) Interdependencies between flame length and fireline intensity in predicting crown fire initiation and crown scorch height. *International Journal of Wildland Fire* **21**(2), 95–113. doi:10.1071/WF11001

Anderson HE 1968. 'Sundance Fire: an analysis of fire phenomena'. Research Paper INT-56. USDA Forest Service, Intermountain Forest and Range Experiment Station. Ogden, UT.

Anderson HE (1983) 'Predicting wind-driven wild land fire size and shape [fire behavior models]'. Research Paper INT-305. USDA Forest Service, Intermountain Research Station, Ogden, UT.

Anderson DH, Catchpole EA, De Mestre NJ, Parkes T (1982) Modelling the spread of grass fires. *The ANZIAM Journal* **23**(4), 451–466.

Billing P, Rawson R (1982) *A Fire Tornado in the Sunset Country*. Forests Commission Victoria, Melbourne.

Bird RB, Codding BF, Kauhanen PG, Bird DW (2012) Aboriginal hunting buffers climate-driven fire-size variability in Australia's spinifex grasslands. *Proceedings of the National Academy of Sciences of the United States of America* **109**(26), 10287–10292. doi:10.1073/pnas.1204585109

Blackmarr WH (1972) 'Moisture content influences ignitability of slash pine litter'. Research Note SE-173. USDA Forest Service, Southeastern Forest Experiment Station, Asheville, NC.

Bond H (ed) (1946) *Fire and the Air War: A Compilation of Expert Observations on Fires of the War Set by Incendiaries and the Atomic Bombs, Wartime Fire Fighting, and the Work of the Fire Protection Engineers who Helped Plan and the Destruction of Enemy Cities and Industrial Plants*. National Fire Protection Association International, Boston, MA.

Bradshaw LS, Deeming JE, Burgan RE, Cohen JD (1984) 'The 1978 National Fire-Danger Rating System: technical documentation'. General Technical Report INT-169. USDA Forest Service, Intermountain Forest and Range Experiment Station, Odgen, UT.

Bunting SC, Wright HA (1974) Ignition capabilities of non-flaming firebrands. *Journal of Forestry* **72**(10), 646–649.

Burrows N, Ward B, Robinson A (2009) Fuel dynamics and fire spread in spinifex grasslands of the Western Desert. *Proceedings of the Royal Society of Queensland* **115**, 69–76.

Butler BW, Cohen J, Latham DJ, Schuette RD, Sopko P, Shannon KS, Jimenez D, Bradshaw LS (2004*a*) Measurements of radiant emissive power and temperatures in crown fires. *Canadian Journal of Forest Research* **34**(8), 1577–1587. doi:10.1139/x04-060

Butler BW, Finney MA, Andrews PL, Albini FA (2004*b*) A radiation-driven model for crown fire spread. *Canadian Journal of Forest Research* **34**(8), 1588–1599. doi:10.1139/x04-074

Byram GM (1954) 'Atmospheric conditions related to blowup fires'. Station Paper SE-SP-35. USDA-Forest Service. Southeastern Forest Experiment Station, Asheville, NC.

Byram GM (1959) Forest fire behavior. In *Forest Fire: Control and Use*. (Ed. KP Davis) pp. 90–123. McGraw Hill, New York, NY.

Byram GM, Nelson RM (1970) The modeling of pulsating fires. *Fire Technology* 6(2), 102–110.

Carrier GF, Fendell FE, Feldman PS (1981) 'Criteria for onset of firestorms'. Report no. ADP001804. Defense Technical Information Center, Fort Belvoir, VA.

Catchpole EA, de Mestre NJ, Gill AM (1982) Intensity of fire at its perimeter. *Australian Forest Research* 12, 47–54.

Catchpole EA, Alexander ME, Gill AM (1992) Elliptical-fire perimeter- and area-intensity distributions. *Canadian Journal of Forest Research* 22(7), 968–972. doi:10.1139/x92-129

Caton SE, Hakes RS, Gorham DJ, Zhou A, Gollner MJ (2017) Review of pathways for building fire spread in the wildland urban interface part I: exposure conditions. *Fire Technology* 53(2), 429–473. doi:10.1007/s10694-016-0589-z

Cetegen BM, Ahmed TA (1993) Experiments on the periodic instability of buoyant plumes and pool fires. *Combustion and Flame* 93(1–2), 157–184. doi:10.1016/0010-2180(93)90090-P

Cheney NP 1981. Fire behaviour. In *Fire and the Australian Biota*. (Eds AM Gill, RH Groves, IR Noble) pp. 151–175. Australian Academy of Science, Canberra.

Cheney NP (2010) Fire behavior during the Pickering Brook wildfire, January 2005 (Perth Hills Fires 71–80). *Conservation Science Western Australia* 7(3), 451–468.

Cheney P, Sullivan A (eds) (2008) *Grassfires: Fuel, Weather and Fire Behaviour*. CSIRO Publishing, Collingwood.

Clark TL, Radke L, Coen J, Middleton D (1999) Analysis of small-scale convective dynamics in a crown fire using infrared video camera imagery. *Journal of Applied Meteorology* 38(10), 1401–1420. doi:10.1175/1520-0450(1999)038<1401:AOSSCD>2.0.CO;2

Coen J, Mahalingam S, Daily J (2004) Infrared imagery of crown-fire dynamics during FROSTFIRE. *Journal of Applied Meteorology* 43(9), 1241–1259. doi:10.1175/1520-0450(2004)043<1241:IIOCDD>2.0.CO;2

Cohen JD (2000) Preventing disaster: home ignitability in the wildland–urban interface. *Journal of Forestry* 98(3), 15–21.

Countryman CM (1964) 'Mass fires and fire behavior'. Research Paper RS-RP-19. USDA Forest Service, Pacific Southwest Forest and Range Experiment Station, Berkeley, CA.

Countryman CM (1971) *Fire Whirls ... Why, When, and Where*. USDA Forest Service, Pacific Southwest Research Station, Berkeley, CA.

Cruz MG, Butler BW, Alexander ME, Forthofer JM, Wakimoto RH (2006) Predicting the ignition of crown fuels above a spreading surface fire. Part I: model idealization. *International Journal of Wildland Fire* 15(1), 47–60. doi:10.1071/WF04061

Cruz MG, Sullivan AL, Gould JS, Sims NC, Bannister AJ, Hollis JJ, Hurley RJ (2012) Anatomy of a catastrophic wildfire: The Black Saturday Kilmore East fire in Victoria, Australia. *Forest Ecology and Management* 284, 269–285. doi:10.1016/j.foreco.2012.02.035

Despain DG, Clark DL, Reardon JJ (1996) Simulation of crown fire effects on canopy seed bank in lodgepole pine. *International Journal of Wildland Fire* 6(1), 45–49. doi:10.1071/WF9960045

Dupuy JL, Maréchal J, Portier D, Valette JC (2011) The effects of slope and fuel bed width on laboratory fire behaviour. *International Journal of Wildland Fire* 20(2), 272–288. doi:10.1071/WF09075

Ellis PFM (2011) Fuelbed ignition potential and bark morphology explain the notoriety of the eucalypt messmate 'stringybark' for intense spotting. *International Journal of Wildland Fire* 20(7), 897–907. doi:10.1071/WF10052

Ellis PFM (2015) The likelihood of ignition of dry-eucalypt forest litter by firebrands. *International Journal of Wildland Fire* 24(2), 225–235. doi:10.1071/WF14048

Emmons HW, Ying SJ (1967) The fire whirl. *Symposium (International) on Combustion* 11, 475–488.

Fang J, Tu R, Guan JF, Wang JJ, Zhang YM (2011) Influence of low air pressure on combustion characteristics and flame pulsation frequency of pool fires. *Fuel* 90(8), 2760–2766. doi:10.1016/j.fuel.2011.03.035

Fendell FE (1986) Crown streets. *Combustion Science and Technology* 45(5–6), 311–315.

Fernandes PM, Cruz MG (2012) Plant flammability experiments offer limited insight into vegetation–fire dynamics interactions. *New Phytologist* 194(3), 606–609. doi:10.1111/j.1469-8137.2012.04065.x

Fernandez-Pello AC (2017) Wildland fire spot ignition by sparks and firebrands. *Fire Safety Journal* 91, 2–10. doi:10.1016/j.firesaf.2017.04.040

Finney MA (1998) 'FARSITE, Fire Area Simulator – model development and evaluation'. Research Paper RMRS-RP-4 USDA Forest Service, Rocky Mountain Research Station, Fort Collins, CO.

Finney MA (2002) Fire growth using minimum travel time methods. *Canadian Journal of Forest Research* 32(8), 1420–1424. doi:10.1139/x02-068

Finney MA, McAllister SS (2011) A review of fire interactions and mass fires. *Journal of Combustion* 2011, 548328. doi:10.1155/2011/548328

Finney MA, Bartlette R, Bradshaw L, Close K, Collins BM, Gleason P, Hao WM, Langowski P, McGinely J, McHugh CW, Martinson E (2003) Fire behavior, fuel treatments, and fire suppression on the Hayman Fire. General Technical Report RMRS-GTR-114. In *Hayman Fire Case Study*. (Ed. RT Graham) pp. 33–179. USDA Forest Service, Rocky Mountain Research Station, Fort Collins, CO.

Forestry Canada Fire Danger Group (1992) 'Development and structure of the Canadian Forest Fire Behavior Prediction System'. Information Report ST-X-3. Forestry Canada, Ottawa.

Forthofer JM (2019) Can scientists predict fire tornadoes? *Scientific American* 60–67.

Forthofer JM, Goodrick SL (2011) Review of vortices in wildland fire. *Journal of Combustion* **2011**, 984363.

Frankman D, Webb BW, Butler BW, Jimenez D, Forthofer JM, Sopko P, Shannon KS, Hiers JK, Ottmar RD (2013) Measurements of convective and radiative heating in wildland fires. *International Journal of Wildland Fire* **22**(2), 157–167. doi:10.1071/WF11097

Fric TF, Roshko A (1994) Vortical structure in the wake of a transverse jet. *Journal of Fluid Mechanics* **279**, 1–47. doi:10.1017/S0022112094003800

Fromm M, Bevilacqua R, Servranckx R, Rosen J, Thayer JP, Herman J, Larko D (2005) Pyro-cumulonimbus injection of smoke to the stratosphere: Observations and impact of a super blowup in northwestern Canada on 3–4 August 1998. *Journal of Geophysical Research, D, Atmospheres* **110**, D08205. doi:10.1029/2004JD005350

Gill AM (1974) Toward an understanding of fire-scar formation: field observation and laboratory simulation. *Forest Science* **20**(3), 198–205.

Goens DW (1978) 'Fire whirls'. NOAA Technical Memorandum NWS WR-129. National Weather Service Office, Missoula, MT.

Haines DA (1982) Horizontal roll vortices and crown fires. *Journal of Applied Meteorology* **21**(6), 751–763. doi:10.1175/1520-0450(1982)021<0751:HRVACF>2.0.CO;2

Hall J, Ellis PF, Cary GJ, Bishop G, Sullivan AL (2015) Long-distance spotting potential of bark strips of a ribbon gum (*Eucalyptus viminalis*). *International Journal of Wildland Fire* **24**(8), 1109–1117. doi:10.1071/WF15031

Hamins A, Yang JC, Kashiwagi T (1992) An experimental investigation of the pulsation frequency of flames. *Symposium (International) on Combustion* **24**(1), 1695–1702.

Harmathy TZ (1978) Experimental study on the effect of ventilation on the burning of piles of solid fuels. *Combustion and Flame* **31**, 259–264. doi:10.1016/0010-2180(78)90138-4

Heinselman ML (1973) Fire in the virgin forests of the Boundary Waters Canoe Area, Minnesota. *Quaternary Research* **3**(3), 329–382. doi:10.1016/0033-5894(73)90003-3

Heskestad G (1991) A reduced-scale mass fire experiment. *Combustion and Flame* **83**(3–4), 293–301. doi:10.1016/0010-2180(91)90076-N

Hicke JA, Johnson MC, Hayes JL, Preisler HK (2012) Effects of bark beetle–caused tree mortality on wildfire. *Forest Ecology and Management* **271**, 81–90. doi:10.1016/j.foreco.2012.02.005

Huffman KG, Welker JR, Sliepcevich CM (1969) Interaction effects of multiple pool fires. *Fire Technology* **5**(3), 225–232. doi:10.1007/BF02591520

Jolly WM, Parsons RA, Hadlow AM, Cohn GM, McAllister SS, Popp JB, Hubbard RM, Negron JF (2012) Relationships between moisture, chemistry, and ignition of *Pinus contorta* needles during the early stages of mountain pine beetle attack. *Forest Ecology and Management* **269**, 52–59. doi:10.1016/j.foreco.2011.12.022

Kok JF, Parteli EJ, Michaels TI, Karam DB (2012) The physics of wind-blown sand and dust. *Reports on Progress in Physics* **75**(10), 106901. doi:10.1088/0034-4885/75/10/106901

Koo E, Pagni PJ, Weise DR, Woycheese JP (2010) Firebrands and spotting ignition in large-scale fires. *International Journal of Wildland Fire* **19**(7), 818–843. doi:10.1071/WF07119

Lareau NP, Nauslar NJ, Abatzoglou JT (2018) The Carr fire vortex: a case of pyrotornadogenesis? *Geophysical Research Letters* **45**, 13107–13115. doi:10.1029/2018GL080667

Latham DJ (1991) Lightning flashes from a prescribed fire influenced cloud. *Journal of Geophysical Research* **96**, 17151–17157. doi:10.1029/91JD01808

Latham D, Williams ER (2001) Lightning and forest fires. In *Forest Fires – Behavior and Ecological Effects*. (Eds EA Johnson, K Miyanashi) pp. 375–418. Academic Press, New York, NY.

Linn RR, Winterkamp JL, Edminster C, Colman JJ, Smith WS (2007) Coupled influences of topography and wind on wildland fire behavior. *International Journal of Wildland Fire* **16**, 183–195. doi:10.1071/WF06078

Linn RR, Winterkamp JL, Weise DR, Edminster C (2010) A numerical study of slope and fuel structure effects on coupled wildfire behavior. *International Journal of Wildland Fire* **19**, 179–201. doi:10.1071/WF07120

Mahesh K (2013) The interaction of jets with crossflow. *Annual Review of Fluid Mechanics* **45**, 379–407. doi:10.1146/annurev-fluid-120710-101115

Martin RE, Pendleton DW, Burgess W (1976) Effect of fire whirlwind formation on solid fuel burning rates. *Fire Technology* **12**(1), 33–40. doi:10.1007/BF02629468

McAllister SS (2021) Effect of reduced plume entrainment on the burning rate of porous fuel beds. *Progress in Scale Modeling* **2**(2), 6. doi:10.13023/psmij.2021.02-02-06

McAllister SS, Finney MA (2014) Convection ignition of live forest fuels. *Fire Safety Science* **11**, 1312–1325. doi:10.3801/IAFSS.FSS.11-1312

McArthur AG, Cheney NP, Barber J (1982) *The Fires of 12 February 1977 in the Western District of Victoria*. CSIRO Division of Forest Research, Canberra. doi:10.4225/08/58712e075f1da

McCarter RJ, Broido A (1965) Radiative and convective energy from wood crib fires. *Pyrodynamics* **2**(1), 65–85.

McRae DJ, Jin JZ, Conard SG, Sukhinin AI, Ivanova GA, Blake TW (2005) Infrared characterization of fine-scale variability in behavior of boreal forest fires. *Canadian Journal of Forest Research* **35**(9), 2194–2206. doi:10.1139/x05-096

McRae RH, Sharples JJ, Wilkes SR, Walker A (2013) An Australian pyro-tornadogenesis event. *Natural Hazards* **65**(3), 1801–1811. doi:10.1007/s11069-012-0443-7

Michaletz ST, Johnson EA (2006) A heat transfer model of crown scorch in forest fires. *Canadian Journal of Forest Research* **36**(11), 2839–2851. doi:10.1139/x06-158

Moore T (2008) *Indians Fire Accident Prevention Analysis*. USDA Fire Service, Pacific Southwest Region, Vallejo, CA.

Mottus B, Pengelly I (2004) An analysis of the meteorological conditions associated with the 1999 Panther River Fire in Banff National Park [abstract]. In *Proceedings of the 22nd Tall Timbers Fire Ecology Conference: Fire in Temperate, Boreal, and Montane Ecosystems*. (Eds RT Engstrom, KEM Galley, WJ de Groot) p. 239. Tall Timbers Research Station, Tallahassee, FL.

Nelson RM (1993) Byram derivation of the energy criterion for forest and wildland fires. *International Journal of Wildland Fire* **3**(3), 131–138. doi:10.1071/WF9930131

Nelson RM, Jr (2002) An effective wind speed for models of fire spread. *International Journal of Wildland Fire* **11**(2), 153–161. doi:10.1071/WF02031

Nelson RM, Jr (2003) Power of the fire – a thermodynamic analysis. *International Journal of Wildland Fire* **12**(1), 51–65. doi:10.1071/WF02032

Peet GB (1967) The shape of mild fires in Jarrah forest. *Australian Forestry* **31**(2), 121–127. doi:10.1080/00049158.1967.10675433

Perrakis DD, Lanoville RA, Taylor SW, Hicks D (2014) Modeling wildfire spread in mountain pine beetle-affected forest stands, British Columbia, Canada. *Fire Ecology* **10**(2), 10–35. doi:10.4996/fireecology.1002010

Peterson DA, Hyer EJ, Campbell JR, Solbrig JE, Fromm MD (2017) A conceptual model for development of intense pyro-cumulonimbus in western North America. *Monthly Weather Review* **145**(6), 2235–2255. doi:10.1175/MWR-D-16-0232.1

Pitts WM (1991) Wind effects on fires. *Progress in Energy and Combustion Science* **17**(2), 83–134. doi:10.1016/0360-1285(91)90017-H

Potter BE (2012a) Atmospheric interactions with wildland fire behaviour – I. Basic surface interactions, vertical profiles and synoptic structures. *International Journal of Wildland Fire* **21**(7), 779–801. doi:10.1071/WF11128

Potter BE (2012b) Atmospheric interactions with wildland fire behaviour–II. Plume and vortex dynamics. *International Journal of Wildland Fire* **21**(7), 802–817. doi:10.1071/WF11129

Rein G (2009) Smouldering combustion phenomena in science and technology. *International Review of Chemical Engineering* **1**, 3–18.

Richards GD (1994) The properties of elliptical wildfire growth for time dependent fuel and meteorological conditions. *Combustion Science and Technology* **95**(1–6), 357–383.

Richards GD (1995) A general mathematical framework for modeling two-dimensional wildland fire spread. *International Journal of Wildland Fire* **5**(2), 63–72. doi:10.1071/WF9950063

Richards GD, Walberg R (1998) The computer simulation of crown fire streets. In *Proceedings of the III International Conference on Forest Fire Research and 14th conferences on Fire and Forest Meteorology*. 16–20 November 1998, Luso, Portugal. pp. 435–440. ADAI – Associacao para o Desenvolvimento da Aerodinamica Industrial, Coimbra.

Rosenfeld D, Fromm M, Trentmann J, Luderer G, Andreae MO, Servranckx R (2007) The Chisholm firestorm: observed microstructure, precipitation and lightning activity of a pyro-cumulonimbus. *Atmospheric Chemistry and Physics* **7**(3), 645–659. doi:10.5194/acp-7-645-2007

Rossa CG, Fernandes PM (2018) Live fuel moisture content: the 'pea under the mattress' of fire spread rate modeling? *Fire (Basel, Switzerland)* **1**(3), 43. doi:10.3390/fire1030043

Rothermel RC (1991) 'Predicting behavior and size of crown fires in the northern Rocky Mountains'. Research Paper INT-438. USDA Forest Service, Intermountain Research Station, Ogden, UT.

Santoso MA, Christensen E, Yang J, Rein G (2019) Review of the transition from smouldering to flaming combustion in wildfires. *Frontiers of Mechanical Engineering* **5**, 49. doi:10.3389/fmech.2019.00049

Schroeder MJ, Buck CC (1970) *Fire Weather: A Guide for Application of Meteorological Information to Forest Fire Control Operations* (No. 360). USDA Forest Service, Washington, DC.

Scott JH, Reinhardt ED (2001) 'Assessing crown fire potential by linking models of surface and crown fire behavior'. Research Paper RMRS-RP-29. USDA Forest Service, Rocky Mountain Research Station, Fort Collins, CO.

Seto D, Clements CB (2011) Fire whirl evolution observed during a valley wind–sea breeze reversal. *Journal of Combustion* **2011**. doi.org/10.1155/2011/569475

Sharples JJ (2008) Review of formal methodologies for wind–slope correction of wildfire rate of spread. *International Journal of Wildland Fire* **17**(2), 179–193. doi:10.1071/WF06156

Sharples JJ, McRae RH, Wilkes SR (2012) Wind–terrain effects on the propagation of wildfires in rugged terrain: fire channelling. *International Journal of Wildland Fire* **21**(3), 282–296. doi:10.1071/WF10055

Silvani X, Morandini F, Dupuy JL (2012) Effects of slope on fire spread observed through video images and multiple-point thermal measurements. *Experimental Thermal and Fluid Science* **41**, 99–111. doi:10.1016/j.expthermflusci.2012.03.021

Simard AJ, Young A (1978) 'AIRPRO: an airtanker productivity computer simulation model'. Information Report FF-X-66. Environment Canada, Canadian Forest Service, Forest Fire Research Institute, Ottawa, Ontario.

Simard AJ, Haines DA, Blank RW, Frost JS (1983) 'The Mack Lake fire'. General Technical Report NC-83. USDA Forest Service, North Central Forest Experiment Station, St Paul, MIN.

Simpson CC, Sharples JJ, Evans JP (2014) Resolving vorticity-driven lateral fire spread using the WRF-Fire coupled atmosphere–fire numerical model. *Natural Hazards and Earth System Sciences* **14**(9), 2359–2371. doi:10.5194/nhess-14-2359-2014

Soma S, Saito K (1988) A study of fire whirl on mass fires using scaling models. In *Proceedings of the First International Symposium on Scale Modeling*. 1988. p. 353. Japan Society of Mechanical Engineers, Tokyo.

Stephens SL, Collins BM, Fettig CJ, Finney MA, Hoffman CM, Knapp EE, North MP, Safford H, Wayman RB (2018) Drought, tree mortality, and wildfire in forests adapted to frequent fire. *Bioscience* **68**(2), 77–88. doi:10.1093/biosci/bix146

Stocks BJ (1987) Fire behavior in immature jack pine. *Canadian Journal of Forest Research* **17**(1), 80–86. doi:10.1139/x87-014

Stocks BJ, Alexander ME, Wotton BM, Stefner CN, Flannigan MD, Taylor SW, Lavoie N, Mason JA, Hartley GR, Maffey ME, Dalrymple GN (2004a) Crown fire behaviour in a northern jack pine black spruce forest. *Canadian Journal of Forest Research* **34**(8), 1548–1560. doi:10.1139/x04-054

Stocks BJ, Alexander ME, Lanoville RA (2004b) Overview of the international crown fire modelling experiment (ICFME). *Canadian Journal of Forest Research* **34**(8), 1543–1547. doi:10.1139/x04-905

Tarifa CS, del Notario PP, Moreno FG (1965) On the flight paths and lifetimes of burning particles of wood. *Symposium (International) on Combustion* **10**(1), 1021–1037. doi:10.1016/S0082-0784(65)80244-2

Taylor SW, Wotton BM, Alexander ME, Dalrymple GN (2004) Variation in wind and crown fire behaviour in a northern jack pine black spruce forest. *Canadian Journal of Forest Research* **34**(8), 1561–1576. doi:10.1139/x04-116

Thomas PH (1971) Rates of spread of some wind-driven fires. *Forestry: An International Journal of Forest Research* **44**(2), 155–175. doi:10.1093/forestry/44.2.155

Thurston W, Kepert JD, Tory KJ, Fawcett RJ (2017) The contribution of turbulent plume dynamics to long-range spotting. *International Journal of Wildland Fire* **26**(4), 317–330. doi:10.1071/WF16142

Tohidi A, Caton S, Gollner MJ, Bryner N (2017) Thermo-mechanical breakage mechanism of firebrands. In *Proceedings of the 10th US National Combustion Meeting*. 23–24 April 2017, University of Maryland, College Park, MD.

Tohidi A, Gollner MJ, Xiao H (2018) Fire whirls. *Annual Review of Fluid Mechanics* **50**, 187–213. doi:10.1146/annurev-fluid-122316-045209

Umscheid M, Monteverdi J, Davies J (2006) Photographs and analysis of an unusually large and long-lived firewhirl. *Severe Storms Meteorology* **1**, 1–13.

Van Wagner CE (1964) History of a small crown fire. *Forestry Chronicle* **40**(2), 202–209. doi:10.5558/tfc40202-2

Van Wagner CE (1977) Conditions for the start and spread of crown fire. *Canadian Journal of Forest Research* **7**(1), 23–34. doi:10.1139/x77-004

Van Wagner CE (1993) Prediction of crown fire behavior in two stands of jack pine. *Canadian Journal of Forest Research* **23**(3), 442–449. doi:10.1139/x93-062

Viegas DX (2004) Slope and wind effects on fire propagation. *International Journal of Wildland Fire* **13**(2), 143–156. doi:10.1071/WF03046

Wade DD, Ward DE (1973) 'An analysis of the Air Force bomb range fire'. Research Paper SE-105. USDA Forest Service, Southeastern Forest Experiment Station, Asheville, NC.

Weise DR, Biging GS (1997) A qualitative comparison of fire spread models incorporating wind and slope effects. *Forest Science* **43**(2), 170–180.

Zhou R, Wu ZN (2007) Fire whirls due to surrounding flame sources and the influence of the rotation speed on the flame height. *Journal of Fluid Mechanics* **583**, 313–345. doi:10.1017/S0022112007006337

9
Measurements in fire behaviour

We introduced the science of fire behaviour in **Chapter 1** by emphasising that experiments and observational data are essential to explaining and thus managing wildland fires. In this chapter, we describe the measurement methods used to collect data on fire behaviour and the underlying physical processes. These measurements must be tailored to the specific research or management questions to be answered and by the characteristics of the environment in which the fires are burning. For field experiments, many measurement methods require people accessing the site which, if precautions are not taken, can alter the fuel properties by sampling and subsequent behaviour of the fire.

To improve understanding and modelling of fire behaviour, we need to quantify the fundamental physical processes of combustion, heat transfer and ignition as they occur in wildlands. Theories and models of wildfire behaviour idealise and simplify these basic processes, but they must be validated by laboratory and field measurements of the processes themselves. In both fire science and management, we need reliable methods for obtaining data that are relevant to wildland fire behaviour, and we must interpret those data appropriately. In this chapter, we discuss how we can obtain data that help us investigate and document the primary fire behaviours and physical processes in both laboratory and field conditions. For each measurement technique, we describe some of the challenges in its use and some possible complications or confounding factors. Some of these measurement techniques are also summarised in the review by Kremens et al. (2010).

Measurements of fire behaviour generally fall into two categories: those that quantify specific physical processes or subprocesses and those that quantify behaviours of the system at the scale of large areas or the whole fire. Both are essential to increasing our understanding because wildfire behaviours reflect the workings of a dynamical system of combustion, heat transfer and ignition. To develop understanding and models of fire behaviour that apply to the broad range of fuel types and environmental conditions in wildlands, we must obtain measurements from field-scale fires that are comparable to measurements from laboratory fires. Quantified descriptions of whole-fire characteristics and their variation are essential for improving fire management, fire planning, fire mitigation and safety on wildland fires. They are also essential for validating model predictions and assumptions about how the underlying physical processes manifest in fire behaviours. It has seldom been possible to obtain extensive data on fuels or fire behaviour on wildfires – that is, fires that are not fully controlled – because predictions of weather and fire behaviour have not been specific enough for researchers to opportunistically gain safe access. This problem may be addressed in the future with technological innovations, particularly in remoting sensing and unmanned aerial vehicle use. In this chapter, we review traditional methods of measuring wildland fire behaviour and also describe emerging techniques, many from application of new technology. While fire science overlaps with many other disciplines, which use methodologies specific to their fields (e.g. atmospheric sciences, ecology and forestry), we focus here on methods with a direct bearing on wildfire behaviours.

Sampling and experimental design

Variability in fuels, weather and fire behaviour make it essential that good *sampling design* be applied to organising and obtaining data, particularly in field-scale experiments. Sampling design is used to structure the collection of a limited number of measurements when it is impossible to measure completely (i.e. to census) an entire population. Considerations important to sampling design in wildland fire include the spatial pattern, precision and timing of measurements.

Measurement methods can be classified as *direct* or *indirect*. Direct measurements describe attributes of an object of interest (i.e. fuel material such as a log, or the height of flames) by measuring the object itself, while indirect measurements describe these attributes by associating them with a related characteristic that is more easily measured (e.g. counting plant stems to estimate biomass). *Precision* describes the repeatability of measurements; it is often expressed as the variation associated with estimates of the characteristics of a distribution; for example, variance and standard deviation describe the variation around an average. Because of the huge variability of fuels and wildland fire behaviours, it is important to measure and control the precision of measurements. Precision differs from *accuracy*, which describes how well a set of measurements estimates the true quantity. To obtain accurate data with maximum precision, it is essential to select appropriate measurement techniques and instrumentation and to use them properly.

Experimental design can determine the quality and applicability of experimental results. In a well-designed experiment – which includes carefully controlled environment and input variables, appropriate ranges of observations, and adequate numbers of replications – the main effects of each variable can be identified and isolated from sources of error. Laboratory experiments can, of course, control the environment and isolate the input variables more easily than field experiments can. The natural complexity of the field environment renders experimental design for use in prescribed fires and wildfires especially challenging. Fortunately, the details of experimental design are well described by many academic resources, covering methods for vegetation ecology, range management, forestry and other fields in natural resources (e.g. Cook and Stubbendieck 1986; Coulloudon *et al.* 1999; Bonham 2013). In many cases, the same methods can be used for field measurements in fire science. Because the topic of experimental design is covered well in these sources, we focus this Chapter on measurement techniques specific to fuels and fire phenomena, not on the broader design in which those measurements would be taken. We also discuss ways in which specific fuel and fire behaviour considerations should influence experimental design.

Fire measurements

Combustion and heat release

Laboratory and field fires offer different opportunities and challenges for measuring fire behaviour, but reliable measurements are essential at all scales. Fuel consumption is important for characterising the post-fire environment. It is also used to indirectly estimate heat release from combustion, even though this provides limited insight into the timing or rates of heat release in flaming spread. Heat release in the flaming phase can be directly measured or estimated from flame zone characteristics. Remote sensing methods are being introduced to provide quantitative information on many of these variables.

Fuel consumption

The complexity of wildland vegetation presents major challenges to the description and measurement of biomass as fuel for fire behaviour (**Chapter 6, Wildland fuel**). Similar challenges apply to measuring the amounts and rates at which those fuels are consumed by fire, even though fuel consumption is one of the primary means of estimating heat release during fires. The amount of fuel combusted has been termed the *available fuel* (Byram 1959), which means its condition (e.g. moisture, temperature, size, spatial arrangement) makes it available for combustion at the time when it is exposed to fire. But the mass of fuel consumed is not directly convertible to heat release because, first, moisture must be evaporated from the fuel before it ignites and, second, combustion itself is rarely complete (leaving char and partially combusted residues that are difficult to measure). In addition, if fuel consumption is to be used for estimating fireline intensity (Byram 1959), then losses of heat to radiation must be subtracted from the heat release of the fuel burned in flaming (Nelson and Adkins 1986). Thus far, fire research has had little success in distinguishing the amount of fuel consumed in flaming from that consumed in smouldering or glowing. Nevertheless, fuel consumption is often the primary method of estimating heat release from field burns and even from laboratory-scale fires.

Fuel consumption measurements are designed to quantify the rate at which fuel is consumed (mass per unit time), the fraction of pre-burn fuel consumed, or the total fuel consumed (mass). The values are expressed as an average over a specific area. Fuel properties and consumption may be quite uniform in small areas in laboratory fires, more heterogeneous for field-scale experiments (10^2 to 10^4 m^2), and highly variable in wildfires over vast landscapes. For most research purposes, measurements of fuel consumption should be able to describe specific fuel components, such as litter and duff, woody fuels of various size and conditions (rotten *v.* sound), grasses, shrubs, and canopy fuels. Ideally, these measurements should be partitioned into the fractions consumed in the flaming and non-flaming phases. With that information, fuel consumption measurements can be used to distinguish the energy released in the flaming front (fireline intensity) from that released in glowing and smouldering combustion behind the front, possibly over long time frames. Data that distinguish flaming from smouldering consumption can also be used to explain and predict smoke production, since because flaming produces fewer particulates than smouldering for a given amount of mass consumed.

Fuel consumption: rate

Measurement of fuel consumption rates (kg m^{-2} s^{-1}) in laboratory fires have long used some form of electronic balance beneath the fuel bed to record changes in fuel weight over time (Anderson and Rothermel 1965; Rothermel 1972; Wilson 1982). In spreading fires, this requires a section of the overall fuel bed to be weighed in isolation from physical contact with anything next to the weighing platform (**Figure 9.1**), since contact with nearby fuels or the fuel bed may bias or add error to the estimate of mass consumed solely on the weighing section. The difficulty of isolating the fuel bed being weighed is perhaps the reason why this technique is mainly used with simple fuel types with small particles, such as wooden cribs, conifer needles or excelsior. Continuous mass-loss measurements with stationary fires, such as wooden cribs, is more straightforward (see an example in **Chapter 4, Burning and heat release rate**). Laboratory experiments that use wind on spreading or stationary fires are more complicated, since airflow impinging on the balance itself may add variability to the measurements.

Continuous weighing of burning fuels produces a weight-loss or mass-loss history of fuel consumption that often exhibits a period of steady burning, indicating flaming combustion and an extended period of slower glowing and smouldering combustion. The rate of mass loss and duration of the flaming period can then be calculated (see

a

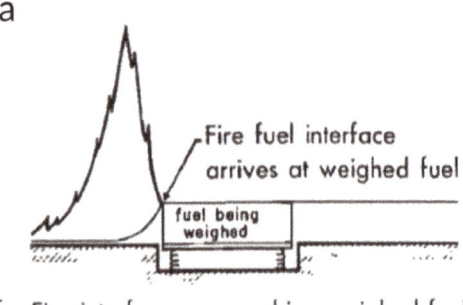

I Fire interface approaching weighed fuel

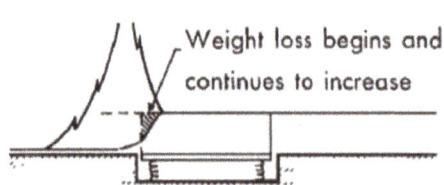

II Fire burning into weighed fuel

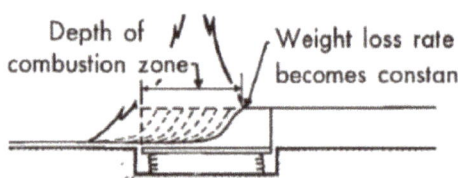

III Steady weight loss rate achieved

b

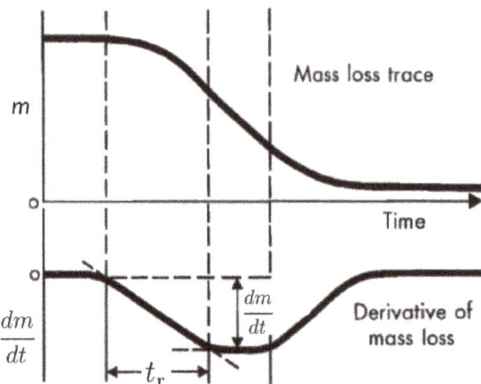

Figure 9.1: Illustration of (a) the method of measuring fuel mass loss in a laboratory fire spreading across a weighing section of a fuel bed and (b) the idealised time-series of mass-loss data obtained from these measurements, where t_r is the flame residence time, dm is the change in mass, and dt is the change in time, used to obtain the mass-loss rate (from Rothermel 1972).

Rothermel 1972; Wilson 1982). The technique of continuously weighing fuels in spreading fires must account for the rate of fire progress across the weighing platform. In the ideal case, a fire will spread steadily with a narrow flame zone across a relatively large weighing platform, producing a mass-loss history adequate for estimating the burning rate of fuel in the flame zone and thus the fireline intensity. In laboratory experiments, the entire fuel bed can be mounted on load cells (Nelson and Adkins 1986), or a separate weighing section can be used (Anderson and Rothermel 1965; Rothermel 1972; Wilson 1982). However, a fire spreading quickly or with a flame zone that is deeper than the length of the weighing platform will record consumption rates that depend on the progress of the fire. In this case, the time trace of fuel consumption will be strongly influenced by the relative position of the flame zone within the overall burning region, so time integration of the data will be needed to estimate the steady consumption rate.

Use of mass-loss data for distinguishing flaming fuel consumption from post-flaming consumption is difficult because it is not clear when flaming ceases unless all of the fuel is consumed in the flaming phase. This challenge is present even in the most controlled laboratory studies and even when burning only dry, fine fuels. The back of a flaming zone is often ragged as fuels transition from flaming to smouldering combustion or are extinguished. Even pine needles and excelsior may not be completely consumed in flaming unless they are very dry and loosely packed. Large woody materials, densely packed fuel beds, and fuels with high moisture contents are even more challenging and can smoulder for a significant time. If a boundary between flaming and smouldering combustion cannot be determined, fuel consumption in smouldering will inflate calculations of fireline intensity and flame dimensions.

Fuel consumption: total

With field-scale fires, it is more common (and feasible) to measure the total mass of fuel consumed per unit area (kg m^{-2}) than to measure the rate of consumption. Total fuel consumption is the difference between the pre-burn oven-dry mass and the mass of post-burn residual material (fuel, char and ash). As mentioned above, this is called the available fuel. It is affected by many fuel bed and particle properties and also by environmental conditions (described in **Chapter 6, Fuel burning rates and consumption**). Under field conditions, direct measurements focus almost entirely on total consumption for two reasons. First, it is difficult to install invasive equipment to measure time-dependent characteristics of fuel consumption without altering the fuel bed properties and subsequent burning behaviours. For example, it would be very difficult to isolate duff and litter or woody fuels in a forest or grass in a meadow on a weighing platform without changing their burning or moisture characteristics. Second, the variability of fuels in many vegetation types would require using these expensive measuring techniques across very large areas, having a very large sample size, or both. While direct measurement is generally used to measure fuels, the sampling process itself can be invasive enough to alter the fuel bed. As samplers walk through and stand in the fuels, they may crush the fine fuels, which would affect fire behaviour within and around the plot.

Under field conditions, fuels of all types are highly variable across a given land area. Sample design can be used to help determine the sampling intensity needed for a given level of precision. The lowest sample variance is achieved if the pre-burn and post-burn measurements are taken at exactly the same locations. This applies to all kinds of fuels – forest floor, dead woody material, grass and shrubs, and canopy fuels. To relocate plots for remeasurement, their positions can be marked with unburnable monuments (i.e. steel posts). Remote sensing techniques, and post-processing of LiDAR (Light Detection and Ranging) returns and thermal and video images, have established routine procedures for geo-rectifying data (Kremens *et al*. 2010). Thermal imaging, in general, has become increasingly effective for quantifying total fuel consumption through measurements of radiative power (Kremens *et al*. 2012; Hudak *et al*. 2016). This is discussed below, when we address more direct measurements of heat release in fires.

For operational prescribed burns, fuel consumption is usually calculated using equations based on the moisture content and pre-burn loading of selected fuels (for examples, see Brown *et al*. 1985; Reinhardt *et al*. 1991; Prichard *et al*. 2006). Equations for consumption of duff, litter and dead woody material in both natural fuels and in *slash fuels* (residues left over from logging operations) are helpful in estimating smoke emissions, the exposure of mineral soil for forestry applications (replanting, seeding, etc.), and the impact of long-duration heating on soils and plants. Other predictive models, both empirical and physically based, have been developed for estimating fuel consumption; the reader is referred to syntheses by Hollis *et al*. (2010, 2011) and Ottmar (2014) for modelling details.

Direct methods of fuel measurement (i.e. *fuel inventory*) have been developed for each fuel component and vertical stratum because of the diversity of fuel sizes, spatial distributions and heterogeneity that characterise wildland fuels (Brown *et al*. 1982). To quantify fuels in wildlands, then, a variety of methods must be combined into a general sampling scheme. Sampling in forests must account for litter and duff, downed woody material, herbaceous and shrubby cover, tree regeneration and canopy fuels. Sampling in shrublands must account for grass and other herbs, litter and probable patchiness of fuels. The limitations described for field-scale fuel characterisation in **Chapter 6, Wildland fuel**, also apply to measurements of fuel consumption. The greatest limitation is that most field-scale inventories can usually provide only estimates of average fuel loading and consumption (kg m^{-2}) over large areas, perhaps many hectares in size. These data are likely to be far less detailed than the fine-scale fuel bed and spatial variability characteristics that are needed to explain and predict the physical processes of fire spread and energy release. However, with the rapid advances in capability and practicality of remote sensing techniques, it may become possible to directly measure many more characteristics of wildland fuels, both before and after burning. We focus here on techniques that are appropriate at relatively fine scales (landscape or smaller) rather than on large-scale regional, continental, or global scales, which can also be addressed with remote sensing technologies.

Forest floor (litter, duff, ground fuel)

Forest floor material in most forests and some shrublands includes recent leaf litter, partially decomposed litter (i.e. *duff*) and buried rotten woody debris. In many forest ecosystems, this material is relatively continuous across the land and often constitutes the highest loading of any surface fuel component. Photo series and their associated measurements have been developed to estimate the quantity and depict the variability of ground fuels and other fuel materials in a variety of forest and vegetation types of the United States (https://www.fs.fed.us/pnw/fera/research/fuels/photo_series/).

The duration of forest floor combustion is often measured, in both laboratory and field experiments, with thermocouples inserted at different depths (Hartford and Frandsen 1992). (For more detailed information on thermocouples, see **Convection** below.) Forest floor combustion can occur in both flaming and smouldering, and it can transition back and forth throughout the long duration of a burn as air temperature, humidity and wind ventilation change. For example, smouldering typically occurs at night and in the morning but may transition to flaming by afternoon, when the ground is exposed to sunlight, higher air temperatures and ventilation from up-valley or up-canyon winds. Smouldering can potentially release heat for many hours, days or even months, over large areas. The duration of burning for duff and litter depends on moisture content, wind conditions, and interactions with combustion of other fuels (Albini and Reinhardt 1995); for duff and organic soils, duration and completeness of burning also depend on its inorganic constituents (i.e. mineral soil) (Frandsen 1987; Hungerford *et al.* 1995). Smouldering can generally take place at much higher moisture contents than flaming combustion.

Consumption of the ground fuel stratum is typically sampled at points within a burn unit by measuring the change in depth after burning. The most common technique, explained by Beaufait *et al.* (1975) and Brown *et al.* (1985), involves the use of metal spikes (*duff pins*) inserted into the ground before burning with the spike heads flush with the upper litter surface (**Figure 9.2a**). Locations of the duff pins must be marked to facilitate relocation after the fire when the vertical distance between the remaining duff or soil surface and the top of the spike is measured. The change in depth is converted to fuel loading (kg m^{-2}) based on correlations of depth with bulk density (kg m^{-3}). The correlation is obtained by destructive sampling for each vegetation or litter type – pine *v.* fir, for example, or duff *v.* litter (see Finney and Martin 1993; Stephens *et al.* 2004; Alexander *et al.* 2004). *Destructive sampling* means physically removing the fuel from a specific volume (e.g. collecting duff 1 cm thick within a sampling frame 0.2 m × 0.2 m), drying it in a laboratory, and then weighing it.

Relationships between depth and fuel mass vary with many conditions – duff depth itself, for example, as well as vegetation composition, moisture regime and productivity of the site, since the balance between litterfall rates and decomposition rates affects the loading–depth relationships (Keane 2008). Variability at scales of a few metres or tens of metres means that some form of stratified sampling is often used to reduce sample variance and the needed sample size. *Stratified sampling* means that sampling of a variable, such as duff depth, is partitioned into separate categories based on known or suspected differences between categories. In stratified sampling, each category or combination of categories is sampled independently. Ideally, stratified sampling would be used to measure the separate depths of pre-burn duff and litter, but the layers cannot easily be distinguished before burning without disrupting the material and altering its

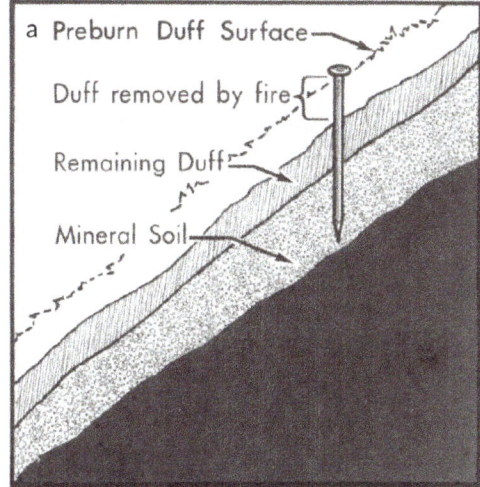

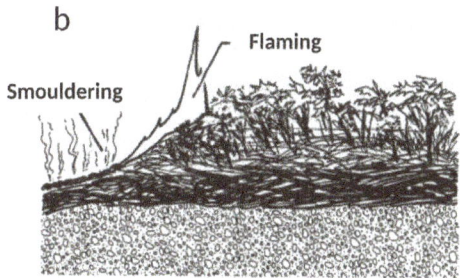

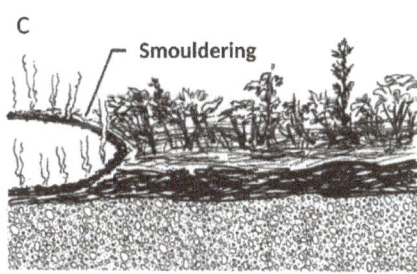

Figure 9.2: Illustrations of (a) duff pin used for measuring consumption of duff on the forest floor by calculating the difference between pre- and post-burn depths (from Beaufait *et al.* 1975); (b) consumption of forest floor by downward flaming; and (c) consumption of forest floor through horizontal spread by smouldering (reproduced with permission from Ryan and Frandsen 1991).

subsequent burning behaviour. Because of this problem, proxy locations (outside the experimental area or away from sample plots) must be used for measuring pre-burn duff and litter depths. Forest floor material is typically deeper underneath tree or shrub canopies where litterfall rates are higher, so

sampling can use distance to the stem of nearby trees or shrubs in order to reduce variance in the sample (Hille and Stephens 2005). The same proxy technique can be used to sample spatial variation in pre-burn moisture content. Topographic attributes may be helpful in stratifying the sample, because topography contributes to moisture variability over large areas (Potts *et al.* 1986). Thus, proxy samples might be taken from ridgelines, midslope, and valley bottoms, as well as different aspects (e.g. north and south).

The duff pin technique is well suited to a fuel structure that burns from the top down, such as loosely compacted pine litter (**Figure 9.2b**), especially if consumption in the flaming zone is nearly complete. However, this is not always the case, especially during smouldering combustion. Dry forest floor materials, including organic soil horizons, are often cracked or have other disruptions (e.g. from burrowing rodents) that allow smouldering to proceed laterally, from beneath the surface (**Figure 9.2c**). This burn progression creates patchy consumption (Ryan and Frandsen 1991; Hungerford *et al.* 1995) and thus increases variation in measurements. In addition, if only the lower layers of the forest floor are consumed, it may bias the resulting calculations, since the bulk density of forest floor material tends to increase with depth. In other words, since recent litterfall is less compact than deeper layers of decomposed material, burning only the lower layers of the forest floor may consume greater loadings than indicated simply by measurements of reduced depth.

Dead and downed woody fuel

Dead and downed woody material distinguishes logs and branches lying on or near the ground from those that remain standing or attached to standing trees (live or dead). Woody material is readily measured by *non-destructive methods* (i.e. not disturbing the sample material by measurement), which avoid altering the fuels and their responses to burning, thus avoiding the need for proxies (discussed above). Advancing technology offers a widening variety of alternatives and improvements in methods for estimating pre- and post-burn fuel characteristics. The reader is referred to the text by Keane (2015) for details on the variety of techniques available for direct and indirect fuel measurement.

The most common direct, non-destructive method for measuring loading and consumption of downed woody fuel material at the plot level has been the *line-intercept method*, also called the *planar-intercept method* (Van Wagner 1968; Brown 1971; Brown 1974; Brown *et al.* 1982, de Vries 1986). A tape or string is attached to a pin that marks the sampling location. The tape is then stretched out at a random angle from the pin. The sampler then counts every fuel particle by size class that intersects the tape, both before and after burning (**Figure 9.3**). The diameters or diameter classes of fuel intercepts are converted to volume, then to mass (based on the density of the fuel material), and finally they are summarised per unit land area based on the length of the transect. The numbers of transects and their lengths can be customised to accommodate spatial variability of fuel materials (Pickford and Hazard 1978). The line-intercept method has undergone much testing and refinement which demonstrate that it can robustly accommodate the variability associated with natural fuel configurations, including sloping topography and elevated or suspended fuel elements. Consumption and loading can be summarised by any of the properties tabulated during the inventory, including fuel size class, rotten or sound wood, and even species (**Figure 9.3**).

The endpoints of each fuel transect should be permanently marked with steel stakes during pre-fire inventories to facilitate relocation and remeasurement after burning. Spatial variation in fuels is very high, especially for large logs. Therefore, if post-fire data are not collected at the same location as pre-fire data, it may be impossible to obtain any statistically reliable conclusions about the amount of fuel consumed by the fire. Field experience also shows that sometimes woody fuels *increase* after burning if trees fall to the ground or roll downhill, if large logs formerly buried by duff or litter

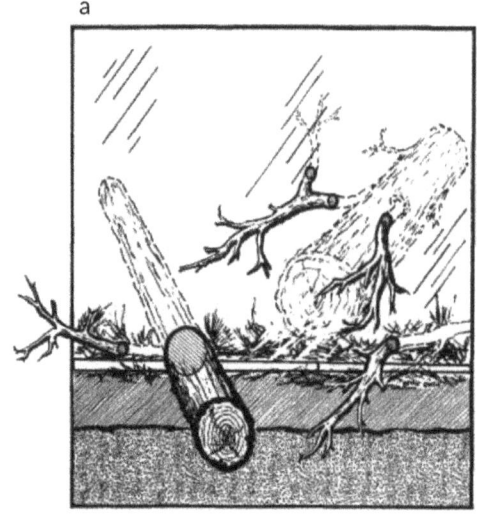

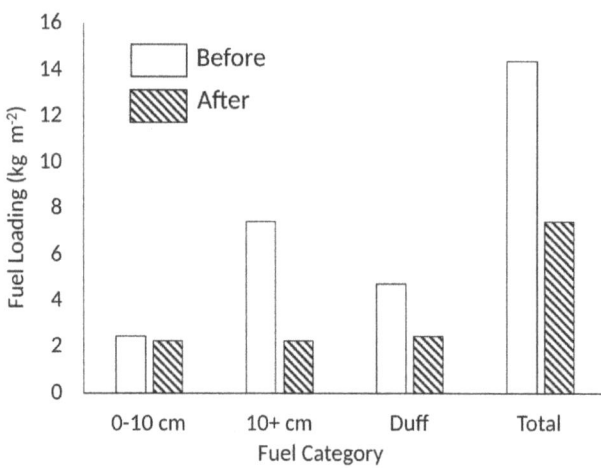

Figure 9.3: (a) Sketch of fuel inventory by line-intercept method, in which the intersections of dead and downed woody fuel are tallied along a transect (from Brown *et al.* 1982) and (b) example measurements of pre- and post-burn fuel loading, which can be used to determine fuel consumption by particle size class (from Reinhardt *et al.* 1991).

become exposed by the fire (Reinhardt *et al.* 1991), or if substantial branchwood falls during the fire. These possible confounding factors make it important to conduct post-fire inventories as soon as possible after burning and perhaps to use pre-burn photographs to account for these processes.

Indirect methods for estimating plot-level pre-burn fuel properties use photographic references ('fuel photo series') for making visual assessments. It is possible to adapt such methods to estimate surface fuel consumption, but this has not been demonstrated. The idea of a fuel photo series was developed in the 1970s (see Maxwell and Ward 1980) to assist fire managers in estimating fuel properties and selecting the fuel models appropriate for use in fire models. Fuel photo series have been implemented widely in the United States (https://www.fs.fed.us/pnw/fera/research/fuels/photo_series/) and also in parts of Australia (Gould *et al.* 2007; Hines *et al.* 2010), and Canada (Lavoie *et al.* 2010). A photo series is developed from detailed fuel inventories of example areas, which are then photographed. Then users visually match the photographs with fuel conditions observed at other locations. For a photo series to be useful for estimating fuel consumption (as opposed to pre-fire loading only), post-burn conditions must also be included (Lavoie *et al.* 2010), but this is uncommon. Visual methods are convenient, but they have been shown to be imprecise in practice, so direct methods are preferable for quantifying fuel loading for fire behaviour research. An alternative is to develop a series of photographs to serve as a guide for evaluating specific fuel components individually, such as shrubs, grass, woody debris etc. Photos of different fuel components can then be visually matched with the appearance of those components in the field (Sikkink and Keane 2008). With this method, the user matches individual fuel components with photographs of that component rather than attempting to match all features of an entire fuel bed, as is done with other photo series.

Large woody fuels can produce large amounts of heat in a wildland fire. Their consumption can be directly measured by wrapping wires around individual logs before and after burning to obtain pre- and post-burn circumferences (**Figure 9.4a**) (Reinhardt *et al.* 1991), which can then be converted to diameter (assuming the logs have round cross-sections) and volume consumed. The moisture content and density of the logs (rotten and sound) must be measured before burning if changes in diameter are to be converted to mass consumption and heat released (**Figure 9.4b, c, d**). In practice, this technique is most appropriate for large pieces of woody debris (> ~10 cm). It has been most commonly used for studies of burning in logging slash. Few details about sampling design have been published, but four criteria for stratification have been suggested: green (recently cut or killed) *v.* dead before the current year, sound *v.* rotten, the amount of bark still attached, and whether the log is in contact with the ground or other logs, or elevated above ground. Many of these properties of large woody material have been shown to influence their consumption in fires (Hollis *et al.* 2010).

Only a few experiments have tried to distinguish the consumption of dead and downed woody fuels in flaming combustion from consumption in the smouldering phase. One of these used an outdoor fuel bed containing slash fuels that were thoroughly inventoried by size class before burning (Brown 1972). Line fires were used to burn the length of the fuel beds; as flaming ceased, smouldering fuels were immediately extinguishing with a spray of water fog. A thorough post-burn inventory then enabled researchers to estimate the proportion of fuel consumed in flaming alone (**Figure 9.5a**). The larger diameter of the fuel, the lower the proportion of mass consumed in flaming and the higher proportion of mass that could have been consumed had smouldering been permitted. A similar finding was reported for laboratory burns consisting of 1 kg baskets of fuels of different sizes (Burrows 2001) (**Figure 9.5b**). Brown's (1972) relationship was subsequently used to estimate flaming consumption and total fuel consumption on a large-scale prescribed burn in logging slash in Ontario, Canada (Ohlemiller and Corley 1994). However, the relationship may not be widely applicable, since differences among tree species and variation in moisture could cause actual fuel consumption to differ substantially from these predictions.

Herbaceous fuel

Like many wildland fuels, herbaceous fuels consist of an immense variety of plant species and their multiple organs (i.e. stems, roots, leaves,

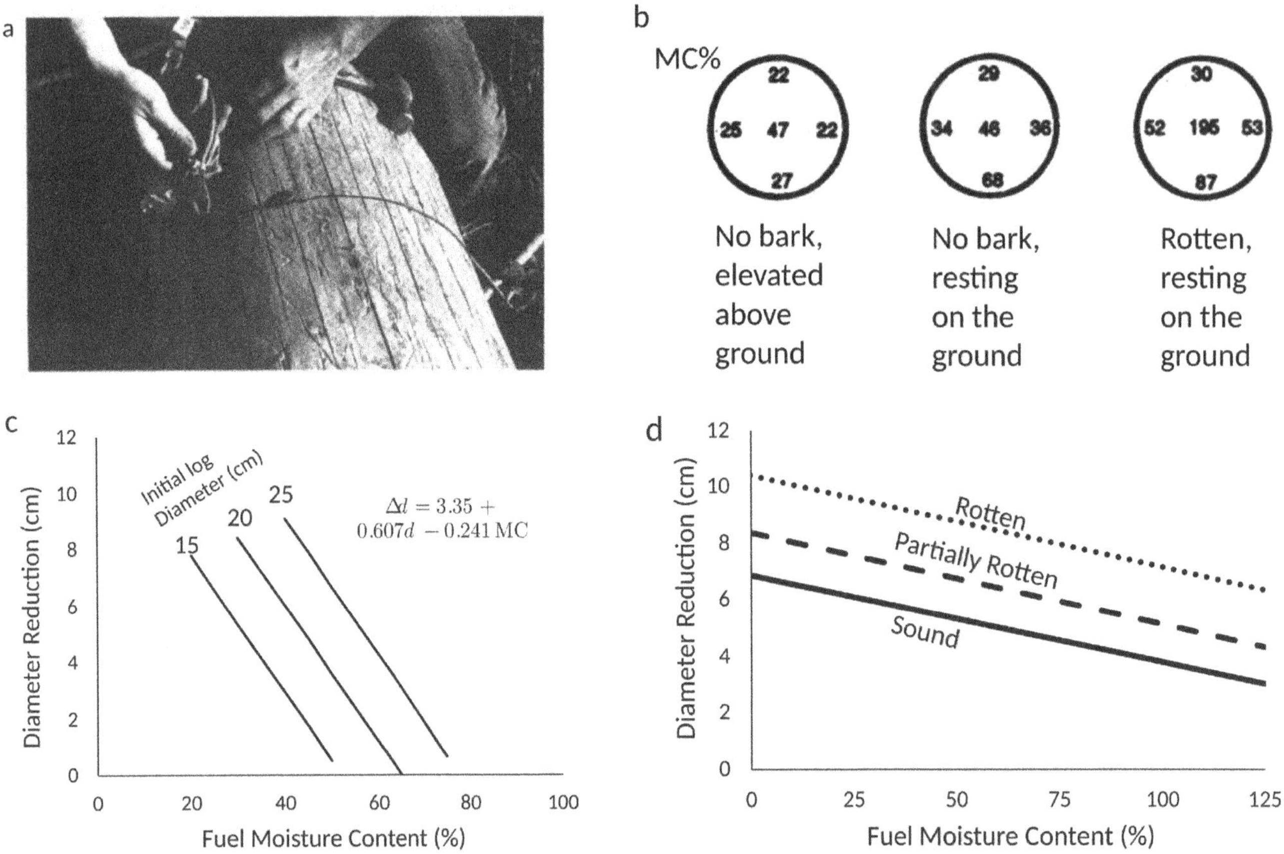

Figure 9.4: (a) Photograph of technique for measuring diameter reduction of large logs by wrapping wire around them; (b) measurements of the moisture variation within large logs shows greater moisture in the interior and for rotten logs and those logs resting on the ground; (c) model of diameter reduction of sound large logs by burning developed from experimental burns in Montana, USA, showing effect of average log moisture content (MC) and pre-burn diameter (d); and (d) variation in diameter reduction for logs based on their average moisture content and degree of decay (from Reinhardt et al. 1991).

$$\Delta d = 3.35 + 0.607 d - 0.241 \, MC$$

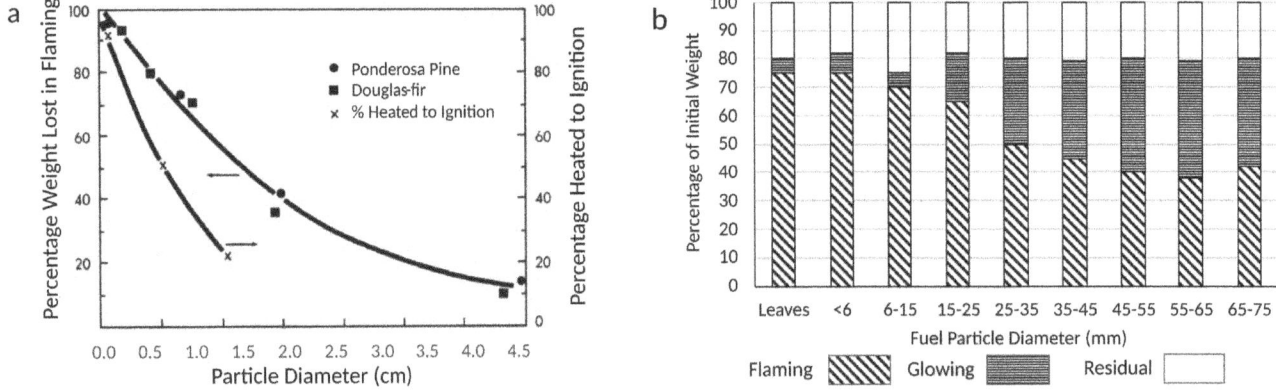

Figure 9.5: Percentage of fuel weight lost (a) in flaming combustion for outdoor experimental burns in *Pinus ponderosa* and *Pseudotsuga menziesii* slash fuels in terms of particle diameter, also showing the percentage of the fuel by particle size heated to ignition (from Brown 1972) and (b) in flaming and glowing combustion from containers with different size classes of woody fuels in laboratory burns (reproduced with permission from Burrows 2001).

fruiting bodies). Fuel-related attributes of grasses and other herbs have been measured for very few species (see **Chapter 6, Wildland fuel**). Unlike woody fuels, most herbaceous vegetation grows anew each year and exhibits strong seasonal and species-specific trends. Throughout the growing season, the aboveground parts of grasses and herbs grow, deteriorate with age, die and decay. Each of these processes affects their particle and bulk characteristics, including moisture, height, loading, density and continuity. Their bulk density is further affected when they are manipulated by mowing or grazing or matted by snow or rain. For these reasons, they must be sampled as close as possible to the time of the fire – both before and after. Although seldom quantified, the spatial variability in grasses and herbaceous fuel properties is important to all of the physical processes of fire spread and behaviour, and new techniques will probably need to be developed to provide such details for future modelling.

Many techniques have been developed for measuring and estimating the properties of herbaceous vegetation (i.e. averages and variations) for ecological and management purposes; the reader is referred to standard texts for details on various techniques (Mueller-Dombois and Ellenberg 1974; Coulloudon et al. 1999; Bonham 2013). The most reliable measurements for our purposes – better understanding the relationships between herbaceous fuels and fire behaviour – come from destructive sampling and *double sampling methods*. There are no reliable substitutes for destructive sampling of grassland fuel loadings and consumption, but destructive sampling obviously alters the fuel bed, so fire behaviour cannot be measured at exactly the same sampling locations. To overcome this obstacle, destructive pre-burn sampling can be performed outside the boundaries of the burn unit if the vegetation is comparable to that within the unit. Destructive sampling is also very time-consuming, so it imposes practical limits on the number of plots that can be sampled. Double sampling provides one way to overcome these obstacles.

In double sampling, researchers first measure *key attributes* of the herbaceous material within a sampling frame ((i.e. *quadrat*, usually < 1 m^2). These are attributes likely to be related to biomass, such as cover, height, density, species composition, or volume (with samples collected in larger, three-dimensional (3D) frames); they may be measured directly or visually estimated. After measuring key attributes, the scientists use destructive sampling (clipping and weighing) to measure biomass from the same quadrat. The clipped and weighed material is converted to dry loading (kg m^{-2}) using sub-sampled moisture contents (see **Fuel moisture** below). After developing correlations between the key attributes and biomass, a second procedure is used to measure *only the key attributes* in new quadrats within the study area, then the correlations applied to estimate biomass. Because the key attributes can be measured much more quickly than destructive sampling permits, double sampling enables scientists to obtain many samples in a short time. This is especially helpful since pre-burn measurements of herbaceous vegetation must be obtained as close as possible to the burn date because of their constantly changing phenology.

Destructive pre-burn sampling can be stratified by cover or species composition. Stratification may produce correlations specific to attributes of individual species, helping to reduce error in measurements obtained during the second stage of sampling. This may be especially helpful if a particular species or group of species (e.g. invasive annual grass species in California, USA) influences fire behaviour differently from other species in the study area.

The second stage of double sampling begins with collection of key attribute data from a large number of plots within the study area. This stage is faster and much less disturbing to the fuel bed than the destructive sampling stage. Various sampling designs can be used to find unbiased ground locations for measuring the key attributes; these include linear transects, belt transects (i.e. having a fixed width) and random points. All sampling design considerations apply to this stage of measurement, including potential stratification of the study area by overstorey or understorey species, disturbance history, successional stage or other features. Sample plots or transect end points must be marked for remeasurement after the fire. Correlations from the first (destructive) sampling stage are used to convert data from the second stage to biomass and to describe the distributions of the data (using mean, mode, variance, etc.).

The *weight-estimate method* is a variant of double sampling. In this method, destructive samples are used to calibrate the observer's visual estimate of fuel mass on a large number of non-destructive plots. This enables the observer to rapidly estimate fuel loadings on study plots, which can greatly increase the number of samples collected in a limited time.

For some species biomass can be estimated by using another indirect measurement technique – *allometry*, the consistent scaling of measurements among parts of an organism. The dry mass of some grass species can be estimated from measurements of height or other characteristics. For example, range scientists have found that the dry weight of bunchgrasses (which grow in distinct tufts, i.e. 'bunches') can be predicted from the diameter and height of the bunch (Nafus *et al.* 2009). Given such a relationship, scientists can measure the diameter and height of bunchgrass plants at sample points on transects, then convert the measurements to fuel mass. One drawback of the method is that grass allometries can vary considerably from year to year based on rainfall, grazing or disturbances (Johnson *et al.* 1988), so they must be redone from new samples each year.

For both quadrat and allometric methods, fuel consumption is estimated by subtracting post-burn measurements from pre-burn measurements. If the vegetation was dormant and very dry at the time of burning, there will usually be little fuel remaining. But if the vegetation was partly green or patchy when burned, or characterised by thick stalks, tufts or deep mats on the ground, considerable biomass is likely to remain. In this case, post-burn measurement will require repetition of the measurement techniques used for pre-burn measurements. If pre-burn measurements were collected with destructive sampling only, then the study could be designed to use paired plots, either systematic or randomised, for pre-burn and post-burn sampling. If this approach is used, samples would be collected before and after fire from alternating quadrats (see Wright and Prichard 2006; Wright 2013) or from one of a pair of plots chosen at random. If the burn is patchy or spatially incomplete, consumption estimates will have high variance; to obtain meaningful information, researchers may need to stratify post-burn data by burned *v.* unburned, or they may need to estimate and record the proportion of biomass burned in each plot. **Figure 9.6** shows results from a study in sagebrush shrubland that stratified samples by the proportion of the area burned at each site.

Shrub fuel

Areas dominated by shrubby vegetation are known variously as shrublands, brushlands or scrublands. Measurement of shrub fuels and their consumption by fire requires specialised techniques to account for their short stature and multistemmed woody growth form. Shrub heights may range from less than one metre to several metres and may exhibit a wide range of coverage and patchiness. Extensive shrublands occur in desert and Mediterranean climates on every vegetated continent, and many of these are characterised by intense, fast-moving wildfires. Well-known examples of wildfire problems are found in the chaparrals of Southern California and Arizona, USA; sagebrush (*Artemisia tridentata*) in the interior Great Basin of the United States; matorral, maquis or garrigue of Mediterranean Europe; matorral of Chile; fynbos of South Africa; and mallee and spinifex in Australia.

Sampling for fuels is especially difficult in some shrublands because the plants can be nearly inaccessible in dense, thorny or allergenic vegetation. In addition, measurements are challenging because of the vertical distribution of fuel particles and leaves, the range of size classes of fuels, and

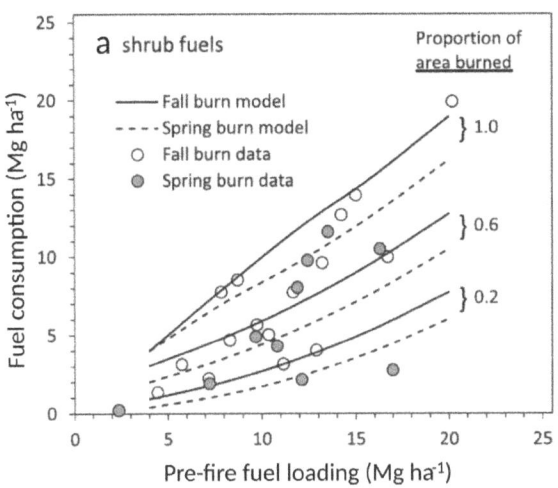 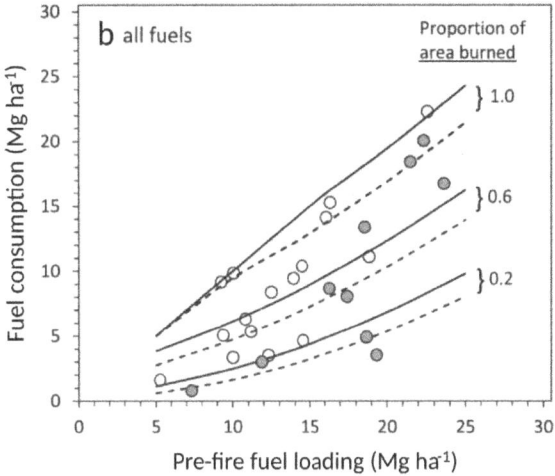

Figure 9.6: Fuel consumption data obtained by destructive sampling and modelled relationships for sagebrush (*Artemisia* species) fires in spring and fall showing differences associated with burn coverage at each of 26 sites (a) for sagebrush only and (b) for all above ground biomass, including grass and litter (reproduced with permission from Wright 2013).

different live and dead fractions. However, just as with herbaceous vegetation, direct destructive sampling and double sampling methods are the most accurate means of quantifying loading and consumption of the fuels. The primary phase of destructive sampling involves clipping or sawing all material from within a 3D quadrat frame ~0.5–1 m² across throughout the entire vertical profile of the plant (Sapsis and Kauffman 1991; Wright 2013). The collected material is then separated by stem size and foliage, weighed in the field or a laboratory, and subsampled for moisture (see **Fuel moisture** below). Live and dead materials are accounted for separately. Once researchers have developed correlations between key attributes (e.g. cover, height, diameter, species) and biomass from destructive sampling, they can proceed with the second step of double sampling – using transects or randomised point locations to measure only the key attributes of intercepted shrubs for estimating fuel loading. As for herbaceous sampling, double sampling can greatly increase the number of plots that can be sampled within a limited time before burning.

An alternative to the quadrat method is to use allometries of shrubs that have been derived from destructive sampling. Allometric equations have been developed for many shrub and small tree species all over the world, primarily for estimating aboveground biomass rather than for fire applications (see Brown 1976; Smith and Brand 1983). Many of these predict branch size distributions, foliar fractions and biomass for individual species as a function of stem diameter, cover and/or height, and they are very applicable to fuel inventory. Before using these equations, however, the researcher should check that they are applicable to the vegetation on the particular study site; if they are not, new equations should be developed.

As with all other fuel types, non-destructive sampling must be redone in the same locations before and after burning to minimise variance. Destructive sampling is often used to obtain post-fire measurements (Sapsis and Kauffman 1991). Even if allometric equations were used to obtain pre-fire measurements, they are unlikely to apply to shrubs after fire, so use of allometry does not alleviate the need for post-fire destructive sampling to determine residual fuel loading by size class. Fernandes *et al.* (2000) provide an example of using shrub allometry for fuel consumption in Portuguese shrublands.

Shrub vegetation rarely grows without accumulating leaf litter under the canopy. For example, a study of fire behaviour in sagebrush (Wright 2013) showed that the fuel load of shrubs alone (**Figure 9.6a**) was less than that for the total fuel complex (shrubs plus litter) (**Figure 9.6b**). For a complete fuel inventory, litter must be sampled or measured using the same methods that are used for measuring duff and litter on the forest floor (see **Forest floor (litter, duff, ground fuel)** above). In locations with discontinuous shrub cover, which are very common in dry ecosystems of the world, grasses and forbs often grow in the interstices between shrub canopies; this also requires inventory (see **Herbaceous fuel** above). An important aspect of fuel measurement in many shrublands concerns the continuity of shrub canopies and grass or litter (often represented by average percentage of cover), which determine the wind and fuel moisture conditions under which fire will spread. These thresholds were discussed in **Chapter 7, Effects of fuel continuity**. Advances in fire spread modelling are likely to require that size distributions for fuel patches and gaps be quantified characterisations of shrub and other fuel types in order to properly represent the spatial variability in physical processes of heat transfer and ignition in discontinuous fuel complexes.

There are an increasing number of examples of the use of LiDAR to estimate shrub biomass in pre- and post-burn conditions (Wang and Glenn 2009; Hudak *et al*. 2020) and these offer the advantage of being spatially resolved at finer resolutions than ground-based inventory techniques. Given the difficulty of direct sampling of shrub fuels, and the limitations in representing spatial fuel patterns, standardised methods for this technology would be very helpful in future fire behaviour and fire ecology studies.

Ladder and canopy fuels

In forests, *ladder fuels* are located between the surface fuels and the lower reaches of a tree canopy. The term describes their function in helping a surface fire spread vertically, as if climbing up a ladder. The term is intuitively appealing, but the physical role of ladder fuels in heating, ignition and vertical fire spread, which facilitate transition from a surface fire to crown fire, has not yet been examined. We do not yet know how much ladder fuel material is needed, and in what vertical and horizontal arrangements, to trigger canopy ignition and spread under specific fire and weather conditions

Ladder fuels traditionally include dead lower tree branches, loose bark flakes or strips, small understorey trees, and vines, lichens and epiphytes – basically any flammable biomass that contributes to vertical continuity. Qualitative descriptions of ladder fuels were discussed in **Chapter 6, Wildland fuel**, but little quantitative information is available on consumption of ladder fuels by fire. One example is a study by Lavoie *et al*. (2010), which inventoried fuels before and after experimental crown fires in jack pine (*Pinus banksiana*) and black spruce (*Picea mariana*) forests. Their results show that dead branches and bark flakes contributed substantially to pre-burn fuel loading in some stands (Lavoie *et al*. 2010); these were presumably consumed completely in crown fire (**Figure 9.7**). Bark fuels for ember source material and vertical flame spread are particularly important in eucalypt forests, and visual estimation methods have been developed specifically for assessing those components (see Gould *et al*. 2011; McCaw *et al*. 2012).

Most experimental studies of crown fire behaviour have relied upon manually intensive methods for estimating canopy loading and consumption at the plot level. Methods for obtaining these estimates combine many forest inventory methods, including a description of the distribution of tree sizes according to diameter, destructive sampling to establish tree biomass allometry for different species, and visually estimating percentages of consumed aerial fuel components. In forests that contain a substantial number of standing dead trees, these must be sampled separately from living trees. Dead branches and foliage on living trees also must be inventoried separately.

Consumption of ladder and canopy fuels is often determined by visually estimating the percentage of the canopy burned and assuming that

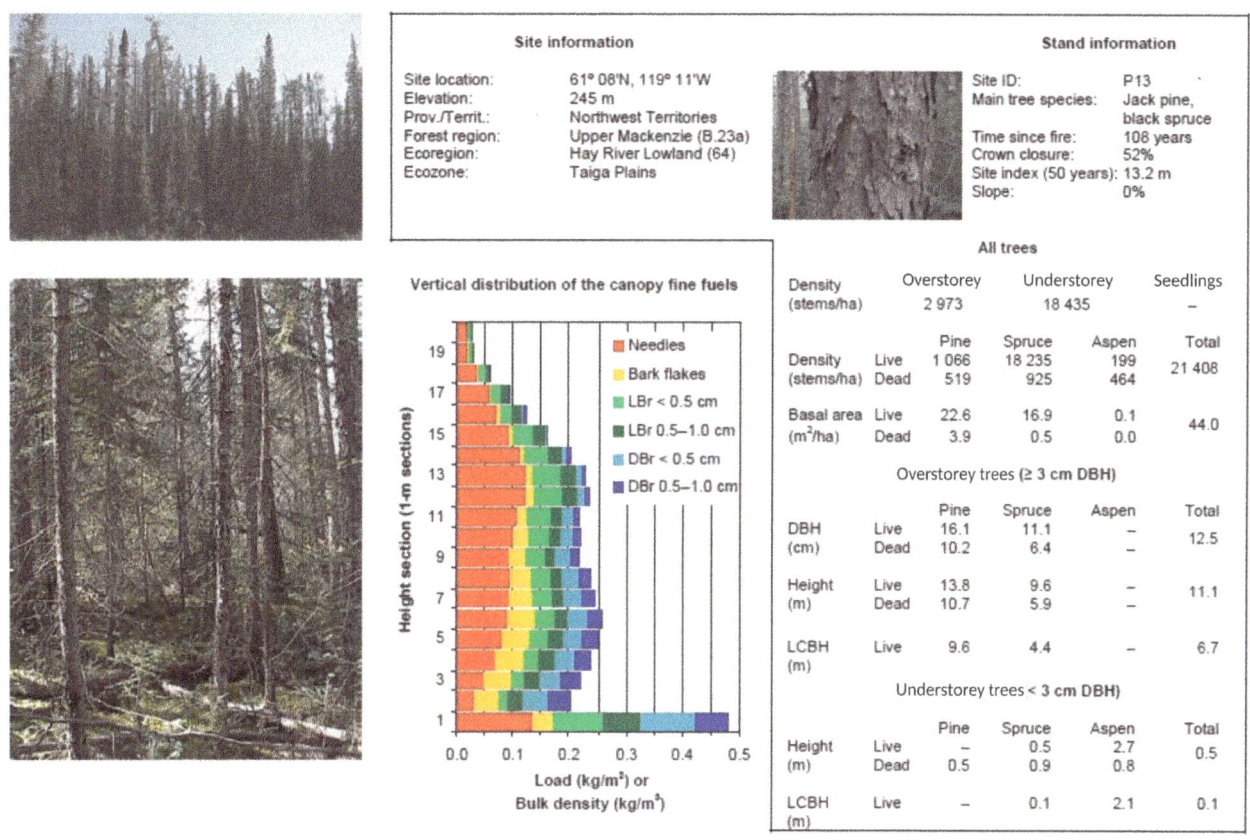

Figure 9.7: Photographs and data showing vertical distribution of fuel in jack pine–black spruce forests 108 years after fire, illustrating the presence and continuity of ladder fuel components, which include bark flakes (from Lavoie et al. 2010). DBH = diameter at breast height, DBr = dead branches, ID = identification, LBr = live branches, LCBH = live crown base height. Note: 1.0 kg/m² = 10 t/ha.

the fire consumed some proportion of the fuel of each size and condition in the burned fraction of the tree crowns. It is commonly assumed that 100% of foliage is consumed on trees burned in crown fires, but observation of charred needles, cones and small branches lofted and transported from crown fires (see **Chapter 8, Spotting and spot fires**) suggests that some unknown fraction of fine fuel is simply evacuated from the site by strong fire-induced winds. The fraction of branch wood estimated to be available for combustion – that is, material of a size and condition able to burn in the fire – has ranged from 100% of particles less than 10 mm in diameter (Stocks 1987, 1989) to 50% of particles less than 6 mm in diameter (Reinhardt et al. 2006a).

Direct estimates of the quantities consumed have been based on custom allometric equations for pre-burn canopy fuel loading (Alexander et al. 2004) and post-burn residual loading. Using these equations, Stocks et al. (2004) found that more than half of the branch material up to 30 mm in diameter was consumed by a crown fire. Another method of estimating pre-burn canopy fuels was developed from destructive sampling and provided for field use as a stereo photo guide (Scott and Reinhardt 2005). This publication includes a photo of the forest stand from the side and a hemispherical image looking upward to assess crown closure, associated with metrics that describe the stand in detail (**Figure 9.8**), including fuel loading throughout the crown profile.

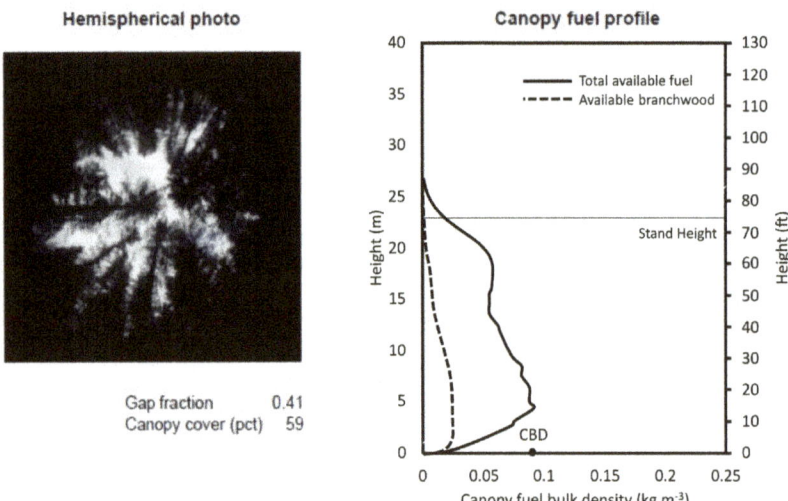

Figure 9.8: Photographs of canopy fuel structure (profile and closure) and graph of foliage and branch wood fuel profile in a stand of *Pinus ponderosa* and *Pseudotsuga menziesii* obtained by destructive sampling (from Scott and Reinhardt 2005).

Forest tree allometries have been developed for many tree species and growth forms (e.g. Brown 1978; Agee 1983; Means *et al.* 1994), but these equations should be used cautiously when estimating fuel loading and consumption until they are evaluated for accuracy in reflecting the specific growing conditions and species mixtures on each site. Reinhardt *et al.* (2006a) reported that custom adjustments of tree allometries greatly improved the precision of estimated pre-burn canopy fuel loading, crown bulk density, and vertical distributions of crown fuel. Keane *et al.* (2005) compared destructively sampled measurements of crown bulk density and branch loading with estimates obtained from a variety of indirect ground-based techniques, finding considerable variability among techniques and widely varying accuracy. The computer program FuelCalc (Reinhardt et al. 2006b) employs published allometries to estimate pre-burn fuel profiles and loading in forest canopies. Even with these approaches, post-burn estimates of fuel consumption require additional post-burn allometries as described above for shrubs and grasses. At this point, no models are available for predicting canopy fuel consumption, and it will be a challenging field to investigate; statistical modelling by Call and Albini (1997) revealed that fuel moisture and particles sizes strongly influence canopy consumption, but their work also showed the extreme variability of fuel consumption in crown fires.

Rapidly developing applications of remote sensing technologies such as LiDAR and Structure from Motion (SFM) have potential to provide comprehensive pre- and post-burn measurements of vertical and horizontal fuel structure, including ladder and canopy fuels (see **Figure 6.15**). LiDAR determines the locations of solid objects by recording the return time of reflected laser light, while SFM uses visible images from multiple angles to construct a model of a 3D surface.

LiDAR and SFM techniques can be employed from ground-based systems or from airborne sensors. Aerial scans provide a way to collect data from extensive land areas, but dense canopies prevent them from detecting fuel strata beneath the canopy and on the ground. Ground-based systems can detect understorey fuel structure but are limited in aerial coverage. For small-scale experiments, both airborne and ground-based LiDAR have been used to estimate fuel consumption on individual trees and changes in the vertical fuel profile (Skowronski et al. 2020). LiDAR has also been used to estimate fuel consumption and its variability over very large-scale wildfires (Alonzo et al. 2017; Garcia et al. 2017). At all scales, this technology has the potential to revolutionise measurements of fuels and fire effects. Once methods are standardised, they will probably be widely used for research on crown fires and other wildland fire behaviour. There are numerous technical challenges in processing data from LiDAR and SFM to estimate fuel quantities (e.g. loading by size class), but they offer potential advantages by producing explicit 3D distributions of foliage and fuel properties (Kremens et al. 2010).

Heat release

As discussed in **Chapter 2, Heat release and fireline intensity** and **Chapter 4, Burning and heat release rate,** heat released from a wildland fire can be quantified in several ways: in terms of a total amount (kJ), a total rate (kJ s^{-1}, or kW), rates for units of area (kW m^{-2}), and rates for the unit length of an advancing line fire (kW m^{-1}). When describing heat release, it is important to explain exactly what is being measured (a rate or an amount); whether the measurement covers flaming combustion, smouldering combustion or both; and, if measuring a rate, the time at which measurements were taken (since the rate of heat release varies over time). Three methods are used to estimate heat release from a fire: (1) using consumption information, (2) directly measuring some of the heat released from the fire, and (3) using laboratory correlations with visual estimates of flame length. Because the rate of heat release is difficult to measure in the field, it is common to calculate total heat release (kJ m^{-2}) from total fuel consumption (described above). Heat release rate in the flaming zone is often estimated from a correlation

Figure 9.9: Illustration and picture of the water can analogue for estimating total heat released in fires (from Beaufait et al. 1975).

of flame length with fireline intensity (kW m^{-1}) or area intensity (kW m^{-2}), which is described below (see **Flame length and height** below).

Of historical interest is the practical method of estimating total heat release in prescribed fires by means of a 'water can analogue' (Beaufait 1966; Beaufait *et al.* 1975). Vented steel cans with a known water volume are placed in the fuel bed before burning; after burning, the total heat released by the fire is estimated from the volume of water boiled away (**Figure 9.9**). The calculation involves summing the heat required to heat the water from ambient to boiling temperature the heat required to convert liquid water to vapour (for the amount evaporated). This approach can be confounded by many factors, including delay between the time of burning and the time of post-burn measurement, and possible heat exchange between the can and its surroundings (which would vary with location). The technique is now little used.

Remote sensing that employs thermal infrared cameras on the ground or on aerial platforms can provide measurements of heat release in terms of thermal radiation across an entire fire or within selected portions of a fire's burning zone (Hudak *et al.* 2016). While satellite-based imaging is applicable at continental and global scales, it does not produce fine enough pixel resolution to be used in explaining or predicting fire behaviour (see Wooster *et al.* 2005 for more information).

Thermal images are produced by an array of detectors that are sensitive to different ranges of wavelengths. The electromagnetic spectrum covers a wide range of wavelengths (*bands*), including thermal bands, which in turn include the visible band (see **Figure 3.11, Chapter 3**). Some thermal detectors have a broad band, which can capture the entire range of radiation given off by a fire, and others have narrow bands, which – for measurements in wildland fires – are used to capture specific ranges within the medium and longer thermal wavelengths. Data from two bands in the thermal spectrum can be used to estimate *radiometric temperature*, which is the temperature sensed as brightness by a radiation sensor (Kremens *et al.* 2010). Before the radiometric temperature can be used to estimate the actual temperature of the heat source, it must be corrected to account for interception by materials between the sensor and the fire (e.g. canopy or vegetation elevated above the burning surfaces).

Thermal images from a stationary platform (a tower or hovering helicopter) or from a short time interval are processed to estimate two primary descriptors of heat release: fire radiative power (FRP: kW or kJ s^{-1}) and fire radiative energy (FRE: kJ), which is simply FRP integrated over time. When FRP and FRE are divided by the area on the ground covered by the pixel, we obtain descriptors of the *density* of fire radiative power (FRPD, kW m^{-2} or kJ m^{-2} s^{-1}) and fire radiative energy (FRED: kJ m^{-2}) *per unit area*. These values are used to derive metrics that describe the actual fire, as long as we know the heat content of the fuel (kJ kg^{-1}) and the fraction of energy released as radiation, along with meeting other assumptions (Kremens *et al.* 2012). When remotely sensed images of FRED are combined with simultaneous spatial reconstructions of fire spread rate, they are theoretically related to Byram's fireline intensity (kW m^{-1}) using **Chapter 2, Eqn [2.11]** (see Johnston *et al.* 2017). The fraction of energy released as radiation that is needed to estimate intensity is difficult to know with certainty. For practical purposes, however, uncertainty regarding the radiation fraction is probably no greater than uncertainty surrounding the heat content of the fuel or the phase of combustion (flaming *v.* smouldering), and these uncertainties are probably less than the uncertainties associated with field estimates of fuel consumption that are described above.

Flame zone properties

Luminous flames are the most visible characteristic of wildfires. The flame dimensions of length, height, width and depth are important metrics of energy release and integral indicators of fire behaviour produced by the dynamical spread system. The energy released by flaming combustion also has critical implications for fire safety, fire management activities, and planning activities for achieving ecological effects. But flames are challenging to quantify, partly because they are highly variable in time and space. This is true even for flames from a stationary source, but it is especially challenging in wildland fires, where they move with the spread of the flame zone through varying fuels and topography, responding to constantly changing weather conditions and the behaviour of the fire itself. Flame dimensions (and hence energy release rates) change through time as fuels are consumed, and flaming can last anywhere from several seconds in light grass to tens of minutes or longer in large woody fuels. As we saw in **Chapter 7, Simple fire spread dynamics**, instantaneous measurements of flame zone and flame dimensions represent a cumulative history of a spreading fire in space and time. Deep, long-burning flame zones display the fire behaviour produced by fuel and weather conditions from perhaps many minutes in the past (or longer).

Understanding of flame zone dimensions is and will remain crucial for understanding fire behaviour, even as remote sensing techniques for measuring them become more widespread. Therefore in this section, we review methods of measuring flame dimensions and their relationships with other fire behaviour characteristics, with particular attention to definitions and sources of uncertainty.

Flame length and height

Flames are highly variable in space and time, and their geometric dimensions in wildland fires are defined imprecisely and used inconsistently. This makes it difficult to compare measurements of flame length from different observers, who may be using different measurement methods. Flame length is related to the energy release rate of a fire, whether it is a line fire (kW m^{-1}) or an area fire (kW m^{-2}) (**Chapter 4, Flame height**), and many empirical correlations have been found between flame length and energy release rate. Unfortunately, differences among these correlations contribute to uncertainty in interpreting them. Despite the challenges with measuring and interpreting flame length, it is a metric widely used in practice for communicating and comparing fire behaviours, partly because it is the most apparent, tangible approximation of energy release rate available in the field and partly because it has a strong bearing on management activities and effects.

The sources of uncertainty and variability in flame length are worth discussing before we consider methods of measurement. First, flames are non-steady in time and space because they are turbulent and their strong buoyancy causes pulsing. Thus it is not clear whether the vertical extent of a reported flame length indicates the longest flames, the tallest flames, or some statistical property such as a maximum or an average. As described in **Chapter 4, Non-premixed or diffusion flames**, the buoyant pulsing of flames generates a distribution of flame presence that can be described by its frequency as continuous and intermittent regions. Buoyancy is also responsible for the peak-and-trough structure of a line fire (**Figures 6.20–6.22**). These complex features are neglected in most idealisations of the shapes of flames. For example, we often caricature flames from a line fire as tilted triangles viewed from the side of the fire (e.g. **Figures 2.9, 2.11**). This perspective ignores the variation produced by the lateral meandering of flame peaks and troughs along the flame front (**Figures 6.20–6.22**).

Second, no precise physical geometry has been determined for flame length, particularly for the base of flames in spreading wildland fires. Flame geometry is often idealised as a tilted triangle in shallow fuels (see **Figures 2.11, 6.19**), but use of the triangle's base to represent the bottom end of the flame length vector, its origin, is ambiguous: it is not clear whether the origin of the flame is on the ground or at the top of the fuel bed, even though the thickness of the bed can vary from a few centimetres to many metres. Flames are often much taller than the depth of the fuel bed (**Figure 2.8**), and in these cases the height of the fuel bed is relatively insignificant compared to the vertical variability of flame size and this source of error can probably be neglected.

It is especially uncertain how to determine flame length for elevated fuel beds, such as those that occur in crown fires. Byram (1959) suggested that flame length be measured from half the fuel height (which, if the crown is fully engulfed, would be half the height of the tree crown) so the calculated flame length would match with observers' impressions. This practice has become convention by repeated use (Johnson 1982; Rothermel 1991), but it has not been examined from a physical basis. Another way to address flame length in crown fires is to use different equations for different fuel beds. For example, Rothermel (1991) suggested using Byram's (1959) equation (**Eqn [4.6]**) for surface fires and Thomas's (1963) equation (**Eqn [4.7]**) for flame length in crown fires.

The physical rationale for defining flame height has been investigated using vertically stacked propane burners to substitute for deep fuel beds (Grumstrup and Finney 2019). This experiment demonstrated that the base of combustion (in wildland settings, often the ground surface) can be justified as the vertical starting point for flame height measurement if flames fill the entire fuel volume (**Figure 9.10**). The explanation involves the dependency of flame height on both the rate of gaseous fuel production and the rate of oxygen diffusion into the gaseous fuels (see **Chapter 4, Non-premixed or diffusion flames**). In vertical fuel beds, a continuous flame volume within the fuel volume means that gas from the decomposition of solid fuel at any height within the bed is entering a collective fuel stream that must rise until it can mix with sufficient oxygen to combust. This would be the case in many brush and crown fires if (or when during the fire) all fuel materials within the vertical profile are engulfed in flame and are thus deprived of oxygen regardless of the heights at which flames are being produced. Thus flame length could be measured from the lowest burning fuels or the ground. However, if

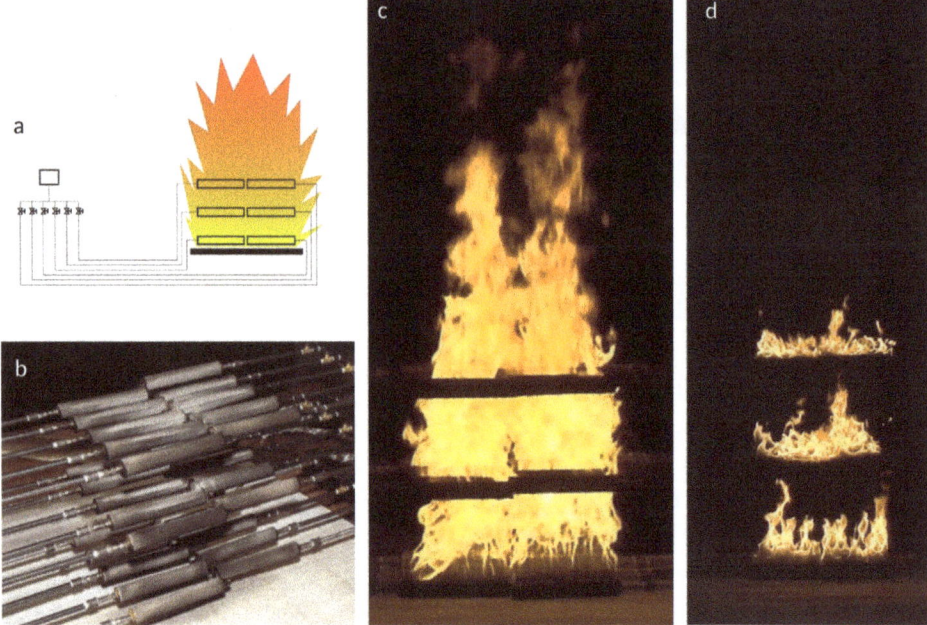

Figure 9.10: (a) Schematic of laboratory experiment using vertical stacked propane burners to determine reference height for flame measurement; (b) photograph of ceramic cylinder burner tubes supplied with propane from the sides; and (c) images and measurements suggesting that flame height for different propane flow rates can be measured from the ground irrespective of the number of vertical burn layers if the fuel bed volume is filled with flame. (d) If separations between burners allow air penetration, then flame height measured from the ground is not an appropriate descriptor.

flames are vertically separated (e.g. if they are attached to individual branches but flames are not connected to the branches above), then oxygen is accessible through multiple gaps in the fuel flow, and flame length should be measured for individual flame layers, not combined through all vertical layers.

The horizontal origin of the flame length vector in wind- or slope-tilted flame zones is often defined as the midpoint of the flame zone depth (**Figure 2.11**). However, this geometry implies that flame length will always be greater than half the flame depth, even if the longest flame originates at or forward of the leading edge of the fire front, which can occur when flames are strongly tilted. Yet for strongly tilted flames, the flame length calculated from one of the various correlation models of fireline intensity can be less than half of the flame zone depth, making graphical depiction and flame length estimation difficult. Other complications associated with defining the flame zone depth described in a later section also affect measurement and interpretation of flame length.

Visual techniques for flame length and height

The most common way to quantify flame length, both in the laboratory and the field, is by visual estimation. For spreading fires, the traditional method is to view the flame front perpendicular to the spread direction (from the side), because the observer must estimate the midpoint of the flame zone – the theoretical origin of flame length. A reference grid of horizontally and vertically marked poles is placed inside the burn within view, to help the observer gauge the length and height of the flames. However, visual estimates – even for stationary fires such as a campfire – show considerable variation among observers (Johnson 1982). This source of error becomes more significant with bigger fires because it is more difficult to see into a large fire and visually estimate where the flames intersect vertical reference markers. It is also more difficult to see the horizontal extent of the flame zone (depth) in larger fires because it is farther away and likely to be obscured by foreground fire and smoke.

Videography of flame height and length offers many advantages over visual estimates because it affords a more objective, repeatable basis for measurements and it can address flame variability. Researchers have used this technique with both thermal and visible-spectrum cameras. When this technique is used, graduated reference poles are located within the view field of the camera, so flame size can be measured or estimated within each frame. Videography can be used for visual estimates in the same way that human observations are used but, in addition, the image sequence can be analysed to quantify flame size and rapid variations that cannot be detected by the human eye. In this case, the image in each frame of the video is processed to distinguish flaming from non-flaming portions. The resulting images are analysed to produce percentages of presence and absence of flames within the entire field of view, which can be used to derive distributions of flame size and position. With thermal cameras, an apparent temperature threshold is used to define the edges of flaming combustion, and with visible-spectrum cameras, a greyscale brightness value serves the same purpose. Videography often neglects variation in flame thickness and emissivity, but it offers greater objectivity than human observations. However, choices regarding the imaging technology (e.g. frame rate, visible *v.* infrared) and processing (e.g. temperature thresholds, brightness, percentile of flame presence/absence) make it difficult to objectively compare measurements from different studies.

A methodology and apparatus for measuring flame dimensions from video analysis with visible reference lengths was described by Adkins (1995) and has been used in both field and laboratory experiments. Videography has been used extensively to examine small-scale laboratory flames, either from stationary sources (burners) or in spreading fires. As Wilson (1982) states, however, the length of flames represented by video and from photographs depends upon the frame rate and the exposure time, which can smear fast-moving flames together, thus inflating their apparent size.

The standard frame rates of commercial cameras (30 to 120 frames per second) seem sufficient for many applications with small and slow-moving flames. Assuming proper exposure, videography of the controlled, small flames in laboratory conditions – particularly laminar flames – affords opportunities for rigorous image processing, which can reveal considerable detail about flame dimensions.

Adkins's (1995) methodology requires that cameras be located external to the fire. Thus it is most useful for narrow laboratory-scale fires and small experimental or prescribed fires in the field. The development of battery-powered in-fire video cameras (**Figure 9.11**) placed within protective boxes allows cameras to be located within the interior of large experimental fires and wildfires (Kautz 1997; Jimenez *et al.* 2007; Butler *et al.* 2010). The camera box is made of aluminium or steel with a single tempered glass window or double-pane glass to protect it from damage from flames. The camera boxes are often wrapped in aluminium foil or fibreglass insulation, which can be disposed of and replaced if damaged. In situ cameras have been used in fires across the spectrum of fuels and fire types – grass fires, understorey fires and crown fires (Taylor *et al.* 2004). As with other techniques for measuring flame dimensions, marked poles in the field of view provide a visual reference for

Figure 9.11: (a) Photograph of in-fire digital camera and insulated aluminium box mounted on a metal pole for field experiments. (b) Image from crop stubble fire with flames ~5 m tall; a 3 m tall pole is visible within the flame zone and a 30 m tower is visible in the background. (c) Image from inside gorse brush fire (*Ulex* spp.) with ~15 m flames; a 3 m tall height pole is visible near the flames as are other instruments including thermocouple arrays and heat flux packages (described subsequently in text). (d) Image from crown fire in lodgepole pine (*Pinus contorta*) (photo by Bret Butler, USFS). (e) 360° camera mounted within a water-filled glass globe. (f) Single image from 360° image of an approaching fire in gorse.

flame size. When a single camera is used, there is some ambiguity in the relative position of flames and reference height. Pickford and Sandberg (1975) developed methods for using multiple cameras external to a fire to measure flame and plume dimensions; these methods could be applied to in-fire cameras as well. If multiple cameras are arranged with overlapping fields of view, they could provide stereo imaging of flames, which would reduce ambiguity in interpreting results.

Standard commercial video cameras offer only a relatively narrow viewing angle, but more recent technology makes it possible to use 360° cameras that automatically correct distortions and join images from multiple lenses (**Figure 9.11e**). From the recorded video, several arbitrary view angles can be explored using commercial software. This technology offers many advantages over single-view video for the purposes of understanding fire spread and characterising flame sizes and dynamics. The challenge in using these cameras is to protect them from heat. Because of their waterproof design, they can be mounted inside water-filled glass globes, which keep their temperatures low enough to function undamaged. An electric pump is needed to circulate cool water in the globe during fire passage.

Point sampling

A different approach to measuring flame height and/or length involves point sampling and is based on the idea of a *passive flame height sensor* (Ryan 1981) (**Figure 9.12**). The original sensors consisted of cotton string soaked in fire retardant salts (e.g. diammonium phosphate) to prevent sustained independent flaming (so that they would burn only when in contact with external flames). Strings were suspended vertically within the burn plot. Heights of burning, charring or singeing were measured after the fire. To convert string measurements to flame heights, treated strings were calibrated with observed flame contact in a separate procedure. A similar technique uses lead solder (Simard *et al.* 1989), which melts at a height consistent with duration of flame exposure. Both techniques have been used on prescribed fires. When these methods are used in conjunction with photographs of fire spread, measurements of flame angle can be used to calculate flame length from flame height measurements (Finney and Martin 1992). The utility of passive string sensors is improved if the strings are pressure-treated so retardant will saturate them uniformly, and if they are supported vertically by a metal guide during passage of the fire (like line supported along a fishing pole). These passive sensing techniques can be deployed in a sampling design throughout an experimental fire to statistically estimate distributions of flame heights or lengths and the plot-level variance. The disadvantage is that flame height must be converted to flame length from the observed or measured flame tilt angles and these may be difficult to obtain from inside a large fire area without the use of in-fire cameras.

Flame depth and residence time

The flame depth refers to the extent of the flaming region behind the fire front. We idealise it as the one-dimensional (1D) linear distance along the ground in the direction of fire travel between the forward leading edge (the ignition interface) and the trailing edge (where visible flaming stops) (**Figure 2.11**). In a steady spreading fire, the flame depth is a function of the *flame residence time*, also known as the *flame reaction time* (Rothermel 1972; Nelson 2003), as expressed by **Chapter 2, Eqn [2.1]**. In fact, Rothermel (1972) defined reaction time for a steady fire as the time required to spread the distance of the flame zone. Residence time has also been defined as the duration of flaming combustion at a point on the ground. This is distinct from the *particle residence time* because, as fuel beds become denser, collections of particles in a bed burn more slowly than individual fuel elements, which lead to longer residence times (see **Chapter 4**, **Burning and heat release rate** and **Chapter 6, Figure 6.12**). The flame residence time, and by association the flame length and depth, are

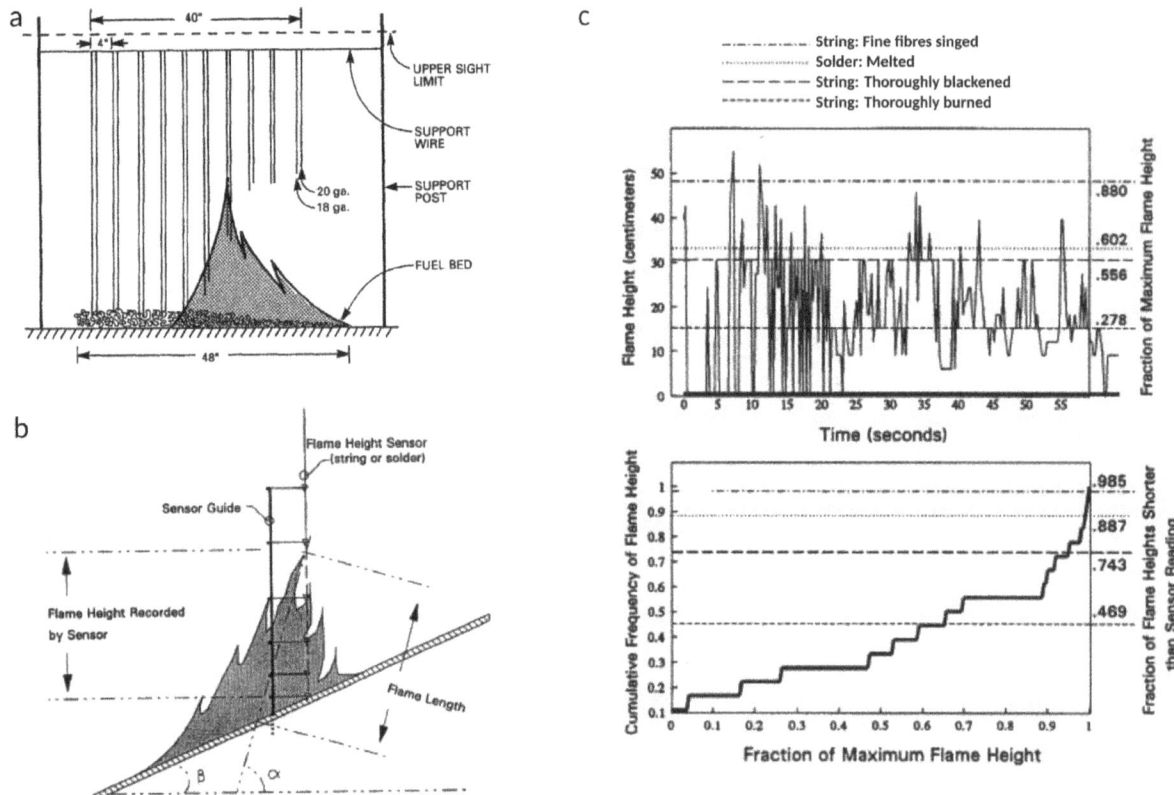

Figure 9.12: Passive flame height sensors used in prescribed fires: (a) on flat ground, solder wire of different diameters melts where contacted by flame (reproduced with permission from Simard *et al*. 1989); (b) on sloping topography, flame angle must be determined to estimate flame length; and (c) calibrations of string and solid sensors are required to compare observed distributions of flame height, which in this case shows the observed time variation in video-recorded flame height compared with measured levels of string damage and solder melting (reproduced with permission from Finney and Martin 1992).

critical indicators of combustion rates and they determine the maximum length of time in which heat can be transferred and fuels can be ignited during flaming. Byram (1959) described a similar role of flame residence time as the 'critical burn-out time' – 'the maximum length of time that a fuel can burn and still be able to feed its energy into the base of the forward traveling convection column'. The same principle was described by Wilson (1982) as an 'effective combustion zone depth'.

Idealised definitions of flame depth and residence time help us describe and communicate visual impressions of line fires and provide some quantitative metrics for fire modelling. As mentioned in the section above on visual techniques, a measurement of flame depth is also integral to determining flame length. In practice, our definitions of the flame zone are ambiguous because of difficulty in distinguishing the leading and trailing edges of a line fire and, by association, the duration of flaming. This conundrum was discussed by Wilson (1982) with a quote from F.A. Albini:

The problem here is that the idealization of the spreading fire as exhibiting two distinct event boundaries, one marking the onset of flaming combustion and one its termination, is not a wholly accurate one. We seek to preserve this idealization of the process for its power as a modeling concept, leaving the experimentalist the ill-defined task of fixing an operational definition of its measurement.

The leading edge of a spreading fire would seem to be definitive at any instant as the location where flaming ignition has just been achieved. Fires in laboratory fuel beds of homogenous single-sized particles might be the ideal case described by Wilson (1982), but at fine scales, the discrete nature of particle ignition requires some spatial averaging to determine a single flame front location. In mixed fuels, say composed of grass and leaf litter, or large logs lying on a bed of pine needles, ignitions of one size of particle often occur first, leaving a gap between the forwardmost ignition and the coherent flame front, which by definition consists of flames merged together from all burning particles. Different heat transfer processes can be responsible for different ignition sequences. For example, flame contact might ignite fine material beyond coarser or denser materials in the fuel bed (or fuel gaps); radiation in crown fires might ignite logs and tree stems before the fine fuels are ignited (**Figure 6.5a, b**).; and, perhaps the most confusing case, embers transported ahead of the flame zone cause isolated ignitions (i.e. spot fires, **Figure 8.4**). All of these behaviours have been observed in spreading crown fires (Taylor *et al.* 2004), and they all challenge our ability to determine the leading edge of a coherent, unambiguous flame zone.

It is difficult to measure the trailing edge of the flaming zone even in simple fuel beds because fuel burnout and waning gas production from the fuel bed cause intermittent flaming that alternates with glowing coals. Even in very simple fuel beds – burning cribs of fuels with uniform size – the mass-loss curves show a long tail of diminishing mass in later stages of burning, making it impossible to clearly distinguish the flaming phase from the non-flaming phase (see **Chapter 4, Figure 4.24**). Cribs, like natural fuel beds composed of larger material, begin to collapse as the mechanical strength of burning fuel particles weakens. The collapsing fuels change the properties of the fuel bed; bulk density often increases as the particles fall together and break into smaller pieces. The patchy randomness of such late-stage combustion can cause intermittent flaming in isolated pockets near the back of the flame zone, creating a ragged trailing edge. Albini and Reinhardt's (1995) model of woody fuel consumption explicitly represents the localised nature of burning of logs and sticks that occurs when heat transfer is enhanced at intersections among particles and between particles and the duff layer. Patchy burning can occur in all kinds of fuels even within the flame zone, as the flame zone deepens and buoyant circulations of flames impede the mixing of oxygen with gaseous fuels (see **Chapter 6, Figures 6.21, 6.22**). Without access to oxygen, combustion is slowed and fuels burn longer, increasing flame residence time and flaming zone depth.

Finally, we must keep in mind that the length of flames and the depth of the flame zone at any given instant are expressions of the history of spread rate and fuel consumption in a particular region of the fire. As we saw in **Chapter 7, Simple fire spread dynamics**, acceleration of spread produces a time-lagged deepening of the burning zone and an associated increase in intensity and flame length. If the linear spread rate is altered by changing wind speed or direction, heterogeneity of fuels, or converging flame fronts, any instantaneous measurements of flame zone dimensions will also be affected, but these do not reflect the steady-state spread rate or the geometry of a line fire.

Direct measurements

Relatively little technology has been used to directly quantify either flame depth or residence time. This may reflect, in part, the issues with defining them and understanding their functional role in fire spread. In spite of their measurement challenges, these concepts have intuitive value for understanding fire behaviour and for modelling. They are so closely related (see **Chapter 2, Eqn [2.1]**) that we can attempt to directly measure flame residence time and then calculate flame depth – or measure flame depth and then calculate residence time. In most cases, direct measurements of these variables are obtained from instrumentation deployed for other purposes, such as measuring fire spread rate.

It is easiest to visually estimate residence time and flame depth by watching a fire in real time or viewing video recordings. Flame depth can be manually measured in slow, low-intensity laboratory fires, as Wilson (1982) demonstrated using a meter stick in rectangular fuel beds in the absence of wind or slope. It is often easier and more objective to use in-fire video rather than visual observations to measure residence time (Despain et al. 1996; Taylor et al. 2004; Wotton et al. 2012), and to use photographs to measure flame zone depth (with markers for horizontal distance and assuming there is direct line of sight into the fire). Finney et al. (2013) used overhead videos of experimental laboratory fires to measure flame depth and residence time in regularly spaced cardboard fuel beds (see **Chapter 6, Figure 6.21b**). The number of rows of fuel elements and hence the depth of the flame zone was clearly delineated on high-definition video. The video also allowed the ignition and burnout times to be estimated with ~1 s precision – and hence residence time could be directly measured and compared with estimates from flame zone depth and spread rate.

In laboratory fires, Catchpole et al. (1998) used a linear array of photocell receptors to measure spread rates and residence times between the start and end of flames crossing a narrow field of view. The horizontal orientation of the photocells (i.e. looking in from the side) along the narrow fuel bed provided consistent, reliable measurements, but the method assumes spread of a linear fire front oriented perpendicular to the photocell view. Linear arrays of thermocouples oriented parallel with the direction of fire travel can also provide point samples of gas temperatures and can thus be used to estimate residence time and flame zone depth. As point samples, thermocouples do not require that a linear front be assumed. However, thermocouples record extensive temperature fluctuations, so time and temperature thresholds must be established to define the beginning and ending of the coherent flame zone (Finney et al. 2013).

Laboratory experiments have sometimes used mass-loss rates to estimate residence times in spreading fires (Anderson 1964; Rothermel and Anderson 1966; Rothermel 1972). From spreading fires, the mass-loss history illustrated in **Figure 9.1** (Rothermel 1972) shows that the 'burning time' is measured from the time mass loss begins to the point at which it becomes constant. The burning time is assumed to be equivalent to the flame residence time. However, the time when mass loss becomes constant may not be exactly the time when flaming ceases, because smouldering can consume fuel continuously even after flaming becomes intermittent (**Figure 4.24**). Because it is difficult to distinguish flaming from smouldering, direct measurements of flame depth may be smaller than those derived from residence time using thermocouples and mass-loss estimates (Wilson 1982).

The difficulties of directly measuring flame residence time and depth in small, highly instrumented fires are mild when compared to the difficulty of measuring them in complex fuels with mixed fuels and widely varying burning rates. Consider the example of an open forest with litter, duff, grasses and woody fuels. In a fast-spreading heading fire, flames will spread first through the fine material, then later ignite the larger, longer-burning fuels. If woody fuels and duff are dry and the wind ventilates the burning zone, a deep flame zone may develop behind the burned-out fine fuels. It may persist but also become patchy if fuel distribution is not uniform. Exactly where is the end of the flame zone in this case? The residence time and flame zone characteristics are not a blended property of the combined fuel complex burning in this way because the ignition and burning of each fuel type occurs at a different place and time. If the fire is backing into the wind, however, a narrower flame zone may develop and proceed slowly by igniting the leaf litter. The short heat transfer distance inherent in slower backing spread may make it somewhat easier to define both residence time and flame depth. The grasses may provide rapid energy release, but will not ignite much ahead of the longer burning litter and woody fuel components.

Thermocouples have been used in field studies to measure flame presence at a single point. Measurements from all such instrumentation are sensitive to the location of the sensors relative to the fuel bed. Sensors placed above the bed probably record shorter duration of flaming than those placed near the top of the fuels or inside the bed. Relatively thin thermocouples (127 µ) placed 0.5 m above ground showed the duration of temperatures exceeding 300 °C from understorey eucalyptus fires to be well correlated with visual observations of flame residence time (Wotton *et al.* 2012).

Flame ionisation detectors have also been used to measure residence time because the electrical resistance of the air between a pair of electrodes decreases when ions in hydrocarbon flames are present (**Figure 9.13**). Flame ions can be reliably distinguished from gases and smoke, so a time series of detections can probably reveal flame residence time. This technique has many applications for flame detection in industrial processes, and we have tested it experimentally in laboratory and field burns (**Figure 9.13**). The utility of this technique in prescribed fires and laboratory fires is still being evaluated, but results generally agree with data from experiments using thermocouple temperature trends.

Methodology for measuring residence time and flame depth in the field is likely to rely increasingly on video from in-fire cameras or overhead video from unmanned aerial vehicles. Video in the visible spectrum may be sufficiently sophisticated for characterising many fires, particularly backing fires and those heading through light fuels with little post-frontal smouldering. Infrared video has an advantage, especially in large fires and heavier fuels, because it can penetrate smoke to some extent. An example of infrared measurements of flame zone depth was produced by McRae *et al.* (2005) (**Figure 9.21a, b**). Where video techniques are used, post-burn processing will be constrained by pixel resolution and will require the analyst to decide

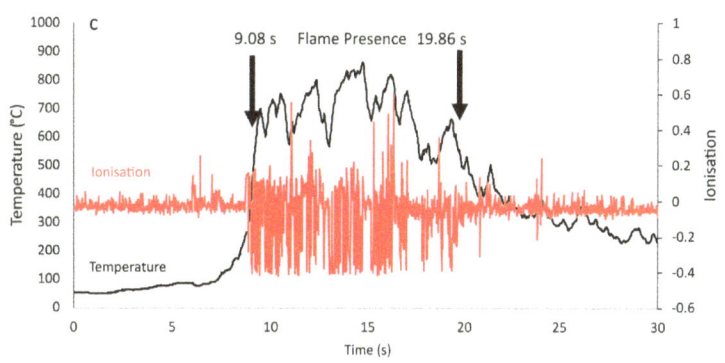

Figure 9.13: Photographs of (a) flame ionisation detector probe and thermocouple wire extending 15 cm above ground for measuring flame residence time in field experiments; (b) data loggers attached to the sensors are protected from heat in a box beneath the sensors during the fire; and (c) comparison of data from paired thermocouple and flame ionisation detector for a prescribed fire in a wheat field with flame presence between ~9.08 and 19.86 s. Photographs by Ian Grob.

how to distinguish the leading and trailing edges of the flaming zone. For example, when viewed from above, wind-blown flames that are tilted far forward of the ignition interface may lead to overestimating the front of the flaming zone, and long-burning litter or discontinuous clumps of flaming grass could lead to estimation of a deeper trailing edge. Overhead video offers an advantage over in-fire video in that it provides a two-dimensional (2D) context for determining the dimensions of coherent flaming activity and relating them to other fire behaviours, and this context may be help scientists better define what constitutes a flame zone and decide how best to measure it.

Heat transfer

Heat transfer is a key physical process that enables wildfires to spread. In wildland fires, heat transfer is the process in which heat produced by combustion of biomass fuel is transmitted by radiation and convection to unburned fuel, thereby raising the temperature of the fuel. When the fuel particles reach a sufficiently high temperature, they break down into flammable gases that are ignitable. Because heat transfer plays a central role in fire spread, measuring its magnitude in laboratory and field fires is critical for improving our understanding of wildland fire behaviour.

Radiation

Radiation heat transfer is the transmission of thermal energy from one place to another via emission and absorption of electromagnetic thermal energy (**Chapter 3, Radiation heat transfer**) and is an important means of heat transfer in spreading wildland fires. It is measured in both laboratory and field studies. Radiation heat transfer in wildland fire research is normally quantified as the *incident thermal radiation flux* – that is, a quantity of electromagnetic energy incident on a surface per unit time per unit area (J s^{-1} m^{-2} or W m^{-2}). For example, a surface fire with flames of a metre or so in height would produce an incident thermal radiation flux of 40 kW m^{-2} on a nearby sensor, while a crown fire would routinely produce values exceeding 100 kW m^{-2}. Data from Frankman *et al.* (2013) show radiation fluxes measured from surface fires and crown fires (**Figure 9.14**). Great care must be taken in interpreting such measurements because the forward radiant heat flux relevant to heating and ignition of fuel particles is confined to the period preceding the arrival of the flame front. In other words, we must know the position of the fire relative to the sensor in order to interpret the meaning or context of the radiant flux recorded. Radiation flux occurring after flame impingement on the sensor will be very high but confounded with convective heating.

If one records the thermal radiation flux on some surface of known area over time, the total flux experienced over that time can be calculated. This will describe the total radiant energy incident on that surface (J m^{-2}). Thermal radiation flux can be measured using remote sensing techniques, including sensing by instruments in aircraft and even spacecraft (Lentile *et al.* 2006). Here we focus on laboratory and field-going instrumentation.

A *radiometer* is a general term for a sensor that measures the rate of electromagnetic radiation arriving at a surface. Fire research typically uses a radiometer that detects radiation in the *thermal* spectrum (**Figure 3.11**). The most common type of radiometer is one that measures the difference in temperature (ΔT) between a radiation-absorbing outer surface and an *isothermal* interior heat sink. 'Isothermal' means that the heat sink in the instrument remains at a constant, uniform temperature. Because the temperature of the interior heat sink is constant, ΔT is an expression of the energy absorbed by the outer surface of the radiometer. It can be converted to thermal radiation flux by using relationships determined through calibration of the sensors.

Two main kinds of radiometers are used to measure ΔT in fires. The *bolometer* uses *thermistors* to detect temperature change, since the electrical resistance of thermistors varies with temperature. Another type of radiometer commonly used in

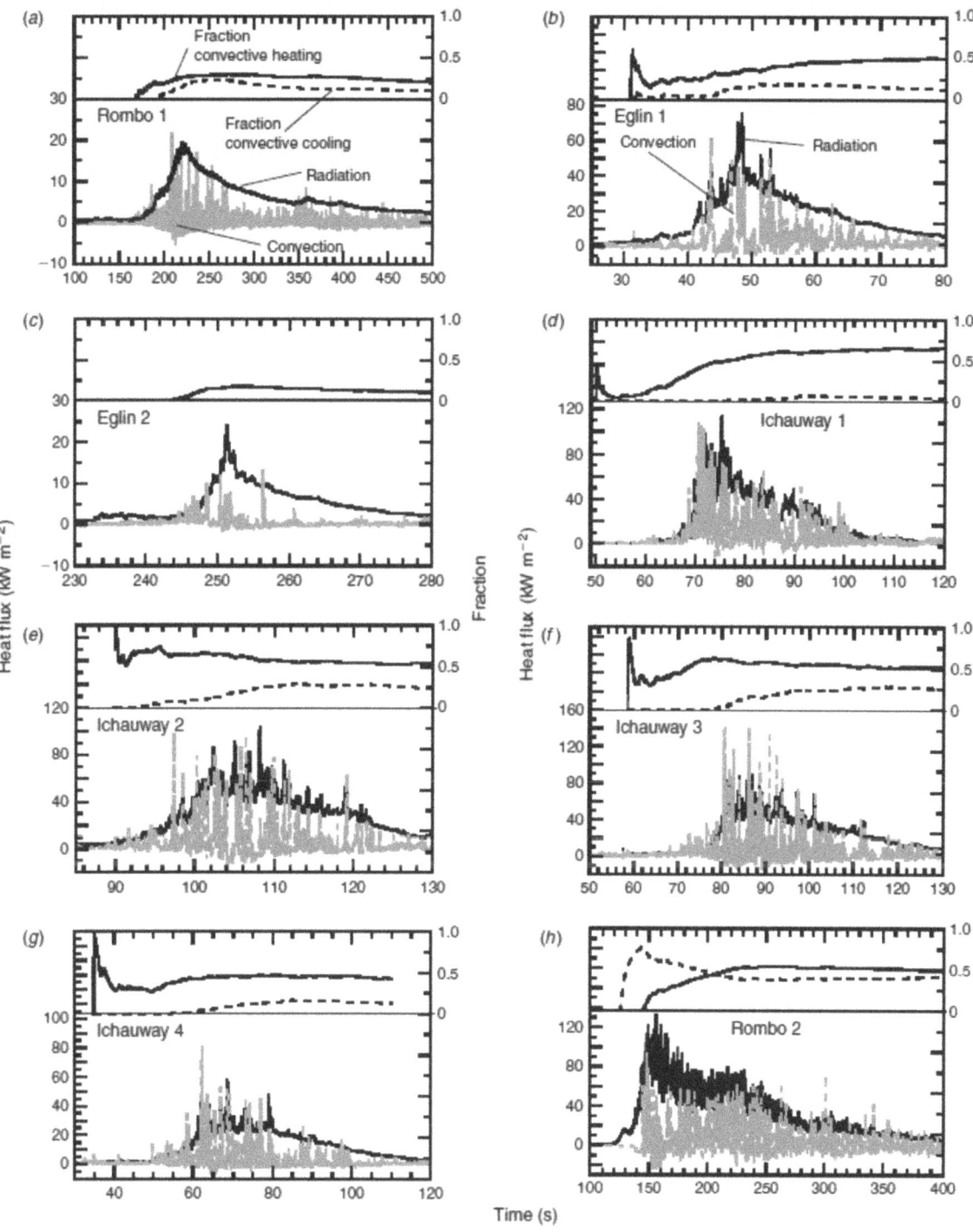

Figure 9.14: Graphs showing radiant and convective heat flux measurements from prescribed fires (reproduced with permission from Frankman *et al.* 2013). A wide-angle radiometer was used in all cases. Panels (a)–(e) are from surface fires; panels (g) and (h) are from crown fires. Note that the measured convective heat flux intermittently drops below zero in many of these fires because of convective cooling of the sensor. The first spike in convection approximates the arrival of the flame front, meaning the impact of radiation flux on fuel heating is afterwards confounded with convective heat flux. Problems with heat flux sensors in representing actual convective flux to small fuel particles is discussed in the following section on convection measurement.

fire is the *thermopile* (including the Schmidt-Boelter and Gardon sensors). Thermopiles use *thermocouples* to measure temperature, since the voltage generated by a thermocouple junction varies with temperature.

The absorbing surface of a radiometer is often coated with a high-absorptivity substance so it will act as nearly as possible to an ideal blackbody. As a near-blackbody, the material should also have no wavelength dependence across the range of thermal radiation, which is ~100 nm to 100 000 nm (see **Chapter 3, Radiation heat transfer**). Many radiometers pass radiation to the sensor through a window of material (usually sapphire) so convective heating cannot influence the sensor. Neither perfect blackbody behaviour nor absolute wavelength independence is physically possible, so calibration of radiometers is always required.

Radiometers have generally been used in wildland fire science for two different purposes, and the purpose determines the most appropriate field of view. The first purpose is *to determine the radiation received by a fuel element in front of an oncoming fire*. For this purpose, a wide-angle radiometer – that is, one with a very wide field of view – should be used, so it will receive radiation emissions from as large an area as possible, similar to the entire *view* that the particle experiences with an approaching fire. A radiometer's field of view is determined by its *acceptance angle*. An ice-cream cone has a narrow acceptance angle (perhaps 30°), while a floor drain has a very wide one (nearly 180°). To determine the radiation a fuel particle surface would receive, the radiometer's field of view should be close to 180°. With this field of view, the sensor will receive emissions from all radiating sources, both cold and hot, emitted by flame gases and solid materials. While measurements from wide-angle radiometers detect emissions from all sources, they do not provide information about the source of emissions; they record incident heat flux regardless of the sources' temperatures, sizes and emissivities.

The second use for radiometers is to determine the source of radiant heat flux. Narrow-angle radiometers are designed for this purpose. A narrow-angle radiometer has an acceptance angle of just a few degrees, so its field of view is assumed to be completely covered by a uniformly radiating source. With this assumption, the radiometric temperature of a specific heat source in a fire can be determined using **Chapter 3, Eqn [3.16]**. However, this equation requires knowing the flame emissivity, which is not detected by the radiometer. One approach to handle this is to assume a value for emissivity, perhaps knowing or measuring the depth of the flame zone (**Chapter 3, Figure 3.14**). If we do this, however, we introduce an important and unverified assumption concerning the source temperature of the radiative heat. Instead, Kremens *et al.* (2010) suggest a method for measuring temperature. They used two radiometers, each fitted with different optical filters. Each filter transmits only a limited range of the wavelengths of thermal radiation. Such a filter is called a *bandpass filter*, and the range of wavelengths over which it transmits is called the *pass band*. Using two co-located radiometers, each with filters having distinct, non-overlapping pass bands and some straightforward post-processing, the flame radiometric temperature can be determined. This value could be used to develop or validate physical models of fire spread.

Convection

Convection describes the transmission of heat energy through the movement of fluids. In wildland fires, convection heats and cools fuel elements (e.g. leaves, branches, logs) by the flow of air and combustion gases (see **Chapter 3, Convection heat transfer**). If we are to characterise fire behaviour and fire effects, and if we are to understand fire behaviour better and develop or validate models of fire behaviour, we need to understand convection. But measuring convection is somewhat more challenging than measuring radiation, because convection works through the interactions of many physical processes.

Just as with the rate of heat transfer by radiation, the rate of heat transfer by convection is a form of *power* – that is, energy per time, expressed as Watts

(W). Since we are interested in the rate of heat moving into and out of surfaces, we often use *convective heat flux* (W m^{-2}) to quantify this power. Convective heat flux cannot be measured or described for an entire fire; that is, there is no way to describe an overall convective heat flux, because convective heat flux is a point quantity – unique to a particular location on a particular surface, the geometry of the flow field and the object, the surface temperature of the object, the gas temperature and thermophysical properties of the adjacent flow. Think of how localised and specific convection heat flux is to the surface of a fuel particle. While convective heat flux is complex, it can be quantified for any fuel element in a laboratory- or field-scale fire by measuring key variables.

To understand what those key variables are, recollect **Chapter 3, Eqn [3.28]**, which describes the convective heat flux to the surface of any object. This equation shows that the heat flux is equal to the product of the heat transfer coefficient h and the temperature difference between the surface of the object T_s and the fluid or gas surrounding the object T_∞. To determine the key variables for measuring convective heat transfer, we must ask what measurements we need to calculate **Eqn [3.28]** for an arbitrary object? The surface temperature of the object (T_s) is initially assumed to be the ambient air temperature before the arrival of flames; it will increase in time due to heating. We can use thermocouples to measure the temperature of the object (T_s) over time, as discussed below. However, the heat transfer coefficient h is more difficult to measure experimentally, and it cannot be analytically derived. Therefore, we calculate h by using a correlation that is appropriate for the shape of the object in question and characteristics of the surrounding flow field (**Chapter 3, Convection heat transfer**). For example, for many fuels we can use the correlation in **Eqn [3.36]**, which produces the heat transfer coefficient for a cylinder. The only measurements required for this equation are the temperature and velocity of the gas flowing over the object (which can be used to calculate other gas thermophysical properties needed). So, while it is not possible to measure a general convective heat flux for a given object in or near a fire, it is possible to calculate the heat flux from measurements of gas temperature and flow velocity. In the following, we briefly discuss measurements of temperature and flow velocity in wildland fire.

Temperature measurement in wildland fire research is most frequently accomplished with thermocouples. The thermocouple is one of the most common devices used for measuring temperature of gases and solids in wildland fire. Our synopsis here covers the general design and strategy behind using thermocouples for fire measurements, but the science and technology of thermocouple uses is considerable, and the reader is referred to authoritative texts (e.g. Kerlin and Johnson 2012) on the subject. A thermocouple consists of two wires of different metal alloys that are welded together on one end in what is called a junction. The thermocouple junction generates a miniscule electrical voltage that changes in a predictable manner with variation in temperature. The voltage signals pass through specialised signal processing circuitry that yields the temperature at the junction. Different types of thermocouples use different combinations of wire alloys. Type K thermocouples, which are made of Chromel/Alumel (nickel-chromium/nickel-aluminium), are perhaps the type most commonly used in fire science because the wire is inexpensive and they have a broad temperature range (approximately –200–1250 °C) with relatively low uncertainty (about ± 2.2 °C). Type K thermocouples weaken with use, however. Repeated and/or long-term exposure to flame temperatures (> 800 °C) will cause the wire to become brittle and prone to breaking. This is particularly problematic for very fine-gauge thermocouples (12.5–50 µ wire). Types R and S thermocouples are platinum-based (platinum-rhodium/platinum) and much more tolerant of flame temperatures, but they are substantially more expensive than type K.

Thermocouples can be selected based on their wire diameter and thus their response time to temperature change, with finer wires more suitable for recording rapid fluctuations of gas temperatures and larger ones for capturing slowly changing trends. Fine-gauge thermocouples are very fragile for field use, and they may not be needed for measuring rate of spread (discussed below); however, they may be needed for measuring the very rapid fluctuations in gas temperatures that occur in convective heat flow. Flows of hot combustion gas and cool ambient air in and around a fire are often turbulent, as we have seen in fine-wire thermocouple measurements of flames (**Chapter 6, Figures 6.23, 6.24**). To characterise these rapid fluctuations, we must measure temperatures at very short time intervals; the rate of sampling must be at least a few times faster than the frequency of the fluctuations, and the sensors (thermocouples) must respond to the temperature changes at a rate just as fast or faster. Large thermocouple junctions (or worse, sheathed junctions) have long thermal response times and respond too slowly for measuring temperatures in the turbulent gases of a fire. Depending on the flame environment and the need for estimating true gas temperatures, measurements (particularly from larger thermocouples) may require correction for radiation losses (see review by Lemaire and Menanteau 2017). The finest gauge (12.5, 25, and 50 µ) bare-wire thermocouples available are needed to capture even the slowest temperature fluctuations; they have response times measuring around 100 ms. These miniscule junctions require careful handling because they are very delicate. Fine-gauge thermocouples commonly break in laboratory experiments and even more commonly in the field – and they are usually unrepairable. The finest-gauge thermocouples are adequate for measuring convection heating in a fine fuel element like a pine needle or blade of grass but, if a more rapid response to temperature change is needed, it may be necessary to use measurement techniques that employ pairs of closely spaced thermocouples (see Ren *et al.* 2020).

Flow velocity measurement in and near fires is challenging because the flow field exhibits wildly changing velocity and direction at any given point in space. Most flow measurement devices are also sensitive to temperature, which affects gas density, which shows similar highly variable behaviour. To measure 3D velocity in the turbulent environment of convection, a sensor must have a fast response rate, and either be insensitive to or account for the changing temperature (density) of the gas flow. In the case of field measurements, these requirements must be met by a package that can tolerate rough handling and the intense heat and wind generated by a passing fire front. Moreover, it would be ideal if the instrument was relatively inexpensive so that the inevitable (and hopefully rare) loss in a fire was not a catastrophic one.

Several devices are used to measure flow velocity in non-fire applications, including various types of anemometers and pitot probes, but none of these are used in fire research. Anemometers used in meteorology and turbulence research are unsuitable for fire applications because they fail to meet one or more requirements for in-fire measurements. Sonic anemometers are accurate but expensive and intolerant of flame temperatures. Cup-and-vane anemometers are robust but respond slowly and measure only a 2D flow field. Pitot probes, which determine velocity based on differences in pressure between the inside and the outside of the end of a tube, are robust to flame temperatures but must be oriented with the flow direction, which is changing constantly and rapidly in fires.

Bidirectional pressure probes have been a mainstay of flow velocity measurement in structural and wildland fire research since they were introduced by McCaffrey and Heskestad (1976). They are very robust and relatively easy to manufacture. They are also very tolerant of fouling by solid debris, remaining functional even when subjected to heavy showers of embers and ash. Bidirectional pressure probes work by measuring differences in pressure between two opposing ports, one generally oriented in the upstream direction and the other in the downstream direction. They require simultaneous measurement of the flow temperature to estimate pressure. They are intentionally insensitive to directional variation in the flow up to ~50° deviation, making them well suited to measuring flow speed in fires when direction is unimportant. Bidirectional probes have been used along with a thermocouple and other sensors since the mid-1990s (**Figure 9.15**) in *fire behaviour packages* developed at the Fire Sciences Laboratory in Missoula, Montana, USA (Butler *et al.* 2010). A package contains two bidirectional pressure probes oriented perpendicular to each other; these measure velocity in two perpendicular directions (i.e. forward and back, up and down). These data are combined with temperature data to calculate convection heat flux (from both radiation and convection) on the surface of an arbitrarily shaped object.

Increased interest in the rapid fluctuations of flow velocity and temperature in wildland fires have led to the development of an instrument called the *heat flux package* (Grumstrup *et al.* 2018). The heat flux package (**Figure 9.16**) measures 3D velocity with a higher sampling rate than the fire behaviour package described above and is robust enough to be used in the field. Instead of the bidirectional pressure probes used in the fire behaviour package, the heat flux package uses disk anemometers (Green and Rogak 1999) to measure flow velocity in each of three directions. These disks are somewhat more sensitive to variation in flow direction than bidirectional probes, so they can measure more precisely the direction that flows are coming from.

For laboratory studies, laser-based diagnostics can be used to obtain point temperature and 3D velocity measurements at sampling rates much higher than the instruments discussed previously. Laser-based techniques capture velocity fields that allow one to visualise variation in the flows in and around flames (**Figure 9.17**). Both *laser-induced fluorescence* (Daily 1997) and *coherent anti-stokes Raman spectroscopy* (Eckbreth and Hall 1979) can measure gas temperature accurately without disturbing the flow field or combustion chemistry. *Particle image velocimetry* (PIV) (Filatyev *et al.* 2007; Morandini *et al.* 2018; Silvani *et al.* 2018) and *laser Doppler velocimetry* (Dibble *et al.* 1984) can produce 2D and 3D measurements of velocity. All of these techniques are challenging to set up, use expensive high-power lasers and require specialists to operate. Perhaps it goes without saying, but these techniques are impractical for field deployment.

Field studies have used ground-based infrared (IR) video to measure flame flow velocities and investigate flame dynamics in surface (Clark *et al.* 2005) and crown fires (Clark *et al.* 1999, Coen *et al.* 2004). Processing of the brightness patterns in sequential images produced estimates of flow velocity field along the fire fronts (**Figures 9.17** and **8.2c**). Video recording in the thermal spectrum has an advantage over recording in the visible spectrum because infrared imagery is less easily obscured by smoke. However, high-quality infrared cameras are expensive and require portable and long-lasting power supplies. Furthermore, as smoke density or the distance through smoky air increases, even infrared cameras may not be able to detect hot convective gases. Thus these techniques are most suitable for capturing thermal flow

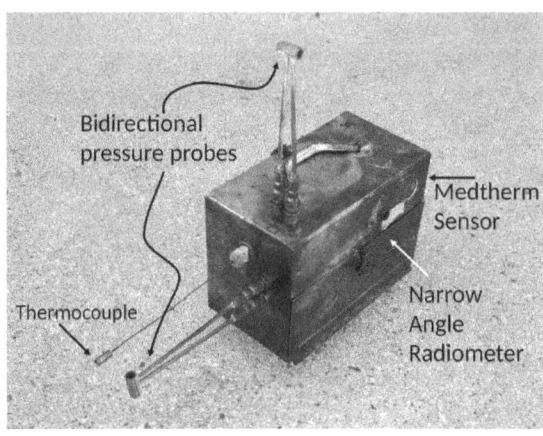

Figure 9.15: Programmable fire behaviour sensor package containing bidirectional pressure probes that measure the velocity of mass flow, a narrow-angle radiometer (NAR) only for radiation, a Medtherm sensor that measures both wide-angle incident radiant flux and total heat flux (radiation + convection), and a thermocouple developed for field deployment in wildland fires (from Butler *et al.* 2010).

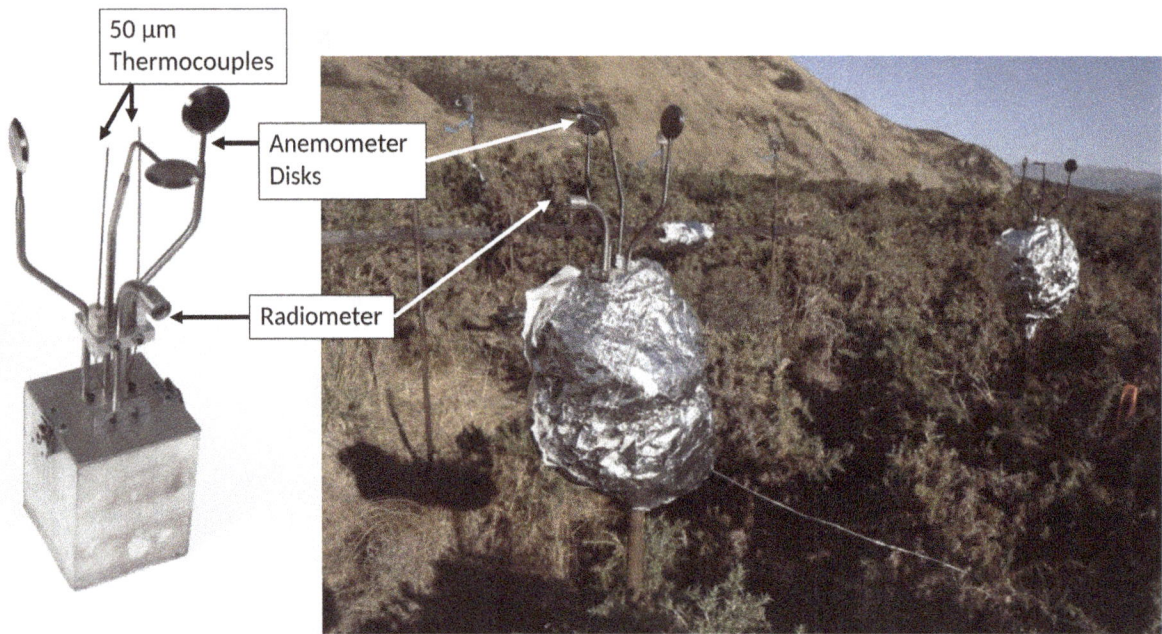

Figure 9.16: This heat flux package made for in-fire measurements contains three disk anemometers that measure pressure differences in each orthogonal direction and thus convective flow velocity. They also include two fine-wire thermocouples for measuring air temperature and a wide-angle radiometer for measuring radiant heat flux (from Grumstrup *et al.* 2018).

dynamics near the edges of a fire rather than in the middle.

Historically, the measurement and interpretation of heat fluxes (radiation and convection) in wildland fire research has been of limited use in understanding the experience of fuel particles during the phase of heating up to ignition. A case in point is the report by many academic articles of 'convective heat flux' measured using heat flux gauges. A pair of gauges is used, one measuring total heat flux (from both radiation and convection such as the Medtherm sensor mentioned previously) and one measuring only radiation heat flux. The authors call the difference between the two signals 'convective heat flux', but it is not. While it seems reasonable that convection heat flux should be equal to the difference between total and radiant heat flux, actual convection heat flux is more complicated than what paired heat flux gauges can measure. Measurements from heat flux gauges are based on properties of the gauges themselves and the flow fields surrounding them. As described above and in **Chapter 3, Convection heat transfer**, however, convection heat flux depends on the geometry of a particular solid object, its temperature, and the local velocity and thermophysical properties of the adjacent flow field. There is no way to describe a generalised or area-wide convection heat flux based on data from paired heat flux gauges, because convective heat flux is a point quantity. Thus the information recorded by heat flux gauges cannot be somehow applied to an arbitrary fuel element like a pine needle or a blade of grass. Pairs of heat flux gauges may be useful for obtaining a *very rough* estimate of the relative contributions of radiant and convective heat flux for particles similar in length to the sensors, but this technique has little utility for measuring convection heat flux to actual fuel particles in fire spread, and because it can be so misleading, its use for this purpose should be actively discouraged.

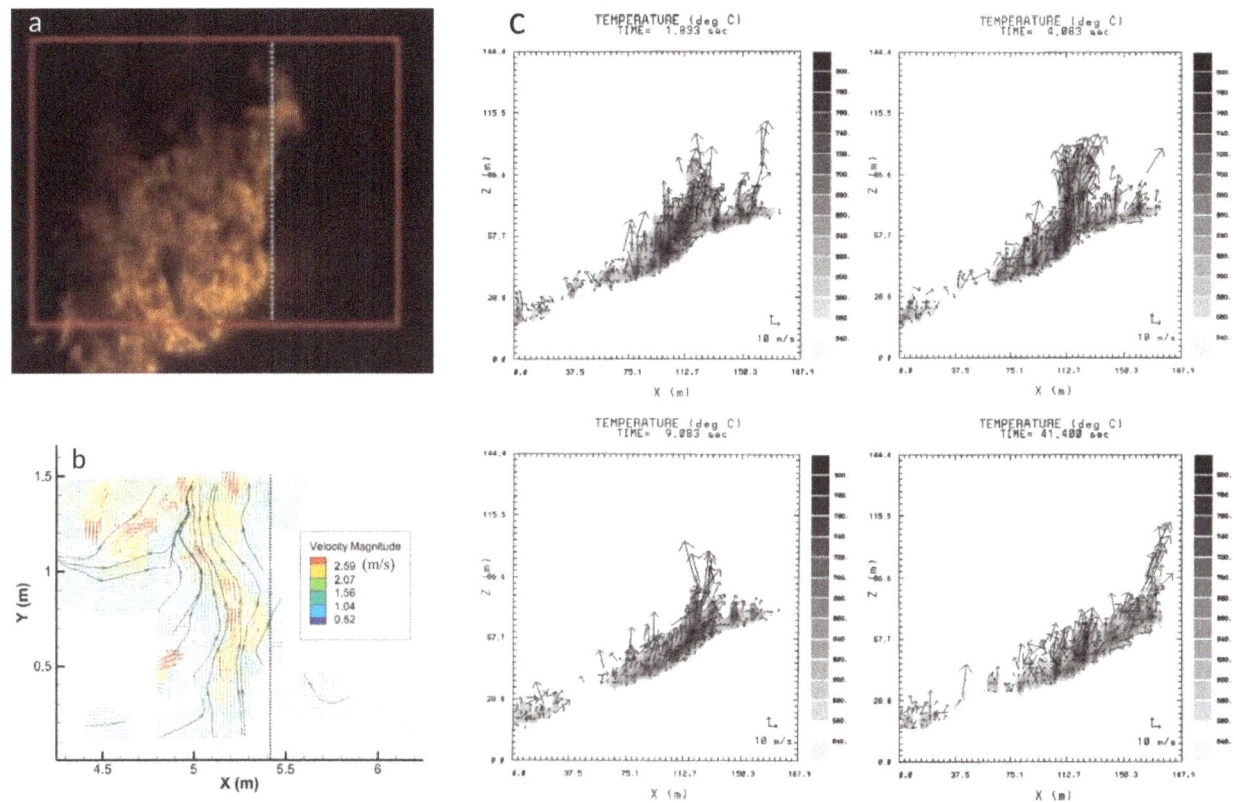

Figure 9.17: Examples of velocity fields of flames obtained from particle image velocimetry applied to a spreading fire in the laboratory, showing (a) the visible image of the flames; (b) the 2D velocity field for a slice through those flames that reveals the converging upward flow and accelerating speed in the flames with height (maximum speed is 4.73 m s^{-1}, see colour legend) (reproduced with permission from Silvani *et al.* 2018); and (c) a ground-based infrared video sequence taken of a spreading crown fire in Alaska that revealed upward flame velocities up to 60 m s^{-1} and downward velocities up to 30 m s^{-1} (the arrow in the bottom right of each panel shows scale of 10 m s^{-1}) and forward bursting of flames (reproduced with permission from Coen *et al.* 2004).

Ignition

Flaming ignition of wildland fuel elements occurs when they reach a temperature sufficiently high to enable thermal decomposition of solid materials into flammable gases at a sufficient rate to sustain combustion (see **Chapter 5, The ignition process**). Time-series measurements of the surface temperature of a fuel element leading up to ignition provide insight into how heating is influenced by convection and radiation.

The most reliable (though painstaking) way to measure the surface temperature of a fuel element is to measure it directly. This can be done using fine-gauge thermocouples embedded on the particle surface. Early studies used relatively thick thermocouples embedded on fuel particles in spreading fires; this technique indicated a rapid but smooth heating trend as the flame front approached (Rothermel 1972; De Mestre *et al.* 1989) (**Figure 9.18**). Use of finer thermocouples (50 μ) by Cohen (2015) provided finer temporal resolution, which revealed that particle surface temperature is very sensitive to intermittent flame contact. This causes a stair-step rise in particle surface temperature up to ignition (**Figure 6.24**). For this research, the fine thermocouples were embedded on the surface of a square wooden stick with a 1 mm side dimension by using a razor blade to make a shallow slit to hold the thermocouple junction. A miniscule amount of thermal

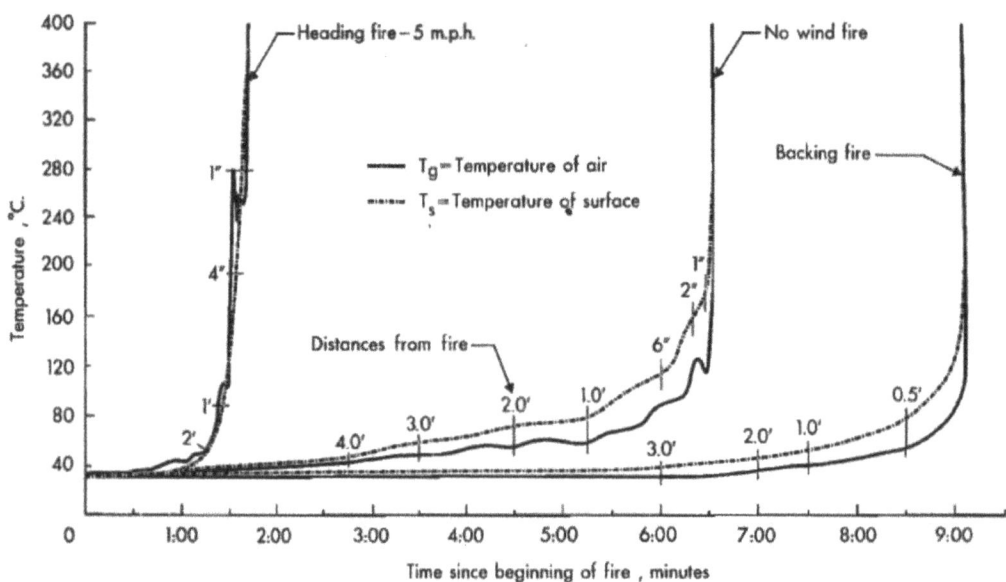

Figure 9.18: Surface heating of fuel particles and adjacent air temperature recorded by thermocouples in spreading fires in the laboratory; data show the rapid rise of particle temperature, primarily by convection, to ignition as the fire front approaches within the final fractions of an inch and second, indicated by measurements in inches or feet along each graph (from Rothermel 1972). Note that the nominal ignition temperature is ~350 °C.

paste ensured good thermal contact between the thermocouple junction and the wood fibres. This technique provided direct, time-resolved measurement of the surface temperatures of fuel elements when they were exposed to radiant heat flux from a gas-fired radiant panel (Cohen and Finney 2010). The technique was also used to measure temperature change from combined convection and radiation heat flux in spreading fires in the laboratory (see Finney *et al.* 2015 and Cohen 2015 for details).

A digital camera can record video or thermal video of a spreading fire if protected inside an insulated, fireproof enclosure. The utility of in-fire cameras for fire research cannot be overstated because they provide video of fire processes that are impossible to see from outside the fire. They are particularly useful for the study of fire spread (discussed below) because they show how fuels are repeatedly blasted by downwashes of flame before ignition. In-fire cameras can be used to identify the sequence in which ignitions occur and thus help identify the physical processes in play. For example, if larger fuel particles ignite before finer ones, it is likely that radiation is sufficient as the heat-transfer process. If smaller particles ignite before larger ones, it is likely that convective heat transfer pre-empts radiant heating. Thermal cameras can also reveal which parts of fuel particles ignite first. For example, **Figure 6.24b** shows the corners and edges of rectangular cardboard heated and ignited in a spreading laboratory fire before the centre of the particle, indicating convective heating was principally important. Heating and ignition along the margins of living leaves can be visualised in laboratory experiments (Prince and Fletcher 2014) and in **Figure 5.14**. In-fire cameras can also show short-range spotting that may be contributing to fire spread and the ways in which canopy fuels are ignited. Action cameras are well suited for these applications because they are relatively inexpensive, reasonably heat tolerant, compact, and record video with good quality. If multiple action cameras are used in a fire, they can be synchronised using GPS-acquired time, which is included on some models.

Rate of spread

The linear rate of spread is the characteristic most sought after for predicting the behaviour of wildland fires. But the rate of spread in fuel beds of discrete particles is a complex quantity to interpret and model because it is not itself a continuous process and does not arise from continuous physical processes (see **Chapter 1, The challenge of validation**). Instead, it emerges from the behaviour of a coupled dynamical system that produces a sequence of discrete ignitions. The ignitions and conditions within the fire are constantly fluctuating and, in field conditions, the environment also adds considerable variability. Fire spread rates are difficult to model because relevant measurements are challenging to obtain, both in the laboratory and under field conditions. Reliable data are essential for understanding the processes that produce fire spread and for validating modelled processes and predictions.

The conditions for spread of wildland fires are constantly changing: the environment varies over time and space (e.g. fuel, wind, temperature, moisture, topography), as do the dynamical behaviours of the fire itself (e.g. acceleration, deceleration, buoyancy, direction of spread, flame instabilities and turbulence). Spread rate also varies because of the discrete nature of ignition. As described in the section above, fires spread through a sequence of ignitions, particle by particle, gap by gap. This rapid but halting pattern introduces wide variability in spread rates over short periods. The leading edge of a fire may not move at all for a few seconds and then leap ahead, producing near-instantaneous spread rates that vary from 0–100 m min^{-1} over a distance of less than 1 metre. The impression of steady spread in homogenous beds of fine fuels is a simplification that probably originates with the limited resolution of the human eye, which can observe fire spread only over intervals of time and distance greater than the scale at which ignition processes actually occur.

Because the physics of fire spread is based in non-steady processes that occur in a non-steady environment, it is challenging to determine the exact position of a fire's front and to measure its rate of travel. With careful measurements, however, we can describe the distributions of fire spread rates over short distances and use statistical distributions to describe the variation in spread rate. Gould and Sullivan (2020) compared the variability in short-interval measurements of fire spread rate with that in long-term averages. The differences indicate that, in order to understand fire spread rates and apply information about them to meet objectives, we must be alert to the ways in which this very important fire characteristic is measured and the sources of error in those measurements.

One-dimensional spread rate

Spread rate measurements have typically focused on the heading portion of the fire; occasionally they have addressed backing fire spread as well. All methods for measuring 1D fire spread are based on a series of point measurements – that is, determination of the time required for the fire front to progress from one known location point to another. They all assume that the fire also travels in a straight line between measurement points.

The simplest, low-technology 1D method for measuring rate of spread is appropriate for all spread directions. It requires an observer to visually record the time the fire front arrives at two points on the ground, which are separated by a known distance. In prescribed fires, the measurement points are often marked by tall poles. Only a stopwatch and a good view of the leading edge of the fire are required to measure rate of spread. Choice of distance is arbitrary, but it is typically selected to cover fairly uniform fuels, and it is usually short enough to justify the assumption that fire spreads the complete distance while wind speeds and directions are stationary. An improvement on this technique makes use of digital video to record the fire's progress. Before digital video was available, filming was done with stopwatches visible in the camera's field of view (Britton *et al.* 1977). Another method was to place firecrackers or small explosives at intervals along a linear transect, with the observer or a microphone audibly detecting the explosions to indicate when the fire arrived (Stephens *et al.* 2008).

More advanced videographic methods have been developed to measure 1D spread rate and flame characteristics in laboratory fires (McMahon *et al.* 1986; Adkins 1995), and these have also been used in prescribed fires (Nelson and Adkins 1986). After a fire, video is processed to measure the progress of the fire's edge past a reference scale within the field of view, and the same reference scale is used to measure the height and angle of flames. Another videographic method uses a longitudinal array of light-detecting photocells with narrow view angle to measure fire progress across a laboratory fuel bed (Catchpole *et al.* 1998).

These 1D observational methods are suitable for small fires of relatively low intensity and are impractical for large and intense fires because the advancing edge of these fires is too remote for an observer to see (or hear) clearly, and the view is likely to be obscured by smoke. A long fire front also makes it difficult to determine the precise time when a fire's edge reaches the predetermined measurement points because of a viewing phenomenon called parallax. This problem can perhaps be addressed with use of multiple cameras (Pickford and Sandberg 1975). All 1D methods also suffer from the assumption that fire spreads in a constant direction rather than wandering laterally or diagonally; this common fire behaviour cannot easily be known from ground-based observations. Because wider fires spread faster than narrow ones (**Chapter 6, Figure 6.31** and **Chapter 7, Figure 7.8**), the limitations of 1D observational techniques to small and narrow flame fronts probably also restrict their usefulness to the lower range of spread rates associated with those fires.

Thermocouple arrays are commonly used in laboratory and field experiments to measure rate of spread (e.g. Wilson 1982; Finney *et al.* 2015; Johnston *et al.* 2018). In this method, linear arrays of thermocouples are oriented parallel to the spread direction. The time when each thermocouple sustains a specified temperature indicating durable flame presence is recorded as the fire front's arrival time, and rate of spread is calculated based on the difference between arrival times. It has been difficult to use this method if the fire's direction of spread is not known when the thermocouple array is installed. The exact arrival time may also be affected by the height above ground that the thermocouples are located, with those a few centimetres above ground showing delayed arrival compared to those at the top of the fuel bed. However, aerial imagery can now be used to record the orientation of the flame zone as it encounters the thermocouple array (**Figure 9.19d**), and this information allows for appropriate interpretation of the fire's progress even if the sensors are not perfectly aligned with the direction of spread. Measurement of fire spread with thermocouple arrays offers an advantage over other point-measurement techniques because time-varying properties of gas and flame temperatures can be recorded in addition to rate of spread (**Figure 9.19**).

Researchers have some options in designing a thermocouple array. It is important to decide whether an accurate estimation of gas temperature and its intermittency is required, or if information on trends is sufficient. For example, thermocouples made of thick wire with a large junction have slow response times and will underestimate temperatures of short-duration flame gases, but a series of them may be sufficient to record time-varying spread. Fine wire thermocouples can record more accurately the intermittent flame contacts and the temperature variation. Spacing between the thermocouples should be determined based on the resolution needed in results. For example, thermocouples spaced a few centimetres apart have been used to record changes in rate of spread in laboratory fires with flames up to 2 m tall (Finney *et al.* 2015). Thermocouples spaced tens of centimetres apart have been used for field applications (see **Figure 9.19**). Arrays deployed in the field require battery-powered data loggers.

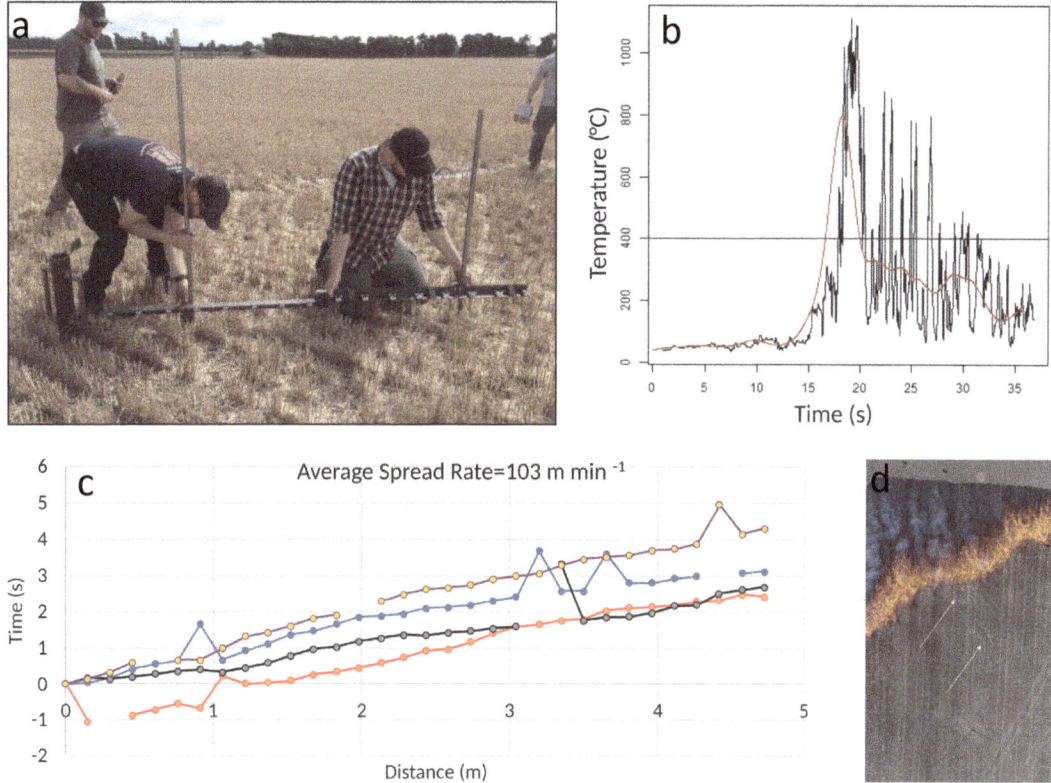

Figure 9.19: (a) Linear arrays of 32 thermocouples (50 μ wire) at 0.15 m spacing installed above the crop-stubble fuel bed before a prescribed fire, parallel with the direction of fire spread (photograph by Ian Grob); (b) record of temperature fluctuations (black line) during flame arrival at 50 μ thermocouple and the smoothed trends red line; (c) fire progress through the fuel bed on each of four thermocouple arrays oriented perpendicular to the spread direction (different colours represent separate arrays); and (d) progress of a fire approaching thermocouple arrays as seen on imagery from an unmanned aerial vehicle (arrows point to thermocouple arrays) (photograph courtesy of University of Canterbury, NZ).

Two-dimensional spread rate and fire growth

To avoid the limitations of 1D measurements of fire spread rate, Simard *et al.* (1984) developed generalised methods for analysing fire movement by using a 2D spatial array of sensors. In their original design, each sensor was a digital timer that started when the fire melted a solder wire in the timer. The sensors were located at known points in various triangular configurations within the fire area. The fire arrival times were analysed to describe directionally specific spread rates throughout the burn unit. The researchers recommended that sensors be located relatively close to each other to minimise ambiguity caused by variations in spread at resolutions finer than the grid. The grid method can be implemented irrespective of fire size or fire type, and it does not require direct observation. The method has been used in experimental crown fire research (Taylor *et al.* 2004) (**Figure 9.20**), slash fires (McRae 1999), understorey burns (McRae *et al.* 2005) and grass fires (Clements *et al.* 2019). Recent implementations of the grid technique have taken advantage of small data loggers and thermocouples to record temperatures at high sampling frequencies (2–50 Hz) in the field, but the methods of analysis remain similar, producing a map of fire arrival times that can be analysed to depict 2D fire movement and its variations.

The most comprehensive approach for measuring fire spread rate, as well as residence time and flame zone depth, involves overhead or aerial videography, either visible or infrared. The technology

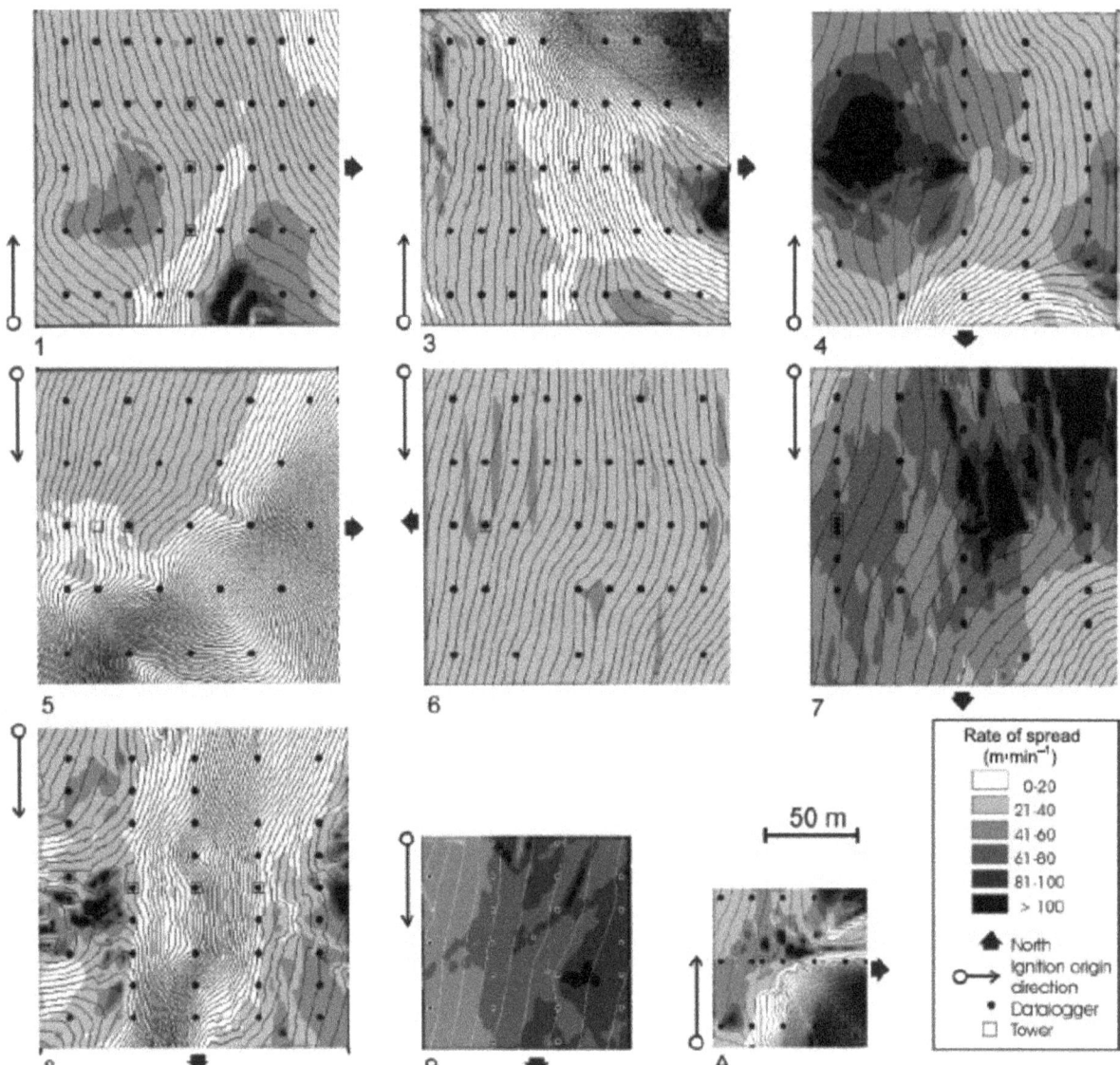

Figure 9.20: Data from grid of ground-based timers (dots) used in crown fire research plots show contours of fire spread that can be statistically reconstructed to describe 2D rate of spread (reproduced with permission from Taylor et al. 2004).

for this approach is advancing rapidly. As the cost of the technology declines and data processing methods are improved, aerial measurement of fire spread will be preferred to ground-based methods.

Infrared cameras are currently used in laboratory experiments for detailed analysis of fire spread (Pastor et al. 2006), and standard high-definition video cameras are routinely positioned to record fire spread from all perspectives, including overhead. The cameras can record at 120 frames per second, which is sufficiently fast for measuring spread rate, flame zone depth and residence time. As unmanned aerial vehicles become more sophisticated and less expensive, scientists are better able to obtain high-quality imagery from overhead sources. Scientific-grade thermal cameras will still require mounting on stationary towers or elevated platforms unless larger aircraft with sufficient payload and power can be acquired. From the resulting data, we can now produce detailed descriptions of fire growth patterns and spread rates (Johnston et al. 2018; Moran et al. 2019).

Contours and vectors describing fire spread in large fires can be derived from imaging systems on fixed-wing aircraft with relatively long time intervals between passes (**Figure 9.21c**). Considerable analysis is required to derive contours of fire movement from these data as explained by Stow *et al.* (2014, 2019). A helicopter (McRae *et al.* 2005) can also be used to obtain overhead data on fire spread, but the resulting data requires complicated post-processing to compensate for aircraft movement (**Figure 9.21a, b**).

Infrared and visible imagery will be essential for exploring how fires spread and flame zone geometry evolve in 2D and 3D. These aspects of fire growth have been little studied because comprehensive data have only recently become available. To help us understand real fire spread in the field – how spread rates vary with curvature of the flame zone, how fires respond to spatially changing topography and fuels, and how multiple flame zones affect fire spread – we must rely on high-resolution overhead video imagery.

Environmental measurements

The coupled, dynamical system of wildland fire operates through the interaction of environmental conditions with the fire itself. Therefore, measurement of environmental conditions is essential in both laboratory and field studies of fire behaviour. Topography is a relatively straightforward characteristic to obtain because it remains static for the duration of a fire. For most small plots, slope and aspect are intentionally chosen to be relatively uniform, but for larger plots or wildfires containing complex variations in canyons or ridges, topography can be reconstructed from digital elevation data. We focus here on fuel moisture and weather factors.

Fuel moisture

The moisture content of living and dead fuels strongly affects fuel consumption, energy release rate, fire spread, flame zone characteristics and combustion behind the flaming zone. While moisture content is sometimes sampled to monitor

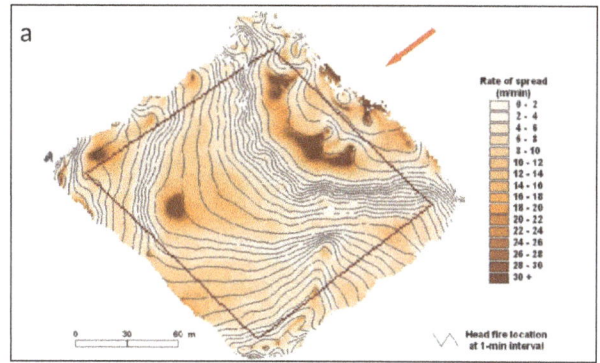

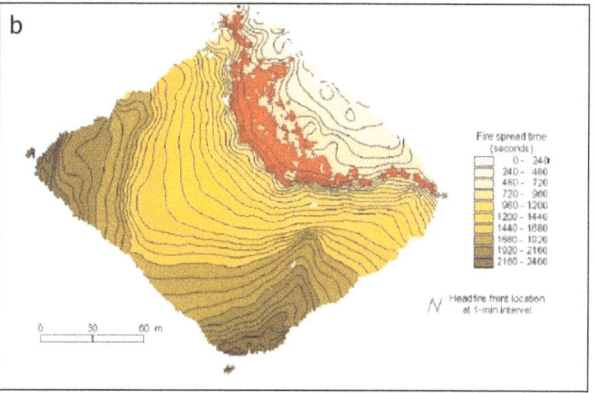

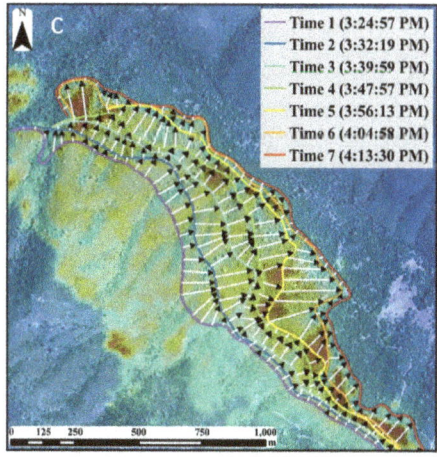

Figure 9.21: Reconstructions of fire spread rates and contours of movement from aerial infrared imagery taken by (a) a helicopter-mounted infrared camera for a small surface fire on an experimental plot, showing (b) time progression contours with superimposed flame zone (red) at ~720 s of elapsed time (reproduced with permission from McRae *et al.* 2005). (c) A fixed-wing infrared imager for large wildfire provides sequential images of fire front position from return overflights that are processed to obtain rate of spread vectors (reproduced with permission from Stow *et al.* 2019).

seasonal trends for fire danger rating, we concentrate here on sampling for experimental or prescribed burns. For this purpose, it is not necessary to sample the moistures of fuels that are not considered 'available' to the fire – that is, not in a condition in which they are likely to ignite if exposed to fire. For example, the large living stems of trees and shrubs are not usually considered available fuel, while their small live branches and bark flakes may be available.

Fuel moisture content is most commonly measured on a *gravimetric* basis and quantified as ratio of the moisture weight in the sample to its dry weight. The moisture weight of the sample is calculated as the difference between fresh weight and dry weight (**Chapter 6, Eqn [6.1]**). For fire applications, moisture is usually not expressed as a fraction of the fresh weight or on a volumetric basis, as is sometimes done for soil. The precision needed for moisture content measurements depends upon the application. Preliminary sampling can be conducted to determine the needed sampling intensity and appropriate levels of stratification. For use in fire behaviour studies, fuel moisture should ideally be measured at a precision of ± 1%. Procedures for sampling and measuring moisture content for a variety of wildland fuel materials are described in Countryman and Dean (1979), Norum and Miller (1984) and (for woody materials) Govett *et al.* (2010).

Fuel moisture sampling is usually stratified to account for the diversity in type, size and spatial distribution of particles in the fuel bed (Norum and Miller 1984; Matthews 2014). In grasslands, for example, it may be important to select samples that represent the main grass species, in both live and dead condition, and to separate the dead material in the thatch and litter from standing stalks, blades and flowering organs. If shrub species are present, then leaves, twigs and large dead stem material should be sampled separately. In forests, samples should be obtained for dead needles of each tree species – particularly if trees have suffered extensive damage from insects or pathogens (Page *et al.* 2013) – and also from branches of different sizes, duff and organic forest floor material, lichens and moss, and large woody material. Live foliage from the current year must be separated from the older foliage of each tree species and separated from cones and other strobili, berries, buds and twigs. Some applications may require sampling available fuel from trees of different sizes and from different positions within the canopy (e.g. low *v.* high branches, interior *v.* exterior). For large land areas or burn units, further stratification of samples may be needed to accommodate variation in topography – for example, to account for the influences of terrain and overstorey shading on fuel exposure to sun and precipitation (Countryman and Dean 1979).

The most common technique used to determine fuel moisture content is *oven-drying* of sampled fuel material. For small materials, a sample typically consists of ~50–100 g fresh weight of just one type of particle in one stratum (e.g. based on size, living *v.* dead, grass *v.* twigs *v.* litter). For large fuel particles, larger, heavier samples are needed. Because logs and deep duff layers have internal moisture variation (see **Chapter 6, Fuel moisture**), it is important to determine before sampling what fuel moisture is important to the study – surface moisture alone, an average value, or a description of the entire gradient (**see Figure 9.4b**). Fuel moisture samples are placed in a sealable container immediately after collection to prevent moisture exchange with the atmosphere. The samples in their sealed containers are weighed to determine fresh weight (i.e. total minus the weight of the container). Then the sample is dried in a laboratory drying oven until its weight is stable (i.e. changing no more than 0.1% between weighings), meaning almost all of the moisture has been removed. Commercial drying ovens typically have forced circulation to provide uniform convective heat exchange. The drying time depends upon the size or quantity of the fuel sample and the temperature of the oven, recommended at 102–105 °C (American Society for Testing and Materials 2010; Matthews 2010). For fine fuels (needles, grass, leaves), drying for a day or two is usually sufficient, although it may take longer if the ambient humidity is high (unless the oven has a humidity control feature). Logs and other large samples may require weeks of drying to reach a stable weight.

Oven drying of living vegetation and materials with volatile organic compounds may release waxes, oils and terpenes in addition to water, which biases the moisture estimates – decreasing the apparent dry weight of the sample and yielding an overestimate of its water content. These effects can be reduced by lowering the oven temperature, but this leaves more water in the sample (Matthews 2010). If high accuracy in fuel moisture is critical to the application, alternative drying techniques may be needed for live fuels and those containing volatile compounds. Techniques could include the use of desiccants, freeze drying or chemical processes.

Microwave ovens have been used to speed drying of fuel samples. This method produces results similar to those from standard ovens for dead material (Norum and Fischer 1980). For live foliage samples, however, uneven heating produced within the oven and disruptions of cell contents may introduce variability and errors in moisture calculations (McCreight 1981). For example, long pine needles and large leaves may not receive uniform heating in a microwave oven and therefore will not be uniformly dried. Also, microwave drying may cause the moisture within living cells to boil, rupturing the cells and expelling biomass as well as water during the drying process. This will cause dry weight to be underestimated and moisture content to be overestimated.

Commercial moisture meters used in industrial applications (e.g. for drying lumber and wood products) can be used to measure moisture content of some wildland fuels. The reader is referred to James (1988) for a discussion of various types used for wood. The accuracy of these measurements depends on instrument calibration and the quality of the sample, and these instruments may not be reliable for rotten or decayed materials or very wet materials (> 30% moisture content). Probe-type resistance or conductance meters have been used for field measurements of fuel particles that have sufficient surface length to insert the probes. The shallow penetration of the probes means that they reflect only moisture in the surface layer of the fuel. If moisture gradients in the fuel are to be examined, large fuels must be sectioned to access interior portions (see **Figure 9.4b**). Several brands of automatic moisture analysers are available for laboratory measurements of fine fuel moisture and can be used in the field with a portable power supply. These analysers have been found suitable and reliable for measuring moisture content of both live and dead materials (Jolly and Hadlow 2012; Rossa *et al.* 2015). Duff moisture can be measured directly with duff moisture meters (Robichaud *et al.* 2004), and these can also produce reasonable estimates of the moisture of live herbaceous fuels (McGranahan 2019).

Because all fuels are continuously responding to atmospheric humidity, fuel moistures must be measured as closely as possible to the time of the experimental fire. This is particularly important for fine dead particles. Logistics of pre-fire sampling must address this issue, ensuring that it is feasible to obtain all samples in a given burn unit within a short time (e.g. < 1 h) before ignition.

Analogue methods have long been used for monitoring fuel moisture content for fire danger rating (Gisborne 1933). The most common is the 10 h fuel moisture stick, which is a commercially produced, pre-weighed and dried assembly of uniform 12.7 mm diameter pine dowels (see Haines and Frost 1978). Fuel moisture sticks are suspended at a standard height above ground so measurements from various locations are comparable. Their moisture content can be readily determined by weighing the sample and correcting for weathering from exposure time. Analogue sticks can be installed on study plots and weighed manually or monitored electronically by weather stations, with data transmitted by satellite or cell phone. Other analogue methods have been developed for research, including custom cribs and sticks of different sizes (Burgan 1987). Fire managers have long used the flexibility of dead pine needles to subjectively assess fuel moisture. Burrows (1991) refined this technique by measuring the angle at which pine needles subjected to bending would break, then calibrating this measurement to moisture content.

Remote sensing techniques are increasingly employed to measure moisture-related characteristics of the soil and living vegetation over broad land areas. These techniques are used primarily for monitoring agricultural crops and fire danger. A variety of sensors have been evaluated, including visible images, synthetic aperture radar, microwave frequencies, near and short-wave infrared bands, and multiband and hyperspectral imagery. Sensors provide estimates of water content per unit land area; this information is then combined with estimates of the dry biomass or the leaf area of vegetation to estimate dry-weight moisture content or relative water content (i.e. percentage of maximum water when leaves are fully turgid). The resolution of these data varies by the sensing platform, with satellite images providing coarser resolutions than aerial or ground-based platforms.

Weather and wind

Wind is often a driving force in wildfire behaviour, but it is very challenging to characterise and predict because it varies on smaller time scales and over finer spatial resolutions than many other environmental influences (e.g. fuel type, fuel moisture, slope). Fire's sensitivity to wind and rapid variation in wind patterns have caused many dramatic, dangerous and sometimes tragic fire events. It is important to measure wind characteristics in as much detail as possible on experimental fires, because other measurements – of weather, fuels, topography and fire behaviour itself – probably cannot be interpreted or modelled without a thorough description of wind speed and direction. Detailed measurements of the wind profile during prescribed fire experiments are beginning to reveal the complex interactions of fire and wind. The reader is referred to details regarding instrumentation in experiments by Clements *et al.* (2007), Clements and Seto (2015), Clements *et al.* (2016) and Clements *et al.* (2019).

While wind speed and direction are often characterised by single, instantaneous measurements, this is a very gross approximation and may not be helpful for understanding fire behaviour. Wind speed and direction change continuously over multiple scales of time and space. Variations occur throughout the year due to seasonal changes. High- and low-pressure weather systems in the midlatitudes alter wind patterns roughly every 4 days. Diurnal variations are caused by changes in the sun angle throughout the day- and night-time cooling after the sun sets; this pattern is complicated in arctic regions because, in early summer, the sun angle remains very low and the sun does not set at all for some days. Storm systems, such as cold front passages and thunderstorms, cause changes in regional wind patterns over intervals of minutes, and turbulence associated with these changes can cause gusts that fluctuate every second or less. Temporal fluctuations in wind speed and direction lead to spatial fluctuations, which are increased by complex terrain. The ground causes frictional drag on the wind, which influences the vertical profile of wind speed and direction (**Chapter 6, Winds**).

For field experimental burns, multiple surface wind sensors are needed to measure the wind field as it varies over time and space. Weather sensors are also usually placed upwind of the fire to measure ambient winds, air temperature and relative humidity (**Figure 9.22**). It is common to measure wind at eye level, 20 feet above the vegetation (in the United States), or 10 m above the vegetation as the international standard. For some research burns, towers 80 m tall or more are erected so sensors can be place at multiple heights to obtain wind profiles (e.g. see Taylor *et al.* 2004; Clements *et al.* 2007; Clements *et al.* 2019). Studies have also used tethered balloons for this purpose.

Measuring winds in complex terrain is especially challenging since terrain features cause significant variability in wind speed and direction. Because complex terrain induces complex wind flow features (e.g. wakes, anabatic and katabatic winds, channelling, mountain waves), multiple wind sensors must be placed in strategic locations to describe the flow field. Butler *et al.* (2015) measured winds over an isolated mountain and in a steep river canyon using ~50 surface wind sensors that traversed all of the terrain features. Their data were used to evaluate microscale wind models (Wagenbrenner *et al.* 2016). An infinite variety of topographic and wind combinations makes it impossible to generalise wind flow patterns and, even if that were feasible, the flow patterns would change in the presence of fire (see **Chapter 6, Topography**). The difficulty of characterising complex wind fields is one reason that experimental studies of fire behaviour are rarely conducted in complex terrain.

It is usually easier to measure wind in laboratory wind tunnels than in the field. The general direction of the wind is known ahead of time, which allows directionally sensitive sensors to be oriented properly. Also, the vertical wind profile is mostly uniform with a very small boundary layer (a few centimetres), so the wind profile is much simpler to characterise in laboratories than in the real atmosphere. This allows researchers some flexibility placing sensors to reliably represent the wind field. Also, the turbulence in wind tunnels is usually very low, so measurements of mean wind speed provides a good representation of the speed in the whole wind field.

Sensors for measuring the wind (i.e. anemometers) are widely available. Each kind has strengths and weaknesses:

- Portable handheld devices are lightweight and can easily be carried by firefighters and observers. These devices include the anemometer in the standard fire belt weather kit and electronic devices like the Kestrel wind meter. A disadvantage is that they are typically measuring winds near the ground as influenced by vegetation and local topography and are thus not generalisable to other areas in complex terrain, and their manual operation means that they are not suitable as instrumentation of long-term trends from specific locations.
- Cup-and-vane and propeller anemometers are less portable but able to record time-averaged horizontal wind speed and direction. Averaging for propeller anemometers usually ranges from 2 to 60 min, but this sensor cannot measure the vertical wind component or fine-scale fluctuations in wind components.
- Sonic anemometers are commonly used in field experiments and can measure 3D winds at high temporal frequency (20 Hz or more). They can also analyse high-frequency wind turbulence and other weather characteristics such as temperature and moisture. Sonic devices are commonly used by atmospheric scientists to estimate turbulent vertical fluxes of momentum, heat and moisture. The downside of these instruments is that they are expensive and more complicated to install but commercial brands can be programmed so that data can be obtained remotely (i.e. via the internet).

An example of comprehensive meteorological measurements associated with prescribed fire research was reported by Clements *et al.* (2019). Vertical arrays of sonic anemometers on towers located within the burn unit recorded changes in the wind profile with fire front passage. A portable weather station provided continuous recordings of winds, temperature and relative humidity (**Figure 9.22**), and a network of cup-and-vane anemometers were arranged around the perimeter of the burn unit. SODAR (sonic detection and ranging) profilers were used to measure wind and temperature in a vertical column up several hundred metres above the sensor (Clements and Seto 2015; Clements *et al.* 2019). LiDAR and radar measurements are also being used to measure winds over large planar areas as the devices sweep the sensor back and forth. Methods of sensing wind high above the ground are also being developed for unmanned aerial devices.

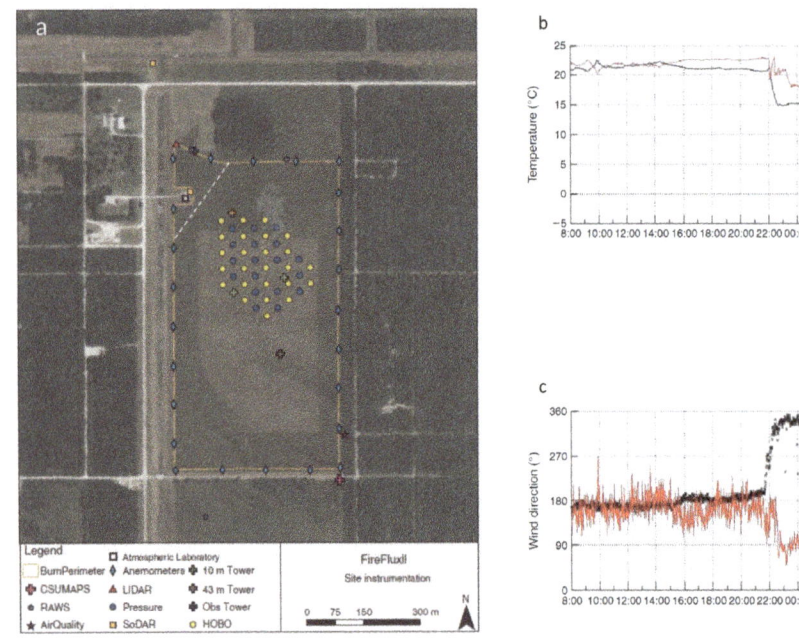

Figure 9.22: (a) Overhead view of monitoring equipment located within and around FireFlux II experimental prescribed burn unit; (b) ambient temperature and humidity from the burn; and (c) wind speed and direction measurements from the burn (reproduced with permission from Clements *et al*. 2019, reprinted with permission from CSIRO Publishing).

Temperature

Air temperature can affect fire behaviour through its impact on fuel moisture and fuel temperature, so it should always be monitored during research studies. Most instrumentation used for weather measurements in fire experiments records temperature, relative humidity and wind characteristics (**Figure 9.22**). Temperature varies significantly over the spatial and temporal scales of a fire, just as wind does but to a lesser degree. Vertical variation in air temperature should be measured if possible, because it strongly influences atmospheric stability, wind and fuel moisture (**Chapter 6, Winds**). This is especially true in complex, mountainous terrain.

The most common types of air temperature sensors are based on electrical resistance. They include thermistors, resistive temperature detectors and thermocouples. These sensors often include technology to reduce errors due to heat by solar radiation – for example, radiation shields and/or air aspiration systems. Placement of temperature sensors in complex terrain should account for topographic influences on air temperature, such as vertical variations between mountains and valleys.

Relative humidity

Relative humidity is a measure of the amount of water in air. It affects fuel moisture and, to a lesser extent, the combustion processes on wildland fires. Relative humidity sensors, called hygrometers, use many different methods to measure this quantity. Most firefighters are familiar with the sling psychrometer, which uses two different temperature measurements (dry bulb and wet bulb) and pressure (estimated through elevation) to produce direct measurement of relative humidity. Older, ingenious methods have used indirect measurement. One method took advantage of the fact that the human hair elongates and contracts due to changes in moisture. Another method used a cooled mirror combined with optical sensors to measure the dew point temperature, from which relative humidity could be computed. Modern sensors measure the change in capacitance or electrical resistance of a substance as it absorbs moisture from the air, or measure changes in thermal conductivity or density of the air itself.

Summary

This Chapter reviews both historical and advanced methods for measuring fire behaviours, physical processes and environmental variables in wildland fire. Laboratory experiments can investigate fire behaviour in detail through the use of careful design and intensive instrumentation. To be useful for understanding and modelling real fires, laboratory studies must be complemented by field-scale experiments with comparable measurements. Laboratory facilities dedicated to studying the behaviour of wildland fires are limited, as are the number of highly instrumented field burns and detailed data from large-scale fires. Experiments in the laboratory and the field are challenged by expense and logistical constraints, because research on fire at any scale requires require custom-designed, robust instrumentation and a wide range of technical expertise to operate it. Detailed study of unplanned wildfires is especially constrained by logistics and safety concerns; in the future, however, the access and technology afforded by remote sensing and unmanned aerial vehicles may greatly increase our ability to obtain detailed, useful data from wildfires. These advancing technologies can create a fruitful, exciting future for wildland fire behaviour research.

References

Adkins CW (1995) 'Users' guide for fire image analysis system – Version 5.0: a tool for measuring fire behavior characteristics'. General Technical Report SE-93. USDA Forest Service, Southern Research Station, Asheville, NC.

Agee JK (1983) Fuel weights of understory-grown conifers in southern Oregon. *Canadian Journal of Forest Research* **13**(4), 648–656.

Albini FA, Reinhardt ED (1995) Modeling ignition and burning rate of large woody natural fuels. *International Journal of Wildland Fire* **5**(2), 81–91. doi:10.1071/WF9950081

Alexander ME, Stefner CN, Mason JA, Stocks BJ, Hartley GR, Maffey ME, Wotton BM, Taylor SW, Lavoie N, Dalrymple GN (2004) 'Characterizing the jack pine – black spruce fuel complex of the International Crown Fire Modelling Experiment (ICFME)'. Information Report NOR-X-393. Natural Resources Canada, Canadian Forest Service, Northern Forestry Centre, Edmonton, Alberta.

Alonzo M, Morton DC, Cook BD, Andersen HE, Babcock C, Pattison R (2017) Patterns of canopy and surface layer consumption in a boreal forest fire from repeat airborne lidar. *Environmental Research Letters* **12**(6), 065004. doi:10.1088/1748-9326/aa6ade

American Society for Testing and Materials (2010) *Standard Test Methods for Direct Moisture Content Measurement of Wood and Wood-base Materials*. American Society for Testing and Materials, West Conshohocken, PA.

Anderson HE (1964) 'Mechanisms of fire spread'. Research Paper INT-8. USDA Forest Service, Intermountain Forest and Range Experiment Station, Northern Forest Fire Laboratory, Ogden, UT.

Anderson HE, Rothermel RC (1965) Influence of moisture and wind upon the characteristics of free-burning fires. *Symposium (International) on Combustion* **10**, 1009–1019.

Beaufait WR (1966) An integrating device for evaluating prescribed fires. *Forest Science* **12**(1), 27–29.

Beaufait WR, Hardy CE, Fischer WC (1975) 'Broadcast burning in larch-fir clearcuts: the Miller Creek–Newman Ridge study'. Research Paper INT-175. USDA Forest Service, Intermountain Forest and Range Experiment Station, Ogden, UT.

Bonham CD (2013) *Measurements for Terrestrial Vegetation*. John Wiley & Sons, New York, NY.

Britton CM, Karr BL, Sneva FA (1977) A technique for measuring rate of fire spread. *Rangeland Ecology & Management/Journal of Range Management Archives* **30**(5), 395–397.

Brown JK (1971) A planar intersect method for sampling fuel volume and surface area. *Forest Science* **17**(1), 96–102.

Brown JK (1972) 'Field test of a rate-of-fire-spread model in slash fuels'. Research Paper INT-116. USDA Forest Service, Intermountain Forest and Range Experiment Station, Ogden, UT.

Brown JK (1974) 'Handbook for inventorying downed woody material'. General Technical Report INT-16. USDA Forest Service, Intermountain Forest and Range Experiment Station, Ogden, UT.

Brown JK (1976) Estimating shrub biomass from basal stem diameters. *Canadian Journal of Forest Research* **6**(2), 153–158. doi:10.1139/x76-019

Brown JK (1978) 'Weight and density of crowns of Rocky Mountain conifers'. Research Paper INT-197. USDA Forest Service, Intermountain Forest and Range Experiment Station, Ogden, UT.

Brown JK, Oberheu RD, Johnston CM (1982) 'Handbook for inventorying surface fuels and biomass in the Interior West'. General Technical Report INT-129. USDA Forest

Service, Intermountain Forest and Range Experiment Station, Ogden, UT.

Brown JK, Marsden MA, Ryan KC, Reinhardt ED (1985) 'Predicting duff and woody fuel consumed by prescribed fire in the Northern Rocky Mountains'. Research Paper INT-337. USDA Forest Service, Intermountain Forest and Range Experiment Station, Ogden, UT.

Burgan RE (1987) A comparison of procedures to estimate fine dead fuel moisture for fire behaviour predictions. *South African Forestry Journal* **142**(1), 34–40. doi:10.1080/00382167.1987.9630281

Burrows ND (1991) Rapid estimation of the moisture content of dead *Pinus pinaster* needle litter in the field. *Australian Forestry* **54**(3), 116–119. doi:10.1080/00049158.1991.10674567

Burrows ND (2001) Flame residence times and rates of weight loss of eucalypt forest fuel particles. *International Journal of Wildland Fire* **10**(2), 137–143. doi:10.1071/WF01005

Butler BW, Jimenez D, Forthofer J, Shannon K, Sopko P (2010) A portable system for characterizing wildland fire behavior. In *Proceedings of the 6th International Conference on Forest Fire Research* [CD-ROM]. 15–18 November 2010, Coimbra, Portugal. (Ed. DX Viegas). University of Coimbra, Coimbria.

Butler BW, Wagenbrenner NS, Forthofer JM, Lamb BK, Shannon KS, Finn D, Eckman RM, Clawson K, Bradshaw K, Sopko P, Beard S (2015) High-resolution observations of the near-surface wind field over an isolated mountain and in a steep river canyon. *Atmospheric Chemistry and Physics* **15**, 3785–3801. doi:10.5194/acp-15-3785-2015

Byram GM (1959) Combustion of forest fuels. In *Forest Fire: Control and Use*. (Ed. KP Davis) pp. 61–89. McGraw Hill, New York, NY.

Call PT, Albini FA (1997) Aerial and surface fuel consumption in crown fires. *International Journal of Wildland Fire* **7**(3), 259–264. doi:10.1071/WF9970259

Catchpole WR, Catchpole EA, Butler BW, Rothermel RC, Morris GA, Latham DJ (1998) Rate of Spread of Free-Burning Fires in Woody Fuels in a Wind Tunnel. *Combustion Science and Technology* **131**(1–6), 1–37. doi:10.1080/00102209808935753

Clark TL, Radke L, Coen J, Middleton D (1999) Analysis of small-scale convective dynamics in a crown fire using infrared video camera imagery. *Journal of Applied Meteorology* **38**(10), 1401–1420. doi:10.1175/1520-0450(1999)038<1401:AOSSCD>2.0.CO;2

Clark TL, Reeder MJ, Griffiths M, Packham D, Krusel N (2005) Infrared observations and numerical modelling of grassland fires in the Northern Territory, Australia. *Meteorology and Atmospheric Physics* **88**(3–4), 193–201. doi:10.1007/s00703-004-0076-9

Clements CB, Seto D (2015) Observations of fire–atmosphere interactions and near-surface heat transport on a slope. *Boundary-Layer Meteorology* **154**, 409–426. doi:10.1007/s10546-014-9982-7

Clements CB, Zhong S, Goodrick S, Li J, Potter BE, Bian X, Heilman WE, Charney JJ, Perna R, Jang M, Lee D (2007) Observing the dynamics of wildland grass fires: FireFlux a field validation experiment. *Bulletin of the American Meteorological Society* **88**(9), 1369–1382. doi:10.1175/BAMS-88-9-1369

Clements CB, Lareau NP, Seto D, Contezac J, Davis B, Teske C, Zajkowski TJ, Hudak AT, Bright BC, Dickinson MB, Butler BW (2016) Fire weather conditions and fire–atmosphere interactions observed during low-intensity prescribed fires–RxCADRE 2012. *International Journal of Wildland Fire* **25**(1), 90–101. doi:10.1071/WF14173

Clements CB, Kochanski AK, Seto D, Davis B, Camacho C, Lareau NP, Contezac J, Restaino J, Heilman WE, Krueger SK, Butler B (2019) The FireFlux II experiment: a model-guided field experiment to improve understanding of fire–atmosphere interactions and fire spread. *International Journal of Wildland Fire* **28**(4), 308–326. doi:10.1071/WF18089

Coen J, Mahalingam S, Daily J (2004) Infrared imagery of crown-fire dynamics during FROSTFIRE. *Journal of Applied Meteorology* **43**(9), 1241–1259. doi:10.1175/1520-0450(2004)043<1241:IIOCDD>2.0.CO;2

Cohen JD (2015) Fuel particle heat exchange during wildland fire spread. PhD thesis. University of Idaho, Moscow, ID.

Cohen JD, Finney MA (2010) An examination of fuel particle heating during fire spread. In *Proceedings of the VI International Conference on Forest Fire Research*. 15–18 November 2010, Coimbra, Portugal. (Ed. DX Viegas). University of Coimbra, Coimbra.

Cook CW, Stubbendieck J (1986) *Range Research: Basic Problems and Techniques*. Society for Range Management, Denver, CO.

Coulloudon B, Eshelman K, Gianola J, Habich N, Hughes L, Johnson C, Pellant M, Podborny P, Rasmussen A, Robles B, Shaver P (1999) *Sampling Vegetation Attributes*. Bureau of Land Management, Denver, CO.

Countryman CM, Dean WA (1979) 'Measuring moisture content in living chaparral: a field user's manual'. General Technical Report PSW-036. USDA Forest Service, Pacific Southwest Research Station, Berkeley, CA.

Daily JW (1997) Laser induced fluorescence spectroscopy in flames. *Progress in Energy and Combustion Science* **23**(2), 133–199. doi:10.1016/S0360-1285(97)00008-7

De Mestre NJ, Catchpole EA, Anderson DH, Rothermel RC (1989) Uniform propagation of a planar fire front without wind. *Combustion Science and Technology* **65**(4–6), 231–244. doi:10.1080/00102208908924051

de Vries PG (1986) Line intersect sampling. In *Sampling Theory for Forest Inventory*. pp. 242–279. Springer, Berlin, Heidelberg.

Despain DG, Clark DL, Reardon JJ (1996) Simulation of crown fire effects on canopy seed bank in lodgepole pine. *International Journal of Wildland Fire* **6**(1), 45–49. doi:10.1071/WF9960045

Dibble RW, Kollmann W, Schefer RW (1984) Conserved scalar fluxes measured in a turbulent nonpremixed flame by combined laser Doppler velocimetry and laser Raman scattering. *Combustion and Flame* **55**(3), 307–321. doi:10.1016/0010-2180(84)90170-6

Eckbreth AC, Hall RJ (1979) CARS thermometry in a sooting flame. *Combustion and Flame* **36**, 87–98. doi:10.1016/0010-2180(79)90048-8

Fernandes PAM, Catchpole WR, Rego FC (2000) Shrubland fire behavior modelling with microplot data. *Canadian Journal of Forest Research* **30**, 889–899. doi:10.1139/x00-012

Filatyev S, Thariyan M, Lucht R, Gore J (2007) Application of simultaneous stereo PIV and double pulsed acetone PLIF to study turbulent premixed flames. In *Collection of Technical Papers 45th AIAA Aerospace Sciences Meeting* **23**, 16005–16011. doi:10.2514/6.2007-1346

Finney MA, Martin RE (1992) Calibration and field testing of passive flame height sensors. *International Journal of Wildland Fire* **2**(3), 115–122. doi:10.1071/WF9920115

Finney MA, Martin RE (1993) Fuel loading, bulk density, and depth of forest floor in coast redwood stands. *Forest Science* **39**(3), 617–622.

Finney MA, Forthofer J, Grenfell IC, Adam BA, Akafuah NK, Saito K (2013) A study of flame spread in engineered cardboard fuelbeds: Part I: Correlations and observations. In *Seventh International Symposium on Scale Modeling (ISSM-7)*. 6–9 August 2013, Hirosaki, Japan. International Scale Modeling Committee, Japan.

Finney MA, Cohen JD, Forthofer JM, McAllister SS, Gollner MJ, Gorham DJ, Saito K, Akafuah NK, Adam BA, English JD (2015) Role of buoyant flame dynamics in wildfire spread. *Proceedings of the National Academy of Sciences of the United States of America* **112**(32), 9833–9838. www.pnas.org/cgi/doi/10.1073/pnas.1504498112. doi:10.1073/pnas.1504498112

Frandsen WH (1987) The influence of moisture and mineral soil on the combustion limits of smoldering forest duff. *Canadian Journal of Forest Research* **17**(12), 1540–1544. doi:10.1139/x87-236

Frankman D, Webb BW, Butler BW, Jimenez D, Forthofer JM, Sopko P, Shannon KS, Hiers JK, Ottmar RD (2013) Measurements of convective and radiative heating in wildland fires. *International Journal of Wildland Fire* **22**, 157–167 doi:10.1071/WF11097.

Garcia M, Saatchi S, Casas A, Koltunov A, Ustin S, Ramirez C, Garcia-Gutierrez J, Balzter H (2017) Quantifying biomass consumption and carbon release from the California Rim fire by integrating airborne LiDAR and Landsat OLI data. *Journal of Geophysical Research. Biogeosciences* **122**(2), 340–353. doi:10.1002/2015JG003315

Gisborne HT (1933) The wood cylinder method of measuring forest inflammability. *Journal of Forestry* **31**(6), 673–679.

Gould JS, Sullivan AL (2020) Two methods for calculating wildland fire rate of forward spread. *International Journal of Wildland Fire* **29**(3), 272–281. doi:10.1071/WF19120

Gould JS, McCaw WL, Cheney NP, Ellis PF Matthews (2007) *Field Guide. Fuel Assessment and Fire Behaviour Prediction in Dry Eucalypt Forest*. Ensis–CSIRO, Canberra and WA Department of Environment and Conservation, Perth.

Gould JS, McCaw WL, Cheney NP (2011) Quantifying fine fuel dynamics and structure in dry eucalypt forest (*Eucalyptus marginata*) in Western Australia for fire management. *Forest Ecology and Management* **262**(3), 531–546. doi:10.1016/j.foreco.2011.04.022

Govett R, Mace T, Bowe S (2010) *A Practical Guide for the Determination of Moisture Content of Woody Biomass*. University of Wisconsin, <http://dnr.wi.gov/topic/ForestBusinesses/documents/BiomassMoistureContent>.

Green SI, Rogak SN (1999) A multiple disk probe for inexpensive and robust velocimetry. *Journal of Fluids Engineering* **121**(2), 446–449. doi:10.1115/1.2822230

Grumstrup TP, Finney MA (2019) Vertical fuel distribution effects on flame lengths in wildfires. Poster presentation at 11th US National Combustion Meeting. 24–27 March 2019, Pasadena, CA.

Grumstrup TP, Forthofer JM, Finney MA (2018) Measurement of three-dimensional flow speed and direction in wildfires. In *Advances in Forest Fire Research 2018. Proceedings of the VII International Conference on Forest Fire Research*. 10–16 November 2018, Coimbra, Portugal. (Ed. DX Viegas) pp. 542–548. doi:10.14195/978-989-26-16-506_60

Haines DA, Frost JS (1978) 'Weathering effects on fuel moisture sticks: corrections and recommendations'. Research Paper NC-154. USDA Forest Service, North Central Forest Experiment Station, St Paul, MIN.

Hartford RA, Frandsen WH (1992) When it's hot, it's hot … or maybe it's not! (Surface flaming may not portend extensive soil heating). *International Journal of Wildland Fire* **2**(3), 139–144. doi:10.1071/WF9920139

Hille MG, Stephens SL (2005) Mixed conifer forest duff consumption during prescribed fires: tree crown impacts. *Forest Science* **51**(5), 417–424.

Hines F, Tolhurst KG, Wilson AAG, McCarthy GJ (2010) *Overall Fuel Hazard Assessment Guide*. 4th edn. Department of Sustainability and Environment, Melbourne.

Hollis JJ, Matthews S, Ottmar RD, Prichard SJ, Slijepcevic A, Burrows ND, Ward B, Tolhurst KG, Anderson WR, Gould JS (2010) Testing woody fuel consumption models for application in Australian southern eucalypt forest fires. *Forest Ecology and Management* **260**(6), 948–964. doi:10.1016/j.foreco.2010.06.007

Hollis JJ, Matthews S, Anderson WR, Cruz MG, Burrows ND (2011) Behind the flaming zone: predicting woody fuel consumption in eucalypt forest fires in southern Australia. *Forest Ecology and Management* **261**(11), 2049–2067. doi:10.1016/j.foreco.2011.02.031

Hudak AT, Dickinson MB, Bright BC, Kremens RL, Loudermilk EL, O'Brien JJ, Hornsby BS, Ottmar RD (2016) Measurements relating fire radiative energy density and surface fuel consumption – RxCADRE 2011 and 2012. *International Journal of Wildland Fire* **25**(1), 25–37. doi:10.1071/WF14159

Hudak AT, Kato A, Bright BC, Loudermilk EL, Hawley C, Restaino JC, Ottmar RD, Prata GA, Cabo C, Prichard SJ, Rowell EM (2020) Towards spatially explicit quantification of pre-and postfire fuels and fuel consumption from traditional and point cloud measurements. *Forest Science* **66**(4), 428–442. doi:10.1093/forsci/fxz085

Hungerford RD, Frandsen WH, Ryan KC (1995) Ignition and burning characteristics of organic soils. In *Fire in Wetlands: A Management Perspective. Proceedings of the 19th Tall Timbers Fire Ecology Conference*. Tallahassee, FL. (Eds SI Cerulean, R Todd Engstrom) pp. 78–91. Tall Timbers Research Station, Tallahassee, FL.

James WL (1988) 'Electric moisture meters for wood'. Gen. Tech. Rep. FPL-GTR-6. USDA Forest Service, Forest Products Laboratory, Madison, WI.

Jimenez D, Forthofer JM, Reardon JJ, Butler BW (2007) Fire behavior sensor package remote trigger design. In *The Fire Environment – Innovations, Management, and Policy Conference Proceedings* [CD-ROM]. Proc. RMRS-P-46CD. (Comps BW Butler, W Cook) pp. 499–505. USDA Forest Service, Rocky Mountain Research Station, Fort Collins, CO.

Johnson VJ (1982) The dilemma of flame length and intensity. *Fire Management Notes* **43**(4), 3–7.

Johnson PS, Johnson CL, West NE (1988) Estimation of phytomass for ungrazed crested wheatgrass plants using allometric equations. *Rangeland Ecology & Management/Journal of Range Management Archives* **41**(5), 421–425.

Johnston JM, Wooster MJ, Paugam R, Wang X, Lynham TJ, Johnston LM (2017) Direct estimation of Byram's fire intensity from infrared remote sensing imagery. *International Journal of Wildland Fire* **26**(8), 668–684. doi:10.1071/WF16178

Johnston JM, Wheatley MJ, Wooster MJ, Paugam R, Davies GM, DeBoer KA (2018) Flame-front rate of spread estimates for moderate scale experimental fires are strongly influenced by measurement approach. *Fire (Basel, Switzerland)* **1**(1), 16. doi:10.3390/fire1010016

Jolly JM, Hadlow AM (2012) A comparison of two methods for estimating conifer live foliar moisture content. *International Journal of Wildland Fire* **21**(2), 180–185. doi:10.1071/WF11015

Kautz J (1997) Appendix C – insulated boxes for protecting video cameras. In *Surviving Fire Entrapments: Comparing Conditions Inside Vehicles and Fire Shelters*. (Ed. R Mangan) pp. 39–40. USDA Forest Service, Missoula Technology Development Center, Missoula, MO.

Keane RE (2008) Biophysical controls on surface fuel litterfall and decomposition in the northern Rocky Mountains, USA. *Canadian Journal of Forest Research* **38**(6), 1431–1445. doi:10.1139/X08-003

Keane RE (2015) *Wildland Fuel Fundamentals and Application*. Springer International Publishing, Cham.

Keane RE, Reinhardt ED, Scott J, Gray K, Reardon J (2005) Estimating forest canopy bulk density using six indirect methods. *Canadian Journal of Forest Research* **35**(3), 724–739. doi:10.1139/x04-213

Kerlin TW, Johnson M (2012) *Practical Thermocouple Thermometry*. 2nd edn. International Society of Automation, Research Triangle Park, NC.

Kremens RL, Smith AM, Dickinson MB (2010) Fire metrology: current and future directions in physics-based measurements. *Fire Ecology* **6**(1), 13–35. doi:10.4996/fireecology.0601013

Kremens RL, Dickinson MB, Bova AS (2012) Radiant flux density, energy density and fuel consumption in mixed-oak forest surface fires. *International Journal of Wildland Fire* **21**(6), 722–730. doi:10.1071/WF10143

Lavoie N, Alexander ME, Macdonald SE (2010) 'Photo guide for quantitatively assessing the characteristics of forest fuels in a jack pine–black spruce chronosequence in the Northwest Territories'. Information Report NOR -X-419. Canadian Forest Service Northern Forestry Centre, Edmonton, Alberta.

Lemaire R, Menanteau S (2017) Assessment of radiation correction methods for bare bead thermocouples in a combustion environment. *International Journal of Thermal Sciences* **122**, 186–200. doi:10.1016/j.ijthermalsci.2017.08.014

Lentile LB, Holden ZA, Smith AM, Falkowski MJ, Hudak AT, Morgan P, Lewis SA, Gessler PE, Benson NC (2006) Remote sensing techniques to assess active fire characteristics and post-fire effects. *International Journal of Wildland Fire* **15**, 319–345. doi:10.1071/WF05097

Matthews S (2010) Effect of drying temperature on fuel moisture content measurements. *International Journal of Wildland Fire* **19**(6), 800–802. doi:10.1071/WF08188

Matthews S (2014) Dead fuel moisture research: 1991–2012. *International Journal of Wildland Fire* **23**(1), 78–92. doi:10.1071/WF13005

Maxwell WG, Ward FR (1980) 'Guidelines for developing or supplementing natural photo series'. Research Note PNW-358. USDA Forest Service, Pacific Northwest Forest and Range Experiment Station, Portland, OR.

McCaffrey BJ, Heskestad G (1976) A robust bidirectional low-velocity probe for flame and fire application. *Combustion and Flame* **26**, 125–127. doi:10.1016/0010-2180(76)90062-6

McCaw WL, Gould JS, Cheney NP, Ellis PF, Anderson WR (2012) Changes in behaviour of fire in dry eucalypt forest as fuel increases with age. *Forest Ecology and Management* **271**, 170–181. doi:10.1016/j.foreco.2012.02.003

McCreight RW (1981) 'Microwave ovens for drying live wildland fuels: an assessment'. Research Note PSW-349. US Department of Agriculture, Forest Service, Pacific Southwest Forest and Range Experiment Station, Berkeley, CA.

McGranahan DA (2019) A device for instantaneously estimating duff moisture content is also effective for grassland fuels. *Fire (Basel, Switzerland)* **2**(1), 12. doi:10.3390/fire2010012

McMahon CK, Adkins CW, Rodgers SL (1986) A video image analysis system for measurement of fire behavior. *Fire Management Notes* **47**, 10–15.

McRae DJ (1999) Point-source fire growth in jack pine slash. *International Journal of Wildland Fire* **9**(1), 65–77. doi:10.1071/WF99006

McRae DJ, Jin JZ, Conard SG, Sukhinin AI, Ivanova GA, Blake TW (2005) Infrared characterization of fine-scale variability in behavior of boreal forest fires. *Canadian Journal of Forest Research* **35**(9), 2194–2206. doi:10.1139/x05-096

Means JE, Hansen HA, Koerper GJ, Alaback PB, Klopsch MW (1994) 'Software for computing plant biomass – BIOPAK users guide'. General Technical Report PNW-340. USDA Forest Service, Pacific Northwest Forest and Range Experiment Station, Portland, OR.

Moran CJ, Seielstad CA, Cunningham MR, Hoff V, Parsons RA, Queen L, Sauerbrey K, Wallace T (2019) Deriving Fire Behavior Metrics from UAS Imagery. *Fire (Basel, Switzerland)* **2**(2), 36. doi:10.3390/fire2020036

Morandini F, Silvani X, Dupuy JL, Susset A (2018) Fire spread across a sloping fuel bed: Flame dynamics and heat transfers. *Combustion and Flame* **190**, 158–170. doi:10.1016/j.combustflame.2017.11.025

Mueller-Dombois D, Ellenberg H (1974) *Aims and Methods of Vegetation Ecology*. Wiley, New York, NY.

Nafus AM, McClaran MP, Archer SR, Throop HL (2009) Multispecies allometric models predict grass biomass in semidesert rangeland. *Rangeland Ecology and Management* **62**(1), 68–72. doi:10.2111/08-003

Nelson RM, Jr (2003) Reaction times and burning rates for wind tunnel headfires. *International Journal of Wildland Fire* **12**(2), 195–211. doi:10.1071/WF02041

Nelson RM, Adkins CW (1986) Flame characteristics of wind-driven surface fires. *Canadian Journal of Forest Research* **16**, 1293–1300. doi:10.1139/x86-229

Norum RA, Fischer WC (1980) 'Determining the moisture content of some dead forest fuels using a microwave oven'. Research Note INT-177. USDA Forest Service, Intermountain Forest and Range Experiment Station, Ogden, UT.

Norum RA, Miller M (1984) 'Measuring fuel moisture content in Alaska: standard methods and procedures'. General Technical Report PNW-171. USDA Forest Service, Pacific Northwest Forest and Range Experiment Station, Portland, OR.

Ohlemiller T, Corley D (1994) Heat release rate and induced wind field in a large scale fire. *Combustion Science and Technology* **97**(4–6), 315–330. doi:10.1080/00102209408935383

Ottmar RD (2014) Wildland fire emissions, carbon, and climate: modeling fuel consumption. *Forest Ecology and Management* **317**, 41–50. doi:10.1016/j.foreco.2013.06.010

Page WG, Jenkins MJ, Alexander ME (2013) Foliar moisture content variations in lodgepole pine over the diurnal cycle during the red stage of mountain pine beetle attack. *Environmental Modelling & Software* **49**, 98–102. doi:10.1016/j.envsoft.2013.08.001

Pastor E, Àgueda A, Andrade-Cetto J, Muñoz M, Pérez Y, Planas E (2006) Computing the rate of spread of linear flame fronts by thermal image processing. *Fire Safety Journal* **41**(8), 569–579. doi:10.1016/j.firesaf.2006.05.009

Pickford SG, Hazard JW (1978) Simulation studies on line intersect sampling of forest residue. *Forest Science* **24**(4), 469–483.

Pickford SG, Sandberg DV (1975) 'Using motion pictures for data collection on prescribed burning experiments'. Research Note PNW-259. USDA Forest Service, Pacific Northwest Forest and Range Experiment Station, Portland, OR.

Potts DF, Ryan KC, Zuuring HR (1986) Stratified sampling for determining duff moisture in mountainous terrain. *Western Journal of Applied Forestry* **1**(1), 29–30. doi:10.1093/wjaf/1.1.29

Prichard SJ, Wright CS, Vihnanek RE, Ottmar RD (2006) Predicting forest floor and woody fuel consumption from prescribed burns in ponderosa pine forests. In *Proceedings of the Third International Fire Ecology and Management Conference: Fire as a Global Process*, 13–17 November 2006, San Diego, CA.

Prince DR, Fletcher TH (2014) Differences in burning behavior of live and dead leaves, part 1: measurements. *Combustion Science and Technology* **186**(12), 1844–1857. doi:10.1080/00102202.2014.923412

Reinhardt ED, Brown JK, Fischer WC, Graham RT (1991) 'Woody fuel and duff consumption by prescribed fire in northern Idaho mixed conifer logging slash'. Research Paper INT-443. USDA Forest Service, Intermountain Forest and Range Experiment Station, Ogden, UT.

Reinhardt ED, Scott JH, Gray K, Keane R (2006a) Estimating canopy fuel characteristics in five conifer stands in the western United States using tree and stand measurements. *Canadian Journal of Forest Research* **36**(11), 2803–2814. doi:10.1139/x06-157

Reinhardt ED, Lutes D, Scott JH (2006b) FuelCalc: a method for estimating fuel characteristics. In *Fuels Management – How to Measure Success*. (Eds PL Andrews PL, BW Butler) pp. 273–282. Proceedings RMRS-P-41. USDA Forest Service, Rocky Mountain Research Station, Fort Collins, CO.

Ren X, Zeng D, Wang Y, Xiong G, Agarwal G, Gollner M (2020) Temperature measurement of a turbulent buoyant ethylene diffusion flame using a dual-thermocouple technique. *Fire Safety Journal*, 103061. doi:10.1016/j.firesaf.2020.103061

Robichaud PR, Gasvoda DS, Hungerford RD, Bilskie J, Ashmun LE, Reardon J (2004) Measuring duff moisture content in the field using a portable meter sensitive to dielectric permittivity. *International Journal of Wildland Fire* **13**(3), 343–353. doi:10.1071/WF03072

Rossa CG, Fernandes PM, Pinto A (2015) Measuring foliar moisture content with a moisture analyzer. *Canadian Journal of Forest Research* **45**(6), 776–781. doi:10.1139/cjfr-2014-0545

Rothermel RC (1972) 'A mathematical model for predicting fire spread in wildland fuels'. Research Paper INT-115. USDA Forest Service, Intermountain Research Station, Ogden, UT.

Rothermel RC (1991) 'Predicting behavior and size of crown fires in the northern Rocky Mountains'. Research Paper INT-438. USDA Forest Service, Intermountain Research Station, Ogden, UT.

Rothermel RC, Anderson HE (1966) 'Fire spread characteristics determined in the laboratory'. Research Paper INT-30. USDA Forest Service, Intermountain Research Station, Ogden, UT.

Ryan KC (1981) 'Evaluation of a passive flame-height sensor to estimate forest fire intensity'. Research Note PNW-390. USDA Forest Service, Pacific Northwest Forest and Range Experiment Station, Portland, OR.

Ryan KC, Frandsen WH (1991) Basal injury from smoldering fires in mature *Pinus ponderosa* Laws. *International Journal of Wildland Fire* **1**(2), 107–118. doi:10.1071/WF9910107

Sapsis DB, Kauffman JB (1991) Fuel consumption and fire behavior associated with prescribed fires in sagebrush ecosystems. *Northwest Science* **65**(4), 173–179.

Scott JH, Reinhardt ED (2005) 'Stereo photo guide for estimating canopy fuel characteristics in conifer stands'. General Technical Report RMRS-GTR-145. USDA Forest Service, Rocky Mountain Research Station, Fort Collins, CO.

Sikkink PG, Keane RE (2008) A comparison of five sampling techniques to estimate surface fuel loading in montane forests. *International Journal of Wildland Fire* **17**(3), 363–379. doi:10.1071/WF07003

Silvani X, Morandini F, Dupuy JL, Susset A, Vernet R, Lambert O (2018) Measuring velocity field and heat transfer during natural fire spread over large inclinable bench. *Experimental Thermal and Fluid Science* **92**, 184–201. doi:10.1016/j.expthermflusci.2017.11.020

Simard AJ, Eenigenburg JE, Adams KB, Nissen RL, Jr, Deacon AG (1984) A general procedure for sampling and analyzing wildland fire spread. *Forest Science* **30**(1), 51–64.

Simard AJ, Blank RW, Hobrla SL (1989) Measuring and interpreting flame height in wildland fires. *Fire Technology* **25**(2), 114–133. doi:10.1007/BF01041421

Skowronski NS, Gallagher MR, Warner TA (2020) Decomposing the interactions between fire severity and canopy fuel structure using multi-temporal, active, and passive remote sensing approaches. *Fire (Basel, Switzerland)* **3**(1), 7. doi:10.3390/fire3010007

Smith WB, Brand GJ (1983) 'Allometric biomass equations for 98 species of herbs, shrubs, and small trees'. Research Note NC-299. USDA Forest Service, North Central Research Station, St Paul, MIN.

Stephens SL, Finney MA, Schantz H (2004) Bulk density and fuel loads of ponderosa pine and white fir forest floors: impacts of leaf morphology. *Northwest Science* **78**(2), 93–110.

Stephens SL, Weise DR, Fry DL, Keiffer RJ, Dawson J, Koo E, Potts J, Pagni PJ (2008) Measuring the rate of spread of chaparral prescribed fires in northern California. *Fire Ecology* **4**(1), 74–86. doi:10.4996/fireecology.0401074

Stocks BJ (1987) Fire behavior in immature jack pine. *Canadian Journal of Forest Research* **17**(1), 80–86. doi:10.1139/x87-014

Stocks BJ (1989) Fire behavior in mature jack pine. *Canadian Journal of Forest Research* **19**(6), 783–790. doi:10.1139/x89-119

Stocks BJ, Alexander ME, Wotton BM, Stefner CN, Flannigan MD, Taylor SW, Lavoie N, Mason JA, Hartley GR, Maffey ME, Dalrymple GN (2004) Crown fire behaviour in a northern jack pine black spruce forest. *Canadian Journal of Forest Research* **34**(8), 1548–1560. doi:10.1139/x04-054

Stow DA, Riggan PJ, Storey EJ, Coulter LL (2014) Measuring fire spread rates from repeat pass airborne thermal infrared imagery. *Remote Sensing Letters* **5**(9), 803–812. doi:10.1080/2150704X.2014.967882

Stow D, Riggan P, Schag G, Brewer W, Tissell R, Coen J, Storey E (2019) Assessing uncertainty and demonstrating potential for estimating fire rate of spread at landscape scales based on time sequential airborne thermal infrared imaging. *International Journal of Remote Sensing* **40**(13), 4876–4897. doi:10.1080/01431161.2019.1574995

Taylor SW, Wotton BM, Alexander ME, Dalrymple GN (2004) Variation in wind and crown fire behaviour in a northern jack pine black spruce forest. *Canadian Journal of Forest Research* **34**(8), 1561–1576. doi:10.1139/x04-116

Thomas PH (1963) The size of flames from natural fires. *Symposium (International) on Combustion* **9**, 844–859.

Van Wagner CE (1968) The line intersect method in forest fuel sampling. *Forest Science* **14**(1), 20–26.

Wagenbrenner NS, Forthofer JM, Lamb BK, Shannon KS, Butler BW (2016) Downscaling surface wind predictions from numerical weather prediction models in complex terrain with WindNinja. *Atmospheric Chemistry and Physics* **16**(8), 5229–5241. doi:10.5194/acp-16-5229-2016

Wang C, Glenn NF (2009) Estimation of fire severity using pre-and post-fire LiDAR data in sagebrush steppe rangelands. *International Journal of Wildland Fire* **18**(7), 848–856. doi:10.1071/WF08173

Wilson RA (1982) 'A reexamination of fire spread in free-burning porous fuel beds'. Research Paper INT-289. USDA Forest Service, Intermountain Forest and Range Experiment Station, Ogden, UT.

Wooster MJ, Roberts G, Perry GLW, Kaufman YJ (2005) Retrieval of biomass combustion rates and totals from fire radiative power observations: FRP derivation and calibration relationships between biomass consumption and fire radiative energy release. *Journal of Geophysical Research, D, Atmospheres* **110**, D24311. doi:10.1029/2005JD006318

Wotton BM, Gould JS, McCaw WL, Cheney NP, Taylor SW (2012) Flame temperature and residence time of fires in dry eucalypt forest. *International Journal of Wildland Fire* **21**(3), 270–281. doi:10.1071/WF10127

Wright CS (2013) Models for predicting fuel consumption in sagebrush-dominated ecosystems. *Rangeland Ecology and Management* **66**(3), 254–266. doi:10.2111/REM-D-12-00027.1

Wright CS, Prichard SJ (2006) Biomass consumption during prescribed fires in big sagebrush ecosystems. In *Fuels Management – How to Measure Success: Conference Proceedings*. Proceedings RMRS-P-41. (Eds PL, Andrews, BW Butler) pp. 489–500. USDA Forest Service, Rocky Mountain Research Station, Fort Collins, CO.

10

Ignition techniques for experimental burning

Greater understanding of dynamical fire behaviour can contribute substantially to improving the safety, usefulness and effectiveness of prescribed fires. *Prescribed burning* is the intentional burning of vegetation to manage fuels and achieve specific ecological objectives. Prescribed burning or experimental burning is also the primary means of obtaining data from wildland fires for research and thus improving understanding, models and other tools for fire management. For these reasons, we use this chapter to discuss the use of various ignition patterns to produce specific fire behaviours. To conduct operational prescribed burns – particularly over large areas and to meet complex land management objectives – one must have extensive formal training, specialised technical qualifications and extensive experience. We do not attempt to cover all of these aspects of prescribed burning here. Instead, we apply the understanding developed throughout this book to the specific problem of how to use particular ignition patterns to produce specific fire behaviours and fire effects.

Prescribed fires that are executed carefully and measured intensively are important for both research and management. Fire scientists use prescribed fires to study fire behaviours beyond the laboratory scale and to explore ranges of fuel and weather conditions that resemble those of full-scale wildland fires. Fire managers use prescribed fires to introduce or maintain fire as an ecological process, manage fuels, prepare sites for regeneration after harvesting, improve wildlife habitat, and meet many other safety- and resource-related objectives.

When we approach wildland fire as a coupled dynamical system, as we have throughout this book, we see that most fire behaviours are manifested at all scales, so we can use the ignition techniques of prescribed burning at small scales to develop greater skill in managing large wildland fires and greater understanding to advance fire behaviour research. We can intentionally use positive and negative feedbacks (**Chapter 7, Positive and negative feedbacks**) to achieve specific outcomes such as measurable fire behaviours, detailed descriptions of two-dimensional fire behaviour, and specific changes in fuels, plant communities, and animal habitat.

Use of experimental burning to investigate fine scale, non-steady fire behaviours can offer insights into the nonlinear dynamics of flame spread that we discussed in **Chapter 6** and partially demonstrated by modelling in **Chapter 7**. These behaviours are sometimes manifested in the large-scale fires discussed in **Chapter 8**, yet they are much different from those most commonly studied and modelled in fire science. For example, most fire behaviour research for the past 80 years worldwide has focused on the steady-state spread of an idealised head fire of constant width from a linear ignition. Yet the success of prescribed burning often requires that practitioners take advantage of non-steady spread and intensity as fires accelerate in two dimensions from point and line ignitions and interact with nearby flame sources from multiple flame fronts. Study of these behaviours in laboratories and the field is at least as important to advancing wildland fire science as understanding steady-state spread.

Thus successful prescribed burning depends not only on competent use of models but also on understanding of fire behaviour that is developed from observation and experience, and on ability to interpret complex and interactive fire behaviours. Many practical references explain various aspects of operational prescribed burning including setting prescription objectives (e.g. Mobley *et al*. 1978; Martin and Dell 1978; Green 1981; Kilgore and Curtis 1987; Wade and Lunsford 1989), so we focus here on the use of prescribed fire for understanding the physical processes of fire behaviour and the environmental factors that produce it. Ultimately, the best way to understand wildland fire is to use it, take responsibility for controlling it, and observe the fire behaviours that result from our actions.

Let us introduce two terms that we use in specific ways for this discussion:

1. We refer to the area of land within a prescribed burn as a *unit*. In practical burning operations, units can have simple or complex topography and mixtures of fuel types, but we assume for this discussion that they are as homogenous as possible because this is best suited to illustrating and studying fire behaviour. We also assume that units are burned under weather conditions that are constant in space and stationary in time, even though this is unrealistic for large units that burn for long periods of time.
2. When we refer to *ignition*, in most cases we are assuming that it is conducted by hand using some kind of torch (e.g. drip torch). Hand ignitions allow practitioners to have a great deal of control over the ignition pattern, observe the fire behaviours that result from their actions closely, and adjust quickly to changing conditions. While our discussion focuses mostly on hand ignitions, fire practitioners can apply the same principles to many other ignition methods (e.g. aerial ignition devices, unmanned aerial vehicles, and ignition devices mounted on motorised vehicles).

Point ignition

A *point ignition fire* (isolated from any other ignition within a unit) is useful mainly for research. Ignition from a single point or small area allows fire to spread in all directions. While this kind of fire would present operational challenges to achieving most fire management objectives, it has great utility for scientists who are studying directional fire spread from a point source and thus the evolution of a fire's shape in relation to wind and slope (**Chapter 2, Figures 2.9, 2.10** and **Chapter 8, Fire shapes and growth patterns**). Only from point ignitions can we obtain measurements to test our assumptions about the ignition location in free-growing fires (**Chapter 8, Fire shapes and growth patterns**). Only from point ignitions can we learn how fires accelerate in two dimensions from point sources rather than from a line-fire front (**Chapter 2, Fire acceleration**). Point ignitions on slopes with various angles and speeds of crosswind can also help us understand the effects of wind-slope interactions on fire shape and spread.

Point ignition means that a small area (less than ~1 m^2) is ignited; then fire spread, shape and intensity are observed as the fire spreads in two dimensions (**Chapter 2, Fire shapes**). These characteristics can best be observed if fuels are uniform and weather conditions are stationary. While wildland fuels are typically variable, several studies have used point ignitions to study fire behaviour in natural fuels; they have been conducted in forest understorey fuels (Lawson 1973), crop stubble or grass (Cheney and Gould 1995; Gould and Sullivan 2020), slash fuels (McRae 1999) and boreal forest woodland (Alexander *et al.* 1990). Because it is difficult to maintain stationary weather conditions, especially with long-duration burns, point-source fire experiments are usually restricted by plot size and wind variability. However, they are probably the only way to study the effects of oscillating wind directions on two-dimensional fire spread and fire shape – knowledge that is essential for developing and validating models of fire growth in both continuous and discontinuous fuels (**Figure 10.1**). The contributions of imagery from unmanned aerial vehicles will be invaluable to this research.

While we need to better understand fire spread in uniform fuels, we also need to learn a great deal about thresholds of fire spread (i.e. spread *v.* no spread), which vary in discontinuous fuel beds and with fire spread direction. From experimental studies of two dimensional fire spread, we can learn how flame characteristics and heat transfer, which vary with the orientation of the fire front, determine whether a fire will spread or not in fuel beds with varying gap sizes, such as discontinuous grass, brush and forest fuels. Similarly, we can learn how fires transition from spread to no-spread as they undergo changes in wind speed and direction, and how flanking and backing fires transition to very wide heading fires when they experience shifting winds (**Figure 8.17**).

Single line ignition

Here we discuss the behaviour of *single line ignitions* that are purposely oriented at different angles relative to wind and/or slope to control the spread rate and fireline intensity (**Figure 10.2**). Ignition is assumed to occur very rapidly (nearly instantaneously) and thus produce a linear flame zone with little influence from a progressing ignition line. For research purposes, line ignitions can help us isolate the effects of the environment and flame zone geometry on fire behaviour. For practical purposes, multiple lines are most commonly ignited progressively and relatively slowly across a given unit (rather than instantaneously); these ignition

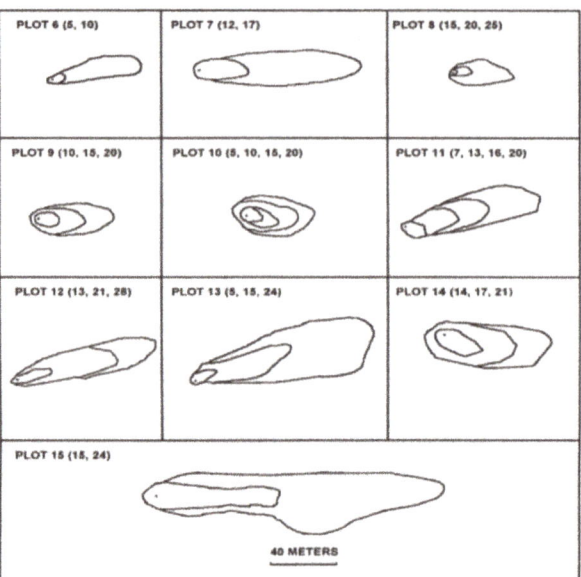

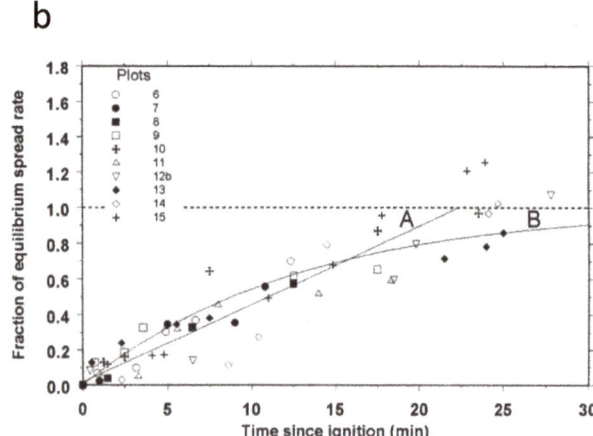

Figure 10.1: Illustration (a) of fire growth recorded for point-source fires in slash fuels, where the ignition point is marked by a small dot near the left edge of the fire and (b) two different models fit to field data showing change in forward spread rate with time relative to steady-state spread (reproduced with permission from McRae 1999).

patterns are discussed below (in the section **Multiple line fires**).

Heading fires

Fires spreading with the wind and/or slope are called *heading fires* or *head fires*. For a given set of environmental conditions, heading fires have higher spread rates and intensities than backing or flanking fires. Heading fires ignited as a continuous line from an upwind edge (**Figure 10.2a**) accelerate rapidly. In prescribed burning, a line ignition is used to produce a heading fire with a uniform flame zone and to cause the fire to rapidly approach a steady state of spread. Line ignitions create the highest possible intensities within a flame zone, which may emulate an arbitrary segment of a wide fireline that is free from lateral edge effects. Wider lines and flame fronts accelerate and spread faster than narrow lines or spot ignitions (Johansen 1987). In research, heading fires are usually designed to develop approximately steady spread rates by the time the fire front reaches sensors or areas to be measured. While the acceleration phase of heading fires is also of interest, for many reasons including understanding feedbacks in the fire spread system and controlling prescribed fire intensities described later with multiple line fires, this is seldom studied by direct measurement, and we do not focus on it here. Prescribed heading fires have been used in many kinds of fuels, including grass (Cheney and Gould 1995), brush (Marsden-Smedley and Catchpole 1995; Fernandes *et al.* 2000; Cruz *et al.* 2013), forest understorey (Fernandes *et al.* 2009; McCaw *et al.* 2012), and boreal forests and woodlands (Alexander *et al.* 1990; Stocks *et al.* 2004).

A wide continuous ignition line, started almost instantaneously, produces the fastest fire acceleration and the most uniform fire front. Small plots (e.g. less than 20 m across) can be ignited very rapidly with hand-held torches. As plot size and the length of the ignition line increase, it may be necessary to use vehicles or to coordinate multiple people igniting separate segments of the line simultaneously in order to achieve nearly instantaneous ignition. **Figure 10.3** shows heading fires in which five people ignited separate segments simultaneously. Recent research in wheat crops suggests that, if only two ignition segments are needed, igniters should proceed from opposite ends of the line towards the centre to produce the fastest spread rates (Cruz *et al.* 2019). Edge-to-centre ignition is probably more effective than centre-to-edge ignition because the interaction of the two merging flame zones

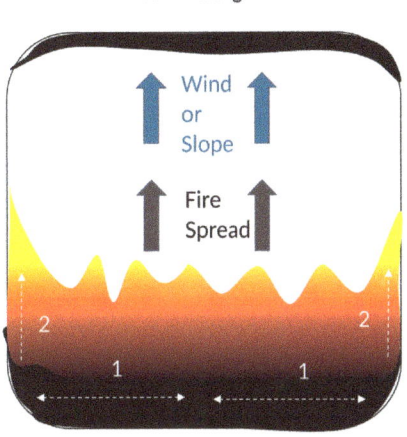

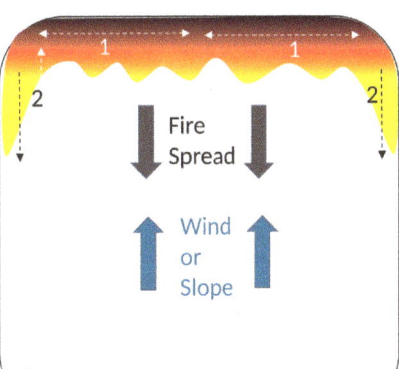

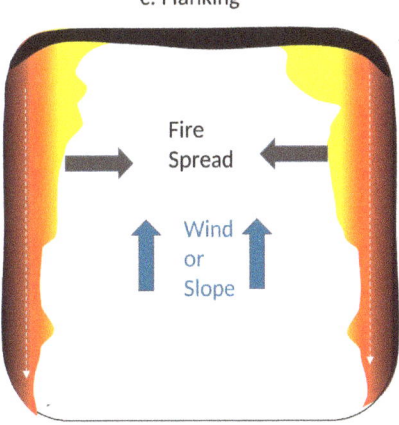

Figure 10.2: Ignition patterns with wind and slope showing (a) heading fire where (1) is the main ignition line with possible lateral ignitions along the sides (2); (b) backing fire with lateral ignitions; and (c) flanking fires along the lateral edges of a burn unit.

Figure 10.3: Aerial photographs of experimental burning in New Zealand with heading fireline ignitions using five people lighting separate segments with drip torches over (a) a 150 m plot edge of cereal crop stubble that produced ~5 m flames and spread rates of ~100 m min^{-1} and (b) a 200 m plot edge of gorse (*Ulex europaeus*) producing ~15 m flames and spread rates of ~70 m min^{-1} (note the vehicles in the lower left). Lateral ignitions progress forward along each side to block wind from wrapping around the flame zone and thus maintain the linear front (photographs by Marwan Katurji and team from the University of Canterbury, New Zealand).

temporarily increases the intensity and local spread rate at the centre of the fire front. Centre-towards-edge ignitions produce a bow-shaped front more quickly, although all initial linear flame zones eventually become bowed in the centre (see **Figure 2.10**).

Where fuels are continuous, it is easier to ignite a heading fire under moderate weather conditions than where fuels are discontinuous, as in many natural vegetation types. Ignition of discontinuous fuels may require stronger winds or lower fuel moistures to produce a heading fire with a continuous flame zone. These considerations have been incorporated into practical guidelines for prescribed burning in discontinuous fuels (see Britton and Clark 1981; Gruell *et al.* 1986; McPherson and Wright 1986; Bunting *et al.* 1987; Burrows *et al.* 2009). If incomplete ignition or fuel discontinuities cause gaps in the fireline, incoming airflow (wind) will be channelled through these gaps and create localised edge effects, which will reduce forward convective heating. This will reduce the acceleration and intensity of the fire front until the flames along the edges of the gaps merge downstream and close the gap – if they do at all. A simple technique for obtaining continuous ignition in discontinuous fuels may be to sweep the torch forward and back across the fuels, igniting a swath of vegetation rather than a single line and thus expanding the depth of the initial flame zone. If gaps in the ignition line are anticipated in fire planning, the plot size can be increased to allow the fire to spread further and close any gaps before it reaches instrumentation.

Edge effects along the sides of a heading fire lift the flames off the fuel bed, locally reducing convective heating in the forward direction (**Chapter 6, Flame front width and shape**). As the fire spreads forward, these edge effects create a continuous drag on the spread rate along its flanks. As a result, all heading fires eventually become bowed or elliptically shaped (**Figure 2.10**). This has been well documented for grass fires (**Chapter 6, Figure 6.31**) and is addressed in the modelling of convective heating (**Chapter 7, Convection heat transfer**). The longer the ignition line used for a heading fire, the greater the distance from the edges and the less effect this drag will have on the fire's acceleration and its eventual spread rate and intensity. As the front of a heading fire becomes increasingly bowed, its curvature affects the orientation of airflows through the flame zone, as evidenced by the direction of longitudinal vortices and peak-and-trough flame structures (**Figure 6.32, Chapter 6, Wind in the presence of fire, Flame front width and shape**). As airflow and convective heating are deflected away from the heading direction of the fire, spread rate and intensity are further reduced. Edge effects must be considered in planning for the size of a burn unit relative to the location of any instrumentation and also in interpretation of measurement data.

There are several ways to measure the behaviour of heading fires and/or to achieve prescription objectives despite the natural elliptical growth pattern of these fires. First, the size of the unit can be scaled to a multiple of the anticipated size of the flames. A rule of thumb is to plan for the width of the unit to be at least 10 times the length of the flames when the fire first reaches instrumentation. This allows the vortex patterns in the flame zone to develop fully and thus maximises convective heating. (Recall that the wavelength of the vortices appears to depend on the length of the flames, as shown in **Figure 6.20d**.) For example, if flames are expected to be 2 m tall, then the flame front must be at least 20 m wide. However, to accommodate the developing curvature of the front, the original ignition line would probably have to be two to four times wider. Second, lateral ignition lines can be extended forward along the edges of the unit to block ambient wind from circumventing the main flame zone and to prevent the edge effects that cause curvature of the fire front (**Figure 10.3**). This appears to be partly effective for extending the time during which the flame front spreads in a straight line, at least in grass fires.

A successful heading fire requires that the downwind (or uphill) edge of the unit be prepared well by blacklining (i.e. pre-burning of fuels along the control line) and/or be attended by a firefighting apparatus. Wide units require more than one person to perform ignition, and the crew must be coordinated to ignite rapidly and uniformly. Because heading fires produce high fireline intensities, they may be difficult to control in fuel types and wind conditions where spotting could occur.

Backing fires

Backing fires are ignited along the downwind edge or at the top of a sloping unit and spread slowly into the wind or downslope (**Figure 10.2b**). Backing fires transfer heat much shorter distances than heading fires, so they produce relatively slow spread rates and low fireline intensities. Convective and radiant heat is transferred only short distances, perhaps only millimetres, because the flames are relatively short and oncoming air and upslope flow attachment tends to bend them away from fresh fuels (**Figure 2.8a**). Oncoming air also cools the fuel particles because it approaches the flame zone across unburned ground. Heat transfer occurs mostly by direct flame contact as flames spread along and among intersecting fuel particles. It is enhanced by oscillations in wind and localised airflows that cause fluctuations in flame direction and tilt angle. Thus backing fires require continuous, horizontal fuels, such as needle litter or grass with thatching, to spread uniformly. Fuel types that tend to be discontinuous, such as bunchgrasses and desert shrublands, may not support backing fire spread with a linear flame zone; instead they may only support spread via sporadic fingers of fire, which produce a ragged geometric edge. Despite this challenge, it is possible to find continuous enough fuels to study backing fire behaviour in the field, at least in small units. For example, Fernandes *et al.* (2009) studied backfires on plots 10–15 m across in a *Pinus pinaster* forest understorey (a combination of litter, grass and shrubs). Unlike heading fires, backing fires do not tend to become curved or bowed as they progress because wind and induced airflow impinge on the flame front without passing through or around the combustion zone.

Fire managers may prefer low intensity backing fires to head fires for burning forest understoreys because backing fires can be conducted with minimal convective heating injury to the forest canopy. For research purposes, backing fires can be conducted on smaller units with shorter ignition lines than those used for heading fires. In fact, small units with short burn times may be more useful than larger-scale burns because they allow scientists to avoid the variability in wind, fuel moisture and other factors that would occur over the course of hours, alter fire behaviours, and thus confound measurements. Smaller units may also help field personnel avoid time constraints for completing a burn.

Flanking fires

Flanking fires are imprecisely defined but generally describe fires that are moving perpendicular to the wind or slope. **Figure 10.2c** shows flanking fires ignited along the lateral edges of the unit. The behaviours of flanking fires have been little studied because they occur over a wide range of directions between the 'pure' directions of heading and backing. The flanks of an idealised elliptical fire can slant over a range of oblique angles relative to wind or slope. Near the head, the flames are angled forward and generally tilt towards the unburned fuel; near the rear, the flames tilt mainly away from the fuel bed. The spread of flanking fires is enhanced by flame impingements from eddies and coherent peak-and-trough structures, both of which migrate with the wind or slope along the flanks. Sometimes, particularly on steep slopes, fire whirls form at regular intervals and progress uphill along the flanks, increasing the rate and variation of flanking spread (see **Chapter 6, Topography**). These tangential flame motions increase in width as they travel, causing the flanking fire to widen in its downwind or uphill portions. Thus ignition of a flanking fire that is perfectly oriented parallel to the wind or slope can gently widen to a fan shape as it progresses.

Flanking fires are mostly overlooked by research, but there is much to be learned from their behaviours about physical fire spread processes. While flames on wind-driven flanks have smaller flames than in heading fires and show little acceleration in spread rate after ignition (unlike heading fires), their heat transfer spread patterns are augmented by cumulative effects of tangential flame motions. These effects suggest important influences at both the flame scale and the fire scale that have been little explored. For example, we do not know if the spread rates and directions of flanking fires are stable once ignited at various orientations to wind or slope, or if instead they vary depending on their context as segments of a larger fire perimeter. Research by Viegas (2002) suggested that the edges of flanking fires appear to change their angle of orientation as they spread. Only more measurements in experiments at multiple scales can address this issue.

Ignition of flanking fires proceeds rapidly from the downwind edge towards the oncoming wind – or from the upslope edge downslope. Ignition may be located parallel to the lateral boundaries of the units or down the middle. Operational use of flanking fires is limited on large units because flanks require a relatively long time to spread wide distances, which allows for unexpected wind shifts to drastically alter the spread pattern. For this reason, flanking fires for management purposes, like heading fires, are mostly executed with multiple ignitions, which is covered in **Multiple line fires** below.

Other line ignitions

Some kinds of line ignition that do not follow strict straight-line patterns are of interest to researchers. For example, line configurations with concave and convex curves and with inverted V shapes have been used at the laboratory and field scales to examine interactions among sections of a flame front caused by changes in induced airflows (Hilton *et al.* 2016; Raposo *et al.* 2018; Sullivan *et al.* 2019). At the field scale, even more complex interactions can be examined. For example, ignition using a series of shapes can create experimental fire whirls by channelling induced airflow through gaps in the pattern (see **Chapter 8, Fire whirls**). We include these variations on line ignition here for the sake of completeness, but no practical or operational applications have been developed for them.

Multiple line fires

In contrast to simple line ignitions (heading, backing or flanking fires), operational burning typically uses multiple ignition lines that advance simultaneously across the unit (**Figure 10.4**). This technique offers practitioners the flexibility to readily adjust the ignition pattern to control the fire's behaviour and effects. The use of multiple line ignitions (i.e. *strips*) enables igniters to interrupt the spread of each strip fire, while it is still accelerating, as it encounters and merges with the backing edge of an adjacent strip (which may or may not still be burning). Thus the technique can

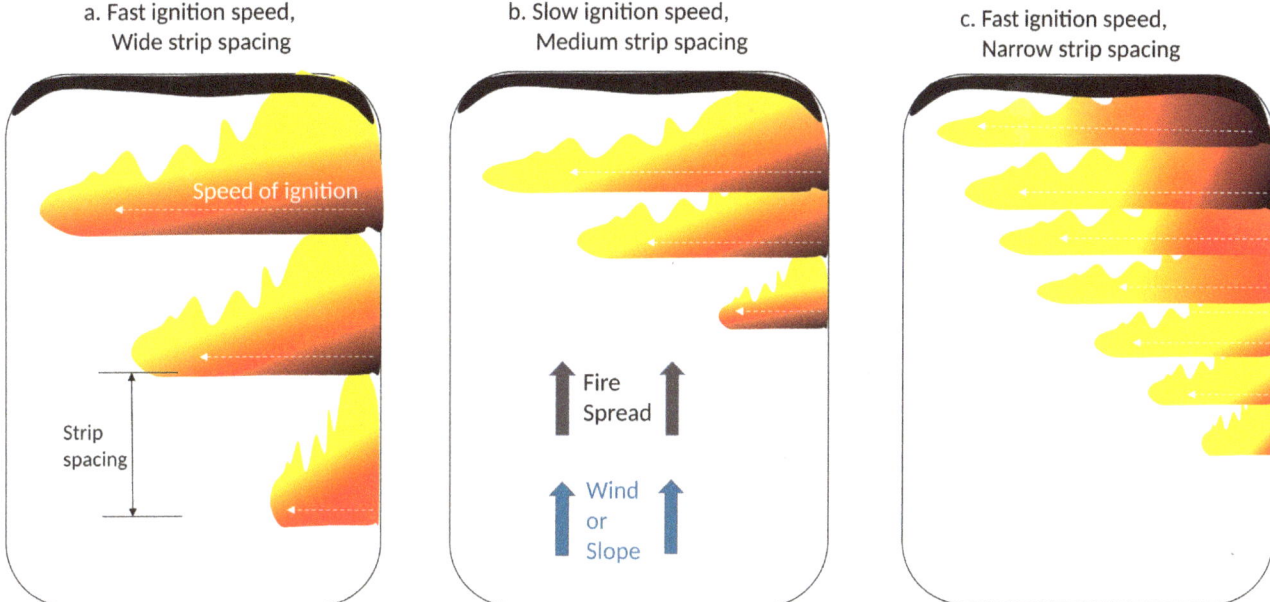

Figure 10.4: Strip head fire ignitions with variations in linear speed of advance and strip spacing, designed to control the intensity and amount of simultaneous active burning across the burn unit: (a) strips spaced widely, progressing rapidly so that higher fireline intensities can build up and creating heterogeneous patterns of intensity; (b) strips ignited slowly at medium spacing so that areas ignited first have nearly finished flaming by the time flames from the adjacent strip fire reach them; and (c) narrowly spaced strips ignited rapidly, so the strip fires merge before they stop accelerating; because flaming is persisting in adjacent strips, this technique increases the proportion of the unit that is burning at one time.

be used to limit the spread rate and intensity reached by any individual strip fire. Multiple ignitions can also allow large units to be burned more quickly than can be done with a single line ignition (which may require a long time to spread across an entire unit), particularly if backing and flanking fires are being used. Strip ignitions may cause considerable variation in intensity, depending on the distances allowed between strips. This variability may confound research into steady-state fire behaviours but may be well suited to characterising non-steady behaviours, including acceleration in rate of spread. One of the disadvantages of using multiple line fires in prescribed burning is that a considerable amount of drip torch fuel may be required to achieve continuous line ignitions over large units.

Strip head fire

A *strip head fire* consists of a series of lines ignited upwind or downslope from a firebreak, so they burn with the wind or upslope towards the firebreak (**Figure 10.4**). Parallel ignition lines are initiated perpendicular to wind or slope. Lines are initiated first near the firebreak and progress downhill or upwind, with each line a specified distance behind the previous line. More than one person is typically involved in igniting a unit. Strip head firing is one of the most commonly used ignition techniques in prescribed burning because it offers great flexibility for controlling fire behaviour and effects. It gives practitioners control over intensity because the fire's forward acceleration from a given ignition line slows as it encounters the backing fire from the adjacent ignition line.

Practitioners consider three main factors when conducting strip head fires:

1. The desired fireline intensity or flame length dictates the distance that separates the ignition lines (i.e. the *strip spacing*). With greater strip spacing, each ignition line can accelerate for a

longer distance, increasing in intensity and flame length, before it spreads into the backing fire along the rear edge of the previous ignition line. There it encounters burned-out fuels, so its rate of spread, flame length and intensity decline. Fires with narrow strip spacing generally burn with lower intensities, so strip depth should be determined based on the fuel type, the spread rate and the maximum desired intensity. However, the determination of strip depth should also be based upon the duration of flaming in each strip, since the flaming zones of adjacent strips will interact if they are still flaming when they merge (see 3 below).

2. The pace of ignition along each strip controls the width of the ignition line itself, and this influences the line fire's rate of spread and intensity. This is very much like how the width of the flame front affects the rate of acceleration and proportion of edge effects for single head fire ignitions discussed above. A person conducting the ignition quickly will create a wider flame front, which will accelerate more rapidly and have a higher intensity than a person conducting the ignition more slowly. If the fire accelerates rapidly, such as in grass or brush fuels, the igniter may need to move quickly to avoid heat from the accelerating fire. In this case, safety considerations may dictate strip depth and the timing of ignitions.

3. The spacing between ignition strips controls the interactions among flame zones and the amount of cumulative area burning. Greater cumulative area burning can change the character of the fire from a series of line fires to an area fire (**Figure 10.4c**). Short distances between igniters and short times between ignitions will release heat from a larger area at one time than longer distances and times. This is more important in long-burning fuel types, such as woody debris and deep duff layers, than in grass fuels, which produce little residual combustion.

These three factors are obviously interdependent, demonstrating once again the nature of wildland fire as a tightly coupled dynamical system. Knowledge of the flame residence time for the particular fuel type, the rates of acceleration, and the maximum intensity and safe proximity of igniters are important interdependent factors. Strip head fires can be used to maintain low fireline intensity in a forest where the convective heating of the canopy foliage is to be minimised. To produce this behaviour, the fires should have a slow rate of ignition and the strip widths should be narrow; these measures will prevent the fire from accelerating and building up intensity beyond the desired levels. In contrast, strip head fires can also be used to cause area-wide fire behaviours similar to those of mass fires (see **Chapter 8, Mass fires**). To produce this behaviour, the strips should be ignited rapidly, which will allow the flaming zones from multiple strips to merge, effectively creating a continuous flaming zone across a large area. In this case, the energy release may be great enough to overcome wind or slope effects, create airflows into the burning zone, and greatly increase air and gas temperatures above the fire. If this technique is used in a forest understorey, it will increase the convective heating of the overstorey foliage and may cause crowning even while the fireline intensity is technically kept low.

Strip head fires are suitable in almost any fuel type where access is practical and safe; this probably excludes very tall, dense shrublands and deep slash fuels unless aerial ignition devices are used. Strip head fires offer practitioners considerable control over fire behaviour and its variability. Practitioners can adjust the strip depth, width, and timing of ignition in response to changes in weather conditions and the fire behaviour they are observing. Deeper strips and slower ignition rates should produce greater contrast between the heading intensity and the ignition line, while shallower strips and faster ignition should create more uniform intensities. Some of the practical considerations discussed for heading fires also apply to strip head fires – for example, wider strips are needed to facilitate spread in moist conditions, and faster winds are needed in sparse fuels. In discontinuous fuels, strip ignitions may not produce any backing spread, so the rear edge of each strip may have stopped flaming by the time the fire from the next strip reaches it.

Flank fire

Prescribed burning with *flank fire* means that an area is treated with lines of fire set into the wind or downslope, which then burn outward at right angles to the wind or slope. This technique requires multiple parallel ignition lines (**Figure 10.5**). Two factors control the fire behaviour resulting from this technique:

1. The spacing between ignition lines determines the distance that fire can freely spread.
2. The rate of ignition of each line (in the upwind or downslope direction) controls the length over which flaming ignition is occurring on a given line at a given moment.

These two factors should be considered to determine the spacing between ignition lines – that is, the distance in which fires can spread freely and flame zones can eventually merge. As flames from each flank ignition approach each other, the buoyancy in the two flame zones induces indrafts of air, which increase flame length in the region where the flame zones meet – the *junction* area (see **Chapter 6, Multiple flame zones and air-flow interactions**). Faster rates of ignition at wider separations create longer junction areas. In the junction areas, intensities are greater than in areas burned close to the ignition lines. If ignition is rapid, the fuel in the areas close to the ignition line is likely to be still burning when the flame zones merge. Thus rapid ignition at wide separations can create area-wide fire behaviours similar to those for strip head fires with rapid ignition and deep strips.

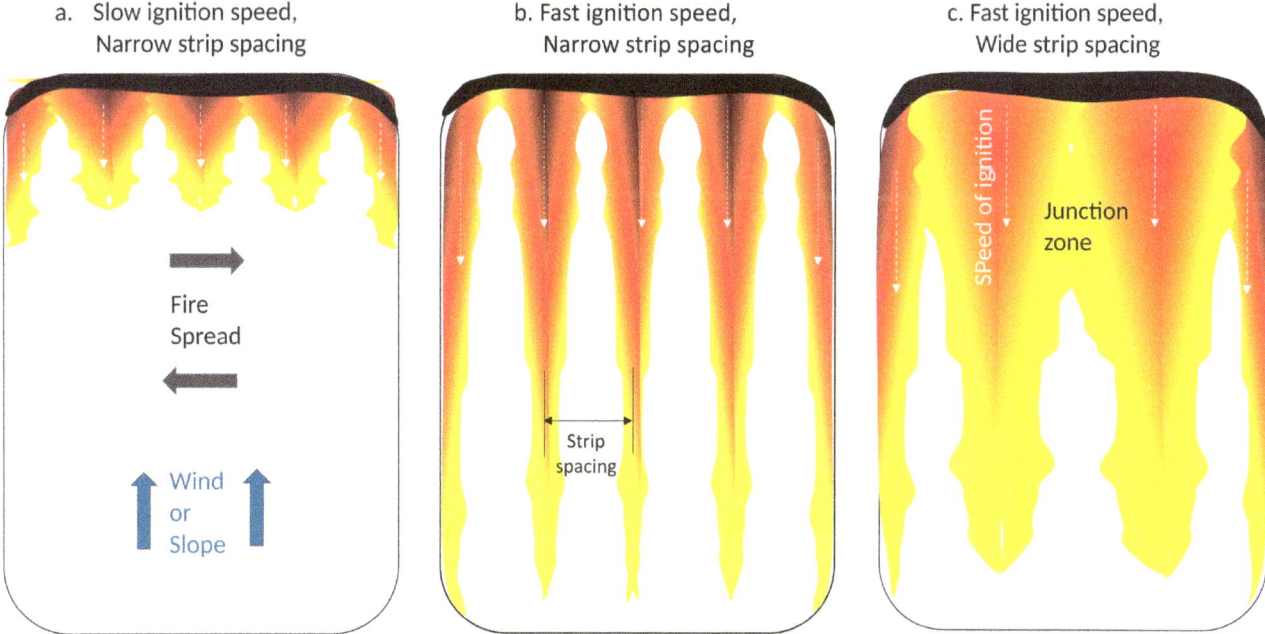

Figure 10.5: Flank ignition patterns with variations in linear speed of advance and spacing between strips, designed to control the intensity, uniformity of intensity, and flame zone interactions in junctions between strips: (a) strips at narrow spacing with igniters progressing slowly into the wind or downslope, so junctions where flame zones merge are narrow; (b) strips at narrow spacing and ignited rapidly, so build-up of fireline intensity is limited except where flame zones merge at junctions, which causes locally high intensities; and (c) strips at wide spacing and ignited rapidly, so junctions are large and the proportion of area burning all at once is increased.

Multiple spot ignitions

Multiple spot ignition (i.e. use of *spot-head fires* (Martin and Dell 1978; Johansen 1984)) is a technique in which points or small areas are ignited in a particular pattern of spacing and timing to control the fire behaviour. This is a versatile technique for managing fire intensity and spread rate. It generally uses less torch fuel than line-fire ignition, and it is well suited to aerial ignition. With this technique, point-source ignitions (small lines or patches) are initiated progressively within broad *lanes* (i.e. rows of successive ignitions perpendicular to the direction of wind or slope) across a unit and then allowed to spread in all directions (**Figure 8.8** and **Figure 10.6**). This technique has several unique features caused by the relatively slow acceleration phase of small ignitions when compared with line ignitions: The slow acceleration of point ignitions keeps intensities relatively low and produces variability in spread and intensity in all directions from each spot fire; and the growth of fire from each ignition stops when it joins the flame zones from nearby ignitions.

Three features of the multiple spot ignition technique provide practitioners with control over fire spread rate and intensity:

1. The size of the spot ignitions themselves. In continuous fine fuels (e.g. continuous dead litter, some grasses and crop residues), with very dry or windy conditions, ignitions the size of a single fuel particle can be sufficient to initiate flame spread, but the fire will accelerate slowly. Small points may take tens of minutes to accelerate to near-steady-state intensity (Johansen 1987), particularly in slash fuels (McRae 1999), while larger ignition areas will accelerate faster. Regardless of ignition size, a spot fire's intensity will increase substantially when it merges with adjacent ignitions. In patchy fuels, practitioners may ignite large

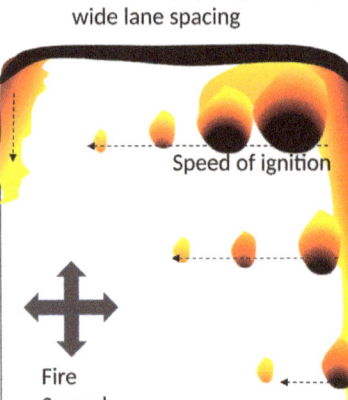
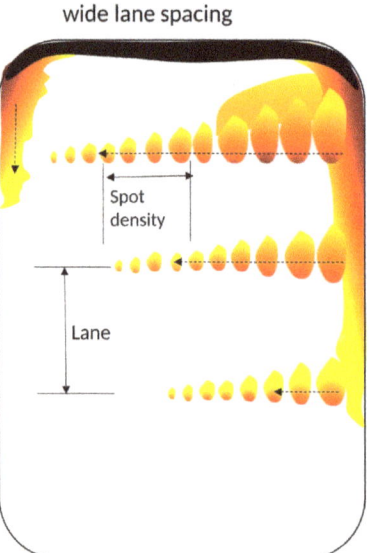
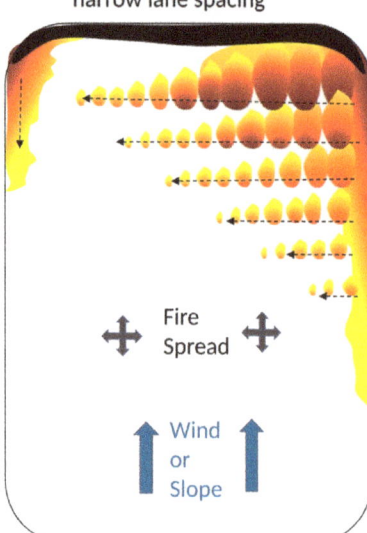

Figure 10.6: Spot ignition patterns with variations in linear speed of advance, density of spots, and forward distance between ignition lanes: (a) wide lanes, low density of spots and slow ignition speed allow burnout of spot interior before it merges with other spot fires at the head and flanks; (b) wide lanes, high density of spots and fast ignition speed increase the continuity of merged spot fires so they form a line fire with high heading intensity; and (c) narrow lanes, high density spots and fast ignition speed cause a large proportion of the area to be on fire simultaneously but limit intensity from each ignition because of its close proximity to neighbouring ignitions.

areas in order to encompass whole fuel patches and span gaps. In moist fuels and low-wind conditions, they may ignite large areas to ensure sustained spread.

2. The density of spot ignitions along a line. Density is used to control the time and distance at which flame zones merge (**Figure 10.6**). A heading and flanking fire front grows from each spot ignition, spreading outward and downwind or uphill. The burning area around each ignition widens until it merges with flames from another spot fire. After this, the merged flame front behaves more and more like a line fire, as local convolutions of edge shape gradually blend into a smoother contour. (This pattern, of course, is altered whenever fuel, topography, or weather patterns become non-uniform.) As airflow stops channelling between separate spot fires, the line fire will probably accelerate in spread and increase in intensity.

3. The speed of ignition. A rapid linear rate of progress will generate multiple flame zones that expand simultaneously, which produces high intensities and may lead to area-wide fire behaviour rather than line-fire behaviour. In contrast, slower ignition speed can produce many small 'doughnuts' of burning spots, each enclosing an area of consumed fuel and taking a longer time to merge with other spot fires, thus producing low relative intensities.

Spot fire ignitions may not be appropriate for prescribed fires in which uniform intensity is desired, and they may not be practical in areas of discontinuous fuels. If a unit has large gaps between fuel patches, spread may be inhibited because the spot fires have relatively low intensities, so individual spot fires cannot merge to form a line fire. Where fuel continuity does permit spread, fire behaviour is likely to vary considerably across the unit. The numerous flame junction zones will have higher intensities than the centres of the spot fires; the developing head fire will spread unevenly because it comprises flame zones from individual spot fires (with relatively low intensity) and junction zones (with higher intensity). The intensities near the origin of each spot fire and along its backing and flanking portions will be relatively low.

Weatherspoon *et al*. (1989) proposed a modified spot fire technique called *tree-centred spot fires*, which could be used to protect individual trees from being damaged by a prescribed understorey fire. With this technique, spot ignitions are initiated near trees that require protection. The flame zone accelerates and grows away from the tree. The fire behaviour near the tree is mainly backing and flanking, so intensities near the tree tend to be low, thus limiting convective heating of the canopy foliage and protecting it from scorching or burning.

Ring fire, centre fire, mass ignition

Unlike most other ignition techniques, the success of *ring fire, centre fire* and *mass ignition* depends heavily upon behaviours at the fire scale; the techniques all manipulate fire-induced airflows, so they aim primarily towards the middle of the unit. In a *ring fire* (**Figure 10.7a**), a circular line ignition is completed quickly around the unit, so it draws the air entrained into the flame zone from the outside towards the middle before substantial outward spread can occur. In a *centre fire*, this pattern of airflow is enhanced by buoyancy generated from flames in the middle of the ring. A centre fire may not be needed for small units or in light winds, particularly if the entire ring can be closed rapidly. But in larger units and stronger winds, the use of a centre fire and progressively larger, concentric rings around it can help offset strong winds and ensure combustion through the whole unit (**Figure 10.7b**). *Mass ignition* requires multiple discrete ignitions, beginning in the middle of the unit and progressing concentrically farther away until the periphery of the unit is reached (**Figure 10.7c**). Mass ignition attempts to draw the competing indrafts from all ignitions towards the middle and join them into a

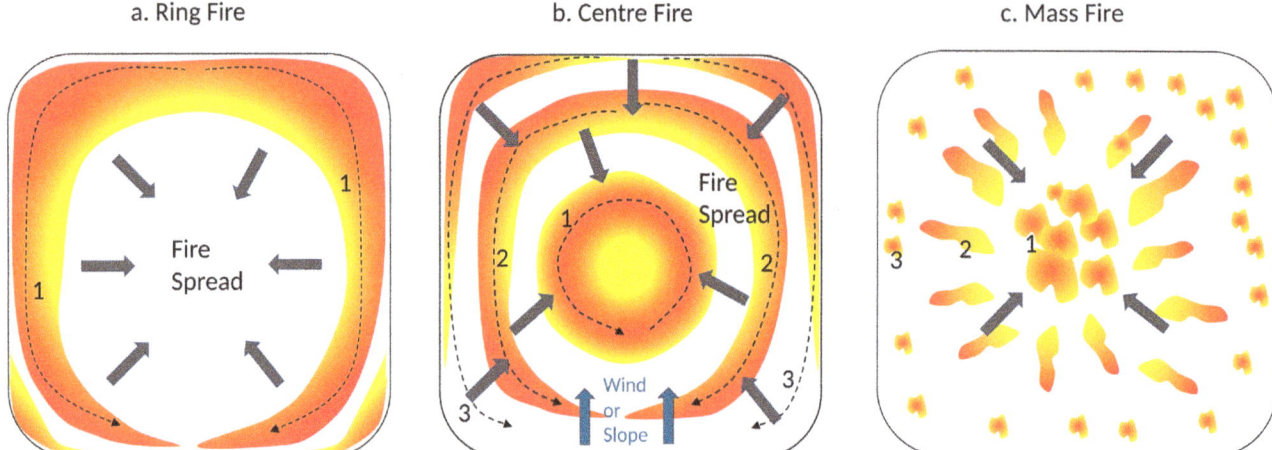

Figure 10.7: Area-based ignition patterns designed to induce airflow towards the middle of the unit: (a) ring fire pattern with no interior ignition causes inward flow of air as soon as the circle is closed; (b) centre fire with interior ignited first, followed by sequential concentric ring ignitions that draw fire primarily towards the middle; and (c) mass fire technique with multiple spot ignitions beginning in the middle of the unit and progressing outward. For large units, these techniques all require aerial ignition.

unified plume that is centred above the unit. All of these techniques require the ignition process to be well coordinated among multiple people or vehicles.

These ignition methods involve complex interactions of ambient and induced airflows at both the local flame scale and the whole-fire scale. As discussed in **Chapter 8, Vorticity** and **Mass fires,** several kinds of fire-scale behaviours, including puffing pool fires, ring fires, and mass fires can be readily demonstrated in the laboratory (see **Figure 8.19**). However, little experimental work has been done to explain how the interplay of fire-line intensity, burning rates and the diameter of the ring or rings can restrict indrafts to the middle of the fire from above. As a fire expands and the fuel in the middle burns out, the fire area becomes large enough that ambient winds can flow into the burned-out portion and impinge on the various portions of the flaming edge, orienting the flames independent of the central plume. Small-scale modelling of these interacting factors would be of considerable value for understanding and predicting such effects in large burns. Scaling to actual large fires would probably require atmospheric stability to aid in the development of tall convection plumes and to control indrafts at the ground surface.

Ring fires, centre fires and mass ignitions are used mainly in slash burns to create a tall plume, which will lift smoke to high altitudes, and to create a strong indraft towards the centre of the unit, which will offset wind and convective heating on the adjacent forest canopy (McRae and Flannigan 1990; Hall 1991; Quintiere 1993). These techniques are also used in brush (Green 1981) and grass fuels where high intensities and rapid spread are desired. The success of these techniques relies on estimating the complex interactions of airflow and flame zone characteristics, and practitioners must adjust the techniques by trial and error based on observations of behaviour and spread. The techniques are best suited for relatively low wind speeds (e.g. several m s^{-1}). Practitioners must evaluate the size of the unit and the fuel type to determine the number of ignition rings and the distance between ignition lines or spots, through which air will be drawn towards the centre. Long-burning fuels with high loading, such as slash, require more time to build intensity and establish buoyancy and indrafts than do grass or brush fuels.

Figure 10.8: Large ring fire conducted on flat terrain in grass, which produced strong indraft with smoke drawn towards the middle of the unit (a) shown for the entire fire area and (b) close-up of the centre portion of the fire (https://www.youtube.com/watch?v=lcLnhs4FXvI&t=80s).

Because access to the centre of these prescribed fires is limited, hand ignition is practical only for small areas (e.g. a few hectares); larger areas require use of aerial ignition devices (Green 1981), such as helicopters or unmanned aerial vehicles, to ignite fast enough to close the circle (**Figure 10.8**).

While ring fires, centre fires and mass ignitions are intended to create strong inward-oriented convection across the unit, they may create some large-fire behaviours that are difficult to control (see **Chapter 8, Mass fires**). These may include lofting of embers by the plume, which causes spotting; vorticity, such as fire whirls, created by channelling airflows between multiple ignition points and burning areas (Countryman 1964; Taylor *et al.* 1973; Green 1981; McRae and Flannigan 1990); and a tall convection column in a crosswind, which may create lee-side vorticity and wake vortices that spread the fire rapidly downwind.

Summary

Effective use of fire for land management and fire research requires understanding of how dynamical fire behaviour is affected by different ignition patterns. In this chapter, we have reviewed a variety of ignition geometries, including points, lines and areas, and we have considered the general kinds of interactions that result from multiple ignitions in each. Manipulation of the rates of ignition, the sizes of ignitions, and the spaces between them offers tremendous control over fire spread, intensity and effects. Control and predictability are essential to conducting experiments if we are to advance our knowledge of physical processes in wildland fires, test model performance based on those interactions, and use fire to accomplish management objectives.

References

Alexander ME, Stocks BJ, Lawson BD (1990) 'Fire behavior in black spruce–lichen woodland: the Porter Lake project'. Forestry Canada, Information Report NOR-X-310. Northwest Region, Northern Forestry Centre, Edmonton, Alberta.

Britton CM, Clark FA (1981) Will your sagebrush range burn? *Rangelands* **3**(5), 207–208.

Bunting SC, Kilgore BM, Bushey CL (1987) 'Guidelines for prescribed burning sagebrush-grass rangelands in the northern Great Basin'. General Technical Report INT-231. USDA Forest Service, Intermountain Research Station, Ogden, UT.

Burrows ND, Ward B, Robinson A (2009) Fuel dynamics and fire spread in spinifex grasslands of the Western Desert. *Proceedings of the Royal Society of Queensland* **115**, 69–76.

Cheney NP, Gould JS (1995) Fire growth in grassland fuels. *International Journal of Wildland Fire* **5**(4), 237–247. doi:10.1071/WF9950237

Countryman CM (1964) 'Mass fires and fire behavior'. Research Paper RS-RP-19. USDA Forest Service, Pacific Southwest Forest and Range Experiment Station, Berkeley, CA.

Cruz MG, McCaw WL, Anderson WR, Gould JS (2013) Fire behaviour modelling in semi-arid mallee–heath shrublands of southern Australia. *Environmental Modelling & Software* **40**, 21–34. doi:10.1016/j.envsoft.2012.07.003

Cruz MG, Hurley RJ, Bessell R, Sullivan AL (2019) 'Fire behaviour in wheat crops'. Client Report No. EP195825. CSIRO Land and Water, , Canberra.

Fernandes PAM, Catchpole WR, Rego FC (2000) Shrubland fire behavior modelling with microplot data. *Canadian Journal of Forest Research* **30**, 889–899. doi:10.1139/x00-012

Fernandes PAM, Botelho HS, Rego FC, Loureiro C (2009) Empirical modelling of surface fire behaviour in maritime pine stands. *International Journal of Wildland Fire* **18**, 698–710. doi:10.1071/WF08023

Gould JS, Sullivan AL (2020) Two methods for calculating wildland fire rate of forward spread. *International Journal of Wildland Fire* **29**(3), 272–281. doi:10.1071/WF19120

Green LR (1981) 'Burning by prescription in chaparral'. General Technical Report PSW-51. USDA Forest Servivce, Pacific Southwest Forest and Range Experiment Station, Albany, CA.

Gruell GE, Brown JK, Bushey CL (1986) 'Prescribed fire opportunities in grasslands invaded by Douglas-fir: state-of-the-art guidelines'. General Technical Report INT-198. USDA Forest Service, Intermountain Research Station, Ogden, UT.

Hall JN (1991) Comparison of fuel consumption between high intensity and moderate intensity fires in logging slash. *Northwest Science* **65**(1), 158–165.

Hilton JE, Miller C, Sharples JJ, Sullivan AL (2016) Curvature effects in the dynamic propagation of wildfires. *International Journal of Wildland Fire* **25**(12), 1238–1251. doi:10.1071/WF16070

Johansen RW (1984) Prescribed burning with spot fires in the Georgia Coastal Plain. *Georgia Forestry Commission – Georgia Forest Research Paper* **49**, 1–8.

Johansen RW (1987) Ignition patterns & prescribed fire behavior in southern pine stands. *Georgia Forestry Commission – Georgia Forest Research Paper* **72**, 1–8.

Kilgore BM, Curtis GA (1987) 'Guide to understory burning in ponderosa pine-larch-fir forests in the Intermountain West'. General Technical Report INT-223. USDA Forest Service, Intermountain Forest and Range Experiment Station, Ogden, UT.

Lawson BD (1973) 'Fire behavior in lodgepole pine stands related to the Canadian Forest Fire Weather Index'. Information Report BC-X-176. Environment Canada, Canadian Forest Service, Pacific Forest Research Center, Victoria, BC.

Marsden-Smedley JB, Catchpole WR (1995) Fire behaviour modelling in Tasmanian buttongrass moorlands. II. Fire behaviour. *International Journal of Wildland Fire* **5**(4), 215–228. doi:10.1071/WF9950215

Martin RE, Dell JD (1978) 'Planning for prescribed burning in the inland northwest'. General Technical Report PNW-GTR-76. USDA Forest Service, Pacific Northwest Forest and Range Experiment Station, Portland, OR.

McCaw WL, Gould JS, Cheney NP, Ellis PF, Anderson WR (2012) Changes in behaviour of fire in dry eucalypt forest as fuel increases with age. *Forest Ecology and Management* **271**, 170–181. doi:10.1016/j.foreco.2012.02.003

McPherson GR, Wright HA (1986) Threshold requirements for burning downed honey mesquite. *Rangeland Ecology & Management/Journal of Range Management Archives* **39**(4), 327–330.

McRae DJ (1999) Point-source fire growth in jack pine slash. *International Journal of Wildland Fire* **9**(1), 65–77. doi:10.1071/WF99006

McRae DJ, Flannigan MD (1990) Development of large vortices on prescribed fires. *Canadian Journal of Forest Research* **20**(12), 1878–1887. doi:10.1139/x90-252

Mobley HE, Jackson RS, Balmer WE, Ruziska WE, Hough WA (1978) A guide for prescribed fire in southern forests. USDA Forest Service, Southern Region, Atlanta, GA.

Quintiere JG (1993) Canadian mass fire experiment. *Journal of Fire Protection Engineering* **5**(2), 67–78. doi:10.1177/104239159300500203

Raposo JR, Viegas DX, Xie X, Almeida M, Figueiredo AR, Porto L, Sharples J (2018) Analysis of the physical processes associated with junction fires at laboratory and field scales. *International Journal of Wildland Fire* **27**(1), 52–68. doi:10.1071/WF16173

Stocks BJ, Alexander ME, Wotton BM, Stefner CN, Flannigan MD, Taylor SW, Lavoie N, Mason JA, Hartley GR, Maffey ME, Dalrymple GN (2004) Crown fire behaviour in a northern jack pine black spruce forest. *Canadian Journal of Forest Research* **34**(8), 1548–1560. doi:10.1139/x04-054

Sullivan AL, Swedosh W, Hurley RJ, Sharples JJ, Hilton JE (2019) Investigation of the effects of interactions of intersecting oblique fire lines with and without wind in a combustion wind tunnel. *International Journal of Wildland Fire* **28**(9), 704–719. doi:10.1071/WF18217

Taylor RJ, Evans ST, King NK, Stephens ET, Packham DR, Vines RG (1973) Convective activity above a large-scale bushfire. *Journal of Applied Meteorology* **12**, 1144–1150. doi:10.1175/1520-0450(1973)012<1144:CAAALS>2.0.CO;2

Viegas DX (2002) Fire line rotation as a mechanism for fire spread on a uniform slope. *International Journal of Wildland Fire* **11**(1), 11–23. doi:10.1071/WF01049

Wade D, Lunsford JD (1989) A guide for prescribed fire in southern forests. Technical Publication R8-TP-11. USDA Forest Service, Southern Region, Atlanta GA.

Weatherspoon CP, Almond GA, Skinner CN (1989) Tree-centered spot firing-a technique for prescribed burning beneath standing trees. *Western Journal of Applied Forestry* **4**, 29–31. doi:10.1093/wjaf/4.1.29

11
Conclusions

Wildland fire is traditionally the domain of land management. Professional foresters and other stewards of our wildlands are concerned with balancing the ecological requirements *for* fire against protecting from threats *by* fire to commercial interests and human lives. Fire ecology and management techniques both depend upon fire behaviour – the physical amounts, durations, and rates of spread and energy release. Thus, the study of fire behaviour is key to achieving a variety of fire management goals.

In this book we have examined the science of wildfire behaviour in terms of physical processes and principles. We've seen that the primary processes of combustion, heat transfer and ignition interact in producing wildfire behaviours through their functions in a dynamical system distinguished by internal feedbacks and external influences of the environment. Principles of this coupled wildfire system are key to explaining even the most mundane behaviours such as the apparently steady spread of a line fire. In this concluding section, we review some of the most important topics covered, the value of the book to researchers with emphasis on some critical research questions, and the principal value of these topics to fire practitioners and land managers.

Key principles and insights

The science of wildfires is distinguished from related disciplines, such as combustion engineering and fires in buildings and industry, by the predominance of discrete fine fuel materials, both living and dead, separated by gaps of ambient air. Furthermore, the range in scales of wildfires and their environmental interactions is immense – extending from millimetres and milliseconds for the physics of combustion and heat transfer to tens of kilometres and months for large wildfires that become atmospheric phenomena. If we hope to someday reliably model or predict the full range of wildland fires with physical fidelity, we're going to have to be clever about how we accommodate the hierarchy of important processes across these scales. This challenge will demand understanding of fire processes beyond the current state of knowledge.

The dynamics, principles and processes of wildland fire behaviour were covered as follows:

- In **Chapters 1** and **2** we reviewed some background of wildland fire science and saw how the science remains incomplete. Decades of experiment-based and empirical research have produced the foundational understanding of fire behaviour that we rely upon today, but additional laboratory and field-scale experiments must be designed to target the missing explanations of physical processes. Wildland fires must spread to be sustained, and we introduced fire behaviour as a dynamical system of physical processes using the simplest and most familiar of all fires – the burning candle. Wildfires involve the same processes, but are so strongly affected by the fuel, topography and weather that these environmental drivers have come to define traditional concepts of spread and intensity in the context of a *line fire*. The line fire is a very useful idea because it simplifies many complexities of wildfire, but it also embodies many assumptions about the environment and the fire itself that limits its application to a wider range of actual fire behaviours.

- To begin explaining *how* fires behave, we outlined in **Chapters 3**, **4** and **5** the technical details of thermodynamics, fluid mechanics, heat transfer, combustion and energy release in flaming and glowing phases, and solid fuel ignition. These physical processes are central to all fires but have specific roles in the wildland context. In particular, the sources of radiation and convection heat transfer from fires of any configuration must be quantified, along with the responses of fuel materials in their vicinity. Fuel particles adjacent to a given fire must experience net rates of heating sufficient for pyrolysis and ultimate flaming ignition. The burning characteristics of diverse arrays of solid fuel materials, both living and dead, determine the rates and quantities of energy release in flaming, glowing and smouldering phases. Fuel variability and influences of moisture and wind make these supremely difficult to generalise and simplify. And all of these processes are interdependent in wildfires where they manifest a myriad of spread characteristics and behaviours.
- From **Chapters 6** and **7** we saw that heterogeneous fuel beds remain challenging to describe for our vast and varied wildlands. In fact, it is so difficult that the salient properties of fuel particles and beds have yet to be defined, or methods devised, to measure actual fuel conditions broadly across wildfire landscapes. Furthermore, the important fuel properties probably change with the kind of fire involved. For example, short flames in backing fires are likely to depend upon finer scale properties of the fuel bed than intense crown fires do. We reviewed the primary influences of weather and topography, detailing examples of how their effects on fire behaviours strongly depend on interactions with the fire itself, particularly involving the property of buoyancy of hot combustion products. We developed a one-dimensional fire spread model to explore the variety of behaviours that arise from the dynamics of interacting physical processes. Though far simpler than actual wildfires, the model offers insights into how the fire system accelerates to approach steady spread under homogeneous conditions of fuels and weather. Similarly, our modelled fire system demonstrated thresholding spread behaviour in discontinuous fuels and with critical limits of wind and moisture when feedbacks were insufficient to maintain spread.
- Many large fires exhibit behaviours that are dependent upon the scale of the entire fire rather than the flame scale. In **Chapter 8** we qualitatively summarised the characteristic behaviours of crown fires, spotting and spot fires, large-fire growth and shapes, pyroconvective storms, manifestations of vorticity, and the limited knowledge of mass fires. We identified some of the many areas that are poorly known for each behaviour and those that remain largely unexplored.
- Advances in wildfire behaviour science will depend upon designing and conducting new experimental studies which, in turn, require us to take reliable measurements of the fire environment and fire characteristics. In **Chapter 9** we surveyed historical and modern techniques for obtaining measurements from laboratory and field-scale fires. Advantages and disadvantages of each technique mean there is no single comprehensive approach and often several are combined in experimental studies.

For field-scale fire experiments, the ignition techniques and their uses in studying different fire behaviour are reviewed in **Chapter 10**. The ability to conduct a variety of field-scale ignitions will be essential to test findings from laboratory research and verify the scaling of physical principles to larger fires.

Principal value to researchers

Our examination of gross-scale wildland fire behaviours and the component physical processes has identified many questions and topics that are poorly resolved or have been left entirely unaddressed by research. This book can offer an entry point to scientists from many disciplines who could contribute fresh ideas and approaches to the persistent challenges of the increasingly important topic of fire behaviour. Without experimental data on these processes and many related subjects, advances in understanding and modelling wildfire spread and behaviour will come very slowly and be greatly limited. We list some of the major unknowns:

- The prevalent uses of line-fire concepts and models in wildfire prediction and training require that we are able to distinguish when the limitations of this approach are exceeded by actual fire behaviours and render calculations either dangerous or misleading. Many fire-scale behaviours are not well approximated simply as large line fires but no research has been directed to this problem.
- Large-fire behaviours described in **Chapter 8** are qualitatively recognisable but are not quantitatively understood. The primary factors explaining atmospheric feedbacks in mass fires and pyroconvective storms or vorticity formation and consequent wildfire behaviours are little explored at any scale.
- Convective heat transfer from the flames to fuel particles is one of the key processes in wildfire spread but is mostly neglected by research. The characteristics of wildfire flames and their adjacent temperature and velocity characteristics in various configurations, including heading, backing and flanking directions, will require concentrated effort to understand and produce scalable methods for modelling.
- What is the wind that affects a spreading wildfire? It is doubtful that the mid-flame or eye-level horizontal wind speed remains unchanged in the presence of different fire configurations, but little research has been directed toward fire interactions with the wind field, either in the laboratory or the field.
- Burning rates and heat release rates are largely unknown for the wide variety of individual fuel components (e.g. duff, woody fuels, live foliage), complex arrangements of fuel materials, and the variety of moisture and wind conditions.
- Our ability to partition flaming from non-flaming phases of combustion in wildland fires is nearly non-existent. These proportions vary greatly, and the ability to transition to and from flaming routinely in wildfires is unexplored. Both factors mean that energy release in wildland fuel complexes is not possible to predict or model under changing environmental conditions or compare between prescribed fires and the largest, most complex wildfires, such as mass fires.
- Living foliage and vegetation clearly can support wildfire spread without the presence of significant dead fuel, but ignition and burning of living leaves and whole plants in the proper context constitutes a giant void in scientific knowledge. We must discover key processes that distinguish live from dead fuel ignition and burning rates and ultimately determine simplified criteria for classifying genera or species by key functional characteristics in fires.
- Inclined topography and incised canyons demonstrably produce flame attachment and rapid uphill fire spread, but these effects are not explained in physical terms. We thus have little reliable basis to incorporate the effects of

steep slopes into fire spread models or to train fire personnel in these behaviours for small or large fires.
- The possible roles of tree stems in the vertical transition of surface to crown fire and the horizontal propagation of crown fires is unexplored. There are no studies which have addressed how heat transfer does or does not change with the density or sizes of tree stems in wind or on sloping topography.
- Threshold conditions in fire spread arise from horizontal and vertical fuel discontinuities (e.g. patchy or clumpy shrubs, or ladder fuels, respectively) and critical values of wind, moisture, or fire size or shape, but these are only addressed by rough empirical observations or anecdotal experience. Thresholds are expressed as dramatic changes in fire behaviour that are potentially perilous to fire personnel because they occur rapidly with minor changes in fire or environmental factors.

Finally, we must be able to use our knowledge of wildfire processes to understand system-level fire behaviours and produce a cogent theory of wildfire behaviour. Theory can serve the needs of science to explain fire phenomena as well as anchor the development of practical models and tools based on that science.

Principal value to managers

Advances in wildfire behaviour science must ultimately be translated to improvements of applied knowledge in order to permit more long-term and large-scale strategic decisions in fire and land management. These functions generally fall into three categories: training and education, planning and mitigation, and operational predictions.

Current training curricula for firefighters, analysts and decision makers are primarily based on concepts employed in modelling systems and thus reflect the limitations of the formulations of those models and tools. The broader goal of education is often compromised for the sake of standardisation and practical necessity. There has been limited official interest from land management agencies in providing a broader service of educating fire personnel throughout their careers, particularly in comparison with producing defined skillsets for practitioners. Universities can provide education for those who attend forestry schools with wildland fire science programs. This book could serve as a resource in university courses for students with a strong interest in the speciality of fire. But many fire personnel do not have degrees from these departments or enter fire management through practical ranks. In fact, most of the active personnel in fire management in the United States have not had, and do not have, such a career path. In the United States, official training is handled through a series of government-sponsored interagency courses, often 1 week each, that build towards credentials recognised in the fire management system. A thorough rewrite of these courses is long overdue and would ideally include foundational science of fire behaviour as contained in this book before any practical modelling is taught. This would give plenty of perspective on the limitations and assumptions of modelling and systems. This is essential to thoughtful analyses and decisions because it allows the use of the vast amount of qualitative information on fires that is beyond modelling capabilities in practice now and in the foreseeable future. Examples include the limitations of fuel description in affecting fire behaviour, fire configurations and interactions with environmental heterogeneity, and recognising fire behaviours that are clearly not addressed by modelling tools.

Planning for wildland fires supports a variety of land management functions and constitutes most of fire behaviour modelling and uses of applied knowledge. For example, community preparedness for wildfires relies upon developing and communicating realistic wildfire scenarios and prioritising areas for mitigation or suburban zoning based on results from intensive Monte Carlo simulations of wildfire risk. Spatial forest planning and fuel treatment design depends upon understanding the efficacy of different treatment prescriptions and being able to evaluate alternative strategic treatment designs for mitigating wildfire impacts. Large-scale prescribed burning utilises fire behaviour calculations for designing prescriptions. Maintenance schedules for repeating prescribed burns depend on understanding fuel dynamics that change wildfire behaviour potential over time. The concern over climate change and carbon storage has also employed modelling to evaluate fire impacts under alternative landscape management schemes. For general purposes of fire mitigation, it has been long recognised that only limited benefits are possible from redoubled efforts in fire suppression, and thus gains in protecting natural resources and communities from wildfire will require that opportunities be expanded for proactively managing fires rather than simply supporting tactical wildfire response. Thus, the science of wildfire behaviour becomes critical to imagining, and implementing, a more strategic, active and holistic approach to wildfire management.

Wildfire prediction to support suppression operations and emergency response has historically been the primary rationale for investments made in the study of fire science – basically to build better and more reliable predictive models. There are two reasons why these applications of fire science constitute a minor fraction of the overall value, however. First is that severe and threatening wildfires represent only a handful of events that occur under the extremes of weather and fuel conditions when all suppression fails, and amount to ~2–3% of all fires in the United States. Thus, only a few dozen large-fire incidents are responsible for burning more than 90% of all burned area in any given year, irrespective of suppression effort. Modelling in support of decisions on this handful of highly visible incidents is performed by a cadre of experienced analysts. The second reason is that large wildfires offer a much narrower decision space for incident management teams than do fires managed for land-management benefit. Large wildfires burning in municipal watersheds or approaching adjacent communities are *always* actively suppressed, with little leeway afforded to decisions on use of firefighting resources or choice of suppression strategies. Such fires acquire political dimensions and even very accurate modelling delivers little prospect for improving the predetermined outcome or actions. Instead, increased reliability of applied fire science can have greater impact when employed to support expanded use of modified suppression responses under the more commonly occurring conditions of mild and moderate weather. The purpose then is to predict when and how we can take advantage of wildfires to achieve landscape-scale ecological benefits and implement sustainable long-term strategies for land and fire management.

In conclusion, the full utility of wildland fire science for sustaining ecosystems and protecting modern communities will ultimately be achieved only when the science is mature enough to simplify the apparent complexities of fire phenomena and allow people to realise how to live compatibly with wildland fire.

APPENDIX A: PHYSICAL QUANTITIES AND UNITS

Introduction and basic concepts

Physical quantities are features of our world that we can measure. Some will be very familiar to you, like length, mass and speed. Others may be less familiar, like energy, power and heat flux. Units are the labels that we attach to physical quantities to communicate magnitude. This appendix will briefly summarise those physical quantities and associated units that are important for understanding wildland fire.

Système international d'unités or the International System of Units was established in 1960 and is the global standard for communicating the magnitude of quantities. The 'International System of Units' is often shortened to the *SI system* or the *metric system*. The Standard (or Imperial) System of Units is often used in the United States for day-to-day transactions and in trades like machining and pipefitting. Metric units and standard units are incompatible, but you can easily convert from one system of units to the other. Modern scientific writings like this book always use metric units because they are the international standard.

The metric system is based on multiples of the number 10. For example, the metre (m) is the fundamental *base unit* of length in the metric system. Imagine a lodgepole pine (*Pinus contorta*) with a height of, say, 20 m. We can measure the diameter of the tree at breast height (DBH) in metres too: let us say 0.32 m. It is easy to misread and awkward to say 0.32 m, so instead we use units of centimetres (cm). There are 100 centimetres in 1 m, so we rewrite the diameter as 32 cm. This is clearer and less likely to be misread. Suppose the tree is felled, bucked, loaded onto a logging truck and hauled to the sawmill, which is 60 000 m away. Here, the metre is an awkward unit in this instance because it is too small to quantify long distances. Instead we use the kilometre (km), which is equal to 1000 m. Now we can rewrite the distance to the sawmill as 60 km.

Notice that both centimetre and kilometre have a prefix (centi- and kilo-) followed by the base unit, metre. The centi- and kilo- prefixes represent the multiplication of the fundamental base unit by 0.01 and 1000, respectively. There are prefixes for all useful multiples of 10, and they are summarised in **Table A.1.** The prefixes in **Table A.1** can be used with any metric unit. For example, 1000 kg is equivalent to 1 Mg.

Note that hecto-, deka- and deci- are used infrequently among scientists and engineers, and appear here for the sake of completeness. The remaining prefixes are in common use.

Scientific notation is another method of writing especially large and small quantities. For example, a hectare of burning timber litter may be producing heat at a rate of 950 million watts. Or the wavelength of light emitted by soot in a flame may be

Table A.1. Metric prefixes, symbol, and magnitudes.

Prefix	Symbol	Magnitude	Written as[1]
giga	G	10^9	billion
mega	M	10^6	million
kilo	k	10^3	thousand
hecto	h	10^2	hundred
deka	da	10^1	ten
-	-	10^0	one
deci	d	10^{-1}	tenth
centi	c	10^{-2}	hundredth
milli	m	10^{-3}	thousandth
micro	μ	10^{-6}	millionth
nano	n	10^{-9}	billionth

[1] Note that the *short scale* of powers of 10 – typical in English-speaking countries – is used here. In the *long scale* – common in French-speaking countries – 10^9 is 'one milliard' instead of 'one billion'.

1.2 millionths of a metre. It is awkward to write and read such numbers: 950 000 000 W and 0.000 001 2 m. Instead we use scientific notation, which is written as a product of the leading numeral and a power of 10. For example, 950 000 000 watts can instead be written as 9.5×10^8 watts, which reads 'Nine point five times ten to the eight watts'. For the wavelength example, 0.000 001 2 m is represented as 1.2×10^{-6} m. The negative exponent here indicates the inverse six power, which is equivalent to one divided by 10 to the power of six ($10^{-6} = 1 \div 10^6 = 0.000\,001$). This is the scientific notation format used in this book. However, when writing scientific notation out by hand, a more abbreviated writing method uses the letter 'e' to represent the power-of-10 part. So, the above examples can be written 9.5e8 and 1.2e-6. Using an upper case 'E' is also acceptable. Scientific notation and metric prefixes are both good methods for communicating particularly large and small quantities. For example, $350\,000$ g $= 3.5 \times 10^5$ g $= 350$ kg $= 3.5 \times 10^2$ kg are all perfectly acceptable.

Base quantities and units

Base quantities are the fundamental physical quantities that we use to quantify and define our world. Most of them will be familiar to you because you use them throughout your day without a second thought. In the following, the names of the units appear, followed by their symbol in brackets.

Mass is a quantity of matter. You can measure the mass of all three phases of matter: solids, liquids, and gases. 'Weight' and 'mass' are sometimes used interchangeably. Strictly speaking, weight is a force (gravity acting on a mass), not a mass. You will weigh less on the moon than on earth because the acceleration of gravity is different, but your mass will be the same. So, in this book we will always use the term *mass* to describe a quantity of matter. Note that the kilogram is the only metric base unit that, for historical reasons, contains a scaling prefix (kilo-).

Metric unit: kilogram [kg]
Standard unit: pound [lb]
Conversion: 1 lb = 0.454 kg

Length is a measure of how far something extends in space or a distance between locations.

Metric unit: metre [m]
Standard unit: foot [ft]
Conversion: 1 ft = 0.305 m
Note: Scientists and engineers often use the term, micron (MI-cron), in place of the unit micrometres, which is equal to one-millionth of a metre. Both are correct, but 'micrometres' can be confused for the measuring device. Unfortunately, and confusingly, both are spelled identically. The length unit is pronounced MI-cro-me-ter and the device is pronounced mi-CROM-eter.

Time refers to how long something takes. The second is the unit for both metric and standard systems. There are, of course, other common units of time as well: minute, hour, day, week, and so on.

Metric unit: second [s]
Standard unit: second [s]
Conversion: 1 s = 1 s

Temperature, fundamentally speaking, is a measure of the average kinetic energy of molecular

Table A.2. Conversions among three temperature scales.

Symbols T_C, T_K, and T_F represent temperature in the Celsius, Kelvin and Fahrenheit scales, respectively.

Convert to	from Celsius	from Kelvin	from Fahrenheit
Celsius: $T_C =$	T_C	$T_K - 273.15$	$(T_F - 32)(1.8)^{-1}$
Kelvin: $T_K =$	$T_C + 273.15$	T_K	$(T_F - 32)(1.8)^{-1} + 273.15$
Fahrenheit: $T_F =$	$11.8 T_C + 32$	$1.8(T_K - 273.15) + 32$	T_F

and/or atomic motion in matter. Colloquially, it is the 'hotness' or 'coldness' of an object. There are two types of temperature scales: absolute and relative. The absolute temperature scales are based on the minimum physically possible temperature. The relative scales are based on an arbitrary offset from the absolute scale. **Table A.2** summarises the equations used to convert among the various temperature scales.

Absolute metric unit: Kelvin [K]
Relative metric unit: Celsius [°C]
Relative standard unit: Fahrenheit [°F]

The **number of elementary entities** like atoms or molecules is represented by the mole. One mole is a standard metric used in chemistry and is equal to $6.022\,140\,76 \times 10^{23}$ entities. The mole is defined as the number of particles (atoms) in exactly 12 g of the carbon-12 isotope. This is a less well-known quantity but represents a key concept in chemistry and combustion. The kilo- prefix is frequently used with the mole: 1000 mol = 1 kilomole = 1 kmol.

Metric unit: mole [mol]
Standard unit: mole [mol]

Before proceeding to derived quantities, a further discussion regarding mass and mole is warranted because the difference between the two is not necessarily obvious. Both quantities communicate an amount of matter, but moles is the number of fundamental particles (e.g. atoms or molecules) comprising the matter, and mass is the overall amount of matter you have.

For example, suppose you have a sealed bottle containing three moles (3 mol) of carbon dioxide (CO_2). The symbol CO_2 means that there is one carbon atom and two oxygen atoms chemically bonded together in a carbon dioxide molecule. Since we know the number of moles of carbon dioxide in the bottle, we can calculate the number of carbon dioxide molecules:

$$\text{Number of } CO_2 \text{ molecules} = (3 \text{ mol})(6.022\,140\,76 \times 10^{23}) \approx 18.066 \times 10^{23}$$

Alternatively, to calculate the mass of 3 mol of carbon dioxide, we need the molecular mass of the molecule: 44.010 g mol^{-1} (consult any chemistry textbook or search online). Given this, the mass of the carbon dioxide in the bottle is the product of the number of moles present and molecular mass:

$$\text{Mass of } CO_2 = (3 \text{ mol})(44.010 \text{ g mol}^{-1}) \approx 132 \text{ g}$$

Moles, while being somewhat obscure outside of chemistry, are easier to understand than mass because mole represents the number of objects, albeit a very large number. Mass is less clear because its formal definition is based on rather esoteric principles of particle physics and cosmology. It is easiest to think of mass as the amount of stuff you have, whereas moles is the number of individual bits of stuff you have.

For example, consider the difference between being informed that (a) fire trucks with a combined mass of 32 000 kg are travelling to the wildland fire or (b) four fire trucks are travelling to the fire. The total mass of fire trucks is unhelpful if you need to know the number of fire trucks that have been deployed. On the other hand, if they must be ferried across a large river, the total mass of the fire trucks *is* required to ensure the ferry has adequate capacity.

Derived quantities and units

Derived quantities are mathematical combinations of base quantities. The units of derived quantities reflect those combinations. Consider, for example, speed, which is defined as a distance travelled per time. If you drive 120 km in 2 h, then your average speed is (120 km)/(2 h) = 60 km h^{-1}. Speed is a derived quantity that represents length divided by time. In base units, the unit of speed is metres per second (m s^{-1}) and, in widely accepted derived units, kilometres per hour (km h^{-1}). In the following, the units of the respective derived quantities will be shown with the named unit and symbol, or with the typical units used if the quantity has no named unit. Derived quantities in standard units are not shown, but they can be found by searching online.

Acceleration is the change in velocity per time.

Unit: m s^{-2}
Base units: m s^{-2}
Note: Acceleration is often a *vector quantity*, meaning that there is a direction associated with the numerical magnitude.

Area is a measurement of the extent of a surface in two or three-dimensions.

Unit: m^2
Base units: m^2
Other units: hectare [ha]
Note: The hectare is an accepted derived metric unit normally used to indicate land area and is equivalent to the area of a square-shaped plot that is 100 m × 100 m. The acre is the standard unit of land area: 2.471 acres is equivalent to 1 hectare.

Density is the mass of matter per volume.

Unit: kg m^{-3}
Base units: kg m^{-3}

Energy is the ability to make changes to surroundings. Both heat and thermal radiation are types of energy that play a key role in wildland fire processes.

Unit: joule [J]
Base units: J = N m = (kg m s^{-2}) m

Force refers to that which pushes on an object (of any phase of matter), causing it to accelerate. It famously features in Newton's Second Law of motion: $F = m\,a$, where F, m and a are force, mass and acceleration, respectively.

Unit: newton [N]
Base units: N = kg m s^{-2}

Heat is a type of energy that can be absorbed or lost by matter, causing a change in temperature or a change in phase.

Unit: joule [J]
Base units: J = N m = (kg m s^{-2}) m

Heat flux is the amount of thermal energy that passes through or onto some area over some unit time due to conduction, convection, and radiation heat transfer.

Unit: W m^{-2}
Base units: W m^{-2} = (kg m^2 s^{-3}) m^{-2}

Heat rate represents energy transferred or produced over some period of time. The terms 'power' and 'heat rate' are identical in meaning, but the latter is often used in fire because it is more descriptive.

Unit: watt [W]
Base units: W = J s^{-1} = (kg m^2 s^{-2}) s^{-1}

Power represents energy transferred or produced over some period of time. The terms 'power' and 'heat rate' are identical in meaning, but the latter is often used in fire because it is more descriptive.

Unit: watt [W]
Base units: W = J s^{-1} = (kg m^2 s^{-2}) s^{-1}

Pressure is a force applied over some area.

Unit: pascal [Pa]
Base units: Pa = N m^{-2} = (kg m s^{-2}) m^{-2}

Speed is the change in position over some period of time. (See note for *velocity*.)

Unit: m s^{-1}
Base units: m s^{-1}
Other units: km h^{-1}

Velocity is the change in position over some period of time in a specified direction.

Unit: m s^{-1}
Base units: m s^{-1}
Other units: km h^{-1}
Note: Velocity, like acceleration, is often a vector quantity, meaning that the numerical magnitude (*speed*) is associated with some particular direction. For example, the symbols u, v and w are all velocities because they specify speed in the x, y and z directions, respectively.

Volume refers to the size of a space in three dimensions.

Unit: m^3
Base units: m^3
Other units: litre [L]
Note: Litre is a derived unit where: 1000 L = 1 m^3. Litres is a common unit for volumes much smaller than 1 cubic metre and especially for quantifying of liquids. Since the lower-case 'l' looks very similar to the number one in typeset text, you will often see the abbreviation for litre written as ℓ or L. However, in the rare instances where volumes in litres appear in this book, the whole word will be spelled out because ℓ and L are symbols representing other unrelated quantities.

APPENDIX B: THERMAL AND PHYSICAL PROPERTIES OF AIR

Thermal and physical properties of air that are relevant to wildland fire are tabulated here. There are five tables corresponding to air properties at standard atmospheric pressure at five selected altitudes: 0 (i.e. sea level), 1000, 2000, 3000 and 4000 m. Note that specific heat, thermal conductivity, dynamic viscosity and Prandtl number do not change with pressure. Also, the values of the properties shown here are for 'real' air and not an ideal gas. As discussed in **Chapter 3**, gases produced by combustion can be treated as air because their compositions are sufficiently similar. Therefore, the properties in the following tables can be used for pure combustion gas, and any mixture of combustion gas and air.

The air properties appearing in the following tables were calculated using the open source CoolProp C++ library: http://www.coolprop.org. See Bell IH, Wronski J, Quoilin S, Lemort V (2014) Pure and pseudo-pure fluid thermophysical property evaluation and the open-source thermophysical property library CoolProp. *Industrial & Engineering Chemistry Research* **53**(6), 2498–2508.

Table B.1. Properties of air at 0 m altitude (standard atmospheric pressure: 101 325 Pa).

temp T, °C	density ρ, kg m^{-3}	specific heat, c_p J kg^{-1} K^{-1}	thermal conductivity k, W m^{-1} K^{-1}	thermal diffusivity α, m^2 s^{-1}	dynamic viscosity μ, Pa s	kinematic viscosity ν, m^2 s^{-1}	Prandtl number Pr
0	1.293	1005.7	2.436×10^{-2}	1.873×10^{-5}	1.722×10^{-5}	1.332×10^{-5}	0.711
10	1.247	1005.9	2.512×10^{-2}	2.002×10^{-5}	1.772×10^{-5}	1.420×10^{-5}	0.709
20	1.205	1006.1	2.587×10^{-2}	2.135×10^{-5}	1.821×10^{-5}	1.511×10^{-5}	0.708
30	1.165	1006.5	2.662×10^{-2}	2.271×10^{-5}	1.869×10^{-5}	1.605×10^{-5}	0.707
40	1.127	1006.9	2.735×10^{-2}	2.410×10^{-5}	1.917×10^{-5}	1.700×10^{-5}	0.705
50	1.092	1007.4	2.808×10^{-2}	2.552×10^{-5}	1.964×10^{-5}	1.797×10^{-5}	0.704
60	1.060	1008.0	2.880×10^{-2}	2.697×10^{-5}	2.010×10^{-5}	1.897×10^{-5}	0.703
70	1.029	1008.7	2.952×10^{-2}	2.845×10^{-5}	2.056×10^{-5}	1.998×10^{-5}	0.702
80	1.000	1009.5	3.023×10^{-2}	2.996×10^{-5}	2.101×10^{-5}	2.102×10^{-5}	0.702
90	0.972	1010.3	3.093×10^{-2}	3.149×10^{-5}	2.146×10^{-5}	2.207×10^{-5}	0.701
100	0.946	1011.2	3.162×10^{-2}	3.306×10^{-5}	2.190×10^{-5}	2.315×10^{-5}	0.700
110	0.921	1012.2	3.231×10^{-2}	3.465×10^{-5}	2.233×10^{-5}	2.424×10^{-5}	0.700
120	0.898	1013.3	3.299×10^{-2}	3.627×10^{-5}	2.276×10^{-5}	2.536×10^{-5}	0.699
130	0.875	1014.5	3.367×10^{-2}	3.791×10^{-5}	2.319×10^{-5}	2.649×10^{-5}	0.699
140	0.854	1015.8	3.434×10^{-2}	3.957×10^{-5}	2.361×10^{-5}	2.764×10^{-5}	0.698
150	0.834	1017.1	3.500×10^{-2}	4.126×10^{-5}	2.403×10^{-5}	2.881×10^{-5}	0.698
160	0.815	1018.5	3.566×10^{-2}	4.297×10^{-5}	2.444×10^{-5}	3.000×10^{-5}	0.698
170	0.796	1020.0	3.631×10^{-2}	4.471×10^{-5}	2.485×10^{-5}	3.120×10^{-5}	0.698
180	0.779	1021.6	3.696×10^{-2}	4.646×10^{-5}	2.525×10^{-5}	3.242×10^{-5}	0.698
190	0.762	1023.3	3.761×10^{-2}	4.824×10^{-5}	2.565×10^{-5}	3.367×10^{-5}	0.698
200	0.746	1025.0	3.825×10^{-2}	5.004×10^{-5}	2.605×10^{-5}	3.492×10^{-5}	0.698
250	0.675	1034.4	4.138×10^{-2}	5.931×10^{-5}	2.797×10^{-5}	4.147×10^{-5}	0.699
300	0.616	1045.1	4.442×10^{-2}	6.903×10^{-5}	2.981×10^{-5}	4.842×10^{-5}	0.701
350	0.566	1056.6	4.737×10^{-2}	7.917×10^{-5}	3.158×10^{-5}	5.577×10^{-5}	0.704
400	0.524	1068.5	5.024×10^{-2}	8.970×10^{-5}	3.328×10^{-5}	6.350×10^{-5}	0.708
450	0.488	1080.5	5.305×10^{-2}	1.006×10^{-4}	3.493×10^{-5}	7.159×10^{-5}	0.712
500	0.456	1092.4	5.580×10^{-2}	1.119×10^{-4}	3.653×10^{-5}	8.004×10^{-5}	0.715
550	0.429	1104.0	5.849×10^{-2}	1.236×10^{-4}	3.808×10^{-5}	8.884×10^{-5}	0.719
600	0.404	1115.1	6.114×10^{-2}	1.357×10^{-4}	3.960×10^{-5}	9.798×10^{-5}	0.722
650	0.382	1125.8	6.374×10^{-2}	1.481×10^{-4}	4.107×10^{-5}	1.075×10^{-4}	0.725
700	0.363	1135.8	6.631×10^{-2}	1.610×10^{-4}	4.252×10^{-5}	1.173×10^{-4}	0.728
750	0.345	1145.3	6.885×10^{-2}	1.743×10^{-4}	4.393×10^{-5}	1.274×10^{-4}	0.731
800	0.329	1154.3	7.135×10^{-2}	1.880×10^{-4}	4.532×10^{-5}	1.378×10^{-4}	0.733
850	0.314	1162.6	7.382×10^{-2}	2.021×10^{-4}	4.668×10^{-5}	1.486×10^{-4}	0.735
900	0.301	1170.5	7.627×10^{-2}	2.166×10^{-4}	4.802×10^{-5}	1.596×10^{-4}	0.737
950	0.289	1177.8	7.870×10^{-2}	2.316×10^{-4}	4.934×10^{-5}	1.710×10^{-4}	0.738
1000	0.277	1184.7	8.110×10^{-2}	2.470×10^{-4}	5.063×10^{-5}	1.827×10^{-4}	0.740
1050	0.267	1191.2	8.348×10^{-2}	2.628×10^{-4}	5.192×10^{-5}	1.947×10^{-4}	0.741
1100	0.257	1197.2	8.585×10^{-2}	2.790×10^{-4}	5.318×10^{-5}	2.069×10^{-4}	0.742
1150	0.248	1202.9	8.820×10^{-2}	2.957×10^{-4}	5.443×10^{-5}	2.195×10^{-4}	0.742
1200	0.240	1208.3	9.053×10^{-2}	3.128×10^{-4}	5.567×10^{-5}	2.324×10^{-4}	0.743
1250	0.232	1213.3	9.285×10^{-2}	3.303×10^{-4}	5.689×10^{-5}	2.455×10^{-4}	0.743
1300	0.224	1218.1	9.516×10^{-2}	3.483×10^{-4}	5.810×10^{-5}	2.590×10^{-4}	0.744
1400	0.211	1226.9	9.975×10^{-2}	3.855×10^{-4}	6.049×10^{-5}	2.868×10^{-4}	0.744
1500	0.199	1234.8	1.043×10^{-1}	4.244×10^{-4}	6.284×10^{-5}	3.157×10^{-4}	0.744
1600	0.188	1241.9	1.088×10^{-1}	4.650×10^{-4}	6.517×10^{-5}	3.459×10^{-4}	0.744
1700	0.179	1248.5	1.133×10^{-1}	5.073×10^{-4}	6.746×10^{-5}	3.771×10^{-4}	0.743
1800	0.170	1254.5	1.177×10^{-1}	5.513×10^{-4}	6.973×10^{-5}	4.096×10^{-4}	0.743
1900	0.162	1260.1	1.222×10^{-1}	5.971×10^{-4}	7.197×10^{-5}	4.432×10^{-4}	0.742
2000	0.155	1265.3	1.266×10^{-1}	6.445×10^{-4}	7.420×10^{-5}	4.779×10^{-4}	0.741
2100	0.149	1270.1	1.310×10^{-1}	6.937×10^{-4}	7.642×10^{-5}	5.138×10^{-4}	0.741
2200	0.143	1274.7	1.354×10^{-1}	7.446×10^{-4}	7.862×10^{-5}	5.509×10^{-4}	0.740
2300	0.137	1278.9	1.398×10^{-1}	7.972×10^{-4}	8.081×10^{-5}	5.891×10^{-4}	0.739
2400	0.132	1283.0	1.442×10^{-1}	8.515×10^{-4}	8.299×10^{-5}	6.286×10^{-4}	0.738

Table B.2. Properties of air at 1000 m altitude (standard atmospheric pressure: 89 874 Pa).

temp T, °C	density ρ, kg m^{-3}	specific heat, c_p J kg^{-1} K^{-1}	thermal conductivity k, W m^{-1} K^{-1}	thermal diffusivity α, m^2 s^{-1}	dynamic viscosity μ, Pa s	kinematic viscosity ν, m^2 s^{-1}	Prandtl number Pr
0	1.147	1005.5	2.436×10^{-2}	2.112×10^{-5}	1.722×10^{-5}	1.501×10^{-5}	0.711
10	1.106	1005.7	2.512×10^{-2}	2.258×10^{-5}	1.771×10^{-5}	1.601×10^{-5}	0.709
20	1.068	1006.0	2.587×10^{-2}	2.407×10^{-5}	1.820×10^{-5}	1.704×10^{-5}	0.708
30	1.033	1006.3	2.661×10^{-2}	2.560×10^{-5}	1.869×10^{-5}	1.809×10^{-5}	0.707
40	1.000	1006.8	2.735×10^{-2}	2.717×10^{-5}	1.916×10^{-5}	1.916×10^{-5}	0.705
50	0.969	1007.3	2.808×10^{-2}	2.877×10^{-5}	1.963×10^{-5}	2.026×10^{-5}	0.704
60	0.940	1007.9	2.880×10^{-2}	3.040×10^{-5}	2.010×10^{-5}	2.138×10^{-5}	0.703
70	0.912	1008.6	2.952×10^{-2}	3.207×10^{-5}	2.056×10^{-5}	2.253×10^{-5}	0.702
80	0.887	1009.3	3.022×10^{-2}	3.377×10^{-5}	2.101×10^{-5}	2.370×10^{-5}	0.702
90	0.862	1010.2	3.092×10^{-2}	3.551×10^{-5}	2.145×10^{-5}	2.489×10^{-5}	0.701
100	0.839	1011.1	3.162×10^{-2}	3.727×10^{-5}	2.190×10^{-5}	2.610×10^{-5}	0.700
110	0.817	1012.1	3.230×10^{-2}	3.906×10^{-5}	2.233×10^{-5}	2.733×10^{-5}	0.700
120	0.796	1013.2	3.299×10^{-2}	4.089×10^{-5}	2.276×10^{-5}	2.859×10^{-5}	0.699
130	0.776	1014.4	3.366×10^{-2}	4.274×10^{-5}	2.319×10^{-5}	2.986×10^{-5}	0.699
140	0.758	1015.7	3.433×10^{-2}	4.461×10^{-5}	2.361×10^{-5}	3.116×10^{-5}	0.698
150	0.740	1017.0	3.500×10^{-2}	4.652×10^{-5}	2.403×10^{-5}	3.248×10^{-5}	0.698
160	0.723	1018.5	3.566×10^{-2}	4.845×10^{-5}	2.444×10^{-5}	3.382×10^{-5}	0.698
170	0.706	1020.0	3.631×10^{-2}	5.040×10^{-5}	2.485×10^{-5}	3.517×10^{-5}	0.698
180	0.691	1021.5	3.696×10^{-2}	5.238×10^{-5}	2.525×10^{-5}	3.655×10^{-5}	0.698
190	0.676	1023.2	3.761×10^{-2}	5.438×10^{-5}	2.565×10^{-5}	3.795×10^{-5}	0.698
200	0.662	1024.9	3.825×10^{-2}	5.641×10^{-5}	2.604×10^{-5}	3.937×10^{-5}	0.698
250	0.598	1034.4	4.138×10^{-2}	6.686×10^{-5}	2.797×10^{-5}	4.675×10^{-5}	0.699
300	0.546	1045.1	4.442×10^{-2}	7.783×10^{-5}	2.981×10^{-5}	5.459×10^{-5}	0.701
350	0.502	1056.6	4.737×10^{-2}	8.925×10^{-5}	3.158×10^{-5}	6.287×10^{-5}	0.704
400	0.465	1068.5	5.024×10^{-2}	1.011×10^{-4}	3.328×10^{-5}	7.158×10^{-5}	0.708
450	0.433	1080.5	5.305×10^{-2}	1.134×10^{-4}	3.493×10^{-5}	8.071×10^{-5}	0.712
500	0.405	1092.4	5.579×10^{-2}	1.262×10^{-4}	3.653×10^{-5}	9.023×10^{-5}	0.715
550	0.380	1104.0	5.849×10^{-2}	1.393×10^{-4}	3.808×10^{-5}	1.002×10^{-4}	0.719
600	0.358	1115.1	6.114×10^{-2}	1.529×10^{-4}	3.960×10^{-5}	1.105×10^{-4}	0.722
650	0.339	1125.7	6.374×10^{-2}	1.670×10^{-4}	4.107×10^{-5}	1.211×10^{-4}	0.725
700	0.322	1135.8	6.631×10^{-2}	1.815×10^{-4}	4.252×10^{-5}	1.322×10^{-4}	0.728
750	0.306	1145.3	6.885×10^{-2}	1.965×10^{-4}	4.393×10^{-5}	1.436×10^{-4}	0.731
800	0.292	1154.2	7.135×10^{-2}	2.119×10^{-4}	4.532×10^{-5}	1.554×10^{-4}	0.733
850	0.279	1162.6	7.382×10^{-2}	2.278×10^{-4}	4.668×10^{-5}	1.675×10^{-4}	0.735
900	0.267	1170.5	7.627×10^{-2}	2.442×10^{-4}	4.802×10^{-5}	1.800×10^{-4}	0.737
950	0.256	1177.8	7.869×10^{-2}	2.611×10^{-4}	4.934×10^{-5}	1.928×10^{-4}	0.738
1000	0.246	1184.7	8.110×10^{-2}	2.784×10^{-4}	5.063×10^{-5}	2.059×10^{-4}	0.740
1050	0.237	1191.2	8.348×10^{-2}	2.962×10^{-4}	5.192×10^{-5}	2.194×10^{-4}	0.741
1100	0.228	1197.2	8.585×10^{-2}	3.146×10^{-4}	5.318×10^{-5}	2.333×10^{-4}	0.742
1150	0.220	1202.9	8.820×10^{-2}	3.333×10^{-4}	5.443×10^{-5}	2.475×10^{-4}	0.742
1200	0.212	1208.3	9.053×10^{-2}	3.526×10^{-4}	5.567×10^{-5}	2.620×10^{-4}	0.743
1250	0.206	1213.3	9.285×10^{-2}	3.724×10^{-4}	5.689×10^{-5}	2.768×10^{-4}	0.743
1300	0.199	1218.1	9.516×10^{-2}	3.926×10^{-4}	5.810×10^{-5}	2.920×10^{-4}	0.744
1400	0.187	1226.9	9.975×10^{-2}	4.345×10^{-4}	6.049×10^{-5}	3.233×10^{-4}	0.744
1500	0.177	1234.8	1.043×10^{-1}	4.784×10^{-4}	6.284×10^{-5}	3.560×10^{-4}	0.744
1600	0.167	1241.9	1.088×10^{-1}	5.242×10^{-4}	6.516×10^{-5}	3.899×10^{-4}	0.744
1700	0.159	1248.5	1.133×10^{-1}	5.719×10^{-4}	6.746×10^{-5}	4.252×10^{-4}	0.743
1800	0.151	1254.5	1.177×10^{-1}	6.216×10^{-4}	6.972×10^{-5}	4.617×10^{-4}	0.743
1900	0.144	1260.1	1.222×10^{-1}	6.732×10^{-4}	7.197×10^{-5}	4.996×10^{-4}	0.742
2000	0.138	1265.3	1.266×10^{-1}	7.267×10^{-4}	7.420×10^{-5}	5.388×10^{-4}	0.741
2100	0.132	1270.1	1.310×10^{-1}	7.821×10^{-4}	7.642×10^{-5}	5.793×10^{-4}	0.741
2200	0.127	1274.7	1.354×10^{-1}	8.394×10^{-4}	7.862×10^{-5}	6.211×10^{-4}	0.740
2300	0.122	1278.9	1.398×10^{-1}	8.987×10^{-4}	8.081×10^{-5}	6.642×10^{-4}	0.739
2400	0.117	1283.0	1.442×10^{-1}	9.600×10^{-4}	8.299×10^{-5}	7.086×10^{-4}	0.738

Table B.3. Properties of air at 2000 m altitude (standard atmospheric pressure: 79 495 Pa).

temp T, °C	density ρ, kg m^{-3}	specific heat, c_p J kg^{-1} K^{-1}	thermal conductivity k, W m^{-1} K^{-1}	thermal diffusivity α, m^2 s^{-1}	dynamic viscosity μ, Pa s	kinematic viscosity ν, m^2 s^{-1}	Prandtl number Pr
0	1.014	1005.3	2.435×10^{-2}	2.388×10^{-5}	1.722×10^{-5}	1.697×10^{-5}	0.711
10	0.978	1005.5	2.511×10^{-2}	2.553×10^{-5}	1.771×10^{-5}	1.810×10^{-5}	0.709
20	0.945	1005.8	2.587×10^{-2}	2.722×10^{-5}	1.820×10^{-5}	1.926×10^{-5}	0.708
30	0.914	1006.2	2.661×10^{-2}	2.895×10^{-5}	1.869×10^{-5}	2.045×10^{-5}	0.706
40	0.885	1006.6	2.735×10^{-2}	3.072×10^{-5}	1.916×10^{-5}	2.166×10^{-5}	0.705
50	0.857	1007.1	2.808×10^{-2}	3.253×10^{-5}	1.963×10^{-5}	2.291×10^{-5}	0.704
60	0.831	1007.8	2.880×10^{-2}	3.437×10^{-5}	2.010×10^{-5}	2.417×10^{-5}	0.703
70	0.807	1008.4	2.951×10^{-2}	3.626×10^{-5}	2.055×10^{-5}	2.547×10^{-5}	0.702
80	0.784	1009.2	3.022×10^{-2}	3.818×10^{-5}	2.101×10^{-5}	2.679×10^{-5}	0.702
90	0.763	1010.1	3.092×10^{-2}	4.014×10^{-5}	2.145×10^{-5}	2.813×10^{-5}	0.701
100	0.742	1011.0	3.161×10^{-2}	4.214×10^{-5}	2.189×10^{-5}	2.950×10^{-5}	0.700
110	0.723	1012.1	3.230×10^{-2}	4.416×10^{-5}	2.233×10^{-5}	3.090×10^{-5}	0.700
120	0.704	1013.2	3.298×10^{-2}	4.622×10^{-5}	2.276×10^{-5}	3.232×10^{-5}	0.699
130	0.687	1014.4	3.366×10^{-2}	4.831×10^{-5}	2.319×10^{-5}	3.376×10^{-5}	0.699
140	0.670	1015.6	3.433×10^{-2}	5.044×10^{-5}	2.361×10^{-5}	3.523×10^{-5}	0.698
150	0.654	1017.0	3.500×10^{-2}	5.259×10^{-5}	2.402×10^{-5}	3.672×10^{-5}	0.698
160	0.639	1018.4	3.566×10^{-2}	5.477×10^{-5}	2.444×10^{-5}	3.823×10^{-5}	0.698
170	0.625	1019.9	3.631×10^{-2}	5.698×10^{-5}	2.484×10^{-5}	3.976×10^{-5}	0.698
180	0.611	1021.5	3.696×10^{-2}	5.922×10^{-5}	2.525×10^{-5}	4.132×10^{-5}	0.698
190	0.598	1023.1	3.760×10^{-2}	6.148×10^{-5}	2.565×10^{-5}	4.290×10^{-5}	0.698
200	0.585	1024.8	3.824×10^{-2}	6.377×10^{-5}	2.604×10^{-5}	4.451×10^{-5}	0.698
250	0.529	1034.3	4.138×10^{-2}	7.559×10^{-5}	2.797×10^{-5}	5.285×10^{-5}	0.699
300	0.483	1045.0	4.441×10^{-2}	8.798×10^{-5}	2.981×10^{-5}	6.171×10^{-5}	0.701
350	0.444	1056.5	4.736×10^{-2}	1.009×10^{-4}	3.158×10^{-5}	7.107×10^{-5}	0.704
400	0.411	1068.5	5.024×10^{-2}	1.143×10^{-4}	3.328×10^{-5}	8.092×10^{-5}	0.708
450	0.383	1080.5	5.304×10^{-2}	1.282×10^{-4}	3.493×10^{-5}	9.124×10^{-5}	0.712
500	0.358	1092.4	5.579×10^{-2}	1.426×10^{-4}	3.653×10^{-5}	1.020×10^{-4}	0.715
550	0.336	1104.0	5.849×10^{-2}	1.575×10^{-4}	3.808×10^{-5}	1.132×10^{-4}	0.719
600	0.317	1115.1	6.114×10^{-2}	1.729×10^{-4}	3.960×10^{-5}	1.249×10^{-4}	0.722
650	0.300	1125.7	6.374×10^{-2}	1.888×10^{-4}	4.107×10^{-5}	1.369×10^{-4}	0.725
700	0.285	1135.8	6.631×10^{-2}	2.052×10^{-4}	4.252×10^{-5}	1.494×10^{-4}	0.728
750	0.271	1145.3	6.884×10^{-2}	2.221×10^{-4}	4.393×10^{-5}	1.623×10^{-4}	0.731
800	0.258	1154.2	7.135×10^{-2}	2.396×10^{-4}	4.532×10^{-5}	1.756×10^{-4}	0.733
850	0.247	1162.6	7.382×10^{-2}	2.576×10^{-4}	4.668×10^{-5}	1.894×10^{-4}	0.735
900	0.236	1170.5	7.627×10^{-2}	2.761×10^{-4}	4.802×10^{-5}	2.035×10^{-4}	0.737
950	0.226	1177.8	7.869×10^{-2}	2.952×10^{-4}	4.934×10^{-5}	2.179×10^{-4}	0.738
1000	0.217	1184.7	8.110×10^{-2}	3.148×10^{-4}	5.063×10^{-5}	2.328×10^{-4}	0.740
1050	0.209	1191.2	8.348×10^{-2}	3.349×10^{-4}	5.192×10^{-5}	2.481×10^{-4}	0.741
1100	0.202	1197.2	8.585×10^{-2}	3.556×10^{-4}	5.318×10^{-5}	2.637×10^{-4}	0.742
1150	0.195	1202.9	8.820×10^{-2}	3.769×10^{-4}	5.443×10^{-5}	2.798×10^{-4}	0.742
1200	0.188	1208.3	9.053×10^{-2}	3.986×10^{-4}	5.567×10^{-5}	2.962×10^{-4}	0.743
1250	0.182	1213.3	9.285×10^{-2}	4.210×10^{-4}	5.689×10^{-5}	3.129×10^{-4}	0.743
1300	0.176	1218.1	9.516×10^{-2}	4.439×10^{-4}	5.810×10^{-5}	3.301×10^{-4}	0.744
1400	0.165	1226.8	9.975×10^{-2}	4.913×10^{-4}	6.049×10^{-5}	3.655×10^{-4}	0.744
1500	0.156	1234.8	1.043×10^{-1}	5.409×10^{-4}	6.284×10^{-5}	4.024×10^{-4}	0.744
1600	0.148	1241.9	1.088×10^{-1}	5.926×10^{-4}	6.516×10^{-5}	4.408×10^{-4}	0.744
1700	0.140	1248.5	1.133×10^{-1}	6.466×10^{-4}	6.746×10^{-5}	4.807×10^{-4}	0.743
1800	0.134	1254.5	1.177×10^{-1}	7.027×10^{-4}	6.972×10^{-5}	5.220×10^{-4}	0.743
1900	0.127	1260.1	1.222×10^{-1}	7.610×10^{-4}	7.197×10^{-5}	5.648×10^{-4}	0.742
2000	0.122	1265.3	1.266×10^{-1}	8.215×10^{-4}	7.420×10^{-5}	6.091×10^{-4}	0.741
2100	0.117	1270.1	1.310×10^{-1}	8.842×10^{-4}	7.642×10^{-5}	6.549×10^{-4}	0.741
2200	0.112	1274.7	1.354×10^{-1}	9.490×10^{-4}	7.862×10^{-5}	7.022×10^{-4}	0.740
2300	0.108	1278.9	1.398×10^{-1}	1.016×10^{-3}	8.081×10^{-5}	7.509×10^{-4}	0.739
2400	0.104	1283.0	1.442×10^{-1}	1.085×10^{-3}	8.299×10^{-5}	8.011×10^{-4}	0.738

Table B.4. Properties of air at 3000 m altitude (standard atmospheric pressure: 70 108 Pa).

temp T, °C	density ρ, kg m^{-3}	specific heat, c_p J kg^{-1} K^{-1}	thermal conductivity k, W m^{-1} K^{-1}	thermal diffusivity α, m^2 s^{-1}	dynamic viscosity μ, Pa s	kinematic viscosity ν, m^2 s^{-1}	Prandtl number Pr
0	0.895	1005.1	2.435×10^{-2}	2.708×10^{-5}	1.721×10^{-5}	1.924×10^{-5}	0.711
10	0.863	1005.3	2.511×10^{-2}	2.895×10^{-5}	1.771×10^{-5}	2.053×10^{-5}	0.709
20	0.833	1005.6	2.586×10^{-2}	3.086×10^{-5}	1.820×10^{-5}	2.184×10^{-5}	0.708
30	0.806	1006.0	2.661×10^{-2}	3.282×10^{-5}	1.868×10^{-5}	2.319×10^{-5}	0.706
40	0.780	1006.5	2.734×10^{-2}	3.483×10^{-5}	1.916×10^{-5}	2.456×10^{-5}	0.705
50	0.756	1007.0	2.807×10^{-2}	3.688×10^{-5}	1.963×10^{-5}	2.597×10^{-5}	0.704
60	0.733	1007.6	2.880×10^{-2}	3.898×10^{-5}	2.009×10^{-5}	2.741×10^{-5}	0.703
70	0.712	1008.3	2.951×10^{-2}	4.112×10^{-5}	2.055×10^{-5}	2.888×10^{-5}	0.702
80	0.692	1009.1	3.022×10^{-2}	4.330×10^{-5}	2.100×10^{-5}	3.037×10^{-5}	0.701
90	0.673	1010.0	3.092×10^{-2}	4.552×10^{-5}	2.145×10^{-5}	3.190×10^{-5}	0.701
100	0.654	1010.9	3.161×10^{-2}	4.778×10^{-5}	2.189×10^{-5}	3.345×10^{-5}	0.700
110	0.637	1012.0	3.230×10^{-2}	5.008×10^{-5}	2.233×10^{-5}	3.503×10^{-5}	0.700
120	0.621	1013.1	3.298×10^{-2}	5.241×10^{-5}	2.276×10^{-5}	3.664×10^{-5}	0.699
130	0.606	1014.3	3.366×10^{-2}	5.478×10^{-5}	2.319×10^{-5}	3.828×10^{-5}	0.699
140	0.591	1015.6	3.433×10^{-2}	5.719×10^{-5}	2.361×10^{-5}	3.994×10^{-5}	0.698
150	0.577	1016.9	3.499×10^{-2}	5.963×10^{-5}	2.402×10^{-5}	4.163×10^{-5}	0.698
160	0.564	1018.3	3.565×10^{-2}	6.210×10^{-5}	2.444×10^{-5}	4.334×10^{-5}	0.698
170	0.551	1019.8	3.631×10^{-2}	6.461×10^{-5}	2.484×10^{-5}	4.509×10^{-5}	0.698
180	0.539	1021.4	3.696×10^{-2}	6.715×10^{-5}	2.525×10^{-5}	4.685×10^{-5}	0.698
190	0.527	1023.1	3.760×10^{-2}	6.971×10^{-5}	2.565×10^{-5}	4.865×10^{-5}	0.698
200	0.516	1024.8	3.824×10^{-2}	7.231×10^{-5}	2.604×10^{-5}	5.046×10^{-5}	0.698
250	0.467	1034.3	4.138×10^{-2}	8.571×10^{-5}	2.797×10^{-5}	5.992×10^{-5}	0.699
300	0.426	1045.0	4.441×10^{-2}	9.976×10^{-5}	2.981×10^{-5}	6.997×10^{-5}	0.701
350	0.392	1056.5	4.736×10^{-2}	1.144×10^{-4}	3.158×10^{-5}	8.059×10^{-5}	0.704
400	0.363	1068.4	5.024×10^{-2}	1.296×10^{-4}	3.328×10^{-5}	9.175×10^{-5}	0.708
450	0.338	1080.5	5.304×10^{-2}	1.454×10^{-4}	3.493×10^{-5}	1.034×10^{-4}	0.712
500	0.316	1092.4	5.579×10^{-2}	1.617×10^{-4}	3.653×10^{-5}	1.157×10^{-4}	0.715
550	0.297	1104.0	5.849×10^{-2}	1.786×10^{-4}	3.808×10^{-5}	1.284×10^{-4}	0.719
600	0.280	1115.1	6.114×10^{-2}	1.960×10^{-4}	3.960×10^{-5}	1.416×10^{-4}	0.722
650	0.265	1125.7	6.374×10^{-2}	2.141×10^{-4}	4.107×10^{-5}	1.553×10^{-4}	0.725
700	0.251	1135.8	6.631×10^{-2}	2.327×10^{-4}	4.252×10^{-5}	1.694×10^{-4}	0.728
750	0.239	1145.3	6.884×10^{-2}	2.519×10^{-4}	4.393×10^{-5}	1.841×10^{-4}	0.731
800	0.228	1154.2	7.135×10^{-2}	2.717×10^{-4}	4.532×10^{-5}	1.992×10^{-4}	0.733
850	0.217	1162.6	7.382×10^{-2}	2.920×10^{-4}	4.668×10^{-5}	2.147×10^{-4}	0.735
900	0.208	1170.5	7.627×10^{-2}	3.131×10^{-4}	4.802×10^{-5}	2.307×10^{-4}	0.737
950	0.200	1177.8	7.869×10^{-2}	3.347×10^{-4}	4.933×10^{-5}	2.471×10^{-4}	0.738
1000	0.192	1184.7	8.110×10^{-2}	3.569×10^{-4}	5.063×10^{-5}	2.640×10^{-4}	0.740
1050	0.185	1191.2	8.348×10^{-2}	3.797×10^{-4}	5.191×10^{-5}	2.813×10^{-4}	0.741
1100	0.178	1197.2	8.585×10^{-2}	4.032×10^{-4}	5.318×10^{-5}	2.990×10^{-4}	0.742
1150	0.172	1202.9	8.820×10^{-2}	4.273×10^{-4}	5.443×10^{-5}	3.172×10^{-4}	0.742
1200	0.166	1208.3	9.053×10^{-2}	4.520×10^{-4}	5.567×10^{-5}	3.358×10^{-4}	0.743
1250	0.160	1213.3	9.285×10^{-2}	4.773×10^{-4}	5.689×10^{-5}	3.548×10^{-4}	0.743
1300	0.155	1218.1	9.516×10^{-2}	5.033×10^{-4}	5.810×10^{-5}	3.743×10^{-4}	0.744
1400	0.146	1226.8	9.974×10^{-2}	5.570×10^{-4}	6.049×10^{-5}	4.145×10^{-4}	0.744
1500	0.138	1234.7	1.043×10^{-1}	6.133×10^{-4}	6.284×10^{-5}	4.563×10^{-4}	0.744
1600	0.130	1241.9	1.088×10^{-1}	6.720×10^{-4}	6.516×10^{-5}	4.998×10^{-4}	0.744
1700	0.124	1248.5	1.133×10^{-1}	7.332×10^{-4}	6.746×10^{-5}	5.450×10^{-4}	0.743
1800	0.118	1254.5	1.177×10^{-1}	7.968×10^{-4}	6.972×10^{-5}	5.919×10^{-4}	0.743
1900	0.112	1260.1	1.222×10^{-1}	8.629×10^{-4}	7.197×10^{-5}	6.405×10^{-4}	0.742
2000	0.107	1265.3	1.266×10^{-1}	9.315×10^{-4}	7.420×10^{-5}	6.907×10^{-4}	0.741
2100	0.103	1270.1	1.310×10^{-1}	1.003×10^{-3}	7.642×10^{-5}	7.426×10^{-4}	0.741
2200	0.099	1274.6	1.354×10^{-1}	1.076×10^{-3}	7.862×10^{-5}	7.962×10^{-4}	0.740
2300	0.095	1278.9	1.398×10^{-1}	1.152×10^{-3}	8.081×10^{-5}	8.514×10^{-4}	0.739
2400	0.091	1283.0	1.442×10^{-1}	1.231×10^{-3}	8.299×10^{-5}	9.084×10^{-4}	0.738

Table B.5. Properties of air at 4000 m altitude (standard atmospheric pressure: 61 640 Pa).

temp T, °C	density ρ, kg m^{-3}	specific heat, c_p J kg^{-1} K^{-1}	thermal conductivity k, W m^{-1} K^{-1}	thermal diffusivity α, m^2 s^{-1}	dynamic viscosity μ, Pa s	kinematic viscosity ν, m^2 s^{-1}	Prandtl number Pr
0	0.786	1004.9	2.435×10^{-2}	3.081×10^{-5}	1.721×10^{-5}	2.189×10^{-5}	0.710
10	0.759	1005.2	2.511×10^{-2}	3.293×10^{-5}	1.771×10^{-5}	2.335×10^{-5}	0.709
20	0.733	1005.5	2.586×10^{-2}	3.510×10^{-5}	1.820×10^{-5}	2.484×10^{-5}	0.708
30	0.708	1005.9	2.661×10^{-2}	3.733×10^{-5}	1.868×10^{-5}	2.637×10^{-5}	0.706
40	0.686	1006.4	2.734×10^{-2}	3.962×10^{-5}	1.916×10^{-5}	2.794×10^{-5}	0.705
50	0.665	1006.9	2.807×10^{-2}	4.195×10^{-5}	1.963×10^{-5}	2.954×10^{-5}	0.704
60	0.645	1007.5	2.879×10^{-2}	4.433×10^{-5}	2.009×10^{-5}	3.117×10^{-5}	0.703
70	0.626	1008.2	2.951×10^{-2}	4.677×10^{-5}	2.055×10^{-5}	3.284×10^{-5}	0.702
80	0.608	1009.0	3.021×10^{-2}	4.925×10^{-5}	2.100×10^{-5}	3.454×10^{-5}	0.701
90	0.591	1009.9	3.092×10^{-2}	5.177×10^{-5}	2.145×10^{-5}	3.628×10^{-5}	0.701
100	0.575	1010.9	3.161×10^{-2}	5.434×10^{-5}	2.189×10^{-5}	3.804×10^{-5}	0.700
110	0.560	1011.9	3.230×10^{-2}	5.696×10^{-5}	2.233×10^{-5}	3.984×10^{-5}	0.700
120	0.546	1013.0	3.298×10^{-2}	5.961×10^{-5}	2.276×10^{-5}	4.167×10^{-5}	0.699
130	0.533	1014.2	3.366×10^{-2}	6.231×10^{-5}	2.318×10^{-5}	4.353×10^{-5}	0.699
140	0.520	1015.5	3.433×10^{-2}	6.505×10^{-5}	2.361×10^{-5}	4.542×10^{-5}	0.698
150	0.507	1016.8	3.499×10^{-2}	6.782×10^{-5}	2.402×10^{-5}	4.734×10^{-5}	0.698
160	0.496	1018.3	3.565×10^{-2}	7.063×10^{-5}	2.443×10^{-5}	4.930×10^{-5}	0.698
170	0.484	1019.8	3.631×10^{-2}	7.348×10^{-5}	2.484×10^{-5}	5.128×10^{-5}	0.698
180	0.474	1021.4	3.696×10^{-2}	7.637×10^{-5}	2.525×10^{-5}	5.329×10^{-5}	0.698
190	0.464	1023.0	3.760×10^{-2}	7.929×10^{-5}	2.565×10^{-5}	5.532×10^{-5}	0.698
200	0.454	1024.7	3.824×10^{-2}	8.224×10^{-5}	2.604×10^{-5}	5.739×10^{-5}	0.698
250	0.410	1034.3	4.138×10^{-2}	9.748×10^{-5}	2.797×10^{-5}	6.815×10^{-5}	0.699
300	0.375	1045.0	4.441×10^{-2}	1.135×10^{-4}	2.981×10^{-5}	7.958×10^{-5}	0.701
350	0.345	1056.5	4.736×10^{-2}	1.301×10^{-4}	3.158×10^{-5}	9.165×10^{-5}	0.704
400	0.319	1068.4	5.024×10^{-2}	1.474×10^{-4}	3.328×10^{-5}	1.044×10^{-4}	0.708
450	0.297	1080.5	5.304×10^{-2}	1.654×10^{-4}	3.493×10^{-5}	1.177×10^{-4}	0.711
500	0.278	1092.4	5.579×10^{-2}	1.839×10^{-4}	3.653×10^{-5}	1.315×10^{-4}	0.715
550	0.261	1103.9	5.849×10^{-2}	2.031×10^{-4}	3.808×10^{-5}	1.460×10^{-4}	0.719
600	0.246	1115.1	6.113×10^{-2}	2.230×10^{-4}	3.959×10^{-5}	1.610×10^{-4}	0.722
650	0.233	1125.7	6.374×10^{-2}	2.435×10^{-4}	4.107×10^{-5}	1.766×10^{-4}	0.725
700	0.221	1135.8	6.631×10^{-2}	2.646×10^{-4}	4.251×10^{-5}	1.927×10^{-4}	0.728
750	0.210	1145.3	6.884×10^{-2}	2.865×10^{-4}	4.393×10^{-5}	2.093×10^{-4}	0.731
800	0.200	1154.2	7.135×10^{-2}	3.090×10^{-4}	4.532×10^{-5}	2.265×10^{-4}	0.733
850	0.191	1162.6	7.382×10^{-2}	3.322×10^{-4}	4.668×10^{-5}	2.442×10^{-4}	0.735
900	0.183	1170.5	7.627×10^{-2}	3.560×10^{-4}	4.802×10^{-5}	2.624×10^{-4}	0.737
950	0.176	1177.8	7.869×10^{-2}	3.806×10^{-4}	4.933×10^{-5}	2.811×10^{-4}	0.738
1000	0.169	1184.7	8.110×10^{-2}	4.059×10^{-4}	5.063×10^{-5}	3.002×10^{-4}	0.740
1050	0.162	1191.2	8.348×10^{-2}	4.319×10^{-4}	5.191×10^{-5}	3.199×10^{-4}	0.741
1100	0.156	1197.2	8.585×10^{-2}	4.586×10^{-4}	5.318×10^{-5}	3.401×10^{-4}	0.742
1150	0.151	1202.9	8.820×10^{-2}	4.860×10^{-4}	5.443×10^{-5}	3.608×10^{-4}	0.742
1200	0.146	1208.3	9.053×10^{-2}	5.141×10^{-4}	5.567×10^{-5}	3.819×10^{-4}	0.743
1250	0.141	1213.3	9.285×10^{-2}	5.429×10^{-4}	5.689×10^{-5}	4.036×10^{-4}	0.743
1300	0.136	1218.1	9.516×10^{-2}	5.724×10^{-4}	5.810×10^{-5}	4.257×10^{-4}	0.744
1400	0.128	1226.8	9.974×10^{-2}	6.336×10^{-4}	6.049×10^{-5}	4.714×10^{-4}	0.744
1500	0.121	1234.7	1.043×10^{-1}	6.975×10^{-4}	6.284×10^{-5}	5.190×10^{-4}	0.744
1600	0.115	1241.9	1.088×10^{-1}	7.643×10^{-4}	6.516×10^{-5}	5.685×10^{-4}	0.744
1700	0.109	1248.5	1.133×10^{-1}	8.339×10^{-4}	6.746×10^{-5}	6.199×10^{-4}	0.743
1800	0.104	1254.5	1.177×10^{-1}	9.062×10^{-4}	6.972×10^{-5}	6.732×10^{-4}	0.743
1900	0.099	1260.1	1.222×10^{-1}	9.814×10^{-4}	7.197×10^{-5}	7.284×10^{-4}	0.742
2000	0.094	1265.3	1.266×10^{-1}	1.059×10^{-3}	7.420×10^{-5}	7.855×10^{-4}	0.741
2100	0.090	1270.1	1.310×10^{-1}	1.140×10^{-3}	7.642×10^{-5}	8.446×10^{-4}	0.741
2200	0.087	1274.6	1.354×10^{-1}	1.224×10^{-3}	7.862×10^{-5}	9.055×10^{-4}	0.740
2300	0.083	1278.9	1.398×10^{-1}	1.310×10^{-3}	8.081×10^{-5}	9.684×10^{-4}	0.739
2400	0.080	1283.0	1.442×10^{-1}	1.400×10^{-3}	8.299×10^{-5}	1.033×10^{-3}	0.738

INDEX

absorptivity 70–1, 80
acceleration 41, 204, 341
accuracy 266
activation energy 89
active crown fires 224
adiabatic flame temperatures 88
adsorption 143
advection 166
aerial fuels 133, 225
aerial imagery 291, 302, 303, 304
air, thermal and physical properties 50, 53, 56, 62, 343–8
air-flow interactions 53
　flame zones 163, 164
　multiple flame zones 100, 182–5
air gaps between fuel particles 137, 141, 149–50
air temperature measurement 310
airborne LiDAR 156, 281
allometric methods
　forest trees 280–1
　grassland fuels 276
　shrub fuels 277
ambient environment radiation heat transfer, 1D fire spread model 197–8
anemometers 296, 309
angular fire shapes 237–8
angular velocity 59–61
area 341
Arrhenius reaction rate equation 89–90
ash 84
atmospheric stability, and pyrocumulus clouds 244
Australia, fuel models 135

autoignition 119
　see also spontaneous ignition
available fuel 22, 36
axisymmetric fire 24, 100–2, 170, 256

backing fires 25, 26, 27, 162, 170
　configuration 179–82
　single line ignition 321, 323
backing spread 24, 26, 27, 29
backing spread rate 34
backing surface fire 23
bandpass filter 294
bark fuels, eucalypt forests 134, 232, 278
baroclinic torque 248, 253
base quantities and units 339–40
beetle-killed forests, effect on crown fire spread 228–9
Bernard cells 164, 165
bidirectional pressure probes 296–7
biomass
　fuel as 132–4
　grass species 276
　ladder and canopy 278
　shrubs 277, 278
black smoke from crown fires 95
black spruce 144, 226, 278, 279
blackbodies 67
　absorption of incident radiation 70
　heat transfer between 71–2
　rate of thermal radiation emission 67
blocking thermal radiation 72–3
'blow-up' fires 160, 245
bodies (thermal radiation emitters and targets) 66

net radiation heat transfer 66–7
　see also blackbodies; greybodies
bolometer 292
bomb calorimeter measurements 85, 87
　and wildland fires 85–6, 99
　woody fuels 110
boundary conditions 145
boundary layers 54–8
　around branches 58, 59
　complex 3D circulations 163
　developing on a flat surface 56–7, 58
　effects of 55
　fires impacting velocities and temperatures in 163
　formation 55–6, 58
　thickness 55, 56–7
　　and convection heat transfer coefficient 78–9
　　fine fuels 78
　see also thermal boundary layer
branches (of plants)
　boundary layer around 58, 59
　conduction heat transfer 63
　as embers 230
　as woody fuel 133–4, 228, 271
brush fires 40, 41, 134
bulk density 105, 139–40
　canopy 224–8, 280
　forest floor 270–1
　fuel bed 22
Bunsen burner
　flame height variables 97
　flames 91, 95
buoyancy 17, 18, 57, 96, 161
　in puffing 95–6, 247–9

buoyancy-dominated jets 253
burn streets 238–42
    explanations 240–2
    variations in shape and completeness 239–40
burning candles 24, 93–4
    as a fire process 15–21
    as a self-sustaining system 15–16
    steps in sustaining 17–18
    see also candle flames
burning rate 102–6
    1D fire spread model 195–6
    charcoal 110, 111
    dead fuels 149–52
    and flame residence time 103, 106
    fuel beds 103–5, 149–50
    and fuel consumption over a period of time 103, 104
    and heat release rate 103
    and heat transfer within fuel particles 105
    and moisture content of fuel 106, 107, 150
    solid fuels 103, 258–9
    thick and thin fuels 105–6
    wind effects 106, 150
burning zone profile 25–7
burnout 24
burnout time 22
Byram's convection number 245–6
Byram's equation for flame length 37, 99
Byram's 'heat engine' 258

Canada, fuel models 135
Canadian Fire Behaviour Prediction System 223
Canadian Fire Weather Index System 146
candle flames
    colours 93–4
    as laminar flames 95
    shape 18–19, 94
    size 19–20
    structure 93, 95
    temperatures 94–5
    turbulent entrainment 243
candles 15–21, 24
    as coupled system 15, 16, 20
    igniting and burning 17–18
canopies, wind profiles 160, 161
canopy fuels 133, 163, 227

characterisation 223
    consumption measurements 278–81
    measurement 278
canyons, fire spread rate 175
carbohydrate molecules 82–3
cellulose 50, 82, 83
    combustion 84–5
Celsius temperature scale 48, 340
centre fires 329, 330, 331
centreline temperature profiles 100–1
centreline vertical velocity 101–2
char 84, 108
char oxidation 86, 108, 109–10
charcoal 109–10, 111
charred firebrands 231–2
charring 86, 108
chemical energy 47, 50
chemical kinetics 89–90
Cladium spp. 242
coherent anti-stokes Raman spectroscopy 297
cold fronts 246, 247, 252
colours
    Bunsen burner flames 91
    non-premixed flames 93–5
combined heat transfer 79–80
combustibility 93, 139
combustion 15, 17, 82–112
    1D fire spread model improvements 218
    flame temperatures 88–9
    heat of combustion 85–8
    and heat release 266
    thermodynamics 84–9
    wildland fires 83–4
combustion dynamics, premixed flames 91–3
combustion reactions 84–5
    chemical kinetics 89–90
combustion triangle 18
commercial moisture meters 307
complex terrain, wind measurements 308–9
concave flame zones 177–8
concentric streets of unburned vegetation 239, 240
concurrent flow conditions 29, 181
conduction heat transfer 47, 50, 61–4
    damaging living tissues of trees above and below ground 64
    one-dimensional 61–2

    pine needles 63, 64
    in wildland fires 62–4, 79
cone calorimeters 86, 87
cones, serotiny 64
configuration factor 70
conifer needles
    conduction heat transfer 63, 64
    effective heat of combustion for dry and live needles 87
    heat transfer coefficient 76, 78, 79
    ignition of live needles 127–30
conifers, foliage moisture content 154
consumability 93, 139
continuous crown fires 224
continuous flame region 98, 100, 101
continuous mass-loss measurements of burning fuels 267–8
convection heat transfer 17, 47, 54, 62, 73–9, 294
    1D fire spread model 198–201, 219
    backing and flaming fires 181–2
    categories 73
    crown fires 226
    enabling both heating and cooling of fuel 79, 80, 119
    flame zone width in 178
    and fuel bed compactness 140–1
    and ignition 125–7
    measurement 294–9
convection heat transfer coefficient 73
    and boundary layer thickness 78–9
    calculations 75–7
    correlations 75, 76, 77–8
    determining 75
    fuel particles 137
    Grashof number (Gr) 74–5, 77, 198
    magnitude 73
    nonlinear relationship with small cylinder diameter 78
    Nusselt number (Nu) 73–4, 198
    Prandtl number (Pr) 74, 76, 77, 199
    proportional to temperature gradient at the surface 78
    for a range of cylinder diameters 76, 77–8
    Reynolds number (Re) 74, 75, 76, 77, 79, 198
convection number 245–6
convective heat flux 141, 150, 201, 207, 293, 295, 298
convective heat loss, for small fuel particles 126–7

convex flame fronts 179
Coriolis force 61
counter-rotating vortex pairs 253–4
coupled systems
    candles as 15, 16, 20
    fire spread 192, 204–9
    fire whirls 250–1
    independencies in 16–17
    wildfires as 17, 20
cribs 104–5, 258, 267, 289
critical heat flux for ignition 119–20
critical heat release rate 117
critical mass flow rate 224, 228
critical mass flux 117, 118
crossflow 169
crown fire spread
    insect-damaged foliage effects 228–9
    processes in 225
    radiation and convection roles 225–7
    tree boles influence 227–8
crown fires 23, 24, 40, 41, 133
    1D fire spread model improvements 219
    behaviour 222–9
    black smoke from 95
    canopy loading and consumption 278, 279
    classification 223–4
    describing and measuring 229
    ember transport 232
    field-based experiments 223
    fireline intensities 222–3
    flame length 38, 99
    ignition 225
    spotting in 229
    turbulent flames 95
crown fuels 133
crowning 177
cumulonimbus flammageniti 243
cup-and-vane anemometers 309
curved streets 239, 240
cyclostrophic flow 250

damped oscillation 207–8
data error 9
dead fuels
    characterisation and classification for predicting fire behaviour 155–7
    comparison of characteristics with live fuels 153
    consumption measurement 271–3
    equilibrium moisture content (EMC) 144–7
    fuel burning rates and consumption 149–52
    moisture content 142, 143–9, 154, 212–13, 307–8
    moisture gradients 148
    responses to changing environmental conditions 143
density 341
dependent crown fires 224
derived quantities and units 341–2
desorption 143
destructive pre-burn sampling 275
destructive sampling
    canopy fuel 278–80
    forest floor 270
    grassland fuels 275
    shrub fuels 277
dewpoint 144
diffusion 50, 143
diffusion flames 91, 93–106
    see also non-premixed flames
dimensionless numbers 73–5, 198–9
direct measurements 266
    downed woody fuel material 272
    flame depth 289, 290, 291
    flame residence time 289–92
    of fuel consumption 269–70
discontinuous fuels, prescribed burns 322, 327, 329
    effect on fire shape 236
    effect on fire spread 213–15
double sampling methods
    herbaceous fuels 275–6
    shrub fuels 277
Douglas-fir 86, 87, 106, 107, 115, 128, 155, 227, 228
downed woody fuels, loading and consumption measurement 271–3
downward flows, in flame zones 166
Drought Code (DC) 146
duff 23, 64, 80, 117, 233, 254, 257, 267, 269
    consumption measurements 270–1
    moisture content 143, 148, 151
    smouldering 107, 110, 112
Duff Moisture Code (DMC) 146

duff pin technique 270, 271
dynamic pressure 54
dynamic viscosity 55–6
dynamical system 191

effective combustion zone depth 288
effective heat of combustion 87
electromagnetic radiation 47, 65
electromagnetic spectrum 65, 282
elementary entities 340
elliptical expansion rates 34, 35
elliptical fire shapes 32–3, 234–7
    area and perimeter 34
    eccentricity 44, 235, 236
    and fire growth patterns 237–8
    and fireline intensity 39, 43–4
    incorporating threshold conditions 238, 239
    and location of the ignition point 237
    and relative fire spread rate 35
    and topographic slope 236
    wind speed effects 235–6
ember lofting and transport 61, 230, 232
embers 229–30, 231
emissivity coefficient 68
emitters 66, 67–8
    thermal radiation from 68–71
empirical modelling 8
energy 47–8, 341
energy balance 49–50
entrainment 102, 174–5, 184–5, 243, 246, 250, 252, 258–9
environmental conditions in the fire behaviour triangle affecting fire spread rate 29–30
environmental measurements 305–11
    see also fuel moisture; relative humidity; temperature; wind
equilibrium moisture content (EMC), dead fuels 144–6
eucalypt species
    bark as ladder fuel 134, 278
    bark ribbons as firebrands 61, 232
    spotting distance 61
evaporation zone (smouldering) 108, 109
experimental burning, ignition techniques 318–31
experimental design 266
extinction points 213

Fahrenheit temperature scale 340
fallen foliage, burning characteristics 139, 140
Faraday, Michael 15, 16, 17
feedbacks in coupled systems 16–17
Fermat's principle 235
fibre saturation point 127
field-scale research 8
    for prescribed burning 318–31
    and validation 8–9
Fine Fuel Moisture Code (FFMC) 146
fine fuels
    convection heating and cooling 79, 80
    ignition time, live foliage 128, 130, 155
    radiation heat transfer 79–80
    thin boundary layer 78, 79
fine-gauge thermocouples 296, 299–300
fire
    as a physical process 15, 47
    as a system 16
    thermal radiation spectrum 65
fire acceleration 41–3
fire area 33–4
fire behaviour 47
    as dynamical system 21
fire behaviour characteristics 28–42
fire behaviour sensor packages 297
fire behaviour triangle 21–4, 29, 132
    disadvantages 21
    environmental variables affecting fire spread rate 29–30
    as useful framework 21
fire characteristics chart 39–41
fire configurations 177–85
fire danger ratings 4–5, 146, 307–8
fire growth 33–5
    and fire shape 234–7
fire growth patterns 237–8, 239, 240
fire measurements 266–92
fire modelling 5–7, 158
fire perimeter 34–5
fire plumes 242–4
fire point 116
fire propagation apparatuses (FPAs) 86, 87
fire protection engineering 3–4
fire radiative energy (FRE) 282
fire radiative power (FRP) 282
fire safety rating 146

fire scale 28
fire science 3
    and the need for experiments 3–4
fire shapes 32–3, 34
    angular 237–8
    elliptical *see* elliptical fire shapes
    and growth patterns 234–8, 239, 240
    local-scale control 235
    patchy or clumpy fuels 237
    and spread rate distribution 35
fire spread 116
    convection *v.* radiation in 180–2
fire spread directions 24–5
fire spread model (1D) 191–219
fire spread rate 9, 16, 25, 29–32, 202
    backing fires 179–80
    definition 29
    environmental conditions in the fire behaviour triangle affecting 29–30
    and flame front width 177, 178
    and flame zone curvature 177–8, 179
    and flow attachment 174
    measuring 301–5
    models 6
    one-dimensional 301–3
    sloping canyons 175
    on sloping topography 29, 173–6
    validation challenges 8–9
fire spread thresholds 29, 212, 213, 215–16
    elliptical fire shapes 238, 239
fire storms 162, 256
fire tornadoes 252–3
fire whirls 59, 61, 176–7, 250–3
    conditions for creating 252
    formation 251, 252
    increased burning rate mechanisms 251
    positive feedback system 251–2
    radial force balance 250
    source of rotation to produce vorticity 252
    v. fire tornadoes 252–3
fire winds 160–8
firebrands 229–30, 231
    charred 231, 232
    raw materials 231–2
    and spot fire ignition 232–3
fireline intensity 36–7, 39
    crown fires 222–3

and elliptical fire shapes 39, 43–4
and flame length 38
fixed-wing aircraft infrared imaging 305
flame angle
    predicting 170
    reduced entrainment effects 184, 185
flame attachment
    canyons 175
    slope-induced 174–5
    and wind 170–2
flame colour 93–5
flame depth 27, 28, 98, 99, 215, 287
    direct measurements 289, 290, 291
    measurement 288, 289
flame detachment distance 199
flame drag 170
flame front width
    1D fire spread model 208–9
    and shape 177–9
flame gases, flow 18–19
flame geometry 170
flame height 28, 96–100
    Bunsen burner 97
    complications 98–9
    and flame zone depth 98
    indirect estimate, wildland fires 99
    measurement 284–7
    point sampling 287
    reduced entrainment effects 185
    visual techniques 285–7
    wildland fires 97–9
flame intermittency 167–8
flame ionisation detectors 291
flame length 28, 37–9, 99, 215
    Byram's equation 37, 99, 196
    indirect estimate, wildland fires 99
    measurement 283–5, 288, 289
    point sampling 287
    reduced entrainment effects 184, 185
    relationship with fireline intensity 100
    Thomas' equation 99–100
    visual techniques 285–7
flame radiation heat transfer, 1D fire spread model 196–7
flame reaction time 27, 287
flame residence time 27, 28, 103, 106, 150–1, 195
    direct measurement 289–92
    measurement 287–8, 289

flame scale 28, 324–30
flame shape 18–19, 94
flame size 19–20, 28, 184
flame speed of premixed flame 92
flame structure 18–19, 164–6
   non-premixed flames 93–5
   premixed flames 91–2
flame temperatures 88–9
   and colour 66, 94–5
   variation with proportions of fuel and air 88
flame tilt, wind effect on 169–70
flame width 178, 211
flame zone curvature, and fire spread rate 177–8, 179
flame zone depth *see* flame depth
flame zone orientation, 1D fire spread model improvements 219
   influence on spread rate 35
flame zone properties measurement 282–92
   see also flame depth; flame height; flame length; flame residence time
flame zones
   and air-flow interactions 100, 163, 164, 182–5
   ambiguity of 288–9
   Bernard circulations 164–5
   downdrafts 165–6
   and flame attachment 170–2
   leading and trailing edges 288–9
   water line, narrow canyons 175
flamelets 95
flames 15, 58
   as bodies 66
   continuous and intermittent flame zones 98, 168, 171–2
   diffusion (non-premixed) 91, 93–106
   premixed 91–3
   thermal radiation emission 68
flaming combustion 82, 86, 93, 106
   ash and char remaining 84
   fuel consumed 99, 272–4
   fuels 82–4
   solid fuels 93
   steps to trigger 120–1
   transition to/from smouldering combustion 107, 110–12
flaming ignition 114–16, 117, 299
   criteria 117–19
   probability of 234

flaming ignition time 117
   predicting 117–19
flaming phase 22, 37
flaming surface fire, life history 24
flaming time 22
   see also flame residence time
flammability
   solid fuels 92–3
   wildland fuels 93, 139
flammability limits 92
flammagenitus 242
flanking fires 176–7, 181–2
   prescribed burning 327
   single line ignition 321, 323–4
flanking spread 24, 26
flash point 116
flashing ignition 116
flow attachment to inclined surfaces 173–4
flow velocity measurements 296, 297, 299
fluid mechanics 46, 53–61
fluids 53
fluid–solid interface 53–4
force 341
forced convection 73, 75, 78
forest fires 40, 41
forest floor material
   consumption measurements 270–1
   see also duff; ground fuels; litter
forest tree allometrics 280–1
Fourier's Law of one-dimensional heat conduction 62
freestream temperature 58, 63
freestream velocity 55
friction 55, 58
Froude number 167
fuel available 36
   phases 22
fuel beds 22
   bulk density 105, 139–40
   burning rate 103–5, 149–50
   characteristics for 'standard case' 204
   compactness 140–1, 149–50
   flame residence time 150–1
   and fuel particles 136–7, 140–2
   optimum packing ratio 104
   porosity factor 105
   properties 136, 139–42
fuel consumption
   in the flaming front 99, 100

   measurements 267–81
   and moisture content 151, 152
   total, measurement 268–70
fuel consumption rates, measurement 267–8
fuel continuity, effects, 1D fire spread model 213–16
   see also discontinuous fuels
fuel cooling, and conduction 64
fuel density
   and ignition time 122
   live fuels 128
fuel descriptions 134–6
fuel flow velocity, and flame height 97
fuel heating, and conduction 63–4
fuel inventory 269, 277
   line-intercept method 272
fuel lean reactants 85, 88, 90
fuel loading, 1D fire spread model 134–6, 207–8
fuel models 134–5
fuel moisture 22, 142–52
   and available fuel 37
   and burning rate 106, 107, 150
   calculated average content 147
   dead fuels 142, 143–9, 154, 212–13
   exchange with the atmosphere 148
   and fire behaviour 37, 128–30, 149
   for fire danger rating 307–8
   and fire spread rate 30, 31
   and heat of combustion 87
   and ignition time 122–4, 130
   live foliage 153–5
   live fuels 128–30
   measurement 305–8
   sampling 306
   seasonal changes 148
   and time-lag classes 146–8
   and total amount of fuel consumed by a fire 151
fuel moisture sticks 307
fuel particles 21
   1D fire spread model 193–5, 206–7
   air gaps between them 137, 141, 149–50
   at equilibrium moisture content 145
   as constituents of fuel beds 136–7
   convective heat loss 126–7
   convective heating, modelling 201
   critical heat flux for ignition 119–20
   fuel bed characteristics 135
   and fuel bed properties 139–42

geometry 137–8
heat transfer and burning rate 105–6
heating and ignition 206–7
heating of the solid 114
ignition process 114–17
material properties 136, 138–9
moisture content 142–52
pyrolysis 114–15
surface temperature measurement 299–300
thermally thick assumption 121, 122
thermally thin assumption 121, 122
fuel photo series 272–3
fuel proportions
and premixed flame speed 92
and flame temperature 88
and heat of combustion 88
and heat release rate 90
and reaction rate 90
fuel rich reactants 85, 88, 90, 103
fuel size, and ignition time 125–6
fuel transects 272
fuel type, wind speed and fire rate spread 30–1
fuel vaporisation 17
fuel(s) 15
characterisation and classification for predicting fire behaviour 155–7
as factor in wildlife behaviour triangle 21
flaming combustion 82–4, 93
heat of combustion 84–8
thermophysical properties 139
time-lag classification 146–7
wildland 21–2

gas conduction 181
gas temperature measurements 297
gaseous fuels 82
burning rate 102
combustion 88
flame height 96–7
premixed, flammability limits 92
reaction rate 89
glowing combustion 106–7, 108
glowing phase 22, 37
glucose 83, 84
glucose molecules 83
Grashof number (Gr) 74–5, 77, 198

grass fires 40–1
grasses 275
allometric methods 276
conduction heat transfer 63
destructive sampling 275
heat transfer coefficient 76, 78, 79
quadrat measurements 275, 276
gravimetric measurement, fuel moisture content 305
greybodies
absorption of incident radiation 70
heat transfer between 71
reflection, absorption and transmission of thermal radiation 70, 71
thermal radiation emission 68
grid method to record 2D fire spread rate 303
ground-based infrared video 297–8
ground-based LiDAR 157, 281
ground fires 23, 24, 108, 134
ground fuels 133, 270–1

hand ignition 319
heading fire spread rate 29
as a function of some environmental variables in the fire behaviour triangle 30
heading fires 23, 25, 26–7, 162, 181
fire front width and shape 177–9
single line ignition 320–3
heading spread 24, 26
definition 29
heading surface fire 23
heat 15, 47, 341
and serotiny 64
heat density 37
heat flux 48, 341
ignition temperatures and fuel behaviour 63–4
spreading fires 180
to communicate incident radiation on any arbitrary object 72
see also convective heat flux; radiant heat flux
heat flux gauges 298
heat flux package 297, 298
heat of combustion 51, 85–8
effective 87
gaseous fuels and air 88
measuring 85–7
wildland fires 85–6, 99

woody fuels 110
heat of vaporisation 52–3, 106
heat per unit area 36, 37
heat rate 341
heat release 35–6, 51
and fireline intensity 36–7
scale of the quantity 36
heat release measurement 266, 281–2
heat release measures, in wildland fires 36
heat release rate 36, 90, 103, 281–2
heat transfer 46–7, 53
and burning rate of fuel particles 105–6
combined 79–80
conduction 47, 50, 61–4, 79
convection see convection heat transfer
measurement 292–9
radiation see radiation heat transfer
heat transfer coefficient (HTC) 73–7, 137
see also convection heat transfer coefficient
heat yield 36, 37
helicopter-mounted infrared cameras 305
hemicellulose 82, 83
herbaceous fuels
biomass measurements 276
destructive sampling 275
double sampling methods 275–6
fuel consumption estimation 276
key attributes 275
measurements 273–6
weight-estimate method 276
horizontal roll vortices 242
horizontal wind speed 158
Huygens' principle 235
hydraulic jump 252
hydrostatic pressure 54, 161, 174, 257, 259
hygrometers 310

ideal gas law 50, 57
ignitability 93, 139
igniting a candle 17
ignition 114–30
1D fire spread model improvements 218–19
centre fire 329–31
critical heat flux for 119–20
crown fires 225

hand ignition 319
mass ignition 329–31
measuring 299–300
multiple line ignition 324–7
multiple spot ignitions 328–9
other line ignitions 324
point ignition 319–20
ring fire 329–31
single line ignition 320–4
types of 119
see also flaming ignition; smouldering ignitions
ignition process 114–17
ignition techniques for experimental burning 318–31
ignition temperature 49, 62, 63, 117–19, 121, 122, 123, 197
ignition time
 factors affecting 122–7
 live fuels 128–30, 154–5
 predicting 120–2
in-fire cameras 144, 287, 291, 300
incident thermal radiation 68
 absorption by blackbody 70
 absorption by greybody 70
 decreases with square of the distance between emitter and target 68–70
 falling on logs from flames 68, 69
 reflection 70
 view factor 70
incident thermal radiation flux 292
independent crown fires 224
indigenous peoples, fire use 4
indirect measurements 66
 allometry for grasses 276
 plot-level pre-burn fuel properties 272–3
 relative humidity 310–11
induced flows 160–8, 250–4, 258–9, 329–31
infinite line fire 25
infrared cameras 297–8, 304, 305
infrared radiation 65
initial fire growth 24–5
initial ignition (candle) 17
insect-damaged foliage, effect on crown fire spread 228–9
instability, in wildland fires 164–5
intermittency 167
intermittent flame region 98, 100, 101, 171–2

International Crown Fire Modelling Experiments 223
irrotational vortices 59, 60–1

jack pine 135, 161, 226, 227, 278, 279
jet point 160
jets in crossflow 253–5
junction zone 184, 327

Kelvin temperature scale 48, 340
kinematic viscosity 56, 74
King's Cross Fire, London Underground train station 174

laboratory experiments 7, 8, 85–7, 104–5, 126–30, 138, 168, 175–8, 249, 268
ladder fuels 133, 134, 223–4
 consumption measurement 278–9, 281
 measurement 278
laminar flames 95
laminar flow 18, 19, 57
laminar zone (boundary layer) 56, 57
large-fire behaviours 222–60
 synoptic weather systems affecting 246, 247
large fires, strong winds caused by 161–2, 250–4, 258–9, 329–31
large woody fuels 110, 225, 228, 283
 consumption measurement 273, 274
laser Doppler velocimetry 297
laser-induced fluorescence 297
latent heat 106, 114, 243–6
leading edge 56, 194–219, 229, 246, 287–9
'leaky-bucket analogy' 49, 121, 141–2
lean flammability limit 92, 115–16
leeward vorticity 227, 241
leftover char and ash zone (smouldering) 108–9
length 339
LiDAR (Light Detection and Ranging) 155–7
 for ladder and canopy fuels 281
 for shrub biomass estimation 278
 for wind measurements 309
lighting a candle 17
lightning 20, 244
lignin 50, 82, 83

line fires
 concept 25–8
 leading and trailing edges 288–9
 multiple 324–7
 segments 29
 see also spreading line fires
line ignitions
 multiple 324–7
 other 324
 single 320–3
line-intercept method 272
linear streets 239–40
liquid fuels 82
liquid wax moving into the flame zone 18
litter 117, 151, 254, 257, 267, 269
 consumption measurements 270
 from shrub fuels 278
live foliage, moisture content 128–30, 153–4
live fuels 127–30, 152–5
 ability to carry fire 153
 characterisation and classification for predicting fire behaviour 155–7
 comparison of characteristics with dead fuels 153
 ignition and burning challenges 127–30, 152–3
 moisture content and ignition time 128–30, 154–5
living vegetation, diversity 152
lodgepole pine 64, 129, 135, 227, 228, 286, 338
logs
 absorbed thermal radiation 80
 conduction heat transfer 64
 convection heat transfer 80
 incident thermal radiation received from flames 68, 69
 seasonal moisture changes 140
 smouldering 107
longitudinal roll vortices 163, 164, 165, 242
loosely packed fuel beds 104
low pressure within the fire 161
 see also hydrostatic pressure

mass 339, 340
mass fires 256–9
 behaviour 257–8
 criteria for 257

dependency on solid fuel burning rates 258, 259
strong surface winds in 257–8
mass ignitions 329–30, 331
mass-loss measurement of fuel consumption 103–4, 117, 267–8
mean gas temperature empirical power-law model 199–201
measurements in fire behaviour 265–311
melting wax 18
methane combustion 84
metric system 338, 339
microwave drying 307
mineral compounds in wood 84
minimum ignition energy 116
mixed convection 73
model error 9
modelling
  and field-scale research 7–8
  see also 1D wildland fire spread model
moisture fuel content see fuel moisture
moisture gradients 148
moisture meters/analysers 307
moles 87, 110, 340
momentum-dominated jets 253
multiple flame zones and air-flow interactions 100, 182–5
multiple line ignitions 324–7
  flanking fires 327
  strip head fires 325–7
multiple spot ignitions 328–9

narrow-angle radiometers 294
natural convection 73, 75, 78
near field 199
near-surface wind profiles 159–60, 162
negative feedback loops 192
negative feedbacks 16
  1D fire spread model 217–18
net heat transfer 49, 119–21
net radiation heat transfer 66, 68
net rate of radiation heat transfer 71–2
neutral stability (wind profiles) 158
Newton's law of cooling 73
no-slip condition 55, 158
no wind, initial fire growth 24, 25
nomenclature xiii–xvi
non-destructive methods
  for shrub fuels 277

for woody material measurement 271–3
non-flaming phase 22
non-premixed flames 91, 93–106
  basic structure and characteristics 93–5
  burning rate 102–6
  colours 93–5
  entrainment 102
  flame height 96–100
  heat release rate 103
  puffing 95–6
  quenching 93, 94, 95
  temperature profiles 100–1
  temperatures 94–5
  vertical gas velocity 101–2
  see also diffusion flames
non-reactive smoke plume
  centreline temperature profile 100–1
  centreline vertical velocity profile 101, 102
non-steady wind, 1D fire spread model 211
nonlinear 191–2
nonlinear feedback 17
Nusselt number (Nu) 73–4, 198

oak twigs, thermal thickness 63
oily rags, autoignition 119
one-dimensional conduction 61–2
1D fire spread rate, measuring 301–3
1D wildland fire spread model 202–16
  effect of dead fuel moisture 212–13
  effects of fuel continuity 213–16
  effects of slope 211–12
  flame front width 208–9
  framework 192–201
  fuel loading 207–8
  fuel particle heating and ignition 206–7
  function 201–2
  improvements 218–19
  non-steady wind 211
  positive and negative feedbacks 216–18
  review of what can be expected from the model 202–4
  simple dynamics 204–6
  wind effects 209–11
operational fire modelling systems 135–6

opposed flow conditions 29, 181
optimum packing ratio 104
oven-drying of sampled fuel material 142, 269, 306–7
oxygen consumption calorimetry 87

packing ratio 104, 140
particle flame residence time 27–8, 150–1, 288
particle image velocimetry (PIV) 297
pass band 294
passive crown fires 223–4
passive flame height sensors 287, 288
phase changes 51–3
photographic methods
  for measuring diameter reduction of large logs 273, 274
  to estimate surface fuel consumption 273
piloted ignition 119
pine needles
  conduction heat transfer 63, 64
  flexibility, to assess moisture content 307–8
pitot probes 296
planar-intercept method 272
plants, structural parts 50–1
plume buoyancy 57
plume structure 18–19
plume vorticity 242, 258
plumes 57
  and pyroconvective atmospheric storms 242–6
  smoke 57–8, 98, 100–1
point ignitions 25–6, 39, 42, 319–20
  multiple 328–9
  see also spot ignitions
point sampling of flame height and/or length 287
point temperature measurements 167, 297, 303
polymers 82, 83
ponderosa pine 64, 144, 156, 228
porosity factor 104–5
positive feedback loop 192
positive feedbacks 16
  1D fire spread model 216–17
power 47–8, 72, 294–5, 341
power-law functions 172, 199–201
Prandtl number (Pr) 58, 74, 76, 77, 199
pre-exponential factor 89
precision 266

preheat zone (smouldering) 108, 109
premixed flames 91–3
    flammability limits 92
    propagating, structure 91–2
premixed gas-phase mixture 115–17
    minimum ignition energy 117
prescribed burn unit 319
prescribed burning 184, 235, 318–19, 321, 322
    discontinuous fuels 322, 327, 329
    flanking fires 327
    multiple line fires 324–7
    multiple spot ignitions 328–9
    strip head fires 325–7
pressure 341
    see also dynamic pressure; hydrostatic pressure
probability of ignition 234
products, combustion 84
propeller anemometers 309
puffing 95–6, 247–50
    and buoyancy 247–9
    demonstration of 249–50
puffing frequency 96, 248–9
pulsating fires 247–50
pyroconvective plumes 242–4
    differences from 'common' atmospheric storms 244
pyroconvective storms 242–6
    processes in development 245
    structure 243
    wind effects 245–6
pyrocumulonimbus clouds 243, 244, 253
    height of 244–5
    processes in development 245
pyrocumulus clouds 242–3, 244
    atmospheric stability effects 244
pyrolysates 83, 109, 115
pyrolysis 20, 83, 89–90, 103, 106, 108, 114–15
pyrolysis and burning zone (smouldering) 108, 109, 111–12
pyrolysis gases 101, 109, 115
pyrolysis rate 89, 90, 97, 103, 111–12, 115
pyrolysis temperature 53, 83, 90, 115

quadrats 275, 276, 277
quenching 93, 94, 95

radial force balance 250
radiant heat flux 150, 191–2, 207, 226, 292, 293

radiation heat transfer 47, 62, 64–73, 172, 292
    ambient environment, 1D fire spread model 197–8
    backing and flanking fire spread 182
    crown fires 225–6
    as dominant mechanism for heating particles to ignition 141
    fine fuels 79–80
    and flame front width 179
    and fuel bed compactness 140, 141
    fuel particles 137–8
    measurement 292–4
    terminology 66
    see also blackbodies; greybodies
radiometers 292–4
    applications 294
radiometric temperature 66, 282, 294
Raleigh-Taylor instability 248
rate of spread see fire spread rate
Rayleigh number 199
reactants 84, 85
reaction rate, combustion reactions 89–90
reciprocity relation for radiation heat transfer 71
reflected thermal radiation 70
reflectivity 70–1
relative humidity (RH) 144
    and equilibrium moisture content 145
    measurement 310–11
remote sensing technologies 155–7, 266, 282, 308
Reynolds number (Re) 74, 75, 76, 77, 79, 198
rich flammability limit 92
Richardson number 198, 199
ridge crests, vorticity-driven lateral spread 255–6
ring fires 249–50, 329, 330, 331
rotational vortices 59, 60
Rothermel's spread rate equation 5, 6, 134–5, 268, 300
rotten wood/logs 122, 143, 233, 270, 273, 274
roughness length 158–9
runaway conditions 192
running crown fires 224

saltation for ember transport 229
sampling designs 266, 275–6

saturation vapour pressure 143–4
scale modelling 10
scientific notation 338–9
scorching 79
self-sustaining system 15–16
sense 61
serotinous cones 64
shear 57
shear flow 57
sheltering 158
shrub fires 41
shrub fuels
    biomass 277, 278
    consumption measurements 276–8
    leaf litter accumulation 277–8
shrublands
    fuel measurement 278
    sampling 276–7
single line ignition 320–4
    backing fires 323
    flanking fires 323–4
    heading fires 320–3
slash fuels 269
sling psychrometers 310
slope angle
    and fire spread rate 30, 31–2, 173
    and flame attachment 174
slope orientation 23, 172
sloping topography
    effects, 1D fire spread model 211–12
    elliptical fire shape 23, 338
    fire spread rate measurement 29, 173–6
    and flame attachment 174
    and flow attachment 174
    initial fire growth 24, 25
    and solar flux 172
    wind profiles 160, 162
smoke plumes 98
    from incense stick 57–8
    non-reactive 100–1
smouldering coals
    net rate of radiation heat transfer to nearby tree trunk 71–2
    thermal radiation emission 67–8
smouldering combustion 106–12
    and charring 108
    emissions 107
    solid fuels 107–8
    transition to/from flaming combustion 107, 110–12

smouldering combustion front 270–1
    sustaining and propagating
        zones 108–9
smouldering ignitions 116–17
    spot fires from 233
smouldering phase 22, 37, 57, 270–1
smouldering stage of wildland fires
    86
SODAR measurements 309
soil heating 64
solar flux 172
solar radiation
    affecting ground surface 173
    and drying of woody fuels 149
    influence on individual fires 172–3
solid body rotation 60
solid fuels 82–4
    burning rates 103, 258–9
    flame height 97–8
    flammability 92–3
    ignition 114–16
    pyrolysis 83, 97
    pyrolysis temperature 83–4
    smouldering 107–8
solid glowing radiation, 1D fire spread
    model 197
sonic anemometers 309
soot 94–5
soot blocking 72–3
sorption curves 143, 145
specific heat capacity (specific heat) 51
    and ignition time 122, 123
    of wood 51–2
spectral emissive flux 65
speed 342
Spinifex vegetation, fires in 237–8
spontaneous ignition 119
    see also autoignition
spot fires 24, 61, 229–34
    dynamics 230–1
    from smouldering ignition 233
    ignition by firebrands 232–3, 234
    probability of ignition 234
spot-head fires 328
spot ignitions 319–20
    multiple 328–9
spotting 61, 229–34
    in crown fires 229
    dynamics 230–1
    probability of 234
spotting distance 61, 230–2
spread rate *see* fire spread rate

spreading fires 167, 168, 169
    flame zone dynamics 166
    heat flux 180
    leading and trailing edges 288–9
    wind speed profiles 162–3
spreading line fires
    and fireline intensity 36
    steady-state assumption 28–9
stable atmospheres 158
steady-state assumption 6, 28–9
steep slopes
    and elliptical fire shape 33
    fire acceleration 43
    initial fire growth 24
Stefan-Boltzmann constant 67
Stefan-Boltzmann equation 67, 70
step-changes 29
stoichiometric proportions 84, 88, 90
stratified sampling 270–1, 275, 306
'stringybark' eucalyptus, spotting
    distance 61
strip head fires, multiple line ignitions
    325–7
strongly coupled 192
Strouhal number 167
structural carbohydrates 83
'structure from motion' (SFM)
    photogrammetry 156, 281
sunlight 65
sun's heat 64–5
surface fires 23–4, 37, 40, 41, 134
    modelling 191–219
surface fuels 133
surface temperature of a fuel element,
    measuring 299–300
sustainability 93, 139
synoptic weather systems affecting
    large-fire behaviours 246
system 191
system behaviour 191–2

tail spread 24
targets 66, 68
    incident energy absorbed 70, 71
    incident energy reflected 70, 71
teardrop shape (candle flame) 18, 94
temperature profiles
    and distance above flames, power-
        law functions 171, 172
    non-premixed flames 100–1
temperature(s) 48, 58, 339–40
    measurement 310

in wildland fires 48–9
    see also flame temperatures
thermal boundary layer 58, 60
thermal cameras 304
thermal conductivity 61
    of common materials 62
thermal diffusion 50
thermal diffusivity 74
thermal energy 47, 51
thermal imaging, to determine heat
    release 282
thermal penetration depth 121
thermal radiation
    blocking 72–3
    incident 68–70
thermal radiation emission
    blackbodies 67
    flames 68, 69
    greybodies 68
    smouldering coals 67–8
thermal radiation flux 197, 292
thermal radiation spectrum 65
    for blackbodies 66
    spectral emissive flux 65, 66
    and visible spectrum 66
    wavelengths 65, 66
thermally thick fuels 63, 121
    burning rate 105–6
    ignition time 121–2, 125–6
    live fuels 128
thermally thin fuels 63, 121
    burning rate 105, 106
    ignition time 121–2, 125–6
    live fuels 128
thermistors 292, 310
thermocouple arrays 302–3
thermocouples 88, 291, 294, 295–6,
    299–300, 310
thermodynamics 45, 50–3
    of combustion 84–9
thermopiles 294
thin boundary layer, fine fuels 78–9
Thomas' equation for flame length
    99–100
3D 'point clouds' 156
3D velocity measurements 297
thunderstorms, common 244
time 339
time-lag classification of fuels 146–8
time lags 191
topography 173–7
    affect on fire behaviour 23

as factor in wildlife behaviour triangle 21
and solar radiation 172
see also sloping topography
torching 23, 24, 223
torching trees
lofted embers from 232
spotting from 231
tornadoes 59
toroidal vortices 57, 248
total fuel consumption measurement 268–70
total heat release 281, 282
transition zone (boundary layer) 56, 57, 58
transitional flow 57
tree boles 227–8, 241
tree-centred spot fires 329
tree crown streets 239
see also burn streets
trench effect 174
trench fires, behaviour 174–5
turbulence 57, 58
turbulent entrainment 243
turbulent flames
Bunsen burner 95, 97
wildland fires 95
turbulent flows 18, 19, 57, 58
turbulent zone (boundary layer) 57, 58
2D fire spread rate and fire growth 303–5
2D velocity measurements 297, 299

ultraviolet radiation 65
unburnt fuel 84
see also fuel consumption
underburning 177
unit area heat release 36
unit energy 37
unstable atmospheres 158
unstable systems 192
uphill flame attachment 174
US National Fire Danger Rating System 234
user error 9

validation 8–9
Van Wagner's classification of crown fires 223–4
vapour pressure 143–4
vapour pressure deficit 144

vegetation
remote sensing to assess moisture content 308
and slope topography 172
and solar radiation 172
velocimetry 297, 299
velocity 342
velocity boundary layer 54, 58, 158, 163
see also boundary layers
ventilation-limited fuel beds 103
vertical gas velocity, non-premixed flames 101–2
videography
1D flame spread rate 302
2D flame spread rate 303–4
flame depth and residence time 291–2
flame flow velocities 297–8
flame height and length 285–7
ignition 300
view factor 70
viscosity 55–6
visible flaming 18
visible spectrum 65, 66
visual techniques, for flame length and height 285–7
volume 342
vortex flows 58–61
vortices 56, 57, 58
counter-rotating vortex pairs 253–4
direction of rotation 61
horizontal roll 242
irrotational vortices 59, 60–1
longitudinal roll 163, 164, 165, 242
rotational vortices 59, 60
toroidal vortices 57, 248
wake vortices 254–5
vorticity 175, 176, 177, 246–56
and baroclinic torque 248, 253
and fire whirls 59, 61, 176–7, 250–3
jets in crossflow 253–5
leeward 227, 241
and puffing 95–6, 247–50
vorticity-driven lateral spread 255–6

wake vortices 254–5
water, phase changes 51–3
'water can analogue', to measure total heat release 281, 282
water potential 143
wavelength 47
weather 157–73

as factor in wildlife behaviour triangle 21
influence on fire behaviour 22–3
see also solar radiation; wind
weight-estimate method 276
wide-angle radiometers 294
wildfire behaviour 28–9
characteristics 29–43
wildfire behaviour science
key principles and insights 333–5
principal value to managers 336–7
principal value to researchers 335–6
wildfire classification 23–4
wildfire spread 20, 191–219
model framework 192–201
model function 201–2
modelled fire spread and behaviour 202–16
system behaviour 191–2
wildfire suppression 4
wildland fire environment 132–85
wildland fire science, since 1900 4–7
wildland fire spread model (1D) 191–219
wildland fires 3
amount of fuel entering the flame zone 19–20
combustion 83–4
conduction heat transfer 62–4
as a coupled system 16, 17, 20
experiment-based research 4
flame height 97–9
flame length 98, 99
flame structure 95
flame temperatures 87
flame width 99
fluid–solid interface 53–4
heat of combustion 85–6, 99
instability 164–5
predicting ignition times 121
smouldering ignition 116–17
turbulent flames 95
wildland fuel particles *see* fuel particles
wildland fuels 21–2, 53, 132–57
chemical characteristics 82–3
flammability 93, 139
ignition and burning 127–30
phases 22
pyrolysis 83
strata types 133–4

wind 157–72
    1D fire spread model improvements 219
    and drying of woody fuels 149
    effect on burning rate 106, 150
    effect on flame tilt 169–70
    effects, 1D fire spread model 209–11
    and flame attachment 170–2
    measurement 308–9
    non-steady, 1D fire spread model 211
    in the presence of fire 160–8
    in pyroconvective storms 245–6
    strong surface winds in mass fires 257–8
wind adjustment factor (WAF) 158
wind direction 23, 308
wind-driven fires
    elliptical shapes 33, 234–6
    ember transport 232
wind profiles 158
    approaching flame zones of spreading fires 162–3
    logarithmic 159, 160
    measurements 308, 309
    sloping topography 160, 162
    throughout the canopy 160, 161
    without fire 158–60
wind speed 22–3, 157–8, 159, 308
    and canopy fuels 163
    effect on elliptical fire shape 235–6
    fuel type and fire rate spread 30–1
    of near-surface winds 159–60
wind tunnels 172, 232, 309
windy conditions
    and fire acceleration 43
    initial fire growth 24, 25
wood
    ash and char after combustion 84
    combustion chemistry 83
    physical properties 139
    smouldering 107–8
    specific heat capacity 51–2
    structural components 83
wood cribs 99, 104–5, 227, 258, 259, 267, 289, 307
    see also cribs
woody fuels 83, 110, 148, 149
    non-destructive measurement methods 271–3